C0-ALU-544

Peter Ax · Multicellular Animals Volume II

The Phylogenetic System of the Metazoa

AKADEMIE DER WISSENSCHAFTEN UND DER LITERATUR · MAINZ

Springer
Berlin
Heidelberg
New York
Barcelona
Hong Kong
London
Milan
Paris
Singapore
Tokyo

Peter Ax

Multicellular Animals

The Phylogenetic System of the Metazoa

Volume II

With 154 Figures

Springer

Professor Dr. PETER AX
University of Göttingen
Institute for Zoology and Anthropology
Berliner Straße 28
37073 Göttingen
Germany

Translated by
Dr. S. KINSEY
06370 Mouans-Sartoux
France

Original edition: Peter Ax. Das System der Metazoa II. Ein Lehrbuch der phylogenetischen Systematik. Gustav Fischer Verlag, Stuttgart, Jena, New York. Akademie der Wissenschaften und der Literatur, Mainz 1999.

Sponsored by Bundesministerium für Bildung, Wissenschaft, Forschung und Technologie, Bonn, and by Niedersächsisches Ministerium für Wissenschaft und Kultur, Hannover

Cover illustration. In front: *Saduria entomon* (Crustacea). Back: *Cheilonereis cyclurus* (Polychaeta). Behind: *Aeolidia papillosa* (Mollusca)

ISBN 3-540-67406-3 Springer-Verlag Berlin Heidelberg New York

Library of Congress Cataloging-in-Publication Data

Ax, Peter, 1927 – (System der Metazoa, English)
Multicellular animals: The phylogenetic system of the metazoa / Peter Ax.
Includes bibliographical references and index.
ISBN 3-540-67406-3 (hardcover)
1.-Metazoa. 2. Phylogeny. I. Title.
QL 45.2A913 1996 591'.012–dc20

Springer-Verlag Berlin Heidelberg New York
a member of BertelsmannSpringer Science+Business Media GmbH

Cover design: Erich Kirchner, Heidelberg
Typesetting: K+V Fotosatz GmbH, Beerfelden
SPIN 10744656 31/3130 - 5 4 3 2 1 0 – Printed on acid-free paper

Preface

Those who wish for permanence in classification must pay the price of stasis – as if forever condemned to confound whales with fish.

M. T. Ghiselin 1981, p. 283

Scientific argument is a debate concerned with the solution of unresolved problems. Before continuing with the phylogenetic system of the Metazoa this foreword gives me the opportunity to discuss some controversial questions, to state selected positions more precisely and to remedy omissions. I would like to draw special attention to serious problems for phylogenetic systematics resulting from the inevitable confrontation with the current rules of nomenclature [1]. In carrying out this debate, I hope not to lose the goodwill of those readers who are experts within the first few pages.

A textbook for students? – Critics who ask that question and answer in the negative probably underestimate the open-mindedness of young people who are not troubled by, or can easily free themselves from, the restraints and arbitrariness of traditional classifications – and to whom systematics is offered as a scientific product in such a form that arguments for every single decision are comprehensible and checkable and can therefore be fully analyzed.

This book certainly does not claim to offer the study of an easy subject matter. However, its contents are possibly more interesting than a study of classes, orders and phyla which do not reflect reality, or learning solely parrot-fashion the 33 carefully numbered orders of insects which, as equally ranked unities, also have no basis in reality.

System – Hierarchical Tabulation – Relationship Diagram

In Nature, only individuals and populations of individuals exist as concrete, material objects. These are the things that we actually observe. In systematics, however, we are far less interested in the genealogical relationships between individuals than in the phylogenetic relationships between species and descendant communities. The latter are conceptional man-made objects (Vol. I, p. 15) [2]. Logically, there can be no phylogenetic sys-

[1] International Code of Zoological Nomenclature. University of California Press, Berkeley, Los Angeles (1985). Fourth edition: The International Trust for Zoological Nomenclature. The Natural History Museum, London (1999)

[2] Multicellular Animals, Volume I. A new approach to the phylogenetic order in nature. Springer, Berlin Heidelberg New York (1996)

tem in Nature which can be grasped as a real tangible object and which can somehow be reconstructed from fragments – although in the literature this is often claimed to be the case.

The phylogenetic system of organisms is a construct of Man – an attempt to represent the immaterial order in Nature which is present as the result of phylogenesis between species and descendant communities as unities with objective common characteristics.

There are two equivalent, alternative possibilities of representing this order – a hierarchical tabulation and a diagram of phylogenetic relationships. It has been pointed out to me that adequate explanations of these terms were lacking in Volume I. I hope to remedy this in the following pages (AX 1984).

Hierarchical Tabulation

Under the name of a monophylum of any rank, we group and indent the subordinate adelphotaxa together. This process is likewise continued for each of the two adelphotaxa, their subordinate taxa and so on. We can thus depict the three first levels of subordination of Arthropoda mentioned in this Volume.

Arthropoda
- **Onychophora**
- **Euarthropoda**
  - **Chelicerata**
    - **Pantopoda**
    - **Euchelicerata**
  - **Mandibulata**
    - **Crustacea**
    - **Tracheata**

The following objection to hierarchical tabulation, as presented above, is sometimes put forward. The indented taxa, which have no categories or numbers assigned to them, allow at most a clear representation of only a small part of the phylogenetic system. In my estimation this objection has no foundation. (1) Generally, only a special restricted part of the phylogenetic system of organisms needs to be dealt with at one time, which is then further clarified in the accompanying text. To stay with the above example – when depicting the high-ranking relationships of adelphotaxa within the Arthropoda – one would hardly want to add the phylogenetic system of the Crustacea or the Tracheata to the tabulation. (2) With the consistent indentation of sister groups we can display – against commonly held opinion – a great deal of information on a single page. For example, from the phylogenetic system of the Insecta, we can show the adelphotaxa relationships of all unities which are treated as orders in traditional classifications (p. 243).

Diagram of Phylogenetic Relationships

The statements contained in the hierarchical tabulation can be augmented by further information and displayed in diagrammatic form – information about stem species and their relationships to terminal taxa, about lineages of species and stem lineages of descendant communities and about the origin of evolutionary novelties (autapomorphies) in these lineages.

I have declared myself in favor of the term "diagram of phylogenetic relationships" for the representation of such information (AX 1984, 1987) as opposed to such terms as dendrogram, phylogram, cladogram or consensus tree – irrespective of what might be reproduced under these terms.

Let us look at the minimal monophylum Cladoceromorpha from the phylogenetic system of the Phyllopoda (Crustacea) (p. 154). The sparseness of the hierarchical tabulation with its indented adelphotaxa *Cyclestheria hislopi* and Cladocera is contrasted with the optical information in the diagram of phylogenetic relationships (Fig. 1). Terminal taxa are single species or supraspecific unities consisting of recent, recent+fossil, or fossil species. In our example, the species *Cyclestheria hislopi* and the species-rich Cladocera (water fleas) are terminal taxa of equal rank with their origin at the splitting of stem species z, which they share. We mark terminal taxa in a relationship diagram with a filled circle and stem species at the time of splitting with an open circle. The lines ascending from stem species z symbolize the lineage of the species *Cyclestheria hislopi* and the stem lineage of the monophylum Cladocera. In the lineage of *Cyclestheria hislopi*, certain groups of setae emerged on the rear segments of the thorax as a presumed autapomorphy (A4)[3]. The stem lineage of a supraspecific taxon can be the lineage of a single stem species as a simple stem lineage or it can be composed of multiple successive stem species and thus form a compound stem lineage (Vol. I, p. 43). One way or another – in the stem lineage of the Cladocera – several autapomorphies (A5)[4] emerged ranging from the closed eye chamber to the reduction of thoracopods to six phyllopods. Finally, we can examine the section between the stem species y and z in our diagram. This is the stem lineage of the Cladoceromorpha in which striking features such as the fusion of the compound eyes or heterogony with an alternation of bisexual reproduction and parthenogenesis evolved (A3)[5].

[3–5] pp. 154–156

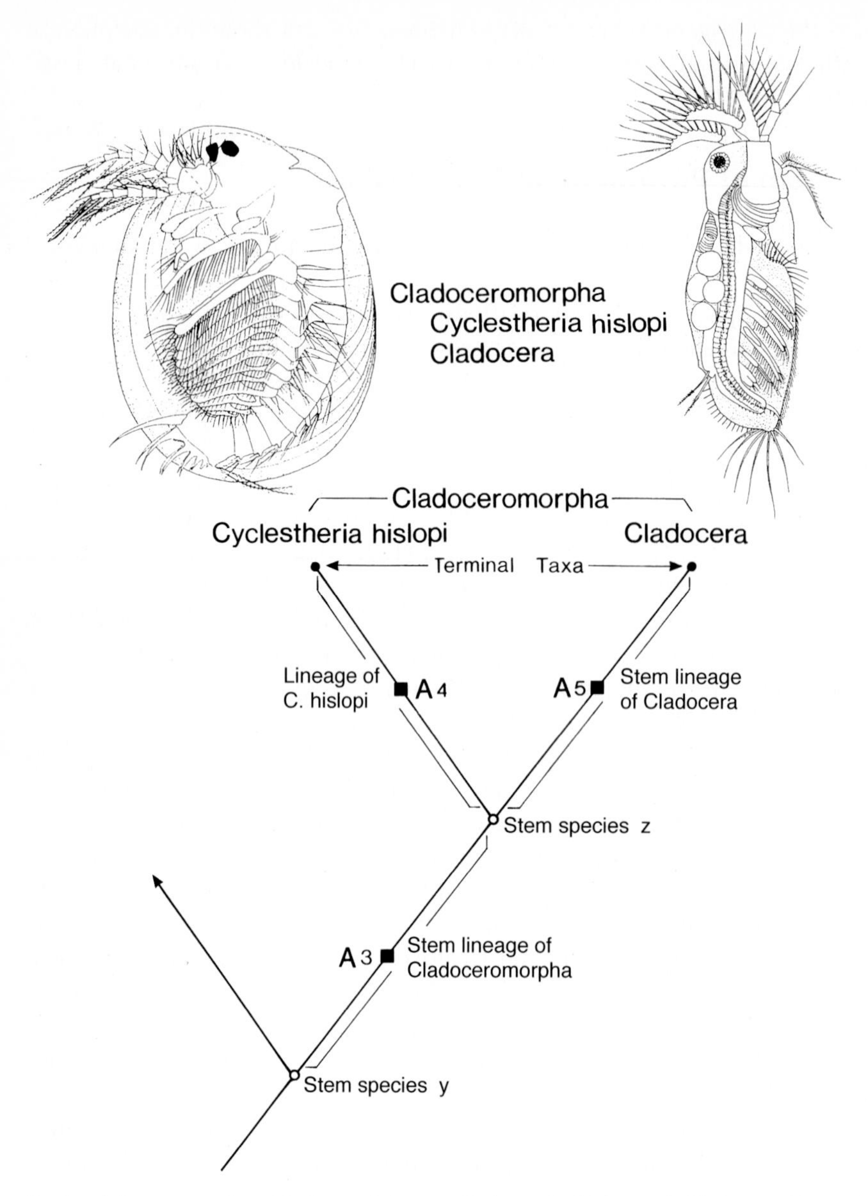

Fig. 1. Cladoceromorpha with the adelphotaxa *Cyclestheria hislopi* and Cladocera. Hierarchical tabulation (*top*) and diagram of phylogenetic relationships. The terms mentioned in the figure such as terminal taxon, stem species, line and stem lineage are explained in the main text. The presentation is taken from the systematization of the Crustacea Phyllopoda (p. 150). The autapomorphies of the monophylum Cladoceromorpha (■ A3), the species *Cyclestheria hislopi* (■ A4) and the monophylum Cladocera (■ A5) are dealt with there. Figures (from Olesen, Martin & Roessler 1997; Olesen 1998): *Left Cyclestheria hislopi*; *right Sida crystallina* (Cladocera)

We symbolize uniquely derived features or autapomorphies with black squares in the lineages of the unities they characterize. In the present example these are autapomorphy ■ 4 in the lineage of *Cyclestheria hislopi* as well as autapomorphies ■ 5 in the stem lineage of the Cladocera and ■ 3 in the stem lineage of the Cladoceromorpha. All the autapomorphies represented by one square are fully explained in the text under the relevant number. Practical aspects (knowledge, extent of describable features) determine the sequence of listing; this makes no assertions about the sequence of evolution of these autapomorphies in the lineage.

Supraspecific taxa are, of course, terminal taxa only at a certain level of the hierarchical system. With regard to the Cladocera, the solid circle denoting the terminal taxon becomes the open circle of a stem species if we want to show the systematization of the taxa included in the Cladocera which originated from the stem species of this unity.

Finally, a word about plotting relationship diagrams. The distances between the symbols for stem species and the angles between stem lineages are kept as uniform as possible – they convey no information as such. They result from the desire to give a clear visual representation of statements within the confines of one page.

The sequence of circles representing stem species in a vertical plane marks their sequence in time. However, the length of the lines connecting the circles does not contain any information about absolute time passed between stem species. Equally, the inclination of the stem lineages does not indicate in any way the speed of the evolutionary processes.

Name – Category – Rules of Nomenclature

Taxa are units of the classification of organisms – either single species or supraspecific assemblages of species. In the phylogenetic system supraspecific taxa must be justified as monophyletic unities. Names of taxa and categories for taxa are two fundamentally different things.

Names of taxa are a necessary **means of communication**. We need the name Insecta if we want to communicate universally about a taxon containing over 1 million species which emerged from one common stem species during geological time. We also need names for every single species of the Insecta – e.g., the name *Cimex lectularius*, if we want to reflect globally on the bite of the bedbug or its position in the phylogenetic system of the Insecta.

This brings us to our first problem. Species have a binomial – just as 250 years ago when LINNÉ first laid down the binomial nomenclature as a general rule in the *Philosophia botanica* (1751; JAHN et al. 1982). Supraspecific taxa receive a uninomial. In the phylogenetic system every supraspecific taxon which can be justified as a monophylum has, in principle, the right to be given its own name – this is valid for unities comprised of only two species as well as for taxa with an infinite number of species.

By creating **categories for taxa** LINNÉ bestowed a second problem on systematics; these categories were established in the first edition of the *Systema naturae* (1735). Categories such as species, genus, family, order, class or phylum are supposed to label the hierarchical ranks of taxa in an encaptic system.

Let us start again with the species taxon. Every species – such as *Cimex lectularius* – can certainly be put in a category species without any problem. However, for our professed goal of denoting their actual rank, this does not make any sense. In the phylogenetic system only sister species can have identical rank. Should one of the two sister species become a species-rich descendant community during the course of phylogenesis, then a single species becomes the adelphotaxon of a high-ranking monophylum. We have just seen such an example with the interpretation of *Cyclestheria hislopi* as the sister species of the whole of the Cladocera (water fleas). The leech *Acanthobdella peledina* is the sister species of all other species of Hirudinea (p. 69), and the worm-shaped *Xenoturbella bocki* may be the adelphotaxon of the entire Bilateria (EHLERS & SOPOTT-EHLERS 1997).

In the categories for supraspecific taxa even the requirements mentioned for species are missing. There are no objective definitions for genera, families, orders, classes ... right up to the level of kingdom, and therefore no possibilities of a justified assignment to taxa. It remains arbitrary to label the taxon Insecta with the category of class or any other term. No scientific procedure exists for using categories to represent the order in Nature. Categories have no place in the phylogenetic system of organisms.

Why these repetitions? It is time to review the **international rules of nomenclature** carefully and bring them up to the level that systematics has reached through its strict orientation towards the processes of phylogeny since the work of WILLI HENNIG (1950, 1966). Rules of nomenclature are still essential as a "universally obligatory statute book" (LAUTERBACH 1996, p. 150). The rules of nomenclature, however, must be adapted to advances in the relevant discipline and not be allowed to gag systematics as a science. The rules of nomenclature urgently need to be freed of elements originating from the time when the order of Nature was understood to be an outcome of a general Plan of Creation and not the result of evolution – this includes all categories as well as the binomial nomenclature connected with the category of genus.

Let us examine the "extortion" exerted by the rules of nomenclature with the help of two examples. In addition to the obligatory genus, the International Code gives clear recommendations to the author creating a new taxon about "class, order, and family to which his new taxon is referred" (1985, p. 231; similar 1999, p. 125).

What happens if suitable taxa with these category ranks do not exist? FUNCH & KRISTENSEN (1995) described a new species of tiny, sessile Bilateria firmly attached to the mouth parts of the lobster *Nephrops norvegicus* (Fig. 2). *Symbion pandora* cannot be placed in any of the high-ranking Bilateria taxa customarily given the category of phylum in conventional classifications. Therefore, a new family, Symbiidae, a new order, Symbiida, a new class, Eucycliophora, and a new phylum, Cycliophora, were estab-

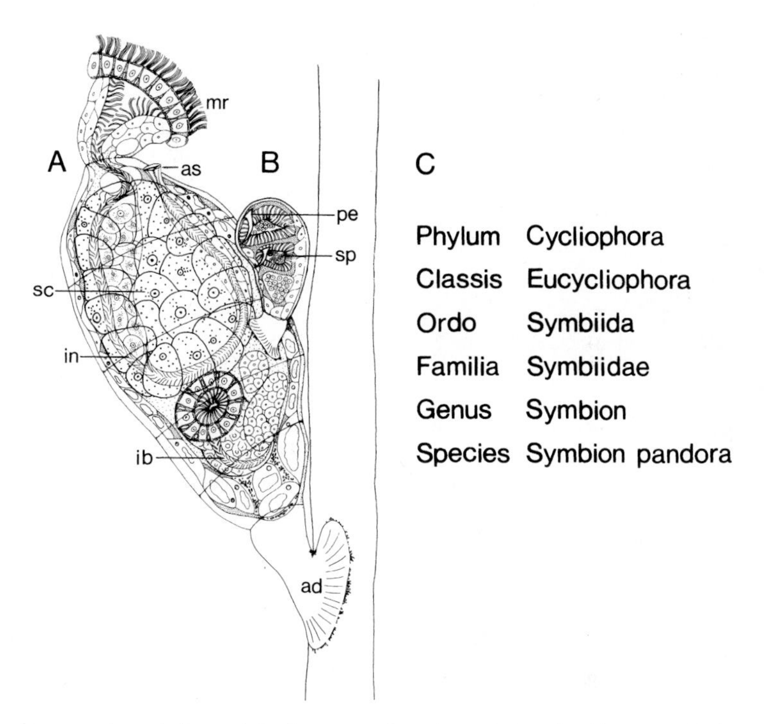

Fig. 2. A, B *Symbion pandora.* **A** Asexual feeding stage (length ca. 350 μm) with terminal adhesive disc anchored onto a chaeta in the mouth region of *Nephrops norvegicus.* Round ciliated oral opening at the entrance to a buccal funnel. U-shaped, ciliated intestinal canal. In the second half of the body an internal bud has already formed mouth and gut. **B** Dwarf male (length ca. 85 μm) attached to the feeding stage. Without mouth or gut. With two cellular bodies that contain spermatids and sperm; each exhibiting a tubular cuticular structure (? penis). **C** Six new taxa which were construed for *Symbion pandora* with categories from species to phylum. The diagnoses identical for all taxa with redundancy is "Acoelomate, marine metazoans with bilateral symmetry and differentiated cuticle. Compound cilia engaged in filter-feeding working as a downstream-collecting system. Sessile, solitary feeding stages with multiciliated cells in the gut, anus just behind a ciliated mouth ring, and internal budding with extensive regeneration of feeding apparatus. Feeding stages alternating with brief, non-feeding, free-swimming stages. Brooding of asexual larva (Pandora larva), male and female. Females brooding chordoid larva with a mesodermal, ventral rod of plate-like muscle cells (chordoid structure) and a pair of protonephridia with multiciliated terminal". *ad* Adhesive disc; *as* anus; *ib* inner bud; *in* intestine with cilia; *mr* ciliated mouth ring; *pe* penis; *sc* stomach cells; *sp* sperm. (Funch & Kristensen 1995)

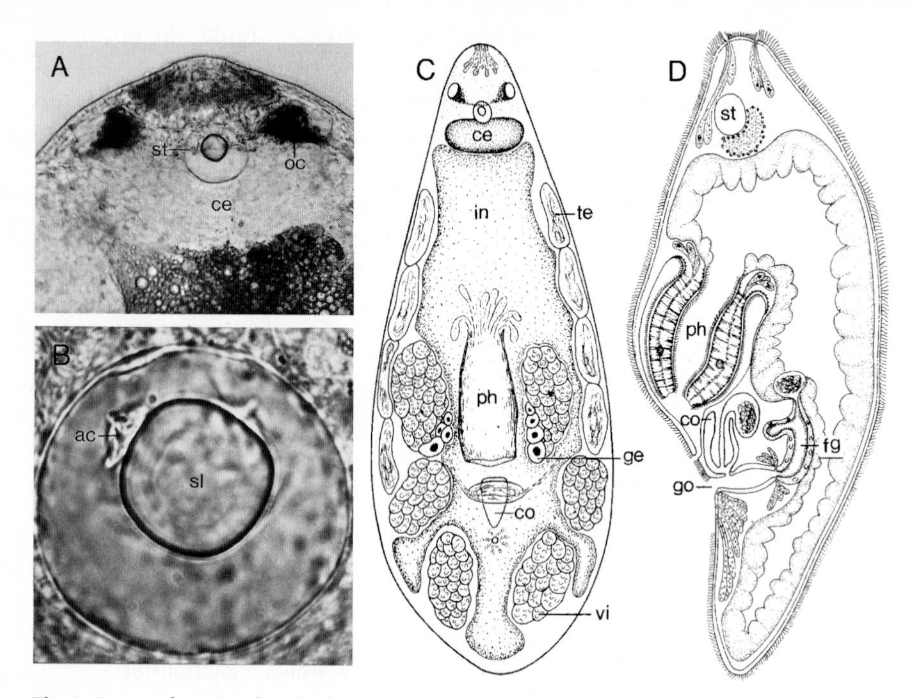

Fig. 3. *Japanoplana insolita* (Lithophora, Plathelminthes). **A** Sensorium of the anterior end with eyes, statocyst and brain. **B** Statocyst with accessory cells on the statolith. **C** Spindle to tear-shaped habitus (length ca. 0.5 mm). Organization from analyses of living animals. Pharynx in the center of the body; behind this, male copulatory organ. Four pairs of lateral testes follicles in the first two-thirds of the body. Three pairs of vitellaria follicles in the second half of the body. One pair of germaria ventrally on the first vitellaria follicle. **D** Internal organization in sagittal section. Pharynx plicatus directed to the rear. Pore of the pharynx pocket before the common opening of the male and female passageways. *ac* accessory cells on the statolith; *ce* cerebrum; *co* copulatory organ; *fg* female genital canal; *ge* germarium; *go* genital opening; *in* intestine; *oc* ocellus; *pe* penis; *ph* pharynx; *st* statocyst; *sl* statolith; *te* testes follicle; *vi* vitellaria follicle; *vs* vesicula seminalis. (Ax 1994)

lished for a single new species – as well as the obligatory genus *Symbion*. All six taxa have one and the same diagnosis – under the umbrella of the rules of nomenclature, systematics becomes an empty formalism.

What really exists in Nature? There are certain individuals attached to *Nephrops norvegicus* that have pelagic larvae in their life cycle, nothing more. A species taxon for them with the binomial *Symbion pandora* needed to be created and the adelphotaxon searched for. All names and categories above the level of the species taxon alone amount to nothing.

Japanoplana insolita (Fig. 3) was established as a species taxon for hitherto unknown populations of Plathelminthes from brackish waters in the north of Japan (AX 1994). On the grounds of the construction of the statocyst with accessory cells attached to the statolith, the new species can justifiably be integrated into the supraspecific taxon Lithophora (SOPOTT-

EHLERS 1985); this unity of Seriata contains several families of traditional classification. However, with a spindle-shaped body just 0.5 mm long, as well as only a few pairs of testes and vitellaria, *Japanoplana insolita* do not resemble the usual organization of the straight to thread-shaped Proseriata with numerous gonad follicles in serial arrangement (Vol. I, Fig. 79). Using the standards of the currently valid rules of nomenclature, I would have to establish a new family within the Lithophora as well as the obligatory genus – however, I will not do this because I refuse to use categories as damaging or useless labels. The binomial *Japanoplana insolita* is no more than a concession to the current valid rules of nomenclature – a concession to introduce the new species to science. Although I neutrally mention *Japanoplana* as a first name and *insolita* as a second name, this is in fact a totally unsatisfactory situation.

The discussion concerning a revised version of the rules of nomenclature has in the meantime been opened (DE QUEIROZ & GAUTHIER 1992; LAUTERBACH 1996; LEE 1996; LIDEN & OXELMAN 1996; SUNDBERG & PLEIJEL 1994 amongst others).

New Taxon – New Name – N. N. (Nomen nominandum)

In practical systematic work, the requirement mentioned above of an assignment of a name to each monophylum should be met in the interest of optimal communication. Obviously, this must be done cautiously – only after thorough justification of the relevant species groups as a monophylum. In some situations it is better to postpone this procedure, for example, if two unities can apparently be interpreted as possible adelphotaxa, but there is as yet no ample evidence for their union under a new name. In such cases we can use the words Nomen nominandum[6] – to be precise "a yet to be named name" – as a temporary solution instead of an ill-founded new name.

The abbreviation N. N. is identical to that of the Nomen nudum that is included in the rules of nomenclature; however, the meanings of **Nomen nominandum** and of **Nomen nudum** have nothing in common. Nomen nudum is an existing, but bare and not available, name because a description or definition for the taxon is missing.

Any author who constructs a new supraspecific taxon with a new name should unequivocally indicate that they have done so. In phylogenetic systematics, which have no categories, this is best done with the additional remark **Taxon novum (tax. nov.)**. I failed to do this in Volume I, where Epitheliozoa, Rhopaliophora and Acrosomata are new taxa with new

[6] Meyers Lexicon. Vol. 15, p. 302. Bibliographic Institute & F.A. Brockhaus AG, Munich (1990).

names. In this, the second Volume, every new taxon is marked distinctly as a tax. nov.

Phylogenetic systematics has been vehemently criticized because it inevitably generates new taxa and taxa names. Those, however, who appreciate the problems concerned with trying to clearly represent the hierarchical division of an extensive system unity from their own work will surely not criticize an unjustified proliferation of taxa and names. To take an example from the group Crustacea – endless sentences and stylistic contortions would be necessary if I had to constantly refer to the common possession of phyllopodia to filter fine particles out of the water as seen in the Anostraca, the Phyllopoda and in the ground pattern of the Malacostraca (realized in the Leptostraca). In place of this long description, I can now mention them once as an autapomorphy of a newly established and newly named taxon Phyllopodomorpha, consisting of the Anostraca, Phyllopoda and Malacostraca (p. 147).

Objectivity

I dedicate this last section to the common misjudgment of the scientific character of statements given in phylogenetic systematics. I am frequently confronted with the opinion that the decisions made in phylogenetic systematics are also subjective and as such no better than the procedure followed in traditional classifications.

This opinion is founded on the following misunderstanding. Statements are made by single individuals as subjects. If we make the noun subject into an adjective, the statements of each single phylogenetic systematist are of course in this sense subjective. But the decisions are not "subjective" in the sense of the common meaning of this word, i.e. "characterized by personal prejudices" or "not of universal validity but random and arbitrary". Phylogenetic systematics has a clear methodology which allows all its statements to be presented in the form of objective – i.e. fully justified and intersubjectively testable – hypotheses. This ranges from decisions of probability between the alternatives of plesiomorphy – apomorphy and synapomorphy – convergence, through the assessment of features as autapomorphies of monophyletic species groups, up to the establishment of the system through the interpretation of relationships between sister groups.

Every hypothesis can be checked at any time and by anyone and thus can be either confirmed or rejected. Should a hypothesis be rejected it must be replaced by one which is empirically better justified (KLUGE 1997). This is objectivity in a historical science oriented towards feature patterns which have evolved over millions of years in innumerable individuals of recent and fossil species.

Phylogenetic systematics determines its position in biology by postulating objective hypotheses about the phylogenetic order in Nature. It is strictly against any forms of speculation which cannot be tested empirically, these unfortunately still inundate phylogenetic literature.

In the figures, I have taken pains to clearly depict autapomorphies of each group. Since these are highlighted in the accompanying legends, the main text should be free from excess references to figures.

Fruitful discussions with Prof. Thomas Bartolomaeus accompanied the preparation of this work. I thank him in particular for the support in his research area – the Annelida and the Mollusca. Dipl.-Biol. Claudia Wolter read the manuscript with a critical eye and suggested many improvements; she prepared original drawings, copied figures from the literature, read the galley proofs and prepared the index. Ms. Renate Grüneberg again managed the computer and undertook the organization of the kinship diagrams and figure lettering. My sincere thanks to both coworkers.

I thank again with pleasure Dr. Carlo Servatius and the Academy of Sciences and Literature in Mainz for their harmonious cooperation.

Last, but not least, I am grateful to Susan Kinsey for her careful translation of the original German version into English.

PETER AX

Contents

Trochozoa

We continue our systematization of the Metazoa with the Spiralia taxon Trochozoa. The Trochozoa are distinguished by a biphasic life cycle in their ground pattern, of which the trochophore larva, which has a prototroch, a pair of ocelli and (?) an apical organ, is the autapomorphy (Vol. I, p. 211) (Fig. 4 → 1).

The Mollusca, Annelida and Arthropoda are the three large, species-rich subtaxa of the Trochozoa and all can justifiably be established as monophyla. There has been relatively little opposition to uniting the Annelida and the Arthropoda into one taxon – the Articulata. However, there is still considerable uncertainty about the kinship relations between the Mollusca and the Articulata, as well as with the smaller unities of the Kamptozoa, Sipunculida and Echiurida. We argue here for a sister group relationship between the Kamptozoa and the Mollusca and between the Sipunculida and the Articulata. The first we unite under the name Lacunifera the second under Pulvinifera. In contrast, the Echiurida are a subordinate taxon of the Annelida.

This all results in the following systematization, the validity of which will be successively demonstrated (Fig. 4).

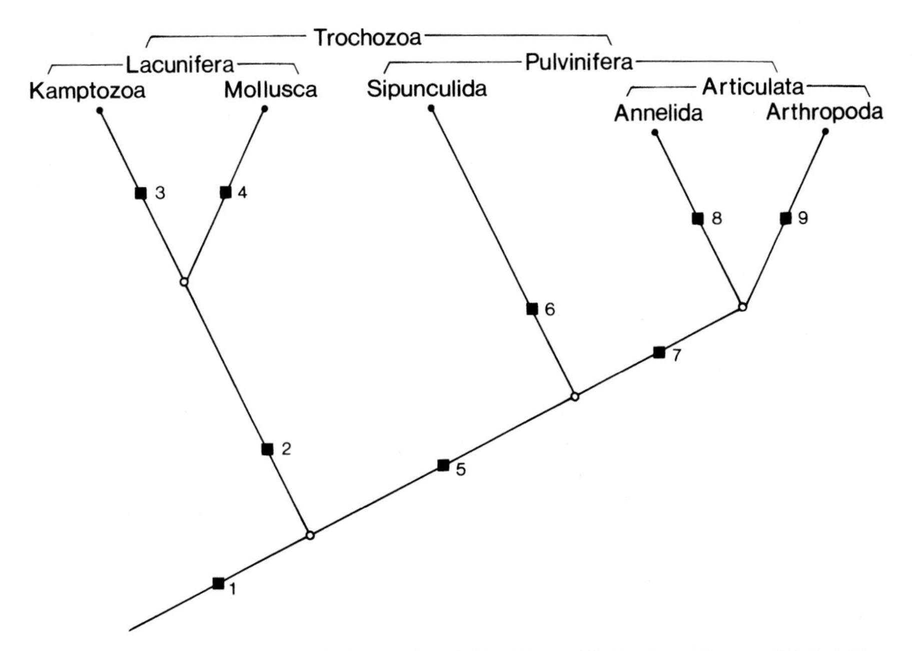

Fig. 4. Trochozoa. Diagram of phylogenetic relationships with the Lacunifera and Pulvinifera as the highest-ranking adelphotaxa

Trochozoa
- **Lacunifera**
  - **Kamptozoa**
  - **Mollusca**
- **Pulvinifera**
  - **Sipunculida**
  - **Articulata**
    - **Annelida**
    - **Arthropoda**

Lacunifera [1]

Because of their trochophore larvae, the Kamptozoa belong in the Spiralia unity Trochozoa. This assertion is now generally accepted. The sister group, however, is as yet unrecognized. Within the scope of phylogenetic systematics, only NIELSEN (1971, 1995) has suggested that an adelphotaxa relationship exists between the Entoprocta (Kamptozoa) and the Ectoprocta (Bryozoa) – albeit with the proviso that "it is difficult to identify synapomorphies of the two phyla" (NIELSEN 1995, p. 183).

In contrast, during the last few years, several arguments for a sister group relationship between the Kamptozoa and the Mollusca have crystallized. (BARTOLOMAEUS 1993a,b, 1997; SALVINI-PLAWEN & BARTOLOMAEUS 1995; HASZPRUNAR 1996). Congruencies exist in four points which can, without argument, be shown to be synapomorphies of the Kamptozoa and Mollusca. Thus they become autapomorphies of the unity Lacunifera.

Autapomorphies (Fig. 4 → 2)

- Dorsal protein-chitin cuticle.
  The covering of the dorsal surface with a cuticle consisting of a protein-chitin complex is observed in both the Kamptozoa and the Mollusca. This feature also distinguishes them from the Sipunculida and the Annelida which have a collagen cuticle.
- Flat ventral surface with cilia.
  In their ground pattern, the Spiralia have a round body completely covered with cilia. The evolution of the dorsal cuticle has restricted ciliation to the ventral surface in the Lacunifera. Here, a flat area with cilia is differentiated in which the mouth and anus are found. The anus has thus moved ventralwards from the presumed primary dorsal position of the Euspiralia (Vol. I, p. 199).

---

[1] tax. nov.

- Foot with creeping sole.
  The state of the ciliated ventral surface can be further specified. A trochophore larva is part of the ground pattern of the Kamptozoa, in which the hyposphere is transformed into a ciliated creeping foot (see below). The obvious similarity with the foot of the Mollusca points to a homology between the two taxa.
- Body cavity with lacunal system.
  Kamptozoa and Mollusca have a compact, acoelomate organization with mesenchyme cells between the epithelia and the organs. This is the plesiomorphous condition of the body cavity of the Bilateria (Vol. I, p. 113). Kamptozoa and Mollusca also possess in their body cavity a special lacunal system bounded by an extracellular matrix for the transport of hemolymph. The assumption that this lacunal system developed only once within a common stem lineage of the Kamptozoa + Mollusca offers the simplest explanation.

Kamptozoa – Mollusca

Kamptozoa

With the distinctive nodding and swaying of their stalks the Kamptozoa (Entoprocta) can easily be distinguished from the polyp colonies of Hydrozoa with which they are often associated on the ocean floor.

Kamptozoa measure only a few mm in length. Amongst the ca. 250 species there are a number of primarily solitary organisms; the majority, however, form colonies. The "Solitaria" (*Loxosoma, Loxocalyx*) have plesiomorphic characteristics throughout and can therefore not be designated as a monophylum. However, the Coloniales (*Pedicellina, Barentsia, Urnatella*) can be seen as a monophyletic unity (EMSCHERMANN 1972).

The Kamptozoa are primarily marine organisms. Only the species *Urnatella gracilis* is found in fresh water.

Fundamental organizational features of the Kamptozoa are at the same time derived characteristics (FRANKE 1993). I list first the autapomorphies which will then be more fully described in the text.

Autapomorphies (Fig. 4 → 3)

- Sessility.
- Division into calyx and stalk.
- Tentacles with five longitudinal rows of ciliated cells and the mechanism of filtration.
- Larva with frontal organ (Fig. 5 D).
- Metamorphosis with torsion of the atrium and intestinal tract.

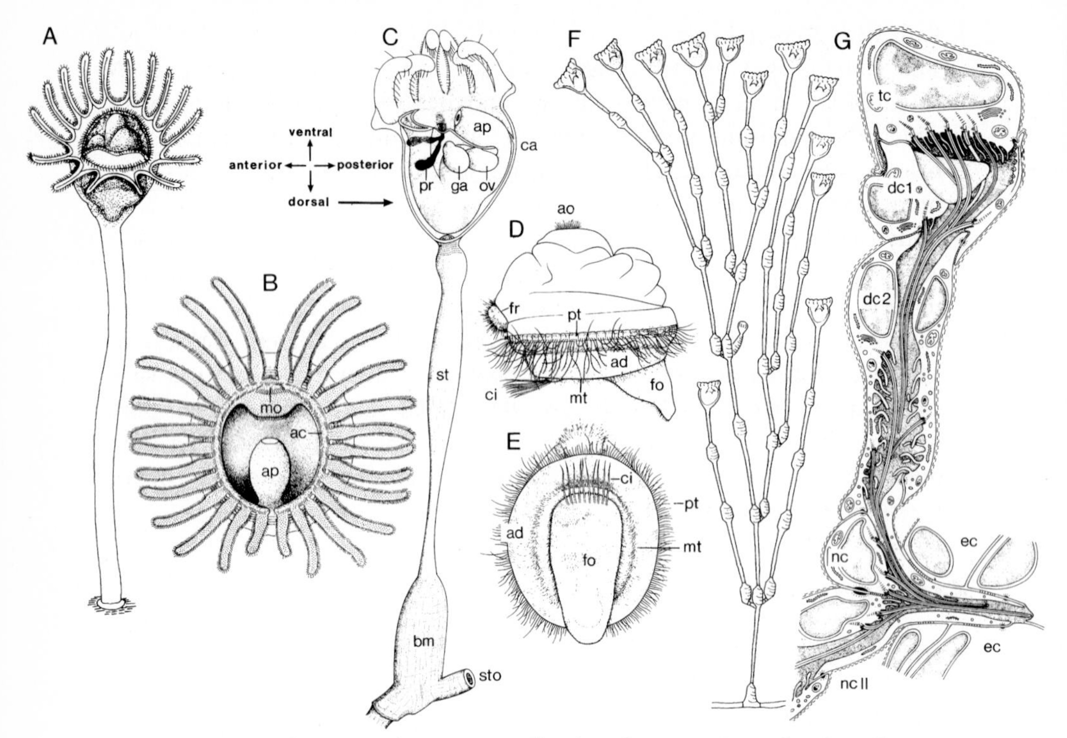

Fig. 5. Kamptozoa (Entoprocta). **A** *Loxosomella phascolosomata.* Example of a solitary species. **B** *Pedicellina cernua.* Colonial species. Inner surface of the calyx=morphological ventral side of the animal. The mouth and the anus mounted on a papilla are surrounded by a crown of tentacles. **C** *Barentsia matsushimana.* Colony-building species. Zooid in lateral view. Half of the tentacle crown removed. End of the stalk with a muscular joint. **D–E** Larva of *Barentsia laxa.* **D** Side view of the larva. From the creeping foot of the hyposphere a bundle of long cilia is visible to the front, to the rear the lobe-shaped end. **E** Ventral view of the larva with prototroch, adoral ciliary zone, metatroch and ciliated creeping foot. **F** *Barentsia ramosa.* Small part of a colony. **G** *Barentsia matsushimana.* Ultrastructure of the protonephridia in longitudinal section. Common opening with the second protonephridium (below). *ac* Peripheral atrial ciliated band; *ad* adoral ciliary zone; *ao* apical organ; *ap* anal papilla; *bm* basal muscle joint; *ca* calyx; *ci* compound cilia of the larval creeping foot; *dc* duct cell; *ec* epithelial cell of the atrium; *fo* foot; *fr* frontal organ; *ga* ganglion; *mo* mouth; *mt* metatroch; *nc* nephropore cell; *nc II* nephropore cells of the second protonephridium; *ov* ovary; *pr* protonephridium; *pt* prototroch; *st* stalk; *sto* stolo; *tc* terminal cell. (**A** Hartwich 1993; **B** Brien 1959; **C,G** Franke 1993; **D,E** Nielsen 1971; **F** Wasson 1997)

With regard to the Kamptozoa, sessility is most definitely a derived characteristic. The division of the organism into calyx and stalk is related to this mode of life. A cup-shaped calyx with a crown of tentacles projects into the water. A slender stalk binds the animal to the substrate. The space which is enclosed by the tentacles is known as the atrium.

The ciliation of the tentacles is very characteristic. Every tentacle has five rows of ciliated cells directed towards the atrium. Two lateral rows of myoepithelium cells with long cilia flank three median rows with short cilia. (NIELSEN & ROSTGAARD 1972; EMSCHERMANN 1996; NIELSEN &

JESPERSEN 1997). These structures carry out the following filtration function. The long lateral cilia pull water up from outside through the tentacles and into the atrium. Fine particles are captured by the short median cilia and transported downwards. On the inner edge of the atrium further ciliated bands are found which sweep the food particles into the mouth.

Mouth and anus both lie within the crown of tentacles – in clear contrast to the Bryozoa (Ectoprocta), in which the anus opens outside the tentacles. The plane between the mouth and anus marks the plane of bilateral symmetry.

The larva of the Kamptozoa has the basic characteristics of the trochophore with an apical organ, and a prototroch and metatroch between which is an adoral ciliated zone (JÄGERSTEN 1972; NIELSEN 1971, 1987). In addition to these features, which are plesiomorphies carried over from the ground pattern of the Trochozoa, there are two larval characteristics that can be discussed as being autapomorphies of the Kamptozoa. (1) The episphere of the larva possesses a frontal organ – primarily paired and with ocelli. This condition is found in the solitary species. In colonial species the frontal sense organ (preoral organ) is unpaired and without ocelli. (2) The larva can crawl over the substrate with a ciliated foot. This is realized by a differentiation of the neurotroch (gastrotroch) of the hyposphere with a complex of long cilia transverse to the anterior (Fig. 5 D, E). The creeping foot is doubtless an apomorphy of the aforementioned trochophore. It must be noted, however, that a creeping foot may be part of the ground pattern of the Lacunifera (Kamptozoa + Mollusca) (see above) which was carried over by the trochophore larvae of the Kamptozoa.

Following attachment of the larva by means of the body wall around the prototroch, an extensive metamorphosis takes place. An invagination of the body wall forms an atrium with which the ventral surface including mouth, anus and foot is closed off from the outside environment. Atrium and intestine then twist themselves around the lateral axis upwards. The original attachment point develops into the stalk. The atrium opens and the tentacles develop and unfold.

Finally I wish to emphasize two plesiomorphies. A pair of unbranched protonephridia belongs in the ground pattern of the Kamptozoa (Fig. 5 G). Each protonephridium consists of four cells – a multiciliated terminal cell, two canal cells and a nephroporous cell (FRANKE 1993). In the freshwater *Urnatella gracilis* strong branching of the protonephridia due to an increase in the terminal organs occurs (EMSCHERMANN 1965).

Furthermore, vegetative propagation may also be a plesiomorphy of the Kamptozoa. In solitary species the ectodermal buds are found on the ventral surface of the calyx between the tentacles; these then detach themselves. In the colonial Kamptozoa the buds of the new zooids develop on stalk bases or stalks of the mother organism. After detachment the remaining connection becomes the stolo. With a steady production of buds, colonies of hundreds of individual zooids can develop (Fig. 5 C, F).

Mollusca

Within this highly diverse group are included the well-known Gastropoda, Cephalopoda and Bivalvia. These are the most species-rich taxa of the Mollusca, which itself is a well-justified monophylum of the Spiralia.

As with the Kamptozoa I will first list the conspicuous autapomorphies. These will then be further explained during the description of the important features of the ground pattern. For these points I refer especially to the works of SALVINI-PLAWEN (1985), HASZPRUNAR (1992), as well as BARTOLOMAEUS & AHLRICHS (1998).

Autapomorphies (Figs. 4 → 4; 6 → 1)

- Dorsal chitin-protein cuticle with calcareous spicules (Fig. 11 D).
- Ventral differentiation into head and foot.
- Mantle with mantle groove.
- Radula apparatus.
- Heart and gonopericardial system.
- One pair of bipectinate gills.
- One pair of osphradia.
- Tetraneural nervous system.
- One pair of muscles for rolling into a ball.

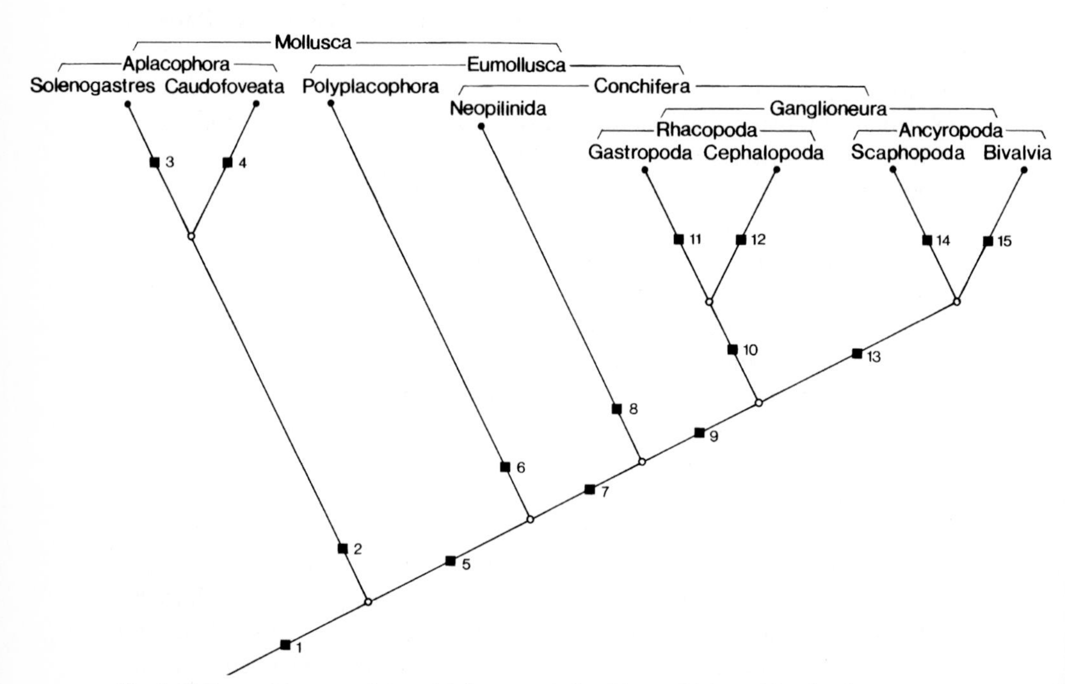

Fig. 6. Mollusca. Diagram of the adelphotaxa relationships at high-ranking levels

The **dorsoventrally flattened stem species** of the Mollusca was possibly a **small organism** only a few mm long. The division into a dorsal chitin-protein cuticle and a ventral foot with a ciliated creeping sole may have derived from the ground pattern of the Lacunifera; the provision of the cuticle with calcareous spicules is, however, an apomorphic peculiarity of the Mollusca. The same can be said for the head with a perioral bulge which is separated from the foot by a ventral transverse groove.

We will now examine the **mantle** of the Mollusca. The dorsal epithelium with cuticle rises over the head-foot area and wraps around to the ventral surface creating a fold. In this fashion the mantle is formed with an outer and an inner epithelium. Between the latter and the foot, the mantle groove is found which encircles the body. A local extension of the mantle groove into the mantle cavity cannot be postulated to be part of the molluscan ground pattern since such a feature is not seen in the Polyplacophora or Neopilinida. This fact leads to the hypothesis that a caudal mantle cavity may have evolved independently in the stem lineages of the Aplacophora (Solenogastres + Caudofoveata) and the Rhacopoda (Gastropoda + Cephalopoda).

The **radula apparatus** (Fig. 7) consisting of a radula and radula support (odontophore) is a unique, apomorphous construction of the Mollusca without any equivalent in any other animals. The apparatus serves to rasp away food particles and in the uptake of detritus. It can also be used in microcarnivorous feeding (Solenogastres, Caudofoveata).

The radula is a flexible band with cuticular teeth in the buccal cavity. When rasping, the odontophore is held vertically and projected in and out of the buccal cavity; in doing so the leading recurved teeth scratch and rasp over the substrate. Teeth and band are destroyed distally and both are continuously renewed proximally within the radula sheath by odontoblasts.

Kamptozoa and Mollusca both possess in their compact, mesenchymic body cavity a lacunal system bounded by an extracellular matrix. The origin of the transport system for hemolymph is hypothesized to have begun in a common stem lineage of the two taxa (p. 3). Therefore this has the status of a plesiomorphy in the ground pattern of the Mollusca.

The Mollusca also have a secondary body cavity with an epithelial covering. The **gonopericardial system** (Fig. 8 A) consists of a pair of gonocoela for the gonads and a pericardium. The pericardium encloses the heart (ventricle) into which two atria (auricles) lead and from which an anteriorly directed aorta (dorsal sinus) opens into the lacunal system. From the gonocoela a pair of gonoducts originate, and from the pericardium two excretory pericardioducts. In the Eumollusca these commonly have separate openings in the mantle groove and it has been suggested that this is part of the molluscan ground pattern. However, it has also been proposed that the original form may be one in which the gonoducts and pericardioducts are united and open through a single pore on each side (HASZPRUNAR

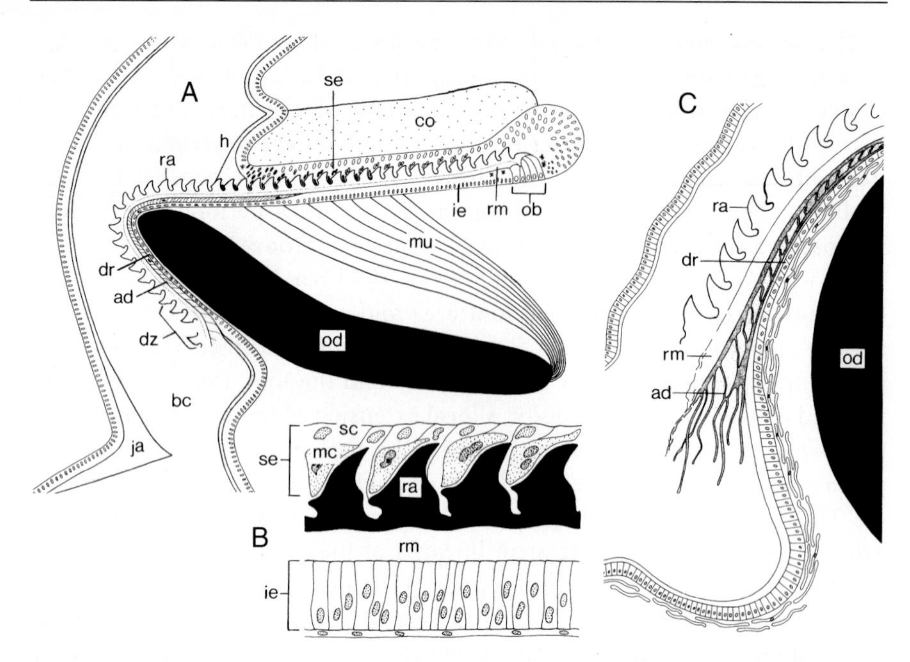

Fig. 7. Example of the Mollusca radula apparatus in the fresh-water snail *Lymnaea stagnalis* (Pulmonata, Gastropoda). **A** Longitudinal section through the radula apparatus with radula sheath (*top*) and odontophore (*bottom*). Radula teeth and radula band develop in the closed end of the sheath from odontoblasts. To the front the upper epithelium forms a cap-like closure of the sheath. In the region of the opening, the lower epithelium forms a special adhesive layer (subradular membrane) beneath the radula membrane and continues into the distal radular epithelium. The transport of the radula is probably based on pseudopodia-like movements of this distal epithelium. The degeneration zone is found at the anterior of the radula in the oral cavity. **B** Longitudinal section through the middle part of the radula sheath. Cells of the upper epithelium harden the radula teeth through secretion of minerals and organic compounds. The lower epithelium supports the radula band. **C** Degeneration zone of the radula. The adhesive layer between the radula membrane and the distal radula epithelium is frayed; the radula teeth fall out. *ad* Adhesive layer (subradular membrane); *bc* buccal cavity; *cb* cuticle of the buccal cavity; *co* collostyle; *dr* distal radula epithelium; *dz* degeneration zone; *eb* epithelium of the buccal cavity; *h* hood (cap); *ie* lower epithelium; *ja* jaw; *mc* mineralizing cell; *mu* muscle; *ob* odontoblast; *od* odontophore; *ra* radula tooth; *rm* radula membrane; *sc* support cell; *se* upper epithelium. (Mackenstedt & Märkel 1987)

1992); "pericardium with one pair of outlets; paired gonad separated rostrally from pericardium with gonoducts separate or somewhere opening into pericardioducts" (SALVINI-PLAWEN & STEINER 1996, p. 31).

The gonopericardial system of the Mollusca is used for the circulation of hemolymph, for excretion and for reproduction. Since in structure and function it is fundamentally different from the coelom of the Pulvinifera (Sipunculida, Annelida), which is a hydrostatic organ used for the purposes of locomotion, the derivation of one from the other is unfeasible. Consequently the gonoducts and the pericardioducts of the Mollusca cannot be interpreted as

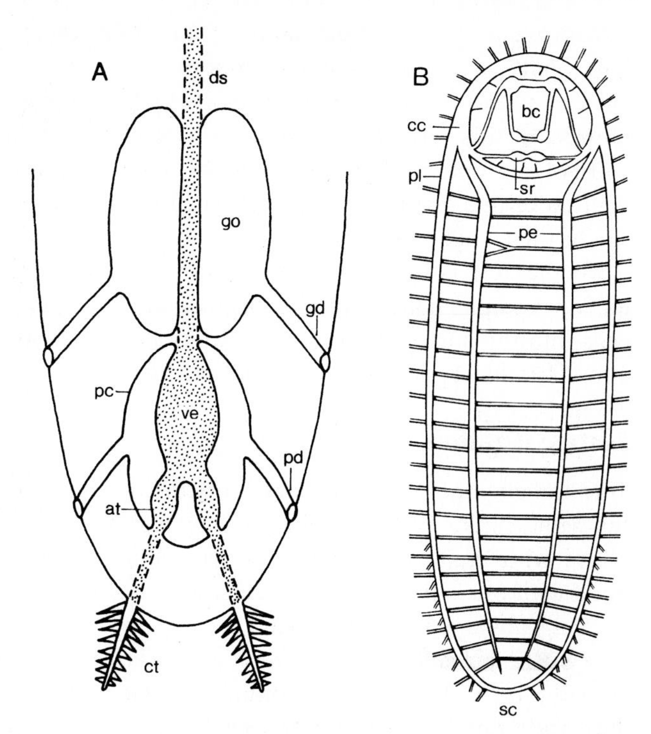

Fig. 8. Mollusca. **A** Basic elements of the gonopericardial system. Paired gonocoela with gonoducts. Unpaired pericard with two pericardioducts. The pericard surrounds the heart (ventricle) and two atria (auricles) leading into the heart. The blood comes from the gills, enters from behind into the atria and is conducted forward from the heart by an aorta (dorsal sinus). **B** Diagram of the tetraneural nervous system (Polyplacophora). Main elements are the cerebral ring, two lateral pleurovisceral cords and two median pedal cords. *at* Atrium (auricle); *bc* buccal ring; *cc* cerebral ring; *ct* ctenidium; *ds* dorsal sinus (aorta); *gd* gonoduct; *go* gonad; *pc* pericard; *pd* pericardioduct; *pe* pedal cord; *pl* pleurovisceral cord; *sc* suprarectal commissure; *sr* subradular ganglion; *ve* ventricle (heart). (**A** Salvini-Plawen 1985; **B** Fischer-Piette & Franc 1960)

being homologous with the metanephridia of the Sipunculida or the Annelida. In the stem lineage of the Mollusca the gonopericardial system obviously evolved independently of secondary body cavities in other Spiralia unities (SALVINI-PLAWEN & BARTOLOMAEUS 1995; BARTOLOMAEUS 1997).

Further autapomorphies in the ground pattern are a pair of distichous pinnate **ctenidia** (bipectinate gills) and a pair of sensory **osphradia** in the mantle groove. The primary function of the ctenidia may be ventilation of the mantle cavity rather than respiration.

The **tetraneural nervous system** (Fig. 8 B) with its cerebral commissure and four nerve cords is a characteristic feature of the Mollusca. The lateral pleurovisceral nerves and ventral pedal nerves are connected by periodic commissures. The pleurovisceral nerves unite terminally via a suprarectal commissure.

With regard to the musculature of the Mollusca, the **muscles for rolling into a ball** are a prominent autapomorphy. These are a pair of longitudinal muscles found in the side of the foot which have mediodorsal antagonists. Muscles of this type make sense as a protection system in the Aplacophora and Polyplacophora, however, they are also found in the Neopilinida. A reduction of this feature can be postulated to have occurred in the stem lineage of the Ganglioneura.

No characteristic features for the Mollusca can be derived from their ontogeny. Ground pattern plesiomorphies are the freely released sperm and eggs into the surrounding waters, external fertilization, spiral-cleavage and a biphasic life cycle with a trochophore larva.

Systematization

Traditionally the Mollusca have been divided into several classes all with the same rank. These are the Aplacophora without shells, the Polyplacophora with eight shell plates, the Neopilinida, Gastropoda and Cephalopoda with one shell and the Bivalvia with a shell composed of two valves.

Using phylogenetic systematics we can free ourselves from the outdated idea of the existence of equal molluscan classes and reject these categories as being obstacles to understanding. A new system can now be presented with wide-ranging agreements over the interpretation of adelphotaxa relationships, in which former classes are placed at different hierarchical levels (LAUTERBACH 1983c, 1984a; SALVINI-PLAWEN 1990; HENNIG 1994; SALVINI-PLAWEN & STEINER 1996; SCHELTEMA 1996; D.L. IVANOV 1996; BARTOLOMAEUS & AHLRICHS 1998).

Mollusca
- **Aplacophora**
  - **Solenogastres**
  - **Caudofoveata**
- **Eumollusca**
  - **Polyplacophora**
  - **Conchifera**
    - **Neopilinida**
    - **Ganglioneura**
      - **Rhacopoda**
        - **Gastropoda**
        - **Cephalopoda**
      - **Ancyropoda**
        - **Scaphopoda**
        - **Bivalvia**

Aplacophora – Eumollusca

The Aplacophora and the other taxa combined within the Eumollusca are identified as the highest-ranking adelphotaxa of the Mollusca. Both are characterized by a series of apomorphous features. They both also have a number of plesiomorphies. They therefore fulfill the criteria for their recognition as sister groups.

The most conspicuous differences are found in the periphery. Through out-group comparison the Aplacophora can be interpreted as primary shell-less Mollusca; their worm-shaped form is, however, an apomorphy. The Eumollusca continue to have as a plesiomorphy the ground pattern division into a broad foot and flat visceral mass. In their stem lineage the shell arises as an evolutionary novelty.

Aplacophora

For the following features an evolution within a common stem lineage of the Solenogastres and Caudofoveata is hypothesized.

Autapomorphies (Fig. 6 → 2)

- Evolution of a vermiform body together with a change in habitat from the sediment surface to within soft sediment.
- Reduction of the broad foot with creeping sole. In the ground pattern of the Aplacophora only a ventral groove with a folded foot remains.
- Papilla cells of the epidermis with a presumed excretory function. Single large cells are pushed with the cuticle into the periphery; they evacuate the contents of apical vesicles to the exterior.
- Mantle cavity as an extension of the mantle groove at the posterior.
- Gonopericardioducts connect the gonads with the pericardium. Separate gonoducts – as in the Eumollusca – are not seen. Instead the gametes are led via the pericardium and the pericardioducts into the mantle cavity.
- Mucus channels (mucus grooves), differentiation products of the mantle cavity, conduct gametes into the water.

Solenogastres – Caudofoveata

Solenogastres

Marine organisms which are either epibenthic or live within the sediment. They exist partly on polyp colonies of Cnidaria and feed on them.

Body length can range from a few mm to 30 cm. *Nematomenia, Dorymenia.*

A prominent feature of the Solenogastres is the foot groove (ventral furrow, ventral fold) which comes from the ground pattern of the Aplacophora. In the longitudinal slit on the ventral side, the foot is found in the form of one or more folds. Mucus glands with openings into an anterior ciliated pit and along the furrow produce a band of mucus on which the Solenogastres glides.

Following separation from the Caudofoveata line, the features listed below must have evolved in the stem lineage of the Solenogastres.

Autapomorphies (Fig. 6 → 3)

- Reduction of the gills.
- Change to hermaphroditism.
- Opening of the gonoducts into voluminous spawn ducts formed from invagination of mucus channels of the mantle cavity into the body.

Caudofoveata

Inhabitants of soft marine sediment out of which the posterior end with gills may project. Body length can vary from between 3 mm to 14 cm. *Chaetoderma, Falcidens.*

The Caudofoveata have a cylindrical body that is, with the exception of the foot shield, fully covered with calcareous scales embedded in the cuticle. The ventral groove is reduced, as is the foot along the ventral side. As a presumed remnant of the foot a shield-like flexible structure behind the oral opening is formed. This functions as a burrowing tool in the sediment. The oral shield is one of the basic autapomorphies of the Caudofoveata.

The Caudofoveata are plesiomorphous in being gonochoristic organisms and in possessing two gill combs (ctenidia) in the mantle cavity.

Autapomorphies (Fig. 6 → 4)

- Oral shield (Fig. 9 E).
- Free hemoglobin in serum.
- Fusion of the pleurovisceral and pedal nerves at the posterior to form a longitudinal stem on both sides.

Eumollusca (Testaria)

Autapomorphies (Fig. 6 → 5)

- Solid calcareous shell in the form of eight plates on the dorsal surface.

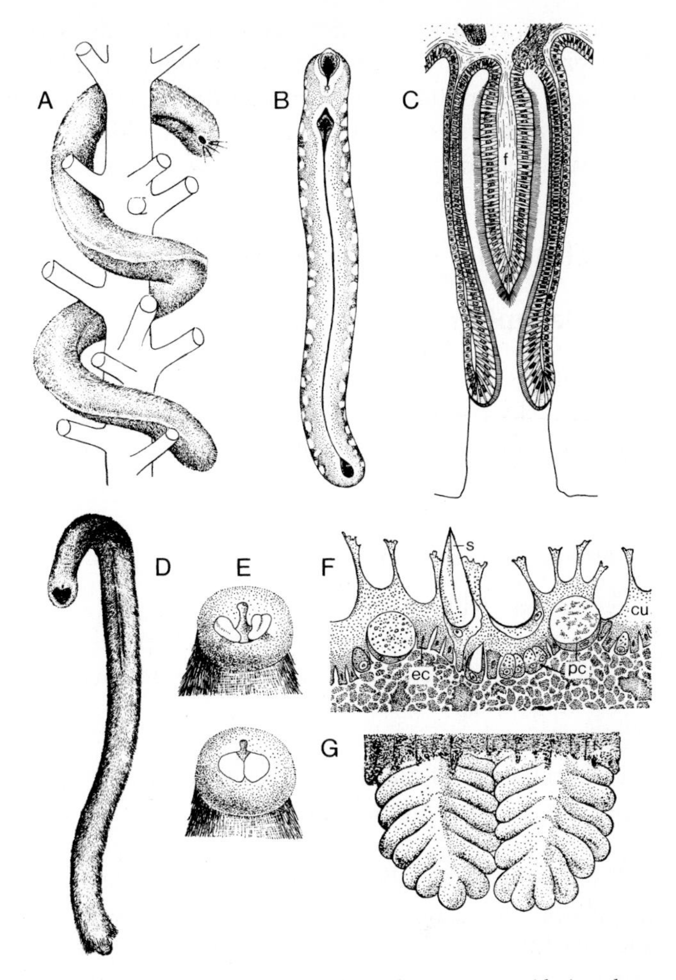

Fig. 9. Aplacophora. **A–C** Solenogastres. **A** *Nematomenia banyulensis* on a Cnidaria colony. **B** *Epimenia verrucosa*. View of the ventral side with foot groove. **C** *Dorymenia antarctica*. Transverse section through the groove and foot. **D–G** Caudofoveata. **D** *Scutopus ventrolineatus*. **E** *Limifossor talpoideus*. Ventral view of the anterior with the partitioned foot shield in various states. **F** *Chaetoderma nitidulum*. Section through the epidermis with calcareous stalk and papilla cells (latter: autapomorphy of the Aplacophora). **G** *Falcidens crossotus*. Posterior ctenidia. *cu* Cuticle; *ec* epidermis cell; *f* foot; *pc* papilla cell; *sp* spiculum. (**A, B, D–G** Salvini-Plawen 1971; **C** Kilias 1993 a)

- 16 pairs of dorsoventral muscle bundles used as foot retractors, two pairs per shell plate. Through contraction of the muscles the body is anchored firmly to the substrate.
- One pair of kidneys – differentiation products of the pericardioducts. The sac-like kidneys are located at the sides of the body.

The number of gills in the stem species of the Eumollusca is disputed. Most probably, a pair of bipectinate ctenidia was inherited as a plesiomorphy from the ground pattern of the Mollusca. Independent additions then occurred in the stem lineages of the Neopilinida, Polyplacophora and Nautiloidea (Cephalopoda).

Polyplacophora – Conchifera

Polyplacophora

The chitons inhabit marine littoral hard substrates, especially the surf zone of rocky coastlines. Their body length usually ranges from between a few mm to a few cm; the maximum reached is seen in *Cryptochiton stelleri* at more than 30 cm.

Autapomorphies (Fig. 6 → 6)

- Tegmentum with aesthetes.
  The outer calcareous layer of the shell plates containing organic substances is traversed by cell processes which are components of the sensory apparatus.
- Articulamentum.
  Differentiation of the hypostracum to anchor the shell plates beneath one another and into the girdle.

Their eight dorsal shell plates create the most noticeable feature of the Polyplacophora – indeed, these plates gave rise to the name of the unity and make them ideally suited for identification. Nevertheless, the eight calcareous plates are not in themselves a reason for establishing the Polyplacophora as a monophylum. It can be hypothesized that a shell consisting of eight plates is more likely to be a feature that was already present in the ground pattern of the Eumollusca as an evolutionary novelty. In comparison to the single shell of its sister group Conchifera, this feature must be viewed as a plesiomorphy of the Polyplacophora.

Two pairs of foot retractors are inserted at every plate. These must also be considered as a plesiomorphy which originated together with the evolution of the eight plates in the stem lineage of the Eumollusca. The eight shell plates overlap each other rather like roofing tiles from the anterior to the posterior. They are framed by the perinotum, a further plesiomorphic feature of the Polyplacophora. The soft girdle with cuticle and calcareous spicules or spines is the remains of the original body wall of the Mollusca. The most usual relationship between the broad plates and narrow girdle is best seen in *Lepidochiton cinereus* or *Tonicella lineata*. Secondary enlarge-

Fig. 10. Polyplacophora. **A** *Lepidochiton cinereus*. Length 1–2 cm. **B** *Tonicella lineata*. Length ca. 2.5 cm. **C** *Katharina tunicata*. Length ca. 9 cm. Shell plates partially overgrown by the girdle. **D** *Cryptochiton stelleri*. Length ca. 32 cm. Shell plates fully overgrown. *Tonicella lineata* for size comparison. **E** Ventral side of *Cryptochiton stelleri*. Separation of foot and head clear. **F** *Cryptochiton stelleri*. View of the ventral side with closely packed gills in the mantle groove. (Originals. **A** Helgoland, North Sea; **B–F** Puget Sound, USA Pacific Coast)

ments of the girdle lead to partial (*Katharina tunicata*) or full (*Cryptochiton stelleri*) covering of the plates (Fig. 10).

What can be said about the layered construction of the shell plates? The terminology for this is not consistent. Here, I orient myself using the work of HAAS (1972, 1981), SALVINI-PLAWEN (1985) and EERNISSE & REY-

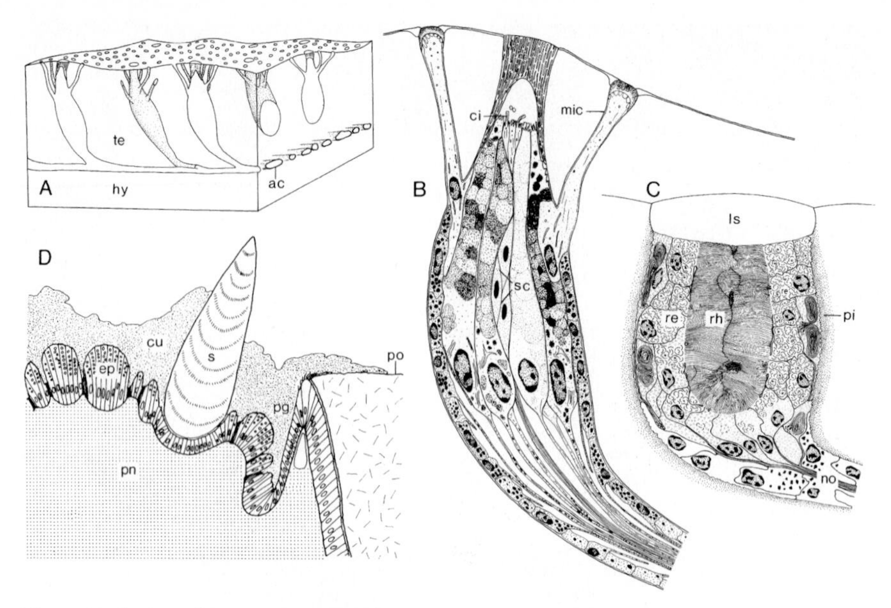

Fig. 11. Polyplacophora. **A–B** *Lepidochiton cinereus.* Aesthetes. **A** The aesthetes radiate inwards at the boundary of the tegmentum and hypostracum in canals of the shell. They ascend in the tegmentum and end at the upper surface in caps. A macraesthete is always surrounded by numerous micraesthetes. **B** Ultrastructure. A macraesthete consists of numerous cells, among these presumed sensory cells with cilia. The associated micraesthetes are cell offshoots of the macraesthetes. **C** *Onythochiton neglectus.* Shell eye with lens, retinula cell and a central rhabdome of microvilli. **D** *Acanthopleura granulata.* Transverse section at the boundary between perinotum (girdle) and shell. Cells of the girdle epithelium secrete the cuticle and calcareous stalk. A girdle pit (periostracum pit) encircles the single shell plate. Cells of the inner edge of the pit produce a very thin organic periostracum (properiostracum, Haas) that covers the exterior of the calcareous plate. The tegmentum is produced laterally from cells of the girdle pit. *ac* Aesthete canals; *ci* cilium; *cu* cuticle; *ep* epithelial papillae; *hy* hypostracum; *ls* lens; *mic* micraesthete; *no* nervus opticus; *pa* epithelial papilla; *pg* periostracum pit; *pi* pigment; *pn* perinotum (girdle); *po* periostracum; *re* retina cell; *rh* rhabdome; *s* spine; *sc* sensory cell; *te* tegmentum. (**A, B** Boyle 1974; **C** Boyle 1969; **D** Haas 1981)

NOLDS (1994). The shell plates of the Polyplacophora consist of three layers – periostracum, tegmentum and hypostracum. The periostracum is a thin organic skin, the tegmentum and hypostracum are calcareous with aragonite crystals.

The periostracum and tegmentum grow through secretion along the edge of the shell by cells of a groove between shell and perinotum. The production of the tegmentum gives rise to surface growth of the shell plates. The hypostracum is the product of cells in the dorsal body epithelium, and leads to thickening of the shell.

The articulamentum and myostracum are distinct differentiations of the hypostracum each with different functions. The articulamentum serves as one of the lateral insertions of the shell plates into the perinotum; it further forms apophyses on plates 2–8, that are inserted as articulation points under the preceding plate. The myostracum is found locally over the attachment sites of muscles.

We can now examine the autapomorphies. The tegmentum as a calcareous layer with living cells is an outstanding feature of the Polyplacophora. Located in the tegmentum are branched canals within which run nerves and cell strands which have a sensory function. The multiple differentiations of the aesthetes include complicated light-sensory organs with lens, retinula cells and rhabdome.

The existence of the articulamentum is probably a further autapomorphy of the Polyplacophora. Its absence in pre-Devonian species supports the idea that the articulamentum with a hinge or joint function was first differentiated in the stem lineage of the Polyplacophora.

Apomorphic for the chitons is the large number of gills – between 6 and 88. However, this feature cannot unequivocally be interpreted as an autapomorphy in the ground pattern of the Polyplacophora, since an increase in the number of gills may have occurred independently several times.

Conchifera

The evolution of a single uniform shell can be seen as the most important development within the stem lineage of the Conchifera. The simplest and most logical explanation for the existence of eight pairs of foot retractors with insertions on the inner surface of the shell is that the single shell of the Conchifera arose from the union of eight separate shell plates. If this were the case one would expect to find congruence in the layered construction of the shells between the Polyplacophora and Conchifera.

The shell of the Conchifera consists of an outer organic layer and two calcareous layers with surface and thickness growth as in the Polyplacophora, which suggests the following equivalencies (HAAS 1972):

Polyplacophora	**Conchifera**
Periostracum	Periostracum
Tegmentum	Ostracum
Hypostracum	Hypostracum

A nacreous layer is not found in the Polyplacophora. In the Conchifera, however, the iridescent inner surface of many shells suggests that this is the original condition. The conversion of the hypostracum to a nacreous layer must therefore be considered to be an evolutionary novelty of the Conchifera.

Autapomorphies (Fig. 6 → 7)

- Single flat shell.
- Hypostracum developed to a nacreous layer.
- A reduction of the perinotum with cuticle and calcareous spines was the consequence of the evolution of a single shell with which the body is fully covered.
- One anterior jaw in the oral cavity.
- Eight pairs of dorsoventral muscles. Reduction from 16 to 8 is a consequence of fusion of pairs of foot retractors (Fig. 12).
- Crystalline style, storage organ for enzymes in the midgut.
- Subrectal commissure between the lateral cords of the nervous system. An apomorphy in comparison to the suprarectal commissure of the Aplacophora and Polyplacophora.

Neopilinida – Ganglioneura

Neopilinida

Before *Neopilina galatheae* was described in the 1950s (LEMCHE 1957; LEMCHE & WINGSTRAND 1959), it had been thought that the "Monoplacophora" were extinct. Since this time around 20 recent species from the deep sea floor have been recovered (WAREN & GOFAS 1996). They are found mostly in water depths of several thousands of meters.

These strongly bilaterally symmetrical creatures are only a few mm long. The flat limpet-shaped shell supports a rostrally directed tip. Head and foot are enclosed by a wide mantle groove. The anus lies medially at the posterior.

The first sensation to be discovered about these animals was the **combination of a single shell with eight pairs of foot retractors.** A distinct correspondence to impressions of muscles, identical in number, in the fossilized shells of certain Conchifera from the Cambrian and Devonian can be seen. Indeed, a single shell and eight pairs of foot retractor muscles are ground pattern features of the Conchifera – in other words, these cannot be used as criteria for creating a monophyletic taxon Monoplacophora within the Conchifera. Fossil Conchifera with the impression of eight or less pairs of retractors can belong principally to three different stem lineages: (1) the stem lineage of the Conchifera, in which these attributes evolved. (2) the stem lineage of the extant Neopilinida; (3) the stem lineage of the Ganglioneura, in which the foot retractors are reduced to a single pair. However, since fossil Monoplacophora may actually be a paraphylum including members of different stem lineages, this name should not be used for a definite monophyletic unity of recent species. Therefore, we use instead the name Neopilinida in reference to the recent species *Neopilina galatheae* (LAUTERBACH 1983c, 1984a,b).

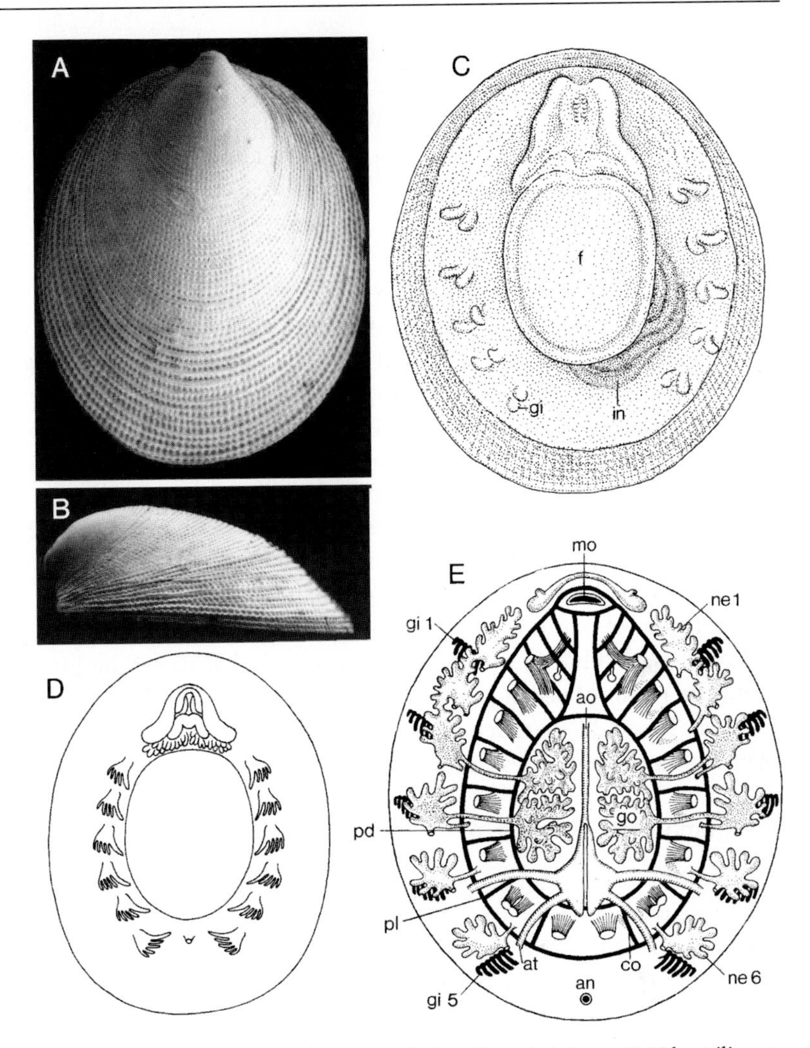

Fig. 12. Neopilinida. **A** *Veleropilina reticulata.* Dorsal view. Length 1.6 mm. **B** *Veleropilina reticulata.* Side view. **C** *Rokopella euglypta.* Ventral view. Length 2 mm. **D** *Vema ewingi.* Schematic ventral view. Six pairs of gills in the mantle groove. **E** *Neopilina galatheae.* Diagram of the organization with serial repetition of six organ systems. Details in text. *an* anus; *ao* aorta; *at* atrium (conduction from the gills to the ventricle); *co* commissure between pleurovisceral and pedal cords; *f* foot; *gi* gills; *in* intestine; *mo* mouth; *ne* nephridium; *pd* pedal cord of the nervous system; *pl* pleurovisceral cord (lateral cord). (**A–C** Warén & Gofas 1996; **D** Lauterbach 1983b; **E** Lemche & Wingstrand 1959; Salvini-Plawen 1981)

A second surprise was the **appearance of several organs in multiple pairs.** This understandably reactivated hypotheses about the origin of the Annelida and Mollusca from a common stem species with metamerism. To date four species have been studied with the following relevant results (LEMCHE & WINGSTRAND 1959; WINGSTRAND 1985; HASZPRUNAR &

SCHAEFER 1997; SCHAEFER & HASZPRUNAR 1997). (The numbers in each case refer to pairs of organs):

	Nephropores[2]	Gills	Gonoducts[3]
Vema ewingi	7	6	3
Laevipilina antarctica	6	5	3 (♂), 2 (♀)
Neopilina galatheae	6	5	2
Micropilina arntzi	3	3	1

Neither in the number of organs within a single species, nor in the comparison of organs between the four species, is there numerical agreement. This would, however, be required if one wishes to hypothesize that the condition seen in the Neopilinida is an expression of metamerism homologous with that of the Articulata. Moreover, coelomic sacs as a basis for this metamerism are lacking in the Neopilinida.

The regular order of commissures between the pleurovisceral and pedal nerve cords is controlled by the arrangement of the eight retractor pairs; these are carried over from the ground pattern of the Conchifera. The serial repetition of kidneys with nephropores, gills, gonads with gonoducts and most probably the existence of two pairs of atria can be interpreted as being the result of secondary increases within the stem lineage of the Neopilinida. To establish exact numbers present in the ground pattern of the Neopilinida more species need to be examined.

Autapomorphies (Fig. 6 → 8)

- Several pairs of kidneys, gills and gonads repeated serially.
- Two pairs of Atria.
- Monopectinate gills.
  Ctenidia with one row of lamellae (Fig. 12 E) in contrast to the bipectinate condition seen in the ground pattern of the Mollusca.
- Absence of renopericardioducts.
  In the Polyplacophora two short tubes lead out of the pericardium into the elongated kidney sacs. The lack of a connection between the pericardium and nephridia, as established by SCHAEFER & HASZPRUNAR (1997), must be secondary in nature.

(Further details in SCHAEFER & HASZPRUNAR 1997).

[2] Maximum six pairs of nephridia. In *V. ewingi* a "surplus" pair of nephropores exist in the anterior body.
[3] Maximum two pairs of gonads with gonoducts, which open into neighboring nephridia, the gametes are expelled through nephropores. In one individual of *V. ewingi* a third pair of gonoducts before the two "normal" pairs has been described. In the ♂♂ of *L. antarctica* the anterior testes have two pairs of gonoducts.

Ganglioneura

The unities described up to now have kept the tetraneural system with two pairs of nerve cords (pedal cords, visceral cords) from the ground pattern of the Mollusca. In the stem lineage of the Ganglioneura – a large monophylum composed of the Gastropoda, Cephalopoda, Scaphopoda and Bivalvia – a ganglioneural nervous system evolved with a **concentration of nerve cell nuclei into specific regions** that we call ganglia. This results in a nervous system with pairs of cerebral, pleural, pedal and visceral ganglia.

The process of differentiation into (nucleus-bearing) ganglia and (anucleate) connectives was largely, but not wholly, realized in the ground pattern of the Ganglioneura. Remnants of nervous systems still retaining nerve cord characteristics were independently eliminated in a number of subtaxa of the Ganglioneura. In some gastropods (*Haliotis, Patella*) ventral nerve cords with commissures originate from already established pedal ganglia (Fig. 13 A).

Further apomorphic peculiarities of the Ganglioneura result from reductions in comparison to the adelphotaxon Neopilinida.

Autapomorphies (Fig. 6 → 9)

- Ganglioneural nervous system.
- One pair of foot retractors, in contrast to the eight pairs seen in the Neopilinida.
- Reduction of the muscles for rolling into a ball.

In their ground pattern the Ganglioneura retain as plesiomorphies one pair of gills and one pair of kidneys.

Rhacopoda – Ancyropoda

Rhacopoda

The following derived congruencies between the Gastropoda and Cephalopoda are the basis for their amalgamation into a monophylum Rhacopoda (Visceroconcha).

Autapomorphies (Fig. 6 → 10)

- Restriction of the mantle groove at the posterior with concomitant development of a cavity. (The displacement of the mantle cavity to the anterior first occurred in the stem lineage of the Gastropoda.)
- Development of a free extrusible head through limitation of the mantle, with the shell on the covering of the visceral mass.

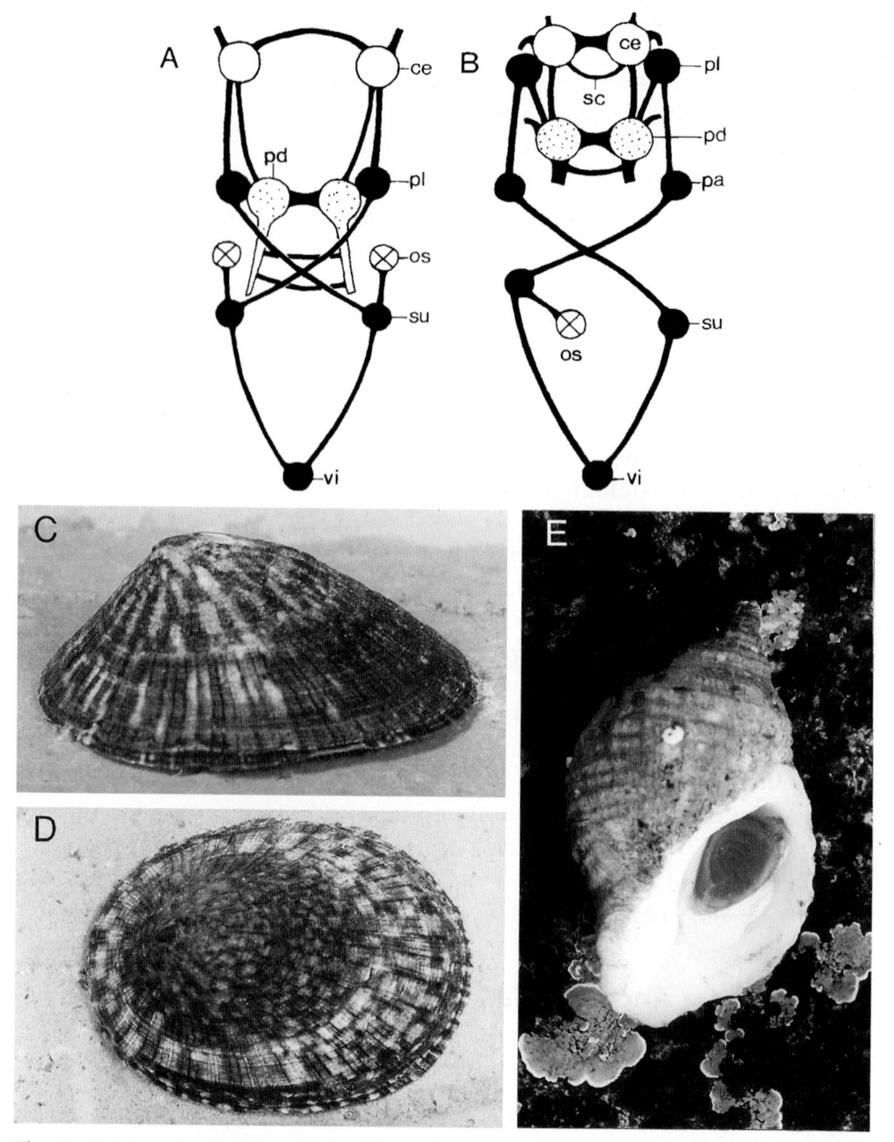

Fig. 13. Gastropoda. **A** Scissurellidae ("Prosobranchia"). Streptoneural nervous system (autapomorphy of the Gastropoda). Paired pleural and subesophageal ganglia in the crossed pleurovisceral cords; visceral ganglia fused terminally. Pedal cords with commissures connected to the pedal ganglia (plesiomorphic feature). **B** Actaeonidae (Opisthobranchia). Parietal ganglia behind the pleural ganglia and the subcerebral commissure (autapomorphy of the Euthyneura). Streptoneury retained in this Opisthobranchia taxon. **C–D** *Acmaea* (Docoglossa, "Prosobranchia"). Cap-shaped shell in side view from the left and from above. **E** *Nucella lapillus* (Muricoidea, "Prosobranchia"). Operculum in the shell opening (autapomorphy of the Gastropoda). *ce* Cerebral ganglion; *os* osphradial ganglion; *pa* parietal ganglion; *pd* pedal ganglion; *pl* pleural ganglion; *sc* subcerebral commissure; *su* subesophageal ganglion; *vi* visceral ganglion. (**A, B** Haszprunar 1988; **C, D** Originals: Puget Sound, USA Pacific Coast; **E** Original: Helgoland, North Sea)

- Evolution of a special muscle system for the rapid extension of foot and tentacles.

Epipodial tentacles in the ground pattern of the Gastropoda have also been hypothesized, and homologized with the arms of the Cephalopoda. This interpretation is, however, questionable. The innervation of epipodial tentacles follows from the pedal ganglia, while the cephalopod arms are supplied from the cerebral ganglia; thereby these nerves pass the pedal ganglia without any synaptic contact (SALVINI-PLAWEN 1981; BUDELMANN & YOUNG 1985; HASZPRUNAR 1988). It is also possible that only one pair of epipodial tentacles belongs in the ground pattern of the Gastropoda (HASZPRUNAR 1988).

Gastropoda – Cephalopoda

Gastropoda

Autapomorphies (Fig. 6 → 11)

- Torsion of the visceral mass and shell above the foot with a twisting of the pleurovisceral nerves (streptoneury) (Fig. 13 A).
- Reduction of the morphological right gonad.
- Operculum (Fig. 13 E).

The torsion by about 180° of the visceral mass with the displacement of the mantle cavity with pallial complex and anus to the proximity of the head appears to be the central process in the evolution of the gastropods.

Inside the body the torsion leads to crossing or twisting of the lateral nerves which connect the pleural ganglia with the visceral ganglia.

The spiral of the shell is not, however, necessarily connected with the torsion of the visceral mass. It has been postulated that the most likely situation in the ground pattern of the Gastropoda is a small cap-shaped shell without a trace of asymmetric coiling in both larvae and adults (HASZPRUNAR 1988). Here the apex of the shell was probably directed anteriorly – as in most of the Docoglossa (Patellacea) (Fig. 13). In this view a spiralization of the shell first appeared within the Gastropoda – either planospiral in one plane (†Bellerophontida) or asymmetrical.

The operculum as a protective covering for the soft body when contracted is a further autapomorphy in the ground pattern of the Gastropoda. The operculum is present in all Gastropoda which have free larvae (HASZPRUNAR 1988).

Plesiomorphies in the ground pattern are the paired development of the pallial organs, ctenidia, osphradia, hypobranchial glands, auricles and kidneys. Within the Gastropoda the unpaired columellar muscle evolved from paired foot retractors.

The **systematization of the Gastropoda** is in progress (HASZPRUNAR 1988; SALVINI-PLAWEN & HASZPRUNAR 1987; SALVINI-PLAWEN 1990; SALVINI-PLAWEN & STEINER 1996 amongst others). There are still several problems to be resolved, however, before a consistent phylogenetic system can be constructed. Therefore, within the bounds of this book, I only deal with the present highest-ranking subtaxa of traditional classification – that is the division of the Gastropoda into Prosobranchia (Streptoneura) and Euthyneura, and of the Euthyneura into Opisthobranchia and Pulmonata.

"Prosobranchia" ("Streptoneura")

The two features of gills situated anterior to the heart and crossing of the pleurovisceral nerves are – as discussed – consequences of the torsion of the visceral mass which occurred within the stem lineage of the Gastropoda. They are therefore autapomorphies in the ground pattern of the Gastropoda, and hence become plesiomorphies within the Gastropoda. In this case they cannot be utilized for further systematization. Since the taxon "Prosobranchia" or "Streptoneura" (both names are used synonymously) is based on these very criteria it is a paraphylum without any logical existence in the phylogenetic system of the Gastropoda.

A series of evolutionary steps that lead from the ground pattern of the Gastropoda to the Euthyneura has been demonstrated by HASZPRUNAR (1988) and SALVINI-PLAWEN & STEINER (1996). The Docoglossa (Patellacea) with a primary cap-shaped shell are interpreted as an early unity. On the other hand it is possible that the Pyramidelloidea are the sister group of the Euthyneura.

Euthyneura

The Opisthobranchia and Pulmonata were united under this name in the nineteenth century (SPENGEL 1881). This combination is durable. In contrast, its use based on the secondary uncrossing of the nervous system (euthyneury) cannot be supported. Firstly, the original streptoneural nerve pattern survives in a number of Opisthobranchia and Pulmonata. Secondly, an euthyneural nerve pattern evolved several times convergently, through the partial detorsion of the visceral mass, through the anterior concentration of ganglia along the pleurovisceral nerves and through a combination of both these processes. There are, however, autapomorphies of the unity Euthyneura. Apart from a special subcerebral commissure and hermaphroditism, HASZPRUNAR (1988) rates the evolution of new parietal ganglia in the pleurovisceral nerves to be of prime importance (Fig. 13 B). These arose by separation from the pleural ganglia.

Opisthobranchia

Within the Euthyneura difficulties soon arise in trying to justify the Opisthobranchia as a valid taxon. Within the stem lineage of the unity Opisthobranchia, SALVINI-PLAWEN (1990) hypothesizes the evolution of a broad headshield as well as a chemical sense organ between the headshield and foot (Hancocks organ) with specific innervation. Since, however, this state is realized in only a few taxa, these structures must have been secondarily lost by the great majority of the Opisthobranchia.

Aeropneusta

SALVINI-PLAWEN (op. cit.) places all euthyneural Gastropoda with additional air breathing in the Aeropneusta having the same rank as the Opisthobranchia. He then divides the Aeropneusta into the **Pulmonata** and the **Gymnomorpha** (*Onchidium*). The homologization of the cutaneous respiration of the Gymnomorpha with the pulmonary respiration of the Pulmonata needs further detailed study. However, the Pulmonata can justifiably be viewed as a monophylum due to the differentiation of the mantle cavity into a pulmonary organ with a localized opening (pneumostome).

Cephalopoda

The stem species of the Rhacopoda can be interpreted as being a benthic organism. In the stem lineage of the Cephalopoda a habitat change to the pelagic realm took place that is linked to profound changes in organization.

Autapomorphies (Fig. 6 → 12)

- Change to floaters in open water with a vertical orientation of the shell and soft body.
- A prerequisite for this was the evolution of a hydrostatic organ. This was achieved through chambering of the apical shell space and the filling of these chambers with gas. From the visceral mass a siphuncle is formed which passes through the chamber walls to the tip of the shell (Fig. 15 A).
- Carnivorous mode of feeding with the capability of catching large objects of prey.
- Connected with this was the evolution of large, beak-like jaws (a parrot-like beak). An apomorphy in contrast with the simple flat jaw structure of other molluscs (Fig. 15 D).

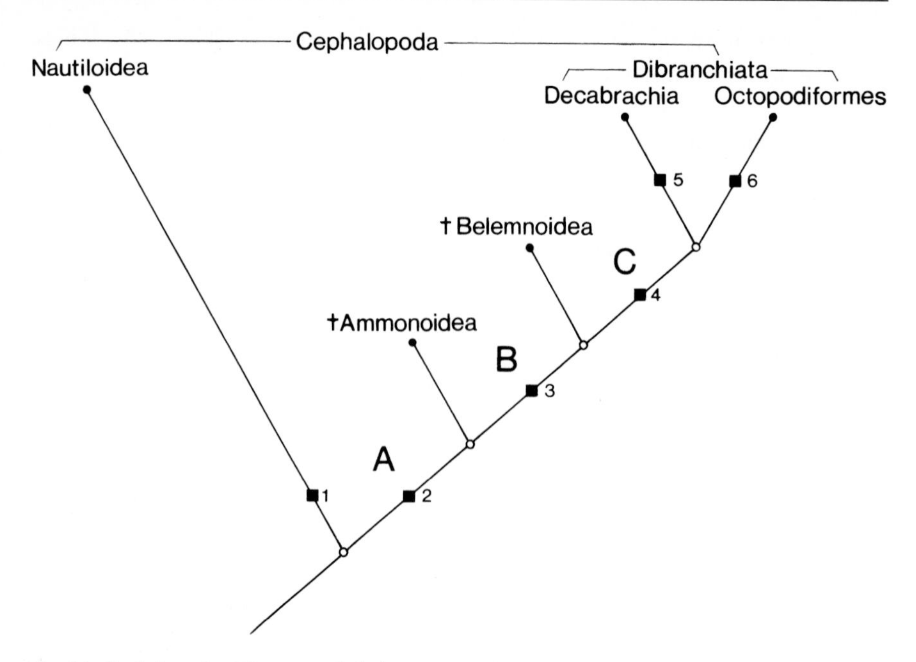

Fig. 14. Cephalopoda. Diagram of phylogenetic relationships placing the †Ammonoidea and †Belemnoidea in the stem lineage of the Dibranchiata (stem lineage concept). *A*, *B* and *C* form the three sections of the stem lineage in which certain autapomorphies of the Dibranchiata originated. Full explanation in the text

- Differentiation of perioral tentacles into arms for catching prey.
- Evolution of a primary two-lobed funnel from the foot for rapid movement using a "jet propulsion" technique.
- Concentration of the nervous system due to fusion of the ganglia to a brain-like center.
- One pair of eversed pin-hole camera eyes (Fig. 15 L).
- Reduction of the crystalline stalk (present in the sister group Gastropoda).
- Connection between the gonocoelom and the pericardium through the gonopericardioduct.
- Differentiation of a storage organ for spermatophores from the distal part of the male gonoduct (Needhams sack).
- Yolky eggs with discoidal cleavage. Direct development through reduction of the trochophore.

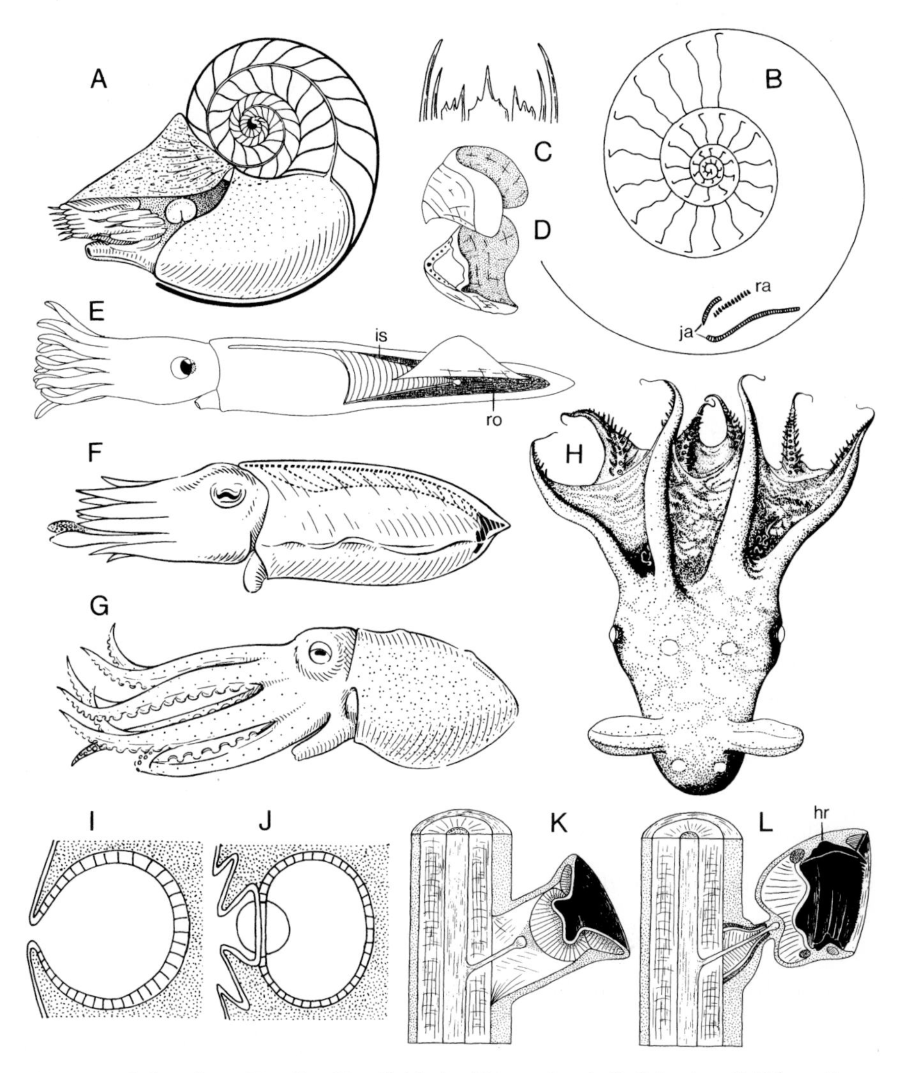

Fig. 15. Cephalopoda. **A** *Nautilus* (Nautiloidea) with exterior shell. Side view. **B** †*Eleganticeras* (Lower Jurassic) (Ammonoidea). Longitudinal section through the outer shell with position of the jaws and radula in the living chamber. **C** †*Eleganticeras* (Ammonoidea). Transverse row of radula with seven teeth (autapomorphy of the Dibranchiata). **D** †*Gaudryceras* (Upper Cretaceous) (Ammonoidea). Upper and lower jaw (autapomorphy of the Cephalopoda). **E** †Belemnoidea. Reconstruction of the soft body; with internal shell. Side view. **F** *Sepia* (Decabrachia). Internal shell a cuttlebone in the visceral mass. **G** *Octopus* (Octopoda). Internal shell found only in the form of two small horny rods. **H** *Vampyroteuthis infernalis* (Vampyromorpha). Aerial view. **I** *Nautilus* (Nautiloidea). Plesiomorphous pin-hole camera eye. **J** *Todarodes* (Decabrachia). Apomorphous lensed eye of the Dibranchiata. **K** Octopoda. Broad unstalked sucker as plesiomorphy. **L** Decabrachia. Stalked sucker with horny ring as an apomorphy. *hr* horny ring; *is* internal shell; *ja* jaw; *ra* radula; *ro* rostrum (**A, F, G** Portmann 1961; **B–E** Ziegler 1983; **H** Kilias 1993 a; **I, J** Mangold-Wirz & Fioroni 1970; **K, L** Kilias 1993 b)

Systematization

A precise phylogenetic systematization of the Cephalopoda (BERTHOLD & ENGESER 1987) underlies the following overview. However, I only accept as sister groups of identical rank, unities which include recent members. Accordingly, three high-ranking adelphotaxa relationships are discussed.

Cephalopoda
- **Nautiloidea**
- **Dibranchiata**
  - **Decabrachia**
  - **Octopodiformes**
    - **Vampyromorpha**
    - **Octopoda**

In the stem lineage concept (Vol. I, p. 43) fossils should be placed sequentially in the stem lineages of monophyla with recent representatives in accordance with the increase in derived features. Within the Cephalopoda the †Ammonoidea and the †Belemnoidea are placed sequentially in the stem lineage of the Dibranchia (Fig. 14). Using these examples, I want to demonstrate once more the usefulness of fossils with incontestable systematization to provide a time sequence for the evolution of certain autapomorphies of monophyletic unities.

Nautiloidea (Tetrabranchiata)

A few species of *Nautilus* are found in the Indo-Pacific. Mostly in water depths of a few 100 m.

Autapomorphies (■ 1 in Fig. 14)

Two pairs of gills, two pairs of atria and two pairs of kidneys. For the ground pattern of the Ganglioneura a simple, single set of these organs can be postulated. Therefore, the duplicate state in the Nautiloidea must be seen as a derived state (doubling). Differentiation of copulation organ (male) and spermatophore store (female) from the inner crown of arms (tentacles).

Hood as a fusion product of the sheaths of the dorsal arms; serves to close across the aperture after withdrawal of the soft body.

Plesiomorphies

The list given here relates to those features that underwent transformation into apomorphous states in the stem lineage of the Dibranchiata. The primary lack of features which were newly evolved within the Dibranchiata is not considered.

- External shell with chambers.
- Large number of 80–90 arms (tentacles, cirri) with adhesive pads for catching prey; without suckers.
- Radula with nine teeth and four marginal plates per row. Probably taken over from the ground pattern of the Cephalopoda.
- Funnel of two unfused lobes.
- Pin-hole camera eyes (without lens).

Dibranchiata (Angusteradulata, Coleoidea)

We use the older common name Dibranchiata, even though it unfortunately refers to a plesiomorphy (two gills).
The †Ammonoidea (sutures, marginal siphuncle) and the †Belemnoidea (a double row of hooks on the arms; HAAS 1989) are fossil monophyla.

Autapomorphies

Ordered according to their appearance in three sections of the stem lineage (■ 2–4 in Fig. 14).

A. Stage between the stem species of the Dibranchiata and the splitting off of the stem lineage of the Ammonoidea (■ 2).
Reduction in the number of arms. The Ammonoidea had little more than ten arms.
Radula with only seven teeth and two marginal plates per row.

B. Stage between the splitting off of the stem lineages of Ammonoidea and Belemnoidea (■ 3).
Final adoption of ten identical arms. Evolution of the ink-sac.
Incorporation of the shell within the body. Overgrowth through the mantle during ontogeny.
Development of a rostrum (calcareous cylinder) at the shells apex (Fig. 15E). Reorientation of the body from a vertical floater to a swimmer with a horizontal position in the water. Evolution of fins.
Close connection between mantle and funnel through cartilage.
Transformation of the pin-hole camera eye into a lensed eye. (Fig. 15J).

C. Stage between the splitting off of the Belemnoidea and the last joint stem species of the recent Dibranchiata (■ 4). Evolution of suckers on the arms – primarily unstalked and without a horny rim (Fig. 15K).
Transformation of one arm pair into hectocotyli used for the transfer of spermatophores.

There are **a large number of further autapomorphies** of the Dibranchiata, that are not proven (or are not provable) in the Ammonoidea and Belemnoidea. The time of their appearance is unclear. Among these are the chromatophores in the epidermis, the closing of the funnel to a tube, poison glands in the foregut, the development of branchial hearts and certain changes in the nervous system such as fusion of the pedal and visceral ganglia, and development of stellate ganglia in the mantle nerves.

Plesiomorphies

One pair of gills, one pair of atria and one pair of kidneys. Carried over from the ground patterns of the Ganglioneura and Cephalopoda.

We now come within the Dibranchiata to the adelphotaxa Decabrachia and Octopodiformes.

Decabrachia

Spirula, Sepia, Loligo, Architeuthis.

Autapomorphies (■ 5 in Fig. 14)

- Lengthening of the fourth arm pair into tentacles for catching prey. Evolution of stalked mobile suckers which have a horny rim (Fig. 15 L).

Plesiomorphy

- Existence of ten arms.

Octopodiformes

Vampyroteuthis, Octopus, Argonauta.

Autapomorphies (■ 6 in Fig. 14)

- Reduction of arms to four pairs.
- Arms joined by a web of skin (interbrachial membrane or velum).
- Statocysts divided in two chambers.

Plesiomorphy

- Stalkless suckers without a horny rim.

BERTHOLD & ENGESER (1987) consider the **Vampyromorpha** and the **Octopoda** as adelphotaxa of the Octopodiformes.

The **Vampyromorpha** have several fossil members but only one living species – the deep-sea *Vampyroteuthis infernalis*. Two small, retractile continuations of the extensive interbrachial membrane are interpreted as transformation products of the second arm pair. Adaptation to the bathyal realm has caused a reduction of the ink-sac and the loss of the capability to change color. Other autapomorphies are the lack of a hectocotylus and the transformation of fins to small lobes.

In the **Octopoda** (*Octopus*, *Argonauta*) the most conspicuous autapomorphy is the sac-shaped body. This evolved together with the reduction of the shell to small horny rods. Furthermore the second pair of arms has been totally lost. Only in the Octopoda is the ink-sac embedded in the mid-intestinal gland.

Ancyropoda

The Cephalopoda conquered the pelagic realm. In contrast, in the stem lineage of the Ancyropoda (Loboconcha) with the adelphotaxa Scaphopoda and Bivalvia a **habitat change** from on the sediment surface to **within soft sediment** took place.

Autapomorphies (Fig. 6 → 13)

- Rostrally extended digging and anchoring foot.
- Lateral extension of the mantle edges and the shell.

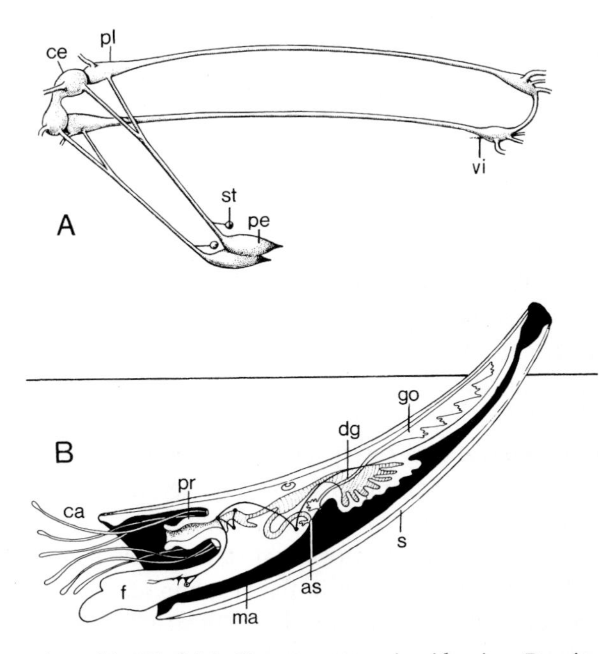

Fig. 16. A *Nucula nucleus* (Protobranchia, Bivalvia). Nervous system in side view. Drawing together of the cerebral and pleural ganglia (autapomorphy of the Ancyropoda). **B** Scaphopoda. Body longitudinal section showing position in sediment. The posterior end projects above the sediment surface. *as* anus; *ca* captacula; *ce* cerebral ganglion; *dg* midgut glands; *f* foot; *go* gonads; *ma* mantle; *pe* pedal ganglion; *pl* pleural ganglion; *pr* proboscis; *s* shell; *st* statocyst; *vi* visceral ganglion. (**A** Franc 1960; **B** Kilias 1993b)

- Concentration of the nervous system with the cerebral and pleural ganglia now lying in close proximity to one another (Fig. 16 A).

A mantle groove encircling the foot remains as a plesiomorphy, in contrast to the restriction of the mantle groove to a cavity at the posterior in the sister group Rhacopoda.

Scaphopoda – Bivalvia

Scaphopoda

Autapomorphies (Fig. 6 → 14)

- Cylindrical tube-shaped shell open at both ends. Product of the mantle which grows together ventrally.
- Proboscis and captacula (capturing filaments) in the oral cavity (Fig. 16 B).
- Absence of gills.
- Water is drawn posteriorly into the mantle cavity.

A species-poor group of marine organisms that live in the upper layer of soft sediment. *Siphonodentalium, Dentalium.*

The Scaphopoda (tusk shells) resemble an elephant's tooth of a few cm in length, dug diagonally into the sediment with the posterior tip overhanging the sediment surface. The **tube-shaped shell** is an important apomorphy of the Scaphopoda. It is secreted by mantle epithelium and achieves its specific form in early ontogeny by fusion of the mantle lobes along their ventral margins.

A further conspicuous feature is the feeding apparatus. The mouth lies at the tip of a mobile proboscis. This oral cone is surrounded by numerous long threads – the captacula. The threads extend into the sediment, glue nutrient particles to the thickened ends and move them by contraction into the oral cavity. *Dentalium entalis*, a North Sea species, specializes on feeding on foraminifers. The retrieved shells are cracked by the radula.

Possession of a radula is an important plesiomorphy that the Scaphopoda have retained in contrast to their sister group Bivalvia.

The gills are reduced in the Scaphopoda and this again is an autapomorphy in relation to the Bivalvia. Oxygen is taken up instead by the mantle epithelium. Water is drawn in through the posterior mantle opening and also exits there. Since in the Bivalvia inhalant water is primarily drawn in from the anterior between the two valves (Protobranchia), the intake of water from the posterior must be a further autapomorphy of the Scaphopoda.

Bivalvia

Autapomorphies (Figs. 6 → 15; 17 → 1)

- Two shell valves with adductor muscles.
- Absence of jaws and radula.
- Fusion of the gill axis to the roof of the mantle cavity.

Bivalves are the only Mollusca with a **two-part shell** divided into a right and left half (Fig. 19). They are also the only Mollusca that have lost the radula – this clearly in connection with the uptake of fine organic particles from soft sediment and water.

To be precise, only in the area of the two calcareous layers ostracum and hypostracum is the shell dorsomedially separated into two. The periostracum, however, wraps over the axis point and binds the two halves together. At this point it forms a two-part ligament of elastic material with an outer tensilium and an inner resilium. The ligament complex and the shell adductors function antagonistically.

The Bivalvia have one pair of bipectinate ctenidia carried over from the ground pattern of the Mollusca and are in this respect more primitive than the Scaphopoda. The first modification to occur within the stem lineage of the Bivalvia was the fusion of the gill axis to the mantle. Since this is not observed in any other unity of the Mollusca it can be seen as an apomorphy of the Bivalvia.

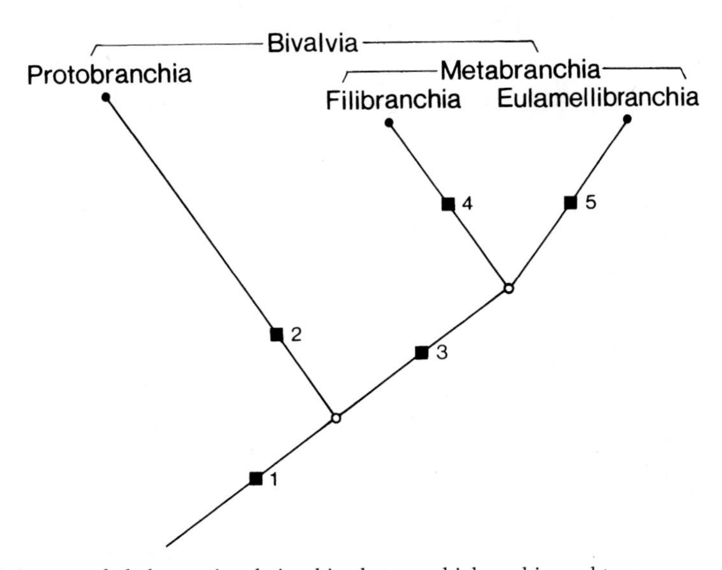

Fig. 17. Bivalvia. Diagram of phylogenetic relationships between high-ranking subtaxa.

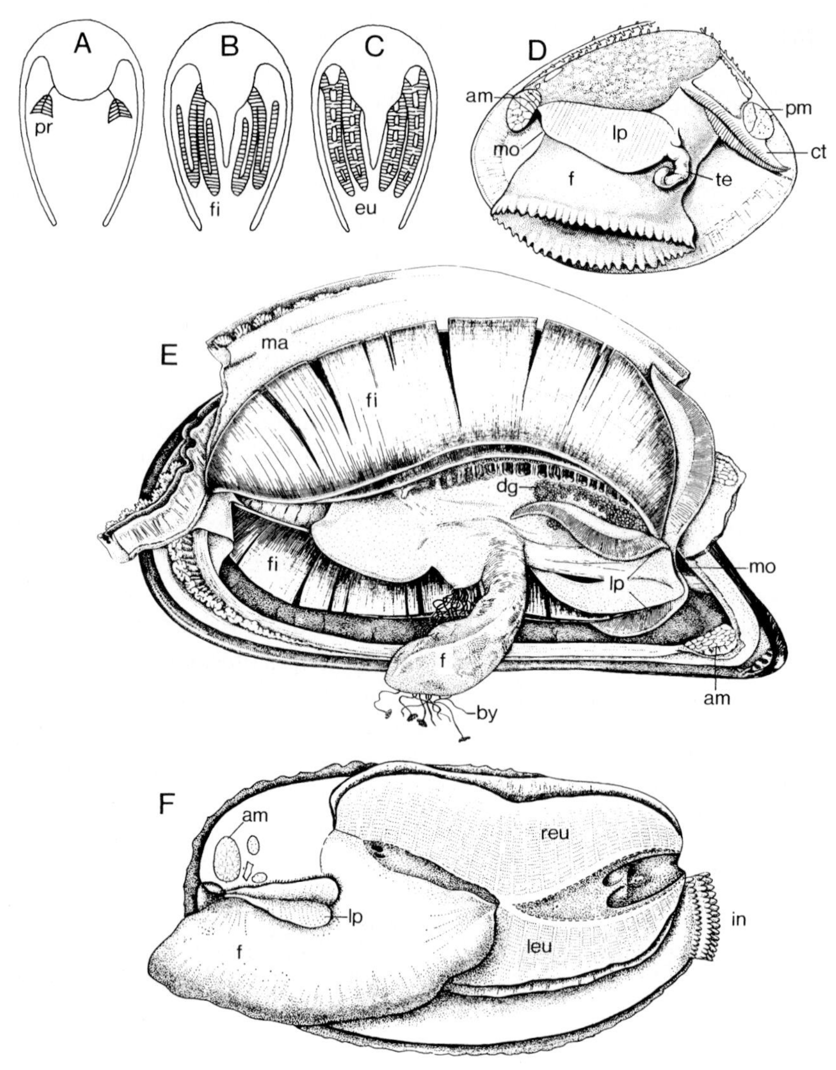

Fig. 18. Bivalvia. **A–C** Different manifestations of the gills, portrayed in schematic body cross sections. **A** Ctenidia of the Protobranchia. **B** Filibranch gills of the Filibranchia. **C** Lamellar gills of the Eulamellibranchia. **D** *Nucula nucleus* (Protobranchia) with ctenidia. Side view from left. Shell and mantle lobes removed from left. **E** *Mytilus edulis* (Filibranchia) with filibranch gills. Side view from right. Shell removed from right. Mantle lobes and gills of the right-hand side raised. Due to the only loosely ciliated connection of the single threads the gills have lightly split into pieces during preparation. **F** *Anodonta cygnea* (Eulamellibranchia) with lamellar gills. Side view from left. Shell and mantle removed from left-hand side. The two gill lamellae of the left-hand side are folded upwards. The single threads are connected by regular longitudinal and transverse bridges and form solid leaves. *am* Anterior closure muscle; *by* byssus threads; *ct* ctenidium; *dg* digestive gland; *eu* eulamellibranch gill; *f* foot; *fi* filibranch gill; *in* ingestion opening; *leu* left eulamellibranch gill; *lp* labial palps; *ma* mantle; *mo* mouth; *pm* posterior closure muscle; *pr* protobranch gill; *reu* right eulamellibranch gill; *te* tentacles of the labial palps. (**A–C** Haas 1935; **D** Franc 1960; **E** Storch & Welsch 1993; **F** Kaestner 1965)

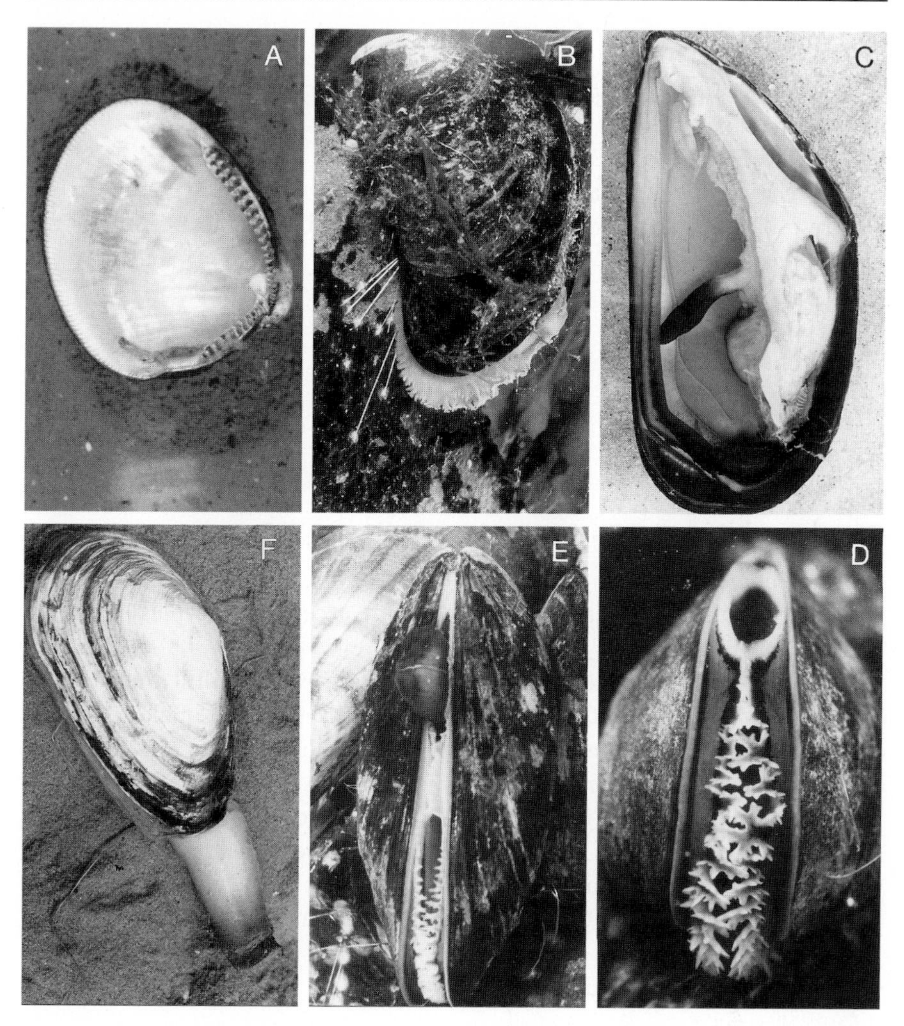

Fig. 19. Bivalvia. **A–C, E, F** Anterior end of the animal to the top. **D** View from behind. **A** *Nucula nucleus* (Protobranchia). Shell with taxodont dentition and nacreous layer (plesiomorphies). **B–E** *Mytilus edulis* (Filibranchia). **B** Anchoring with byssus threads in rocky tidal flat. **C** Left half removed. Byssus gland in the swelling behind the finger-like foot section. Two gill lamellae to the right of the foot. **D** Posterior end. Above egestion opening, below ingestion slit with papillae-like extended mantle edges. **E** Ventral view. Finger-like foot section projects upwards out of the shell gap, below is the ingestion opening. **F** *Mya arenaria* (Eulamellibranchia). Ingestion and egestion areals greatly elongated and united to form a siphon. (Originals: **A, B** Helgoland, North Sea; **C** Arcachon, Atlantic; **D, E** Sylt, North Sea; **F** Cuxhaven, North Sea)

Systematization

"The Bivalvia are notorious for the different classifications and phylogenetic scenarios proposed which emphasize single organ systems....." (SALVINI-PLAWEN & STEINER 1996, p. 42).

The relationship diagram of these authors was based on a study of 42 features. We deal here with the two uppermost hierarchical levels of the phylogenetic system.

Bivalvia
- **Protobranchia**
- **Metabranchia**
  - **Filibranchia**
  - **Eulamellibranchia**

Protobranchia

Nucula, Solenomya.

The Protobranchia present almost without exception the plesiomorphic traits of the Bivalvia.

Firstly, they have distichous, pinnate ctenidia which fulfill the primary function as respiratory organs. Connected with this the intake of water from the anterior into the mantle cavity is a plesiomorphy.

Furthermore, the collection of organic particles from the sediment must be the original mode of feeding in the Protobranchia. Two tentacles (palp probisci), elongations of the labial palps, can be extended between the shells to collect organic matter from the sediment surface and carry it via the cilia to the mouth. Only additionally are particles filtered from the water by the gills.

The foot has a flattened sole.

Autapomorphies (■ 2 in Fig. 17)

SALVINI-PLAWEN & STEINER (1996) hypothesized that an adoral sensory organ and a diagonal orientation of the gills in the mantle (Fig. 18D) may be special features of the Protobranchia.

Metabranchia

All other bivalves can with good justification be grouped together as a monophylum being the sister group of the Protobranchia. In the confusing nomenclature which exists for the highest-ranking taxa of the Bivalvia it appears to me that the name Metabranchia is the best alternative to the name Protobranchia.

Autapomorphies (■ 3 in Fig. 17)

Transformation of the ctenidia to filibranch gills for filtering organic material out of the water. The intake of water into the mantle cavity is transferred to the posterior part of the body.

The single lamellae of the ctenidia evolved into long filaments, connected to one another by cilia. This resulted in the formation of large sheet-like filibranch gills that are, however, relatively easily divided up into their component parts. The V-shaped filibranch gills hang down into the mantle cavity and then rise dorsalwards; however, they do not fuse with the mantle epithelium.

Inhalant water from outside the shell is drawn in from the posterior across the gill filaments and filtered. The water then flows through the interior spaces created by the gill filaments and out of the mantle cavity to the posterior.

As well as filibranch gills, characteristic byssus glands also evolved in the stem lineage of the Metabranchia. In the ground pattern of the Metabranchia the production of adhesive threads is limited to larval and juvenile stages. Finally, in the stem lineage of the Metabranchia, the broad sole of the foot has been lost.

Filibranchia

Mytilus, Lithophaga, Pinctada, Pinna, Ostrea, Pecten

Autapomorphies (■ 4 in Fig. 17)

- Extension of byssus production to the adult organism.
- Existence of an abdominal sensory organ.

It must be emphasized that the Filibranchia cannot be justified as a monophylum due to the presence of filibranch gills. These belong in the ground pattern of the Metabranchia. In comparison with the lamellar gills of the sister group Eulamellibranchia, filibranch gills are a plesiomorphy of the Filibranchia.

Eulamellibranchia

Unio, Anodonta, Margaritifera, Sphaerium, Pisidium, Dreissena (fresh water). *Cerastoderma (Cardium), Tridacna, Ensis, Macoma, Scrobicularia, Mya, Pholas, Barnea, Zirfea, Teredo* (marine).

Autapomorphies (■ 5 in Fig. 17)

Evolution of eulamellibranch gills as solid lamellar gills with the following new features:

1. Longitudinal tissue bridges connect the single gill threads with one another.
2. Cross-connections between the descending and ascending leaves of the gills.
3. The ascending leaves of the Eulamellibranchia fuse distally to the mantle epithelium.

The limiting of byssus formation to the juvenile phase must be seen as a plesiomorphy, as this was taken over from the ground pattern of the Metabranchia.

The **Septibranchia** (*Poromya*) have a horizontal septum in the mantle cavity interrupted by slits in place of the gills. In comparison with all states of the gill already discussed this is an apomorphy. The Septibranchia are a subordinate subtaxon of the Eulamellibranchia (SALVINI-PLAWEN & STEINER 1996).

Pulvinifera[4]

A **cuticle with collagen fibers** and a **secondary body cavity which functions as a hydrostatic cushion** are two features of the Sipunculida and the Annelida (incl. Echiurida and Pogonophora). It is hypothesized that they are apomorphies which evolved only once within the limits of the Spiralia. By means of these features we can create a monophylum, the Pulvinifera, with the Sipunculida and the Articulata (Annelida + Arthropoda) as adelphotaxa.

Autapomorphies (Fig. 4 → 5)

- Cuticle with collagen fibers (Fig. 20).
  The cuticle is interspersed by a multilayered network of collagen fibers. The layers cross one another at right angles. The fibers are embedded in a matrix which peripherally forms an epicuticle.
- Coelom as hydrostatic organ.
  In the ground pattern of the Pulvinifera an undivided, unpaired, continuous coelom that traverses the whole length of the body.

The state of an undivided, unsegmented coelom is realized in the Sipunculida. In the segmentation of the Annelida there are numerous pairs of coelomic sacs in linear succession; these are separated from one another by mesenteries and dissepiments. Both forms of the secondary body cavity have a cellular lining, both develop in the spiral-cleavage from the 4d mi-

[4] tax. nov.

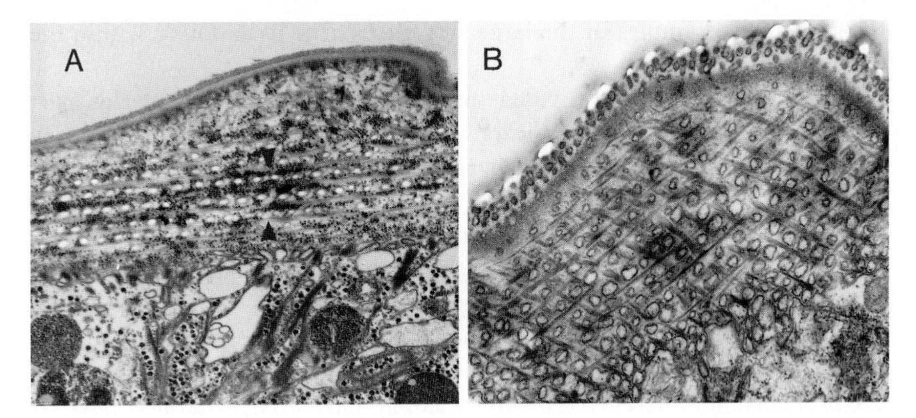

Fig. 20. Cuticle in the ground pattern of the Pulvinifera. With a meshwork of collagen threads. The layers of threads cross one another at right angles. Section through the epidermis and cuticle (electron microscope photographs). **A** *Phascolion strombi* (Sipunculida). **B** *Scoloplos armiger* (Polychaeta). (Originals: T. Bartolomaeus. **A** Marseille, Mediterranean; **B** Sylt, North Sea)

cromere and both are fluid-filled cushions utilized in movement. All this points clearly to homology, whereby the undivided coelom of the Sipunculida must be the plesiomorphy (p. 40) and the segmentation of the Annelida the apomorphic alternative.

The lining of the coelom in the Sipunculida consists of an outer muscle-free peritoneum beneath the somatic musculature and a myoepithelium of epithelial muscle cells around the intestine; an independent visceral musculature does not exist (Vol. I, Fig. 45 D). This differentiation is also widespread in the Annelida (BARTOLOMAEUS 1994). Correspondingly, it could be the primary state in the ground pattern of the Pulvinifera.

There is obviously a close functional connection between the flexible cuticle with crossed collagen fibers and the elastic coelom. Together they create a hydrostatic skeleton (WESTHEIDE 1996) which acts as a support for the action of the body musculature. In the evolution of the Arthropoda they are then both abandoned. In the stem lineage of the Arthropoda the flexible collagen cuticle is replaced by a stiff chitinous cuticle, and the coelom transformed into a new body cavity with a pericardial septum.

Sipunculida – Articulata

Sipunculida

Around 200 species of marine organisms inhabit the sea bed (RICE 1993). They are primarily hemisessile inhabitants of sandy and muddy sediments. Secondary biotopes are empty mollusc shells and limestone crevices. *Sipunculus nudus* (length 25 cm) and *Golfingia minuta* (1.5 cm) from the

North Sea are examples of the large size differences to be found within the taxon.

The peanut worms are divided into a slender retractable anterior part (introvert) and a somewhat thicker posterior part (trunk). The boundary between the two can be marked by an abrupt change in the cuticle ornamentation.

The oral opening lies within a tentacle crown. The long U-shaped intestine has a spiraled descending and ascending branch. The anus is situated dorsally in the anterior part of the trunk.

In the organization of the Sipunculida there are no signs of segmentation. The coelom fills the introvert and trunk uniformly; there are neither dissepiments nor mesenteries. The coelomic cavity is penetrated only by the retractor muscles of the introvert.

Only one pair of metanephridia lie in the anterior of the coelom. The nervous system with a single nerve cord on the ventral side also shows no trace of metamerism. In summary then, the Sipunculida must be seen as a unity of primary unsegmented species.

Just as significant is the absence of "annelid-like chaetae". This can be interpreted as a plesiomorphy of the Sipunculida. There is no evidence either from the organization of the Sipunculida or in established kinship hypotheses to suppose that chaetae have been secondarily lost.

Finally, in connection with the construction of the body cavity, the lack of a circulatory system must also be an original state.

A derived peculiarity of the Sipunculida is the tentacle coelom to the anterior. This is a ring vessel in front of the large body cavity. Canals run from the ring structure to the tentacles. Such a tentacle coelom is missing in *Golfingia minuta* (BARTOLOMAEUS 1994). Until a broader comparative analysis has been carried out, it must remain undecided as to whether this organ is an autapomorphy in the ground pattern of the Sipunculida or first evolved within the unity.

Autapomorphies (Fig. 4 → 6)

- Intestinal canal with ascending branch and opening of the anus at the anterior part of the trunk (Fig. 21 B, C).
- Unpaired nuchal organ (Fig. 21 D).
  Ciliated epithelial thickening at the anterior within or below the tentacle crown. The nuchal organ consists of ciliated and aciliar supporting cells as well as bipolar primary sensory cells. The organ is innervated by paired nerves from the anterior of the brain.
  Fundamental differences to the paired nuchal organs in the Annelida (p. 47) argue against a homology (PURSCHKE et al. 1997).
- Ciliated urns (Fig. 21 E).

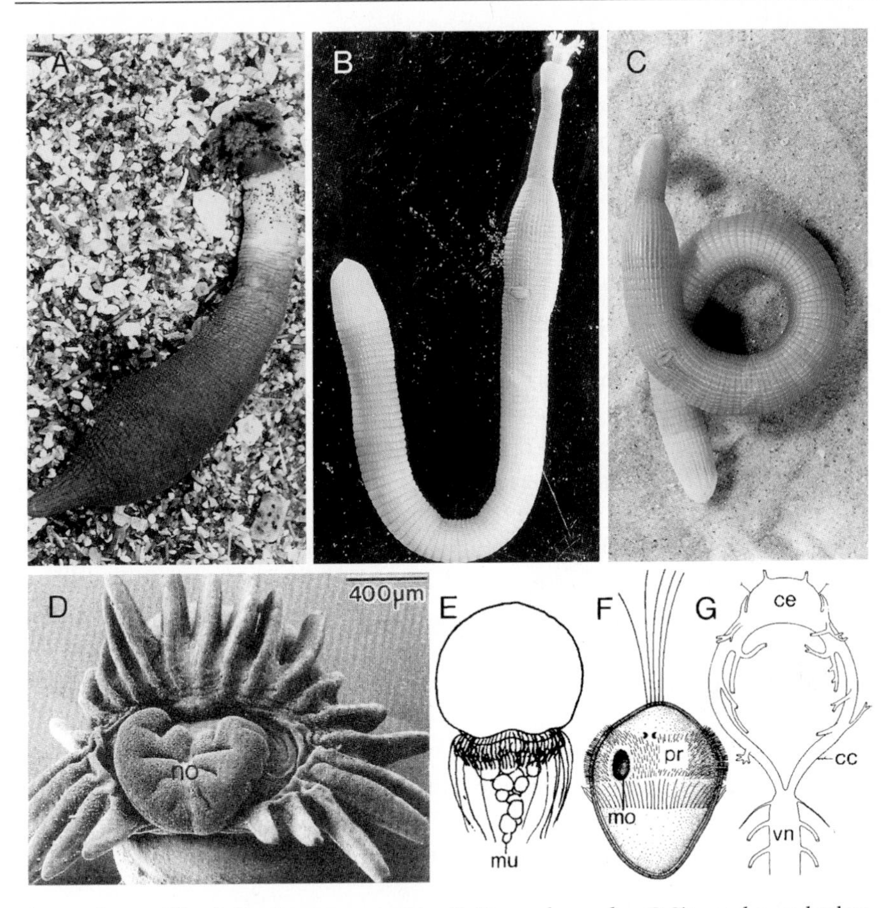

Fig. 21. Sipunculida. **A** *Dendrostoma pyroides*. **B** *Sipunculus nudus*. **C** *Sipunculus nudus* boring into the sediment with the introvert. In both photographs the dorsal anus in the foremost trunk section is visible (autapomorphy). **D** *Phascolosoma*. Anterior end with tentacles and unpaired nuchal organ (autapomorphy). **E** *Sipunculus nudus*. Swimming ciliated urn with two large cells. Anterior cell a transparent bubble, posterior cell a dish with a crown of cilia. Small cells are anchored centrally on the basal ciliated cell. From these and from the margin of the ciliated cell mucus is produced. **F** *Golfingia*. Trochophore larva. **G** *Golfingia minuta*. Central nervous system with brain, paired pharyngeal connectives and unpaired ventral nerve cord. *cc* Circumesophageal connective; *ce* cerebrum; *mo* mouth; *mu* mucus; *no* nuchal organ; *pr* prototroch; *vn* ventral nerve cord. (A–C Originals: A Puget Sound, USA Pacific Coast; **B, C** Arcachon, French Atlantic Coast; **D** Purschke, Wolfrath & Westheide 1997; E Bang & Bang 1962; F, G Tétry 1959)

In the coelomic fluid there are numerous free cells varying in nature. The multicellular ciliated urns, which agglutinate cell debris and other foreign particles, are unique. Mobile urns are only known from *Phascolosoma* and *Sipunculus* (RICE 1993). Urns anchored to the peritoneum are more common. Accordingly, it is possible that they belong to the ground pattern of the Sipunculida.

- ? Absence of protonephridia.
  In the trochophore larva of the Sipunculida protonephridia are said to be absent. Since in young lecithotrophic larvae they are difficult to detect by light microscopy a thorough ultrastructure investigation needs to be carried out to clarify matters.

Articulata

Autapomorphies (Fig. 4 → 7)

- Segmentation (metamerism).
  Division of the body into a row of similar segments (metameres) with a repetition of certain paired organs in the longitudinal axis. These are the coelomic sacs with mesenteries and dissepiments, nephridia, ganglia in a ventral ladder-type nervous system and the lateral canals of a closed circulatory system between the dissepiments. At the anterior and posterior are sections without a segmental character – anterior the prostomium of the Annelida (=acron of the Arthropoda) and posterior, the pygidium of the Annelida (=telson of the Arthropoda).
- Teloblastic growth.
  Formation of segments in a growth zone at the posterior, in front of the pygidium. That the processes in the Annelida and Arthropoda are homologous receives confirmation through a similar mode of expression of the polarity gene engrailed (p. 79) during early segmental development and neurogenesis (WEISBLAT et al. 1993; SCHOLTZ 1997).
- Longitudinal muscles in bands.
  The longitudinal muscles of the primary, uniform body wall musculature of the Spiralia are separated into a few bands of muscles (ROUSE & FAUCHALD 1997). Since this state is seen in both the Annelida and the Arthropoda it can indisputably be interpreted as a further derived characteristic in the ground pattern of the Articulata.

The **segmentation of the body** and the **teloblastic genesis of the metameres** form a unique feature complex of the Articulata (Euarticulata NIELSEN 1997), which within the Spiralia is only found in the Annelida and Arthropoda. Following the principle of the most parsimonious explanation, we postulate a single evolution of this phenomenon in a lineage common only to the Annelida and Arthropoda, even when molecular analyses do not support this interpretation (GHISELIN 1988; EERNISSE et al. 1992; EERNISSE 1997 amongst others).

The Articulata hypothesis is supported, however, by the latest studies of coelomogenesis of the Onychophora (BARTOLOMAEUS & RUHBERG 1999). According to these authors, the segmental ordering of coelomic cavities during ontogeny, and their formation by schizocoely from the entire mesoderm, form an apomorphy for the Annelida and Arthropoda indicating their kinship as sister groups[5].

The question of what caused the evolution of metameres is a totally different matter. If the stem species of the Articulata was a "naked" organism, then the hypothesis that the division of the uniform coelom of the Pulvinifera into segmental sacs represents an efficiency increase in peristaltic movement through sediment is plausible (CLARK 1964, 1969). Unfortunately, the hypothesis cannot be proved by methods of phylogenetic systematics.

For the ground pattern of the Articulata we postulate first of all segmentation, teloblastism and longitudinal muscles bands, features that are to be found in both the Annelida and the Arthropoda. The next step in the discussion is whether further features that are only seen in subgroups of the Articulata possibly belong in the ground pattern of the unity. Let us therefore examine the Articulata more closely.

Annelida – Arthropoda

Annelida

The question of whether the unity Annelida is a monophylum or not is one of the most hotly disputed in phylogenetic systematics. In the case of a positive answer, the division of the Annelida by conventional classification into two equally ranked taxa, the Polychaeta and Clitellata, must be analyzed.

As with the Articulata, we argue using the principle of parsimony. Therefore we use solely those features that are found only in each of the groups to be reviewed. Obviously, before these features are accepted as autapomorphies, they must be critically checked and proven. In this way we can hypothesize that the Annelida, the Polychaeta, and of course the Clitellata, are monophyletic unities.

We begin with those features which can be shown to be derived characteristics of the Annelida.

---

[5] Addition during translation. A. Schmidt-Rhaesa, T. Bartolomaeus, C. Lemburg, U. Ehlers & J.R. Garey. Journal of Morphology 238, 263–285 (1998) examine a recent competing hypothesis according to which the Arthropoda are supposed to belong to a possible monophylum Ecdysozoa comprising all molting Metazoa. Contradictions in the molecular support (18S rDNA) lead J.W. Wägele, T. Erikson, P. Lockart & B. Misof. J. Zool. Syst. Evol. Research 37, 211–223 (1999) "to be cautions in accepting the Ecdysozoa hypothesis, particularly since a strong argument for the alternative hypothesis Articulata can be made from morphological data".

Autapomorphies (Figs. 4 → 8; 25 → 1)

- Capillary chaetae with β-chitin.
  The chaetae of the Annelida are formed in ectodermal pouches or chaetal follicles, which are sunk into the body from the epidermis. Each chaeta is the product of a single cell. This chaetoblast is the basal cell in the chaeta follicle. On the surface directed into the lumen long microvilli are differentiated, these secrete around themselves a tube of fibrillar material containing β-chitin. When the chaeta grows beyond the level of the microvilli, cavities remain in a rigid honeycomb structure. Wall cells of the follicles secrete a scleroproteinaceous skeleton which holds together the tubes of the chaetae (Fig. 22).
  From the numerous different forms of chaetae that have been observed, a simple capillary chaeta belongs in the ground pattern.
- Bundles of chaetae.
  The presence of chaetae in two dorsolateral and two ventrolateral groups per segment is interpreted as being a further characteristic feature (BARTOLOMAEUS 1994). Chaetae bundles in which the chaetae lie close together at the base and then fan out probably belong in the ground pattern. This situation is realized in the Clitellata.

The evolution of segmental chaetae as support and anchoring devices for on or in the sediment may represent an optimization of peristaltic locomotion. Unfortunately this is also a "causal" evolutionary hypothesis that is impossible to prove. We have still to establish precisely why we postulate the **evolution of chaetae in the stem lineage of the Annelida**. For this a comparison with several taxa is necessary.

1. The chaetae of the Annelida have nothing to do with the nearly homonymous structures of the Arthropoda. The setae of the **Arthropoda** are cuticle structures with α-chitin which do not have their genesis in follicles. Furthermore no hypothesis exists which postulates an adelphotaxa relationship between a subtaxon of the Annelida and the Arthropoda. In other words, there is no evidence for a reduction of Annelida chaetae in the stem lineage of the Arthropoda. We must trace the Arthropoda back to an Articulata stem species without chaetae.
2. Chaetae with very similar ultrastructure and which have their genesis in follicles are found on the free edges of the mantle lobes of **Brachiopoda** (STORCH & WELSCH 1972; GUSTUS & CLONEY 1972; ORRHAGE 1973; LÜTER 1998) and on the skin of dibranchiate **Cephalopoda** (BROCCO et al. 1974; BUDELMANN et al. 1997). Obviously there is no segmentation here. Even leaving this circumstance aside, when comparing the Annelida, Cephalopoda and Brachiopoda, established kinship hypotheses force us to the conclusion of a convergent evolution

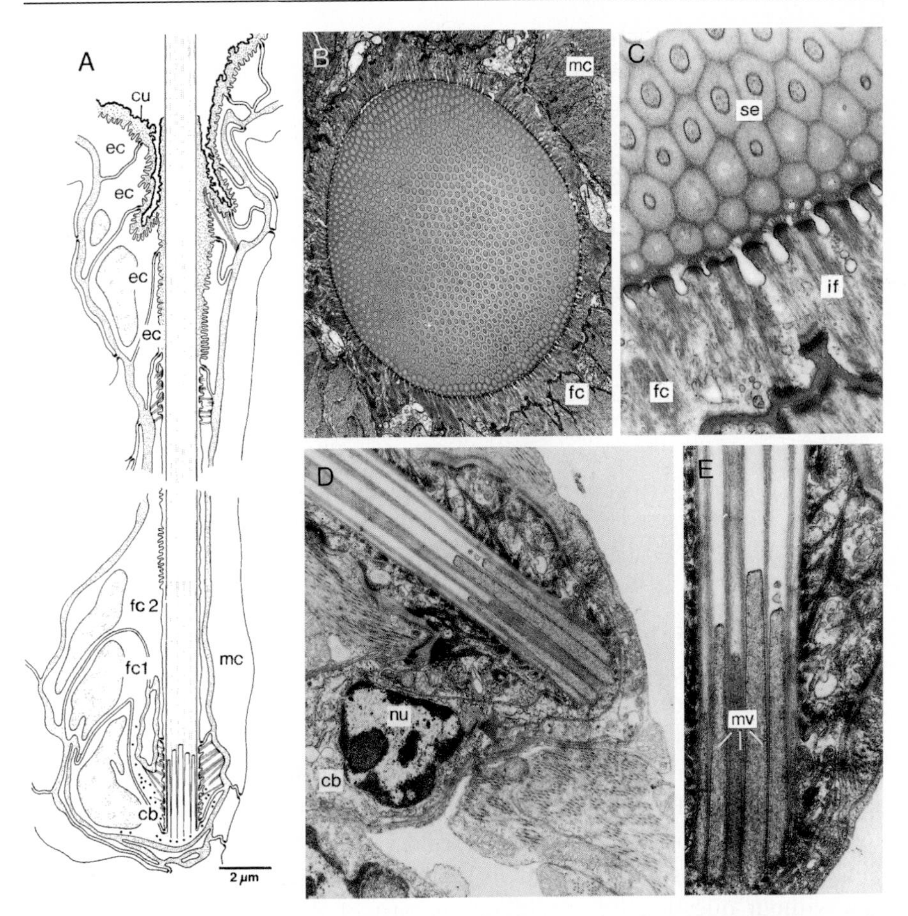

Fig. 22. Annelida. Ultrastructure and genesis of the chaetae. **A** Capillary chaeta of *Pectinaria koreni* (Terebellida). Longitudinal section. Demonstration of chaeta formation from a chaetoblast and the anchoring of the chaeta in the body. Nucleus of the building cell at the side of the proximal end of the chaeta; microvilli project into the chaeta. Distalwards, the chaeta is at first surrounded by two follicle cells, whose cuticle is attached to the chaeta. **B** Acicula from *Eulalia viridis* (Phyllodocida). Transverse section showing the regular system of tubes as the product of single microvilli. The chaeta is surrounded by follicle cells and musculature. **C** Part of the acicula of *Eulalia viridis*. Anchoring of the chaeta through intermediary filaments of the follicle cells. **D** *Pectinaria koreni*. Formation of a chaeta from a chaetoblast. **E** View of a chaeta of *Pectinaria koreni* to show the microvilli in the interior of the tubes. *cb* Chaetoblast; *cu* cuticle; *ec* ectodermal cell; *fc* follicle cell; *if* intermediary filament; *mc* muscle cell; *nu* nuclei of the chaetoblast; *se* chaeta. (**A, D, E** Bartolomaeus 1995; **B, C** Bartolomaeus et al. 1997)

of chaetae. These are the placing of the Annelida in the Articulata, the Cephalopoda in the Mollusca, and the Brachiopoda in the Radialia. Whoever wishes to present an alternative argument must be able to trace the three taxa back to a common stem species with chaetae and then explain the consequent multiple reduction of chaetae.

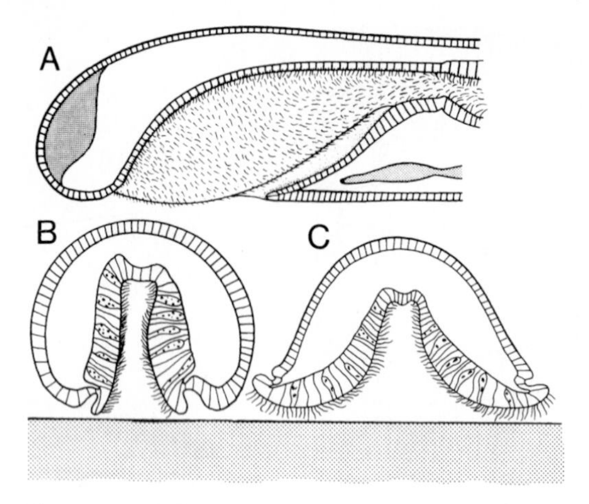

Fig. 23. Annelida. Diagram of the dorsolateral ciliary folds in the foregut. **A** Sagittal section. Resting position. **B** Transverse section. Resting position. **C** Transverse section. Folds everted during nutrient uptake. (Purschke & Tzetlin 1996)

3. The chaetae of the Annelida and the **Echiurida** are identical. If they are indeed homologous, then it is justifiable to place the Echiurida in the Annelida (p. 56).

A further feature which must be mentioned is that of the "ciliated folds in the foregut" (Fig. 23). In numerous Polychaeta ciliated folds or fields form the dorsolateral walls of the foregut (PURSCHKE & TZETLIN 1996). Since these are also widespread in the Clitellata, dorsolateral ciliated folds can without question be considered as part of the ground pattern of the Annelida. The folds are organs of microphagous feeding. They can be extruded to collect organic particles from the sediment, they then convey these to the esophagus. PURSCHKE & TZETLIN (1996) interpret this to be the original mode of feeding in the Annelida. I hesitate, however, to judge the dorsolateral ciliated folds as an autapomorphy of the Annelida. Finally, the stem species of the Articulata must have also eaten and there is no convincing argument against the assumption that it was already a microphagous organism with ciliated folds in the foregut. This is then possibly another derived feature of the Articulata.

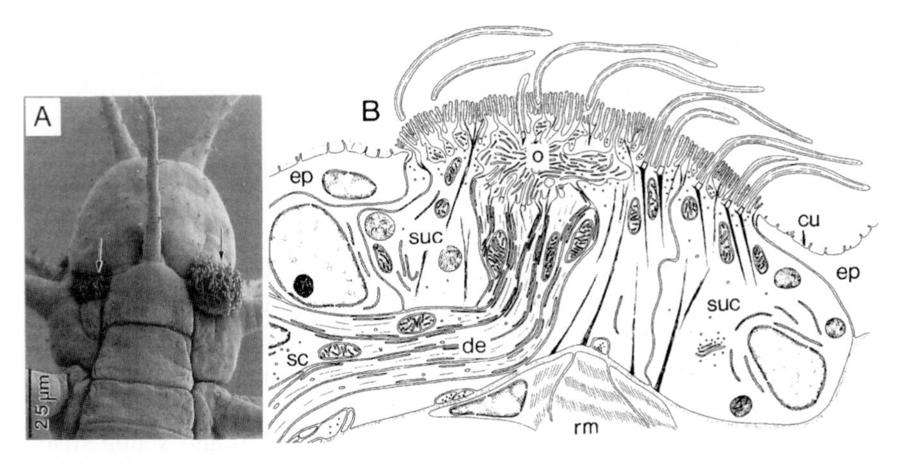

Fig. 24. Polychaeta. Paired nuchal organs. **A** *Microphthalmus* (Hesionidae). Dorsal view. Arrows point to the large nuchal organs. **B** *Nerillidium troglochaetoides* (Nerillida). Reconstruction of a nuchal organ in cross section. *cu* Cuticle; *de* monociliary dendrite of the sensory cell; *ep* epidermis cell; *o* olfactory chamber; *rm* retractor muscle; *sc* sensory cell; *suc* multiciliated supporting cell. (Purschke 1997)

Polychaeta – Clitellata

Polychaeta

Autapomorphies (Fig. 25 → 2)

- Parapodia.
  Two dorsolateral and two ventrolateral projections of the body bearing chaetae carried over from the ground pattern of the Annelida. In the simplest case these are primarily four bulge-like structures in which chaetae are arranged in rows (BARTOLOMAEUS, pers. comm.). These rows are an apomorphy in contrast to the appearance of bundles of chaetae in the ground pattern of the Annelida.
- Nuchal organ.
  Paired chemoreceptors at the posterior of the prostomium, composed of bipolar primary sensory cells, ciliated supporting cells and a retractor muscle (Fig. 24). The sensory cells are usually furnished with a cilium and several microvilli, which project into a subcuticular cavity (olfactory chamber). The nuchal organ is innervated by paired nerves from the posterior part of the brain (PURSCHKE 1997).
- A pair of pygidial cirri.
  The original state which is carried over by the Aciculata. In the Scolecida an increase to two or more pairs of caudal cirri has occurred.

Parapodia, pairs of **nuchal organs**, and **cirri on the pygidium** only exist in the Polychaeta. Consequently we postulate their **evolution in the stem lineage of the unity Polychaeta.** This is the most parsimonious explanation, the validity of which will now be tested.

With regard to the parapodia, I see no methodically grounded justification for the hypothesis of an earlier development in the stem lineage of the Annelida or even the Articulata (WESTHEIDE 1996a, 1997). On the other hand, it is not necessary to postulate the reduction of the parapodia in the stem lineage of the sister group Clitellata. We put forward the simplest hypothesis which is that the Clitellata carry over from the ground pattern of the Annelida the original state of four groups of bundled chaetae without parapodia.

Due to their widespread appearance within the taxon and subtle structural congruencies, the nuchal organs of the Polychaeta have been homologized (PURSCHKE 1997). They are absent, however, in the Clitellata and Arthropoda. As with the parapodia, there are three basic possibilities – evolution in either the stem lineage of the Articulata, the Annelida, or the Polychaeta. In the latter, no further explanations concerning the reduction of nuchal organs in the Clitellata and Arthropoda are needed. Therefore this hypothesis must be favored.

Finally, in the Clitellata or Arthropoda, there are also no equivalents of the pygidial cirri. A repetition of the previous argument can be spared here.

The three features mentioned above can, without conflict, be interpreted as autapomorphies of a monophylum Polychaeta.

Systematization

Fortunately, the first outline of a phylogenetic system for the Polychaeta has now been attempted (ROUSE & FAUCHALD 1997). We present here the highest-ranking adelphotaxa kinships, even when the authors state that "this study must be regarded as one of the initial steps in a new phase of polychaete systematics" (1997, p. 161). Some open questions still remain in presenting relevant justifications for monophyla.

Polychaeta
- **Scolecida**
- **Palpata**
  - **Aciculata**
  - **Canalipalpata**

An overview of all the "families" of the Polychaeta that have been integrated into the phylogenetic system, and those that have yet to be integrated, is presented in a separate study (FAUCHALD & ROUSE 1997).

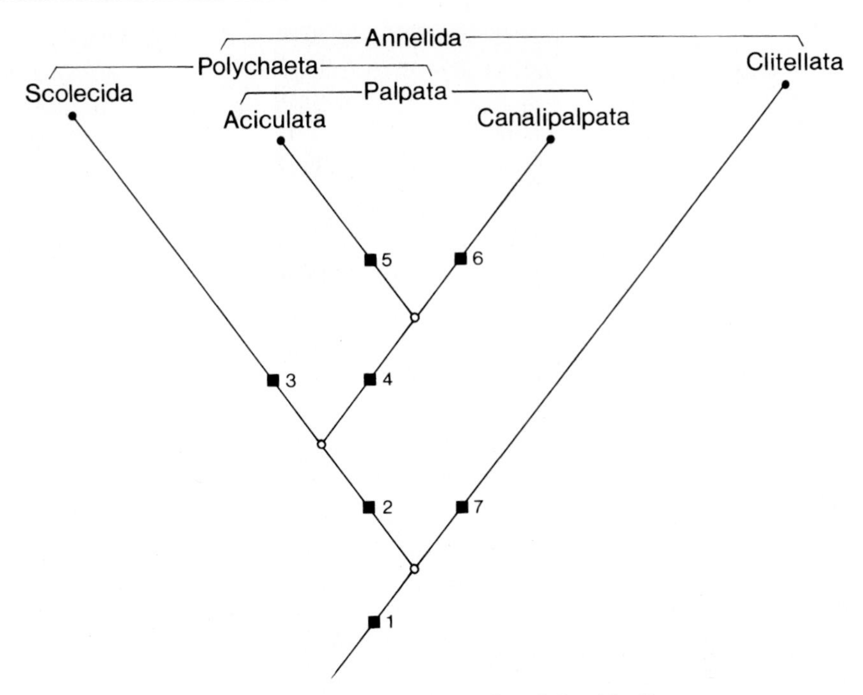

Fig. 25. Polychaeta. Phylogenetic system in the form of a relationship diagram

Scolecida – Palpata

Scolecida

Autapomorphy (Fig. 25 → 3)

- Two (or more) pairs of cirri on the pygidium.
  In the ground pattern there are a minimum of four caudal cirri. This is a doubling in contrast to the ground pattern of the Polychaeta.

That the unity Scolecida is a monophylum has only been weakly justified (ROUSE & FAUCHALD 1997). A plesiomorphy is the absence of appendages (antennae, palps) on the prostomium.

Nine "families" from the conventional classification are placed in the Scolecida. Well-known taxa include *Arenicola, Capitella, Ophelia, Travisia, Scoloplos* (Fig. 26).

The name Scolecida was derived from the name Scoleciformia by ROUSE & FAUCHALD. This name was originally introduced by BENHAM (1896) for certain Polychaeta. This step is somewhat unfortunate as the name Scolecida was a well-known term for a series of simply organized "worms" (HENTSCHEL & WAGNER 1996) outside the Polychaeta.

Fig. 26. Polychaeta. **A** *Arenicola marina* (Arenicolidae, Scolecida). **B** *Ophelia limacina* (Opheliidae, Scolecida). **C** *Travisia pupa* (Opheliidae, Scolecida). **D** *Scoloplos armiger* (Orbiniidae, Scolecida). In the Scolecida antennae and palps are absent on the prostomium. This lack is interpreted as a plesiomorphy. **E** *Halosydna brevisetosa* (Polynoidae, Phyllodocida, Aciculata) with elytra = plate-shaped differentiated dorsal cirri from the parapodia. **F** *Nereis diversicolor* (Nereidae, Phyllodocida, Aciculata). (Originals: **A, D** Sylt, North Sea; **B** Arcachon, Atlantic; **C, E** Puget Sound, Pacific; **F** Cuxhaven, North Sea)

Palpata

Autapomorphy (Fig. 25 → 4)

- Sensory palps on the prostomium (Fig. 27 A, B).
  The evolution of two palps with a sensory function can be hypothesized to have occurred in the stem lineage of the unity Palpata. The palps are found on the prostomium and have a ventral orientation.

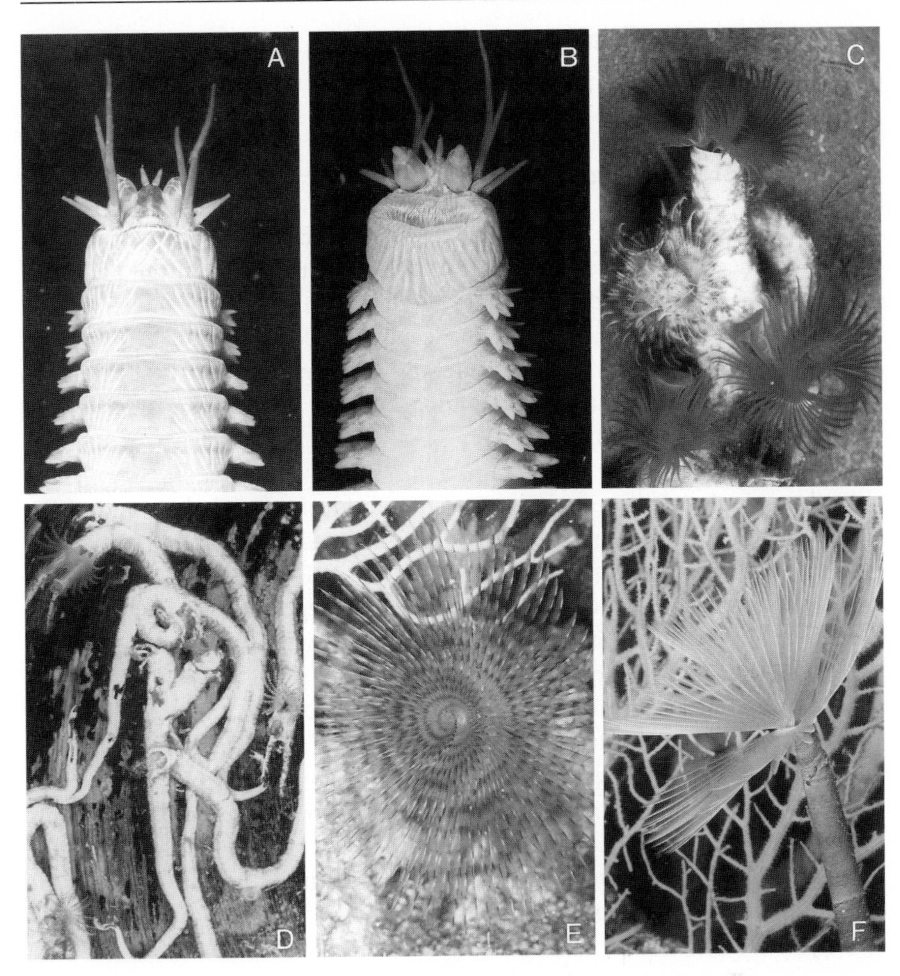

Fig. 27. Polychaeta. **A–B** *Nereis virens* (Nereidae, Phyllodocida, Aciculata). **A** Anterior end, dorsal view. **B** Ventral view. Two small antennae terminal on the prostomium, laterally the two cone-shaped sensory palps. Peristomium and following segment fused, without parapodia but together with four pairs of cirri. Axial pharynx in **B** partially everted. **C** *Serpula vermicularis* (Serpulidae, Sabellida, Canalipalpata). **D** *Hydroides norvegica* (Serpulidae, Sabellida, Canalipalpata). **C–D** Animals in calcareous tubes on hard substrates. **E–F** *Spirographis spallanzani* (Sabellidae, Sabellida, Canalipalpata). Aerial and side view. Animal in soft tube in sediment with funnel-shaped crown of tentacles. (Originals: **A, B** Sylt, North Sea; **C** Puget Sound, USA Pacific Coast; **D** Helgoland, North Sea; **E, F** Banyuls sur Mer, Mediterranean)

Aciculata – Canalipalpata

Aciculata

Autapomorphies (Fig. 25 → 5)

- Locomotory parapodia with aciculae (Fig. 27 A, B).
  Lobe-shaped parapodia with dorsal notopodium and ventral neuropodium. In each of the four branches there is a minimum of one aciculum as a strong, modified chaeta. The deeply embedded aciculae remain in the chaeta follicle and do not generally emerge through the epidermis. The aciculae are skeletal rods to which muscles for the movement of the parapodia are attached.
- Compound chaetae.
  In the Aciculata compound chaetae composed of a basal shaft and a distal appendage exist. The two parts are connected by a joint.
- Parapodial cirri.
  A pair of dorsal cirri on the notopodia of a segment. A pair of ventral cirri on the neuropodia.
- Antennae on the prostomium.
  In the ground pattern of the Aciculata there are three prostomial antennae with innervation from the anterior part of the brain. The antennae point primarily rostroventrally.
  Nereis is an example of the reduction in the median process, here only the two lateral antennae remain (Fig. 27 A, B). It should be noted that the antennae in the Polychaeta obviously have no connection with the homonymous antennae of the Arthropoda.

A plesiomorphy of the Aciculata is a pair of pygidial cirri – taken over from the ground pattern of the Polychaeta.

The Aciculata of ROUSE & FAUCHALD (1997) are better known under the old name Errantia, a term that could still be used today. Their classical counterpart is the non-monophyletic "Sedentaria"; this name, however, is now being phased out. The Phyllodocida and the Eunicida have been presented as subordinated adelphotaxa of the Aciculata.

The **Phyllodocida** have a long axial pharynx that can be extruded terminally (Fig. 27 B). This organ is an **autapomorphy** of a unity which according to ROUSE & FAUCHALD (1997) contains 19 "families" from conventional classification. Well-known taxa include *Aphrodite, Lepidonotus, Anaitides, Hesionides, Nereis, Nephthys, Tomopteris* (pelagic).

In the **Eunicida** the pharynx lies ventrally beneath the esophagus; this may be (?) an alternative **apomorphy** to the manifestation of the axial pharynx of the Phyllodocida. Dorsal gills on the parapodia are probably a further apomorphic peculiarity of the Eunicida. Six "families" according to ROUSE & FAUCHALD (1997). Well-known taxa include *Eunice* (palolo worm), *Lumbrineris, Ophryotrocha.*

Canalipalpata

Autapomorphy (Fig. 25 → 6)

- Grooved palps with a feeding function on the prostomium.
  Transformation of the two palps with sensory function (ground pattern Palpata) into two organs which serve in nutrient uptake. The sensory palps with a circular outline become transport bands with a U-shaped cross section. The cilia of the furrows transport nutrient particles to the mouth. The evolution of grooved palps is related to the change to a stationary life style with conversion to microphagy.

The highest-ranking subtaxa of the Canalipalpata according to ROUSE & FAUCHALD (1997) are the Spionida, Terebellida and Sabellida, with unclear relationships between one another. Due to specific agreement in the structure of the hooked chaetae (uncini), BARTOLOMAEUS (1995) interprets the Terebellida, and a unity of Sabellida + Pogonophora (see below), as sister groups.

In the **Spionida** the palps have moved to the peristomium (segment following the prostomium). A further characteristic feature is the elongation of the nuchal organ to form posterior appendages. Seven "families" from conventional classification. Well-known taxa include *Polydora, Pygospio, Scolelepis, Chaetopterus, Magelona.*

For the **Terebellida** ROUSE & FAUCHALD (1997) highlight two **autapomorphies**. The gular membrane is a muscular dissepiment in the anterior of the body; it serves to raise coelomic pressure to extrude the anterior bodily appendages. A thickening of tissue in the dorsal blood vessel above the esophagus in which blood cells are formed. BARTOLOMAEUS (1995) interprets the existence of perioral primary non-retractile tentacles for microphagous feeding as a further derived peculiarity. Coelomic canals traverse the tentacles. According to ROUSE & FAUCHALD (1997) eight "families" from conventional classification. Well-known taxa include *Lanice, Pectinaria.*

The **Sabellida** with a fusion of prostomium and peristomium are only weakly justified as a monophylum by ROUSE & FAUCHALD (1997). Excluding the Pogonophora (see below) a further **autapomorphy** may be the perioral funnel-like crown of tentacles (BARTOLOMAEUS 1995). From two semicircular formed tentacle bearers (homologous with the palps) feather-like tentacles arise, each with two rows of filaments (pinnulae) per feather for filtering suspended fine particles and for respiration (Fig. 27 C–F). Five "families", according to ROUSE & FAUCHALD (1997), are from conventional classification. Well-known taxa include *Owenia, Sabellaria, Sabella, Spirographis, Serpula, Hydroides, Spirorbis.*

We now come to the remarkable **Pogonophora**, which are mostly very long deep-sea dwellers, only a few millimeters thick. In his classic monograph, A.V. IVANOV writes, "We cannot in the long run escape the conclusion that pogonophores are an independent phylum of the Deuterostomia" (1963, p. 139). Today, there is no longer any doubt that the Pogonophora belong in the Polychaeta and within this constitute a subtaxon of the Canalipalpata. ROUSE & FAUCHALD (1997) place them as a low-ranked unity Siboglinidae (first "family" name from CAULLERY 1914) in the Sabellida, whereas for BARTOLOMAEUS (1995, 1998) they form the adelphotaxon of the Sabellida.

The Pogonophora are characterized by a **whole series of unusual derived characteristics.** In the division of the body, following on from the prostomium with palps, there is a extended peristomium (first segment) and an extremely long second trunk segment. The latter forms by far the largest part of the whole body. Only the short, thickened posterior end is comprised of numerous "normal" segments (Fig. 28 B, C). Adult Pogonophora have neither mouth nor anus. In the second trunk segment the trophosome is found as a transformation product of the intestine. It contains chemolithotrophic bacteria, which probably form the only nutritional source of the Pogonophora. The palps are not conveyor belts for food particles and have therefore lost their ciliated furrows (ground pattern of the Canalipalpata). The number of palps varies from one in *Siboglinum* to more than 200 appendages in *Spirobrachia*. The Pogonophora live in tubes secreted by gland cells of the epidermis. These tubes characteristically contain β-chitin. Nuchal organs are absent. *Siboglinum, Lamellisabella, Spirobrachia, Riftia* (1.5 m long), *Lamellibrachia.*

We are now back at the beginning of the systematization. With the above division of ROUSE & FAUCHALD (1997), the first steps towards a comprehensive phylogenetic system of the Polychaeta have been taken. However, the task is not yet complete. From the "basic classification" (1997, p. 73) 29 "families" are still excluded although a rough placement is in many cases possible. Yet unsolved problems include the relationship of the fresh-water Aeolosomatidae and Potamodrilidae. They are conventionally placed in the Clitellata – a measure, which according to BUNKE (1986), has no justification. ROUSE & FAUCHALD place the Aeolosomatidae and Potamodrilidae in the Polychaeta albeit for the present as taxa incertae sedis.

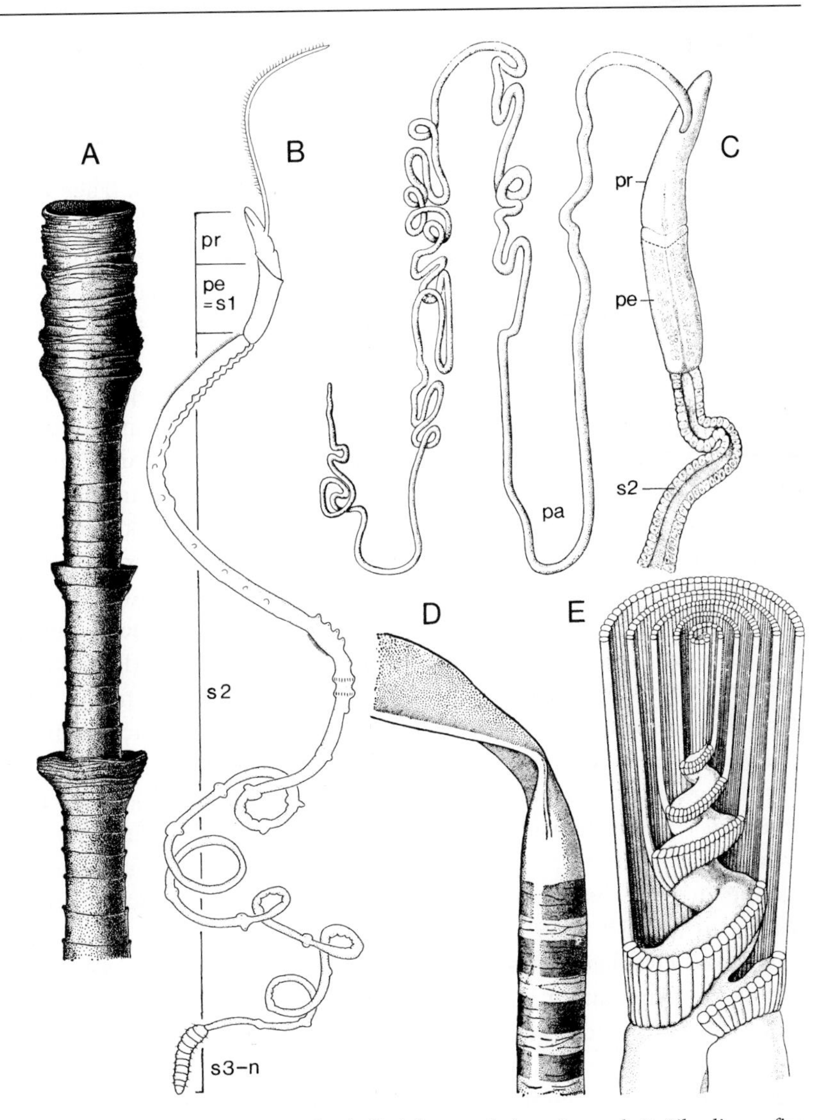

Fig. 28. Pogonophora. **A** Tube of *Lamellisabella johanssoni*. Anterior end. **B** *Siboglinum fiordicum* with one palp on the prostomium. **C** *Siboglinum pellucidum*. Anterior end. **D** *Oligobrachia dogieli*. Tube with membranous top. **E** *Spirobrachia grandis*. Numerous palps on a spirally twisted support structure. *pa* Palp; *pe* peristomium; *pr* prostomium; *s* trunk segment (**A, C–E** Ivanov 1957, 1963; **B** Purschke 1996)

Echiurida

The final problem dealt with here concerns the position of the Echiurida. I will argue for their inclusion in the Polychaeta.

The spoon worms comprise around 150 marine species. They are semi-sessile organisms living in soft sediments (*Echiurus, Urechis*) or in crevices of rocky substrates (*Bonellia*). The body is divided into a large prostomium and a sac-like unsegmented trunk. With a trunk length of 15 cm the spatulate prostomium of *Echiurus echiurus* is around 4 cm. The forked prostomium of *Bonellia viridis* can be up to 1.5 m long.

The muscular **prostomium** is used primarily for microphagous feeding. On the prostomium of *Echiurus echiurus* the stream of fine particles in a ventral ciliated groove can be particularly well observed (Fig. 29 B). In *Bonellia viridis* there is also a constant flow of nutrient particles along the long band to the mouth at the anterior end of the trunk. *Urechis caupo* acquires its food by forming a mucus network that is extended in a U-shaped tube. This is a derived behavior within the Echiurida.

The **trunk** is lined by an **undivided coelom** with somatic peritoneum and visceral myoepithelium (PILGER 1993; BARTOLOMAEUS 1994). The coelom shows no sign of segmentation and is developed from two mesodermal bands which form into coelomic sacs. These produce at first dorsal and ventral mesenteries which later largely disappear.

In the nervous system a large circumesophageal ring develops from the brain area in the episphere of the trochophore. This runs through the prostomium to its tip. In the hyposphere of the trochophore a pair of ventral nerve cords is formed, these soon fuse, however, to form a single nerve cord (KORN 1960).

A closed circulatory system with a short dorsal blood vessel in the first half of the trunk and a long ventral blood vessel belong in the ground pattern of the Echiurida. The blood flows out of the dorsal vessel into a median vessel of the prostomium; this branches anteriorly into two lateral canals through which the blood is led back to the ventral vessel of the trunk.

One or two pairs of metanephridia in the anterior of the trunk also belong to the ground pattern. These serve to discharge gametes. There can be up to 40 serially arranged nephridia.

A pair of anal sacs in the posterior part of the body with numerous ciliated funnels form the excretory apparatus of the Echiurida. In *Bonellia viridis* 8500 ciliated funnels per anal sac have been observed (PILGER 1993).

The Echiurida are gonochoristic. The simple unpaired gonads lie in the posterior part of the trunk. In the ground pattern male and female are identical. The gametes are released via the metanephridia into the water where external fertilization takes place. The extreme sexual dimorphism of

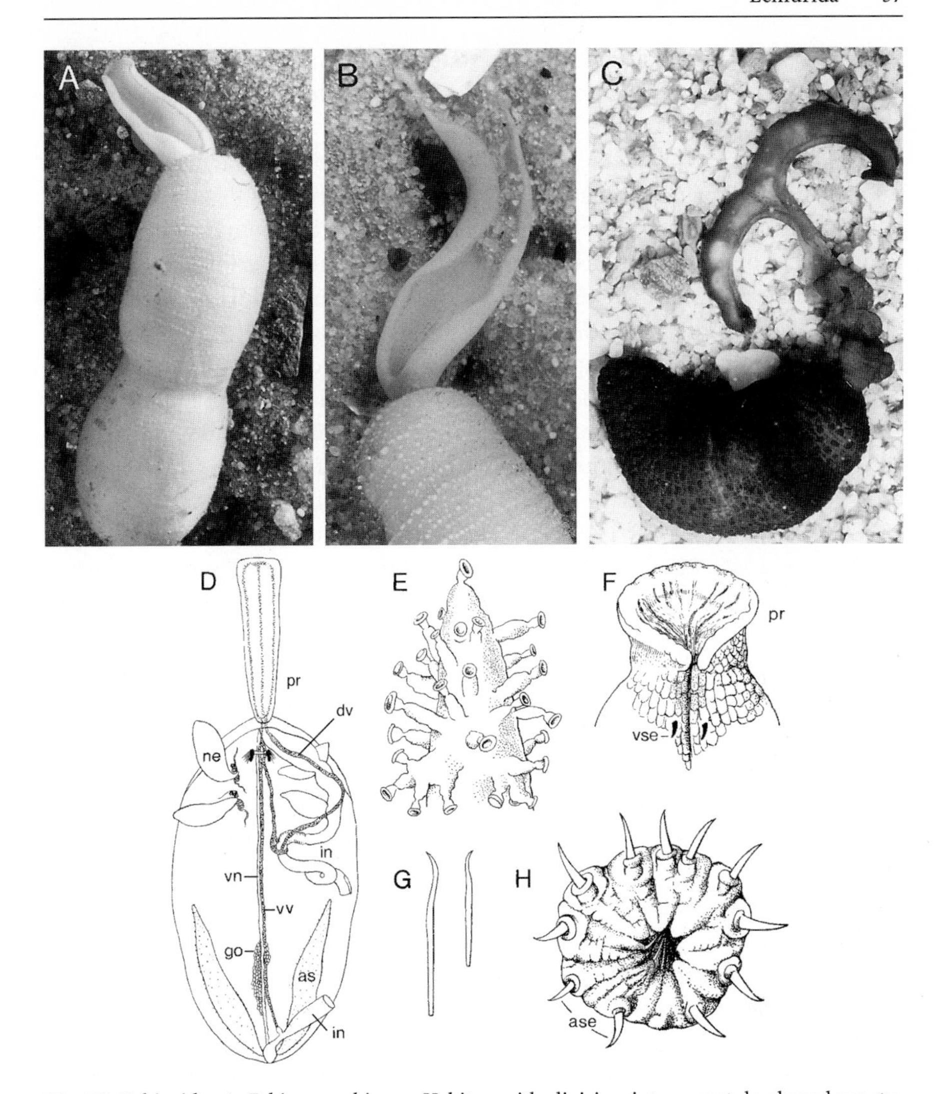

Fig. 29. Echiurida. **A** *Echiurus echiurus*. Habitus with division into a spatula-shaped prostomium and a sac-shaped trunk. **B** *Echiurus echiurus*. Transport of sand grains in the ciliated ventral groove of the prostomium in the direction of the mouth. **C** *Bonellia viridis*. Female. Prostomium, over 1 m long with a large fork at the anterior. The prostomium is strongly contracted against the trunk. **D** Diagram of the organization of the Echiurida. Dorsal view. Midsection of the coiled intestine not shown. **E** *Bonelliopsis alaskana*. Anal sac. Tip of a branch with ciliated funnels. **F–H** *Urechis caupo*. **F** Prostomium and beginning of the trunk. Prostomium reduced to a small collar. Tips of the ventral chaetae project out of the skin. **G** Hook-like ventral chaetae. Up to 1 cm long. **H** Posterior end with anus and a ring of anal chaetae. *as* Anal sac; *ase* anal chaetae; *dv* dorsal blood vessel; *go* gonad; *in* intestine; *ne* nephridium; *pr* prostomium; *vn* ventral nerve; *vse* ventral chaeta; *vv* ventral blood vessel. (**A, B** Originals: Helgoland, North Sea; **C** Original: Banyuls sur Mer, Mediterranean; **D–H** Stephen & Edmonds 1972)

Bonellia viridis is an apomorphous state within the Echiurida. The millimeter-long male lives in the nephridia of the female and the fusion of eggs and sperm cells also occurs here.

The **existence of chaetae** is the most important characteristic in deducing the phylogenetic relationships of the Echiurida.

One pair of large mobile hooks can be placed in the ground pattern of the Echiurida. These chaetae are found ventrally in the anterior region of the body. Further, in the taxa *Urechis* and *Echiurus* around 10–15 small anal chaetae are found. In *Urechis* these form a regular ring at the posterior, but in *Echiurus* they are staggered and so form two wreath-like structures. Although only realized in these two taxa, we postulate that the anal chaetae also belong in the ground pattern of the Echiurida. This will now be proven.

The chaetae of the Echiurida are in their ultrastructure, development, and composition (with β-chitin) identical to those of the Annelida (STORCH 1984; PILGER 1993; BARTOLOMAEUS, pers. comm.).

That there is a kinship between the two taxa is, in principle, undisputed on the grounds of other congruencies – development of the prostomium from the episphere of the trochophore, early coelomogenesis, paired origin of the ventral nerve cord. Due to these facts, the hypothesis of a homology between the chaetae of the Echiurida and Annelida is the most parsimonious and, at the same time, only justifiable explanation. I see no grounds for an assumption of convergence.

If we test the relationship in detail, we are confronted with two competing hypotheses. The Echiurida form either the adelphotaxon of the Articulata (ROUSE & FAUCHALD 1995, 1997) or are a subtaxon of the Annelida (NIELSEN 1995; EIBYE-JACOBSEN & NIELSEN 1996). Assuming a homology of the chaetae results in the following consequences.

1. In a **sister group relationship Echiurida–Articulata**, only a few chaetae would have developed initially on an unsegmented trunk with later increment due to the development of segments in a metameric organism. Furthermore a reduction of chaetae in the stem lineage of the Arthropoda would then be imperative. Neither assumption can be sustained (see p. 44).
2. If the **Echiurida are a subtaxon of the Annelida**, then the few chaetae on the anterior and posterior ends could be the remains of a reduction process from organisms with segmental chaetae. In the absence of directional peristaltic locomotion these would be superfluous. This is the case seen in the Echiurida if they are interpreted as having a secondary ametameric sac-like trunk as an adaptation to an overwhelmingly sedentary life style within the sediment.

In conclusion, I favor the second hypothesis, although the kinship within the Annelida is still unclear. The adelphotaxon of the Echiurida is unknown. Autapomorphies of the Polychaeta (parapodia, nuchal organ, cirri on the pygidium) are absent (? secondarily). On the other hand the Echiurida also have none of the apomorphies of the Clitellata (see below). Therefore their ordering within the Polychaeta is the most economical measure. Correspondingly, I divide the autapomorphies of the Echiurida into two groups: (1) features independent of the question of whether the Echiurida are the sister group of the Articulata or belong in the Annelida and (2) further features applicable if the Echiurida are a subtaxon of the Polychaeta.

Autapomorphies

First group

- Large prostomium as an organ of microphagous feeding (Fig. 29 A–C).
- Two anal sacs as excretory organs with numerous ciliated funnels (Fig. 29 E).

Second group

- Secondary ametameric sac-like trunk with a uniform peripheral muscle layer. Undivided coelom.
- Restriction of the chaetae to the anterior and posterior parts of the trunk (Fig. 29 F–H).
- Absence of parapodia, of nuchal organ on the prostomium and of cirri on the pygidium.

Clitellata

The Clitellata are the sister group of the Polychaeta. They are easily justified as a monophylum due to a whole series of derived characteristics.

Autapomorphies (Fig. 25 → 7)

- Hermaphroditism.
  All Clitellata are hermaphrodites in contrast to the gonochoristic state in the ground pattern of the Annelida. A single evolution of hermaphroditism in the stem lineage of the unity Clitellata is hypothesized.
- Clitellum.
  Gland-rich section of the epidermis in the anterior (Fig. 30 A). Extension over a few segments, not before the female genital openings (JAMIESON 1983 a; BUNKE 1986).
  In the ground pattern of the Clitellata the function of the clitellum is the secretion of mucus during "copulation". The strengthening secretion binds the sexual partners together during mutual transfer of sperm to the receptacula seminis (spermathecae). After separation of the partners

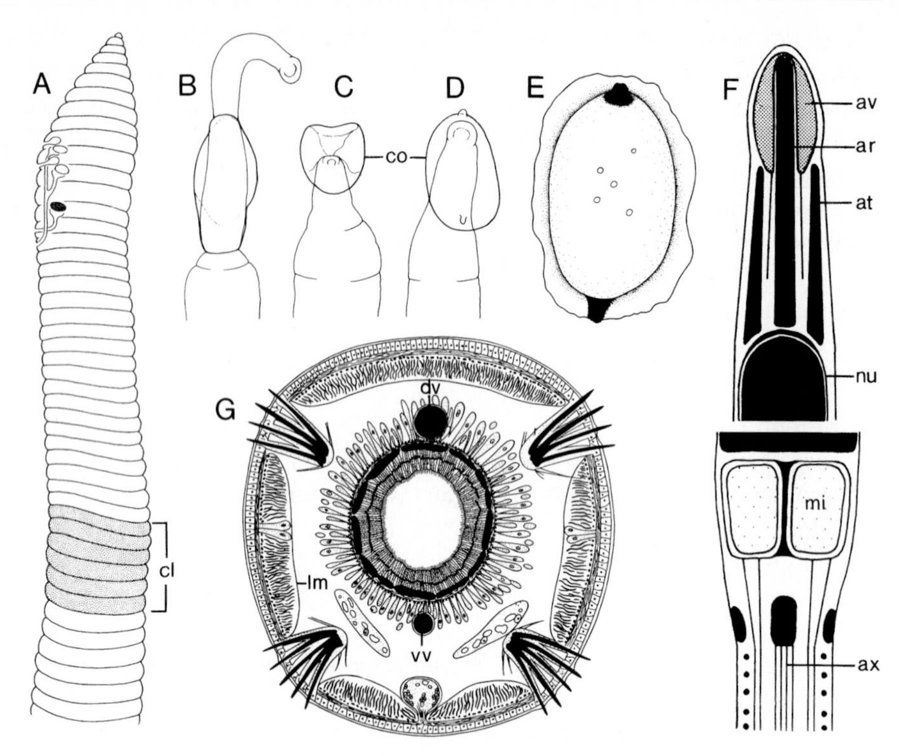

Fig. 30. Clitellata. Ground pattern features. **A** *Lumbricus* (Lumbricidae). Clitellum (autapomorphy) in segments 32–37. **B–E** *Erpobdella octoculata* (Hirudinea). Cocoon as the product of the clitellum (autapomorphy). **B+C** After secretion the anterior body with clitellum retreats backwards out of the cocoon. **D** Working of the cocoon with the anterior end after deposition. **E** Cocoon with eggs in albumen fluid. **F** Sperm in the ground pattern of the Clitellata. Autapomorphies: Acrosome tube and acrosome rod in the tip; mitochondria between nucleus and axoneme. **G** *Enchytraeus albidus* (Enchytraeidae). Cross section. Demonstration of Clitellata plesiomorphies: Division of the body longitudinal musculature into bands; carried over from the ground pattern of the Articulata. Four groups of chaetae without parapodia; taken over from the ground pattern of the Annelida. *ar* Acrosome rod; *at* acrosome tube; *av* acrosome vesicle; *ax* axoneme; *cl* clitellum; *co* cocoon; *dv* dorsal blood vessel; *lm* longitudinal muscle band; *mi* mitochondrium; *nu* nucleus; *vv* ventral blood vessel. (**A** Westheide 1996b; **B–E** Westheide 1980; **F** Jamieson 1992; **G** Avel 1959)

the clitellum produces a cocoon with nutrients for the embryos. Eggs and sperm fuse outside the body in the cocoon.
Different evolutionary transformations occur in the Hirudinomorpha: binding of the sexual partner with clitellum mucus does not take place. Sperm transfer occurs via the penis (Branchiobdellida) or spermatophores (ground pattern of the Hirudinea) (p. 67). Internal fertilization.

- Location of the gonads.
  On a few successive segments. Male before the female.
- Relocation of the brain.
  Withdrawal from the prostomium into the anterior part of the trunk.

- Ultrastructure of the sperm (JAMIESON 1983b, 1992).
  Acrosome tube: a duct in front of the nucleus in the elongated acrosome. It carries the primary acrosome vesicle and contains the acrosome rod (perforatorium). This is not observed in the sister group Polychaeta. Mitochondria are found in the middle section interpolated between the nucleus and the point of the axoneme (Fig. 30F).
- Absence of trochophore larva.
  Reduction of the larva through a change to direct development in the stem lineage of the Clitellata.

Plesiomorphies

A different view is presented for the absence of bodily appendages. A body without antennae, palps and parapodia but with four groups of capillary chaetae could have been carried over by the Clitellata from the ground pattern of the Annelida (pp. 44, 48).

Traditional classification has for a long time divided the Clitellata into two unities, the Oligochaeta and the Hirudinea. To construct a phylogenetic system these must be rejected. In the latest efforts to produce a systematization of the Oligochaeta, including computer analysis and the principle of parsimony (BRINKHURST & NEMEC 1987; ERSEUS 1987; JAMIESON 1988; BRINKHURST 1994 amongst others), I find only one attempt to fulfill the basic requirements of phylogenetic systematics, namely, the attempt to establish the Oligochaeta themselves as a monophylum.

Class Oligochaeta: "Euclitellates[6] with testes anterior to ovaries and lacking jaws. With spacious coelom. Setae not jointed; absent in Achaeta. Sperm acrosome lacking anterior extension seen in Hirudinea" (JAMIESON 1988, p. 395). In contrast to the intention of the author, this "diagnosis" does not contain apomorphies in comparison with the Hirudinea. The attempt has foundered. Understandably, there is also no hypothesis about sister groups on the first subordinate hierarchic level, which in the case of a monophylum should be possible.

The "Oligochaeta" are a paraphylum; their members form a "basal stock" of the Clitellata with constantly reoccurring plesiomorphic features compared with the Hirudinea. "The unnatural taxon 'Oligochaeta' should simply be abandoned for lack of information content" (SIDDAL & BURRESON 1995, p. 1059). However, lack of information is not the only problem, a taxon Oligochaeta is in addition a serious hindrance to understanding kinship.

As the elimination of the "Turbellaria" cleared the way for a consequent phylogenetic systematization of the Plathelminthes (Vol. I, p. 145), so should the elimination of the Oligochaeta enable a phylogenetic system for

[6] Clitellata

the Clitellata to be constructed and to precisely establish first the adelphotaxon of the Hirudinomorpha. It should then be possible to trace the sister group of this unity and so on.

We start by unifying the Branchiobdellida – often interpreted to be "Oligochaeta" – with the Hirudinea to a monophylum Hirudinomorpha. This taxon alone has been well enough justified to be included in this book.

Hirudinomorpha[7]

Branchiobdellida and Hirudinea possess a series of significant congruencies. Their assessment as synapomorphies, or as due to convergence, is still at present under dispute (SAWYER 1986; BRINKHURST & GELDER 1989; HOLT 1989; PURSCHKE et al. 1993; SIDDAL & BURRESON 1995). I interpret the following features as synapomorphies between the Branchiobdellida and the Hirudinea – and therefore as autapomorphies of the monophylum Hirudinomorpha which they comprise.

Autapomorphies (Fig. 31 → 1)

- Tubular muscle cells.
  Differentiation of the muscle cells into a closed outer layer with radially arranged flat myofibrils and a central axis of sarcoplasma with nucleus (Fig. 32 A).
- Ganglia with neuron packets.
  The structure of the ganglia with arrangement of the neurons in packets around the axon is a very peculiar characteristic of the Hirudinea (Fig. 32 B). In principle the same pattern exists in the Branchiobdellida even though only four neuron packets (two lateral pairs) are found in contrast to the six packets of the Hirudinea. Therefore the state seen in the Branchiobdellida is without doubt the plesiomorphy (SAWYER 1986).
- Epizoa (ectoparasites) on fresh-water animals.
- Existence of a posterior sucker.
- Displacement of the anus to the anterior, dorsally before the sucker.
- Secondary annulation of the segments. A minimum of two annuli (Branchiobdellida).
- Male genital pores fused.
- Sperm ultrastructure.
  A long corkscrew-like acrosome (FERRAGUTI & LANZAVECCHIA 1977; FERRAGUTI 1983) may be a further synapomorphy between the Branchiobdellida and the Hirudinea.

---

[7] tax. nov.

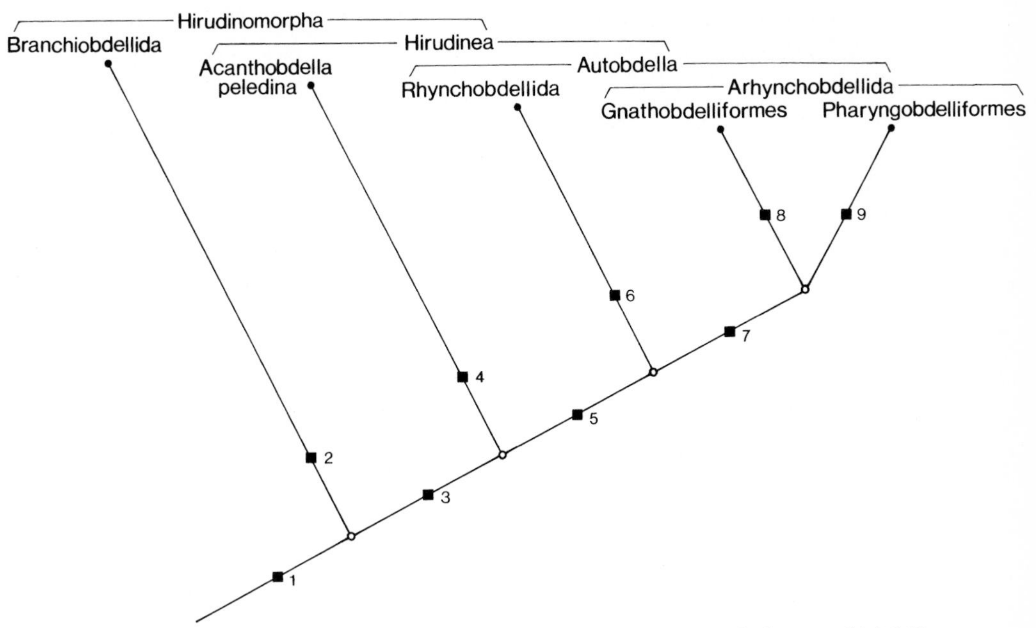

Fig. 31. Hirudinomorpha. Diagram of the phylogenetic relationships with the Branchiobdellida and Hirudinea as the highest-ranking adelphotaxa.

The structure of the **tubular muscle cells** and the ganglia with **neuron packets** form singular features of the Branchiobdellida and Hirudinea without comparison in the other taxa of the Clitellata. There are no established kinship hypotheses which can compel the interpretation of convergence, and we do not accept purely unjustified speculations. We must therefore use the principle of parsimony and postulate a single evolution in a common stem lineage of the Branchiobdellida and Hirudinea. This interpretation also throws light on further congruencies. Obviously their life style as epizoa or ectoparasites and, connected with this, the sucker, the dorsal anus and even the secondary annulation could be the result of convergent evolution. However, there is also no justification for this. When two convincing autapomorphies are already present then it is simplest and most sensible to see the aforementioned features as having developed once in the stem lineage of a unity Hirudinomorpha. This interpretation is internally consistent; there is no other taxon in the Clitellata that comes into question as the adelphotaxon of the Hirudinea and that can lead to a different assessment of certain features of the Branchiobdellida.

A further congruence between the Branchiobdellida and the Hirudinea is the position of the clitellar region, on segment (9) 10–12. Since the sister group of the Hirudinomorpha is not yet known, it must remain for the moment unassessed.

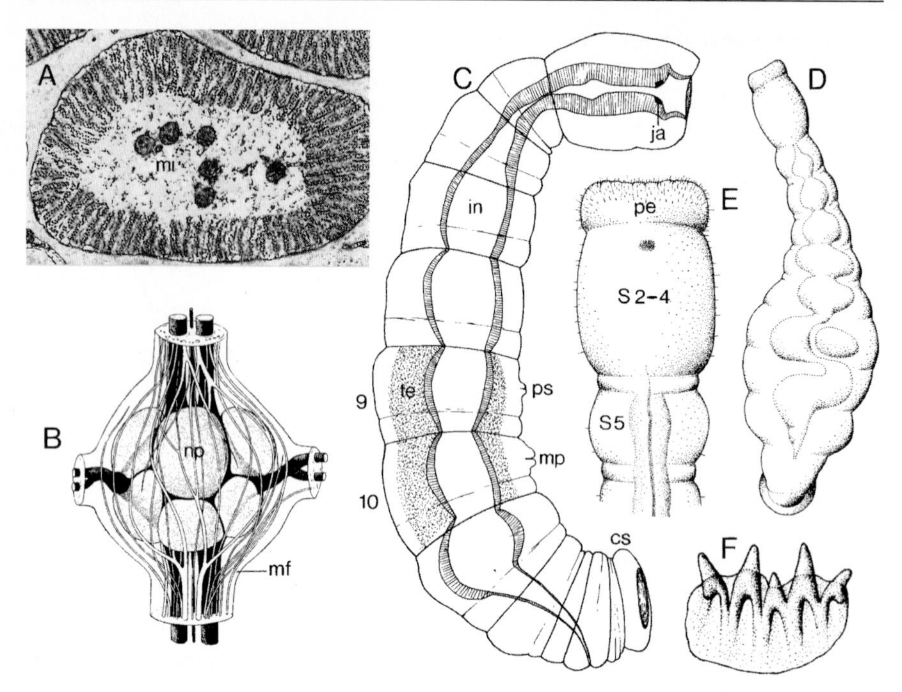

Fig. 32. A–B Autapomorphies of the Hirudinomorpha. **A** *Erpobdella* (Pharyngobdelliformes). Longitudinal muscle cell in cross section. Myofibrils form a wreath at the periphery; sarcoplasma with mitochondria in the center. **B** *Pontobdella muricata* (Rhynchobdellida). Ganglion from the center of the body with six neuron packets. **C–F** Branchiobdellida. **C** *Cambarinicola.* Side view with intestinal canal. **D–F** *Xironogiton instabilia.* **D** Habitus in dorsal view. **E** Anterior end with peristomium and three fused segments. **F** Dorsal jaw. *cs* Caudal sucker; *in* intestine; *ja* jaw; *mf* muscle fiber; *mi* mitochondrium; *mp* male genital pore; *np* neuron packet: *pe* peristomium; *ps* pore of the spermatheca; *s* segment; *te* testis. (**A** Lanzavecchia & de Eguileor 1976; **B** Harant & Grassé 1959; **C** Pennak 1978; **D–F** Franzén 1962)

Systematization

Hirudinomorpha
 Branchiobdellida
 Hirudinea
  Acanthobdella peledina
  Autobdella
   Rhynchobdellida
   Arhynchobdellida
    Gnathobdelliformes
    Pharyngobdelliformes

Branchiobdellida – Hirudinea

Branchiobdellida

Inhabitants of the Holarctic region. Epizoa on fresh-water Crustacea. Predominantly on the body surface and on the jaws of Decapoda, they apparently only seldom change to ectoparasitism feeding on blood.

150 species ranging from a few mm to 1.2 cm in length.

Branchiobdella.

We present a brief outline of their organization before going into more precise details of the derived characteristics of the Branchiobdellida.

Chaetae and prostomium are absent. The common description of 15 segments needs to be more fully examined (SAWYER 1986). A mostly broad head section is said to be formed from four segments (S 1–4). A ring-shaped peristomium with division into upper and lower lip (FRANZÉN 1962; HOLT 1965) is followed by three fused segments. The trunk consists of 11 segments (S 5–15) of which the last metamere is differentiated into a sucking disc (BRINKHURST & GELDER 1989). The trunk segments are divided by a furrow into two rings. The anus lies dorsally in segment 14.

Two pairs of testes in segment 9 and 10; a penis opens ventrally in segment 10. Female organs with unpaired spermatheca in segment 9, two ovaries and two genital pores in segment 11.

Autapomorphies (Fig. 31 → 2)

- Absence of chaetae.
  Convergent reduction in comparison with the Autobdella.
- Subfeatures of the bodily division.
  Absence of prostomium. Strong reduction in the number of segments to a constant 15. Head formed from four segments (Fig. 32 C).
- Two cuticular jaws in the pharynx (Fig. 32 F).
  One dorsal and one ventral jaw. Convergence to the three jaws of the Gnathobdelliformes.
- Reduction of the nephridia to two pairs.

Hirudinea

With the alternating adhesion of the anterior and posterior suckers, leeches have developed their characteristic mode of locomotion that we describe simply as leech-like, an expression used for analogous movement in other animals. Only one of the ca. 300 leech species has no anterior sucker. *Acanthobdella peledina* forms the adelphotaxon of the Autobdella (Euhirudinea), a grouping of all other “real” leeches with two suckers.

We begin by justifying the monophylum Hirudinea as a whole. For the deduction of the apomorphous ground pattern characteristics of the Hirudinea we must make a decision on a first point in which the adelphotaxa differ significantly. This is the different number of segments, caused by differing constructions of the posterior sucker.

The Hirudinea have without doubt carried over the prostomium and the peristomium as the first chaetaeless segment from the ground pattern of the Clitellata. In the Autobdella prostomium and peristomium are incorporated in the formation of the anterior sucker. This is totally missing in *Acanthobdella peledina*, and prostomium and peristomium are either reduced or fused with the second segment (PURSCHKE et al. 1993). We can, with justification, establish the existence of prostomium and peristomium in the ground pattern of the Hirudinea because two ganglia more than the actual number of body segments are present. *Acanthobdella peledina* and the Autobdella both have a supraesophageal ganglion formed from two pairs of neuromeres, of which the first is assigned to the prostomium and the second to the peristomium (1st segment). In *Acanthobdella* 29 ganglia pairs follow for segments 2–30, in the Autobdella 32 ganglia pairs for metameres 2–33 (LIVANOW 1931).

Which then belongs in the **ground pattern** of the Hirudinea, the morphological number of **30 or 33 segments**?

An answer can be deduced from the construction of the posterior sucker. In *A. peledina* and in the Autobdella the sucker begins in each case with metamere 27. In *A. peledina* the sucker is a concave bulge of the last four segments with a vertical orientation in relation to the longitudinal axis of the body. The adhesive disc of the Branchiobdellida has a comparable orientation; however, it is composed of only one segment. In the Autobdella the posterior sucker originates from a ventral flattening of the last seven segments and stands parallel to the longitudinal axis of the body (LIVANOW 1931).

Just using the comparison between *A. peledina* and the Autobdella, a derivation of the *A. peledina* sucker through reduction of the last three body segments is hard to conceive. For the probable case that the Branchiobdellida form the sister group of the Hirudinea, the interpretation of the direction of evolution becomes convincing. The original state is represented by the Branchiobdellida with an adhesive disc formed from one segment. In the stem lineage of the Hirudinea three new segments were incorporated into the still vertical anchor apparatus; this is realized in *A. peledina*. The extension of the body by three extra segments then led to the evolution of the sucker of the Autobdella having an orientation parallel to the longitudinal axis. In conclusion, a stem species of the Hirudinea with prostomium, peristomium and 29 further segments can be postulated, whereby the last four segments formed a posterior sucker.

Since we are concerned with numbers, I would like to draw attention to a small difference in the position of the genital pores – again counting the peristomium as segment 1.

	♂ – Pore	♀ – Pore
Branchiobdellida	10	11 (2 pores)
A. peledina	11	12
Autobdella	10	11

From the in-group Hirudinea the Autobdella correspond to the out-group Branchiobdellida. Accordingly the site of the genital opening in *Acanthobdella peledina* should be the derived state within the Hirudinea.

With this we come to the apomorphous features of the Hirudinea (cf. PURSCHKE et al. 1993).

Autapomorphies (Fig. 31 → 3)

- Body division.
  Prostomium + segments 1–30 (= peristomium+29 further segments).
  Posterior sucker from four segments (S 27–30).
  Stronger annulation through more annuli per segment.
- Nervous system with 31 pairs of ganglia.
  Supraesophageal ganglion (brain) with two ganglia – subesophageal ganglion with four ganglia – ventral nerve cord with 21 free ganglia – posterior ganglion mass from the four ganglia of the sucker segments.
- Ganglion structure with six packets of neurons.
  Each ganglion is made up of six discrete neuron packets. Two packets lie behind one another laterally at each side, and between these ventrally lie the remaining packets. This unique structure allows the ganglia to be counted exactly with a corresponding determination of the number of segments.
- Oblique muscles.
  Two layers of diagonal muscles between the circular and longitudinal muscles of the body wall musculature. The myofibrils of both layers cross each other at right angles.
- Strongly developed parenchymatous connective tissue.
- Transformation of the coelom.
  The transformation begins with constriction by the connective tissue and loss of the mesenteries between the coelomic sacs of a segment.
- Absence of spermatheca.
  No sperm transfer via a penis and storage in a spermatheca (in comparison with the Branchiobdellida).
- Spermatophores.

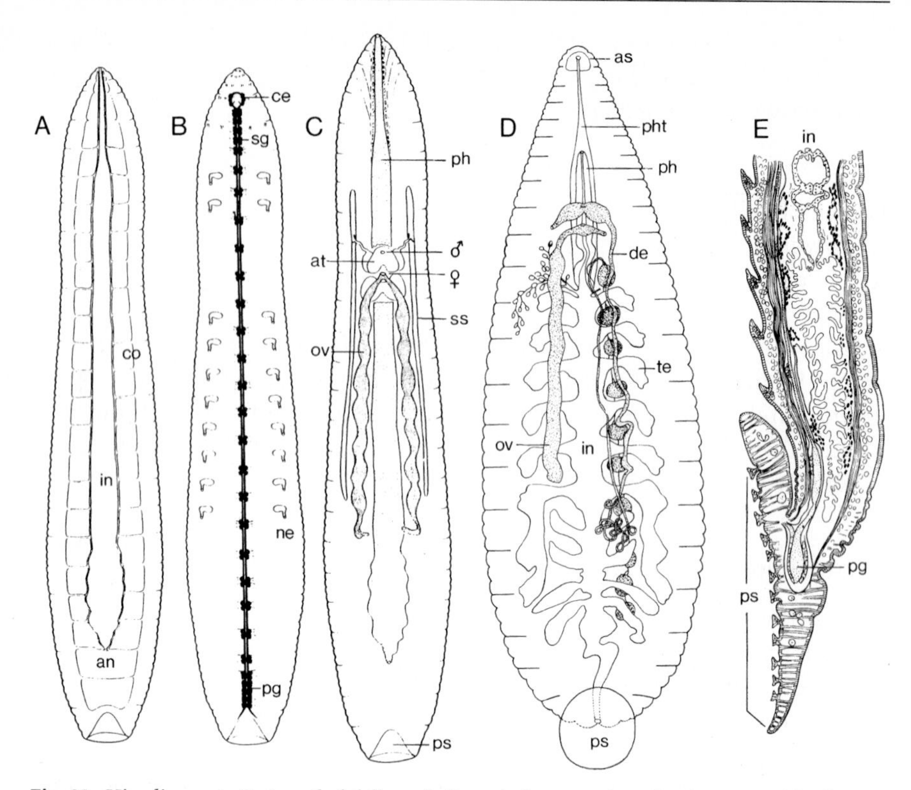

Fig. 33. Hirudinea. **A–C** *Acanthobdella peledina*. **A** Segmental coelomic sacs with dissepiments in the longitudinal axis of the body. **B** Nervous system and position of the nephridia. **C** Genital organs with long sperm-sacs and ovarian tubes. **D** *Glossiphonia complanata* (Rhynchobdellida). Derivation of the testes follicles from the state of the sperm-sacs in *Acanthobdella peledina*. **E** *Branchellion torpedinis* (Rhynchobdellida). Longitudinal section through the posterior body. Posterior sucker of seven segments with a position parallel to the long axis of the body (autapomorphy of the Autobdella). *an* Anus; *as* anterior sucker; *at* atrium; *ce* cerebrum; *co* coelomic sac; *de* ductus ejaculatorius; *in* intestine; *ov* ovarian tube; *pg* posterior ganglia mass; *ph* pharynx; *pht* pharyngeal pocket; *ps* posterior sucker; *sg* subesophageal ganglion; *ss* sperm-sac; *te* testis. (**A–C** Purschke, Westheide, Rohde & Brinkhurst 1993; **D** Pennak 1978; **E** Harant & Grassé 1959)

Sperm transfer through spermatophores that are placed on the surface of the partner (Fig. 34C).

- Unpaired female genital opening.

  Derived state in comparison to the Branchiobdellida.

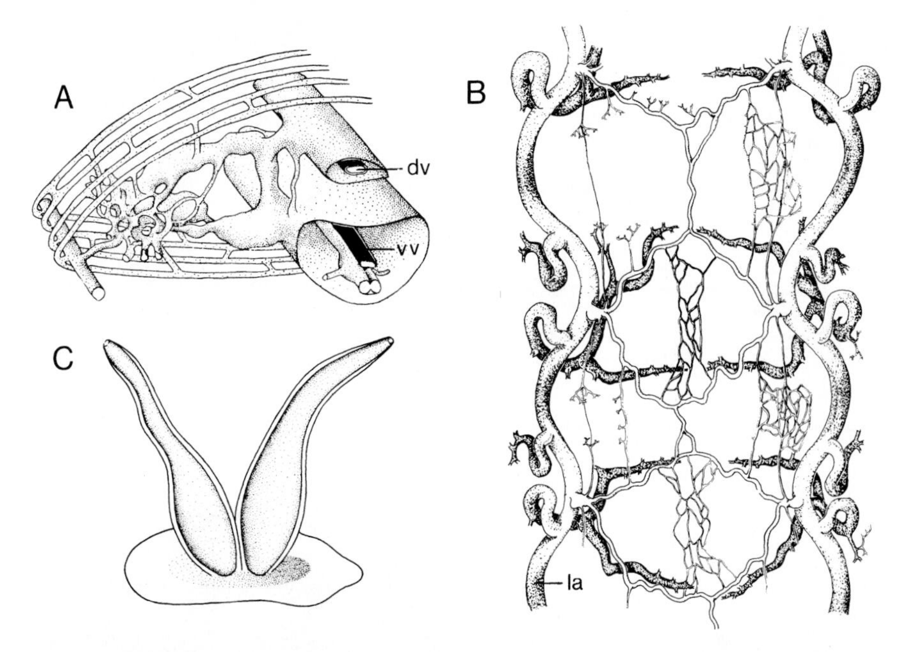

Fig. 34. Hirudinea, Autobdella. **A** *Glossiphonia complanata* (Rhynchobdellida). Canal system of the coelom and primary blood circulatory system. **B** *Hirudo medicinalis* (Arhynchobdellida). Coelomic canals as a secondary blood circulatory system. In both lateral vessels the blood flows from the back to the front; they contract alternately (heart function). **C** *Erpobdella octoculata* (Pharyngobdelliformes). Spermatophore with a basal plate and two tubes. *dv* Dorsal blood vessel; *la* lateral vessel; *vv* ventral blood vessel. (**A** Harant & Grassé 1959; **B** Boroffka & Hamp 1969; **C** Westheide 1980)

Acanthobdella peledina – Autobdella

Acanthobdella peledina

As a single species, *Acanthobdella peledina* forms the adelphotaxon of the Autobdella. *A. peledina* is a fresh-water inhabitant found in the north of Scandinavia and Russia as well as in Alaska. As a permanent ectoparasite on cold-water fish their nutrition comes from skin tissue and blood (SIDDAL & BURRESON 1995).

With a mass of plesiomorphies *A. peledina* is very important for the understanding of the evolution of leech organization. Through the existence of eight chaetae on each of the segments 2–6 the development of the Hirudinea from chaeta-bearing Clitellata is clearly demonstrated. The absence of an anterior sucker documents the evolution of the anchor apparatus in two steps. *A. peledina* is the only leech with well-developed segmental coelomic sacs – even if these are already constricted by connective tissue and a strong musculature. Further signs of a transformation are the incomplete

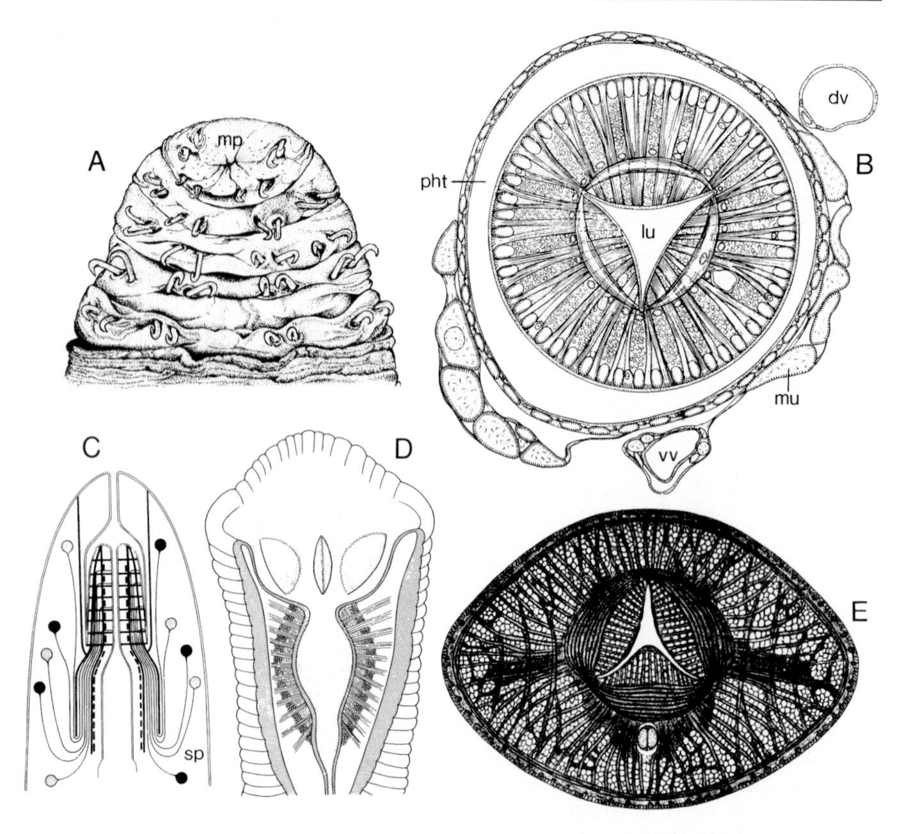

Fig. 35. Hirudinea. **A** *Acanthobdella peledina.* Ventral view of the anterior end. Four pairs of chaetae are found in each of the segments 2–6. **B–E** Construction of the Autobdella pharynx. **B–C** Subtaxon Rhynchobdellida. Pharynx differentiated into a eversible proboscis surrounded by a pharyngeal pocket (proboscis sheath). **B** *Haementeria costata.* Cross section through the proboscis in the proboscis sheath. Dorsal and ventral vessels of the primary blood circulatory system shown. **C** *Glossiphonia complanata.* Longitudinal section through the anterior. Diagram. **D–E** Arhynchobdellida. Pharynx not eversible. **D** *Hirudo medicinalis* (Gnathobdelliformes). Anterior end. Ventral area opened. Mouth surrounded dorsolaterally by the anterior sucker. Three cuticular jaws in the oral cavity. Pharynx developed as a sucking organ. **E** *Erpobdella octoculata* (Pharyngobdelliformes). Cross section through the anterior end with pharynx. Due to twisting of the pharynx one edge is directed upwards. *dv* Dorsal blood vessel; *lu* lumen of the pharynx; *mp* mouth pore; *mu* musculature; *pht* pharyngeal pocket; *sp* salivary glands; *vv* ventral blood vessel. (**A** Westheide 1996b; **B** Harant & Grassé 1959; **C, D** Westheide 1981; **E** Scriban & Autrum 1934)

dissepiments and the widespread loss of mesenteries. *A. peledina* has a well-developed primary circulatory system that stems from the space of the primary body cavity. Long sperm-sacs (testes) and egg tubes extend over several segments (Fig. 33 C). The sperm-sacs are of special interest. The serial arrangement of testes follicles in the Autobdella can be derived from regular constrictions of the sperm-sacs in *A. peledina*, these themselves are a plesiomorphy which can be traced back to the ground pattern of the Clitellata (PURSCHKE et al. 1993).

Let us now look at the derived characteristics of *Acanthobdella peledina.*

Autapomorphies (Fig. 31 → 4)

- Absence of prostomium and peristomium.
  Prostomium and segment 1 (peristomium) are reduced or fused with segment 2 (see above). From the 30 segments in the ground pattern of the Hirudinea only 29 remain.
- Anchor function of the chaetae.
  The 40 chaetae (plesiomorphy) on the flat ventral surface of the anterior are arranged around the mouth and are used for anchoring purposes (PURSCHKE et al. 1993). Arrangement and function of the chaetae are an apomorphic state (Fig. 35 A).
- Lack of nephridial funnels in comparison with the Autobdella

Autobdella (Euhirudinea)

Autapomorphies (Fig. 31 → 5)

- Body division.
  Prostomium + segments 1–33 (=peristomium + 32 further segments).
  Posterior sucker from seven segments (S 27–33).
- Nervous system with 34 pairs of ganglia.
  During the course of the development of the sucker, the number of ganglia pairs is increased in the posterior ganglia mass from four to seven.
- Complete absence of chaetae.
  Convergent chaeta reduction in the stem lineage of the Branchiobdellida (p. 65) and the Autobdella.
- Anterior sucker.
  The reduction of the last chaetae at the anterior made the way free for the evolution of a second sucker. The pit-shaped organ around the ventral oral opening is formed from the fusion of the prostomium, peristomium and four further segments.
- Coelom as a canal system.
  Fusion of the metameric coelomic sacs due to disintegration of the dissepiments. Evolution of the coelom into a system of longitudinal canals and transverse connections (Fig. 34 A).
- Serial testes follicles.
  Evolution from long sperm tubes as realized in *Acanthobdella peledina* (see above).

Rhynchobdellida – Arhynchobdellida

In conventional classifications, the Rhynchobdellida, Gnathobdelliformes and Pharyngobdelliformes are three rank equivalent groups of leeches with two suckers. The latter two, however, form a closer monophyletic unity of primary proboscis-less leeches. These Arhynchobdellida are the adelphotaxon of the Rhynchobdellida.

As their name suggests, the construction of the foregut plays a central role in the systematization of the Autobdella.

A pharynx with a triradiate lumen belongs in the ground pattern, whereby primarily one plane was probably directed dorsalwards and one corner pointed down.

Rhynchobdellida

Autapomorphy (Fig. 31 → 6)

- Proboscis.
  The evolution of the pharynx to an eversible proboscis is the fundamental autapomorphy by which the Rhynchobdellida can be justified as a monophylum (Fig. 35 B, C).
  The muscular organ is surrounded by an ectodermal sheath. For nutrient uptake the proboscis is everted out of the oral opening, searches for a possible puncture site and penetrates deep into the tissues of its prey. Fluid nutrition is sucked up by the action of the radial muscles.

A plesiomorphy is, as with *Acanthobdella peledina*, a primary circulatory system. The main elements are dorsal and ventral longitudinal blood vessels together with oblique connections in the anterior and posterior ends.

Glossiphonia, *Theromyzon*, *Piscicola*.

Arhynchobdellida

Autapomorphy (Fig. 31 → 7)

- Secondary circulatory system.
  In connection with further division of the coelom into finer canals up to capillaries, the original primary circulatory system is fully abandoned. Blood now flows directly in the canals of the coelom. The blood with hemoglobin becomes coelomic fluid. Muscular parts of the lateral canals form the circulatory motor (Fig. 34 B).

In comparison with the Rhynchobdellida the non-eversible pharynx is a plesiomorphy. The Pharyngobdelliformes with a twisted pharynx (see below) and the Gnathobdelliformes with jaws in the oral cavity are well-justified monophyla of the Arhynchobdellida. They may also be sister groups. This is, however, still uncertain because the placing of some jawless taxa without twisted pharynges (*Cylicobdella* amongst others) in the Gnathobdelliformes (Hirudiniformes) (SAWYER 1986) seems to be somewhat problematical.

Gnathobdelliformes (Hirudiniformes)

Autapomorphies (■ 8 in Fig. 31)

- Jaw.
  Three semicircular-formed cuticular saw blades within the oral cavity (Fig. 35D). Between the teeth of the free edges salivary glands open; they produce hirudin which is responsible for inhibiting blood clotting.
- Direct sperm transfer.
  Spermatophores are an apomorphy in the ground pattern of the leeches. They are, however, reduced in the stem lineage of the Gnathobdelliformes. A penis evolved from the area of the genital opening. Due to the absence of a spermatheca it is introduced directly into the genital pore.
  Hirudo medicinalis, Haemopis sanguisuga, Xerobdella lecomptei.

Pharyngobdelliformes (Erpobdelliformes)

Autapomorphy (■ 9 in Fig. 31)

- Spirally twisted pharynx.
  The triradiate pharynx lumen has anteriorly a dorsal and two ventrolateral planes. Soon afterwards the pharynx is twisted 60° anticlockwise, so that the dorsal plane is turned to a laterodorsal position and one edge points upwards (Fig. 35E). *Erpobdella.*

The Pharyngobdelliformes can devour large prey. Correspondingly, the pharynx constitutes one third of the body length.

Arthropoda

The species-poor Onychophora from the tropics form the tiny adelphotaxon to the more than one million arthropod species with articulated appendages which are found worldwide. We unite these under the name Euarthropoda. Within the Euarthropoda the highest-ranking sister groups are the Chelicerata and the Mandibulata. The latter is divided up into the

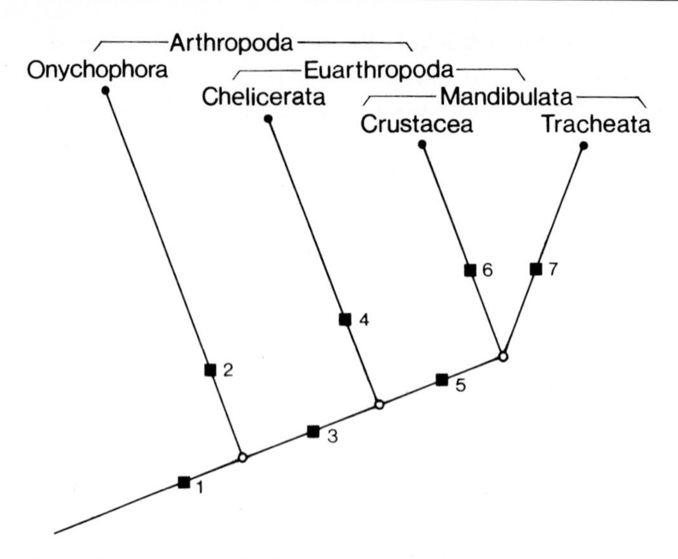

Fig. 36. Diagram of the high-ranking sister group kinships within the Arthropoda

Crustacea and the Tracheata. The relevant relationship diagram (Fig. 36) has its equivalent in the following hierarchical tabulation.

Arthropoda
- **Onychophora**
- **Euarthropoda**
  - **Chelicerata**
  - **Mandibulata**
    - **Crustacea**
    - **Tracheata**

Already the first adelphotaxa relationship is of fundamental importance for the understanding of the evolution of the Arthropoda. Due to the Onychophora feature pattern we can, and must, divide the customary "arthropod features" into two sets. The first set of evolutionary novelties developed within the common stem lineage of the Onychophora and the Euarthropoda. They alone form the autapomorphies that establish the Arthropoda as a monophylum. Of particular importance are the chitin cuticle and ecdysis during growth. A second set of new features, which evolved later in the stem lineage of the Euarthropoda, include the division of the appendages into articulated limbs – from which the name Arthropoda is derived. These features do not, however, contribute to the characterization of the Arthropoda as a whole but are rather autapomorphies of the subtaxon Euarthropoda.

The facts outlined above determine our order of presentation. We will deal first with the autapomorphies of the Arthropoda, then the feature pattern of the Onychophora before finally discussing the apomorphic peculiarities of the Euarthropoda.

Autapomorphies (Figs. 4 → 9; 36 → 1)

- *α*-Chitin-protein cuticle with ecdysis.
  The cuticle with its antiparallel arrangement of neighboring chitin chains and their covalent connection to protein is secreted by the ectoderm. Once produced the cuticle cannot be enlarged. Therefore, to grow, an individual must molt at intervals and produce a new cuticle.
- Absence of locomotory cilia.
  The loss of mobile cilia on the surface of the body is a consequence of the evolution of a hard chitin cuticle.
- Cephalization.
  The head of the Arthropoda is formed from the acron (= prostomium of the Annelida) and probably the first three segments with appendages. The antennae of the Onychophora are homologous with the antennae in the ground pattern of the Euarthropoda (p. 77). However, it is not possible to homologize the jaws and the oral papilla of the Onychophora with certain postantennal appendages of the Euarthropoda.
- Uniramous locomotory appendages whose construction cannot be further defined.
- Body cavity with pericardial septum.
  A dorsal pericardial septum spans the body horizontally. The septum separates a pericardial sinus around the heart from the rest of the body cavity, the perivisceral sinus (Fig. 37C).
- Heart with ostia.
  Hemolymph circulates in arteries that exit from the heart and in sinuses of the body cavity. Veins are absent. The return flow into the heart is effected by ostia, which are segmental slits in the heart wall.
- Nephridia with sacculi.
  The transitory coelom cavity found during ontogeny in the Onychophora becomes the sacculi of the nephridial system (BARTOLOMAEUS & RUHBERG 1999) (Fig. 37C). The sacculi with podocytes are the site of primary urine production.
- Alteration of the spiral-quartet-cleavage (SCHOLTZ 1997) and loss of the trochophore larva.
  1. The original spiral-cleavage is replaced with a radial position of the cleavage products.
  2. In the ground pattern there is a superficial cleavage in which the blastomeres are arranged at the periphery of the embryo. It remains

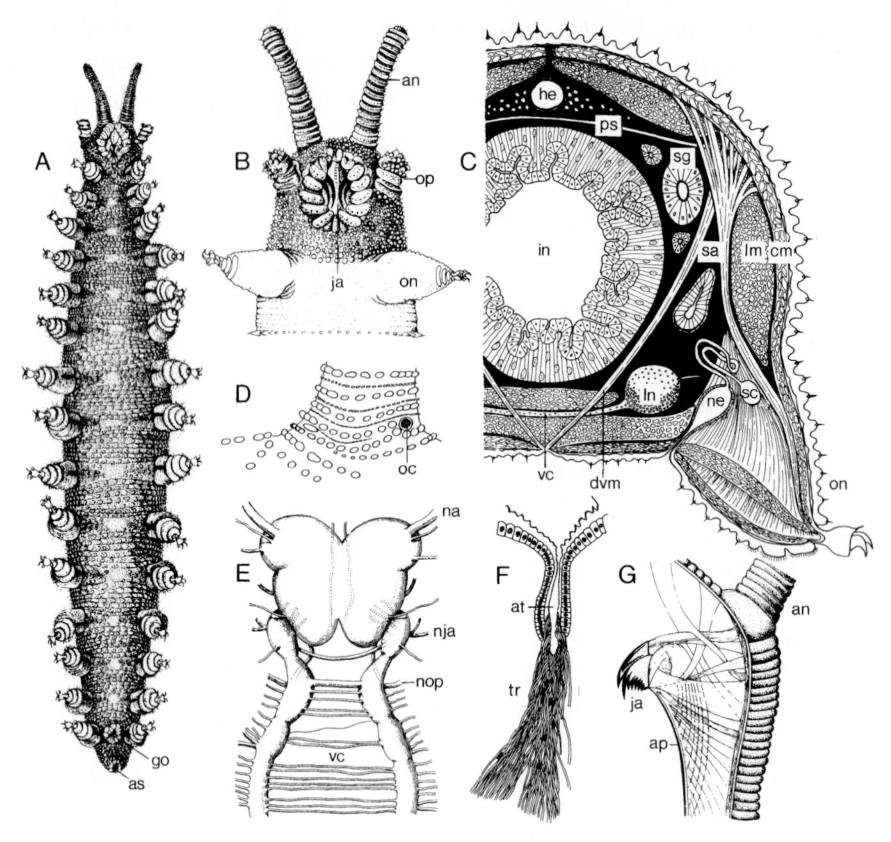

Fig. 37. Onychophora. **A** *Peripatopsidae* sp. Habitus. **B** *Heteroperipatus engelhardi.* Head and first trunk segment. Ventral side. **C** Cross section through the mid-body section. **D** *Peripatus horsti.* Base of the right antenna with lensed eye. **E** *Peripatus jamaicensis.* Brain and longitudinal nerves to demonstrate the innervation of antennae, oral-hooks (jaws) and oral papillae. **F** *Peripatopsis capensis.* Tuft of trachea. Atrium and tracheal capillaries. **G** *Peripatus amboinensis.* Oral-hook (jaw) with muscle supply. *an* Antenna; *ap* apophyse; *as* anus; *at* atrium; *cm* circular musculature; *dvm* dorsoventral muscles; *go* genital opening; *he* heart; *in* intestine; *ja* oral-hook (jaw); *lm* longitudinal muscles; *ln* longitudinal nerve; *na* antennal nerve; *ne* nephridium; *nja* nerve of the oral-hook (jaw); *nop* nerve of the oral papillae; *oc* ocellus; *on* oncopodium; *op* oral papilla; *ps* pericardial septum; *sa* salivary gland; *sc* sacculus; *sg* slime gland; *tr* tracheal capillaries; *vc* ventral commissure. (**A** Pflugfelder 1968; **B** Ruhberg 1985; **C** Ruhberg 1996; **D,E** Cuénot 1949; **F,G** Moritz 1993 a)

unclear as to what special mode leads to this result – an intralecithal cleavage with cleavage products not separated by cell membranes or a holoblastic cleavage.

3. The superficial cleavage leads to a blastoderm stadium with a central mass of yolk.
4. The trochophore larva from the ground pattern of the Articulata is lost in the stem lineage of the Arthropoda.

Onychophora – Euarthropoda

Onychophora

The terrestrial Onychophora with ca. 160 species inhabit a girdle around the equator. In the southern hemisphere they extend through the tropics and subtropics and reach into temperate regions. The discontinuous distribution seen today is explained by the hypothesis of their existence on the formerly connected land mass of Gondwanaland (STORCH & RUHBERG 1993).

Peripatus, Peripatopsis.

The Onychophora are between 5 mm and 15 cm long. As soft-bodied organisms with a thin chitin cuticle they live hidden in protected terrestrial biotopes with high humidity.

In their body division a head section is weakly divided from the trunk. On the head there are three pairs of very different appendages. (1) Long antennae point forwards; they bear sensory equipment. (2) Pointed, claw-like jaws are located deep in the oral cavity; to cut up prey they are pushed forwards. (3) Short oral papillae sit laterally behind the antennae; they are connected to extensive slime glands that produce sticky secretions for catching prey and for defense (Fig. 37 B).

The morphological interpretation of these appendages has met with difficulty. In the adult Onychophora no division of the brain can be recognized (SCHÜRMANN 1987, 1995). Reports of findings of the sections proto-, deuto- and tritocerebrum in the embryo (PFLUGFELDER 1948, 1968) need to be more fully investigated. Independently of this the innervation of the antennae from the brain allows a homologization with the antennae in the ground pattern of the Euarthropoda (= first antennae of the Mandibulata). In contrast, the jaws and oral papillae are not connected to the brain. They are supplied through esophageal connections which link the brain with the two ventral nerve cords (Fig. 37 E). In the Mandibulata this is the area that contibutes to the formation of the subesophageal ganglion and thereby innervates the mandibles and two pairs of maxillae. A homologization of the claw-like jaws and the oral papillae with certain head appendages of the Mandibulata is not possible; we can, however, interpret their special construction and function as derived peculiarities of the Onychophora.

The body of the Onychophora contains a maximum of 43 pairs of uniform walking legs. The short, lobopod legs end in a pair of claws; it is from these claws that the Onychophora derive their name. Their innervation follows through the segmental neuromeres of the ventral nerve. The expression of the body appendages in the form of short, claw-bearing walking legs must be seen as a further autapomorphy, which developed in the stem lineage of the Onychophora during the transition to land.

The ventral nerves of the nervous system show the following peculiarities: (1) they are unusually far apart from one another; (2) the single neuromeres are connected to one another through 9–10 commissures; and (3) Ganglionic swellings are only weakly formed in the neuromeres (Fig. 37E). The deviations mentioned in (1) and (2) from the "normal" ladder-like nervous system of the Annelida and Arthropoda possibly belong to the characteristic features of the Onychophora. We now list these in full.

Autapomorphies (Fig. 36 → 2)

- Terrestrial mode of life.
  The stem species of the Arthropoda was an aquatic organism. The transition of the Onychophora to terrestrial living must be interpreted as a case of convergence to the evolution of terrestrial Euarthropoda.
- Air-breathing organs.
  In connection with the change to land came the development of tracheae; however, these cannot be homologized with the tracheae of the Euarthropoda. Tufts of long, unbranched tracheae spring from a short atrium (Fig. 37F). The associated spiracles are irregularly distributed over the body in large numbers.
- Claw-like jaws.
  Sickle-shaped blades open in the oral cavity. Long apodemes serve as attachment points for musculature which anchor them in the anterior (Fig. 37G).
- Oral papillae.
  Stubby appendages without claws. The orifice for the large slime glands (Fig. 37B).
- Walking legs in the form of short oncopodia.
- Lensed eyes.
  A large lensed eye lies dorsally on both of the antennae bases (Fig. 37D). Parallel-oriented retinal cells carry receptor processes with microvilli. The sensory cells have rudimentary cilia. The highlighting of general similarities with Annelida and Euarthropoda (EAKIN & WESTFALL 1965) remains irrelevant for the purposes of phylogenetic systematics.
  Homologization with the light sensory organs of other Articulata is not possible. Consequently, an independent evolution in the stem lineage of the Onychophora can be hypothesized.
- Salivary glands.
  Pairs of salivary glands open into the oral cavity; they produce an enzyme for extraintestinal digestion of prey (RUHBERG 1996) which is cut up by the jaws. Morphologically, the salivary glands are transformed nephridia of the oral papillae segment.

In comparison with the Euarthropoda, the ground pattern of the Onychophora includes a number of plesiomorphies. I would like to highlight the following two features.

1. The Onychophora have a layered body wall musculature. From the exterior to the interior this consists of a sequence of circular muscles, diagonal muscles and thick bundles of longitudinal muscles (Fig. 37C); the latter originated in the stem lineage of the Articulata.
2. The nephridia with sacculi are original in their extension over the whole trunk and through the existence of a nephrostome with cilia.

Euarthropoda

Autapomorphies (Fig. 36 → 3)

- Skeleton formed from plates.
  Evolution of solid, flexible chitin plates. There is a minimum of a dorsal tergite and a ventral sternite per segment.
- Disintegration of the body wall musculature.
  In connection with the formation of the exoskeleton, the whole body musculature is transformed into separate cords with cross-striated musculature.
- Arthropodium.
  Jointed appendage with two branches. From the basal unpaired protopodite an endopodite (inner branch) and an exopodite (outer branch) arise.
- Cephalization.
  Molecular marking of cells that the polarity gene engrailed[8] expresses in sequential segments, has obviously ended the long dispute over the number of segments in the head of the Euarthropoda (SCHOLTZ 1995, 1997). According to the molecular findings, no preantennal segment exists between the acron and the antennal segment.
  The head of the stem species of the Euarthropoda probably consisted of an acron, an antennal segment and three postantennal segments with biramous appendages similar to the legs of the trunk[9] (Fig. 38). The compound eyes, the labrum (upper lip) and the protocerebrum belong to the acron. The labrum is not a derivative of an appendage, but rather the ventrally and caudally moved tip of the acron. The deutocerebrum is

---

[8] The expression of the polarity or segmentation gene results in a cross-striation of cells at the posterior edge of a segment. The marking is carried out with a monoclonal antibody against the gene product.

[9] Interpretation for the ground pattern of the †Trilobita (MÜLLER & WALOSSEK 1987; WALOSSEK & MÜLLER 1990).

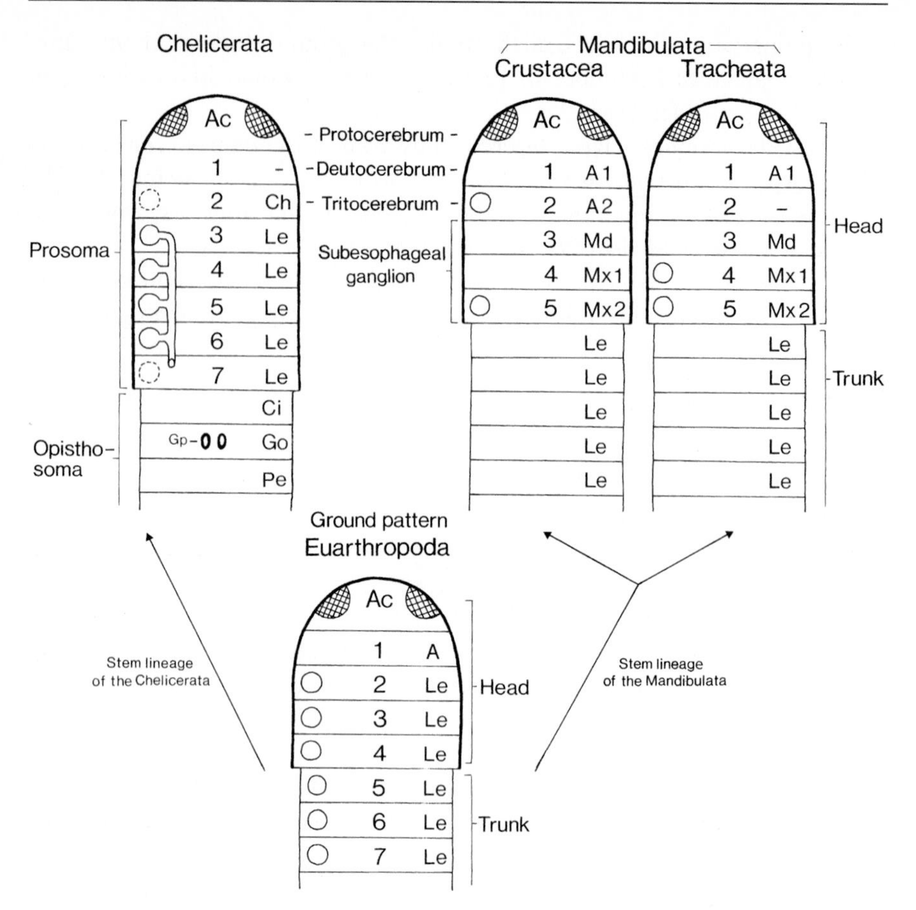

Fig. 38. Diagrams of the anterior in the ground pattern of the Euarthropoda, the Chelicerata (Xiphosura) and the Crustacea and Tracheata as adelphotaxa of the Mandibulata. **Euarthropoda.** Head of acron + four segments. One pair of uniramous antennae on segment 1 and three pairs of biramous legs with endo- and exopodites on segments 2–4. **Chelicerata.** Prosoma of acron + seven segments. Antennae reduced. Chelicerae are differentiation products of segment 2 appendages. Five pairs of uniramous walking legs on segments 3–7 follow. **Mandibulata.** Head of acron + four or five segments. Pairs of first antennae, second antennae, mandibles, first and second maxillae on segments 1–5. Segment 5 possibly convergently fused in the Crustacea and Tracheata. Crustacea with the ground pattern provision. In the Tracheata the second antennae are reduced. **Nephridia.** For the ground pattern of the Euarthropoda one pair in each of the segments 2–7 can be hypothesized. Expression of four pairs in the ground pattern of the Chelicerata (segments 3–6); reduction to two pairs in the ground pattern of the Crustacea (segments 2 and 5) and to two pairs in the ground pattern of the Tracheata (segments 4 and 5). *A* Antenna; *A1* 1st antenna; *A2* 2nd antenna; *Ac* acron; *Ch* chelicera; *Ci* chilarium; *Go* genital operculum; *Gp* genital pore; *Le* locomotory appendage; *Md* mandible; *Mx1* 1st maxilla; *Mx2* 2nd maxilla; *Pe* plate-shaped appendage; *circle* nephridia; *dashed circle* embryonic nephridium; *hatched circle* compound eye.

the ganglion of the antennal segment and the tritocerebrum is the ganglion of the first postantennal segment.
- Faceted eyes (compound eyes).
  One pair of eyes in the acron. Composed of numerous ommatidia. Further explanation follows in the text.
- Median eyes.
  Four separate ocelli in the middle of the head.
- Sensory cells with stereocilia.
  In the sensilla the sensory cells bear stiff cilia in which the two central tubuli are missing ($9 \times 2+0$ pattern) (PAULUS 1996).
- Six pairs of nephridia.
  Reduction of the segmental excretory organs to six pairs on segments 2–7 (Fig. 38).
- Ventral germ band.
  Concentration of the early developmental processes to the underside of the germ. Blastomeres arrange themselves on the prospective ventral side of the embryo in a longitudinal band. They supply the material for paired semicircular head lobes and some anterior segments. The rest of the body develops from a growth zone (SCHOLTZ 1997).
- Larva with three pairs of appendages.

Head, eyes and nephridia must be more closely examined with regard to elementary modifications of the ground pattern within the Euarthropoda.

In the **arrangement of the head** from the stem species of the Euarthropoda, the Chelicerata and the Mandibulata have taken separate paths (Fig. 38).

In the stem lineage of the Chelicerata three trunk segments were united with the original Euarthropoda head to form a new entity. This is the prosoma of our terminology which we contrast with the opisthosoma – the rest of the body. The antennae of the first segment are reduced; the chelicerae evolved from the appendages of the second segment as three-part grasping organs. The following five appendages (segments 3–7) became uniramous structures for locomotion.

In the stem lineage of the Mandibulata a subdivision of the head took place into a pregnathal section with differentiation of a second antennae pair, and a gnathal region with the evolution of mandibles and two pairs of maxillae. The ganglia of the corresponding segments 3, 4 and 5 form the subesophageal ganglion. It has lately been argued that in the stem species of the Mandibulata it is possible that apart from the mandibular segment only the first maxillary segment was incorporated into the head (LAUTERBACH 1980; WALOSSEK & MÜLLER 1990; SCHOLTZ 1997). The transformation of the appendages of the fifth segment into the second maxillae and the joining of this segment to the head would then have occurred in a convergent fashion in the Crustacea and the Tracheata. The ex-

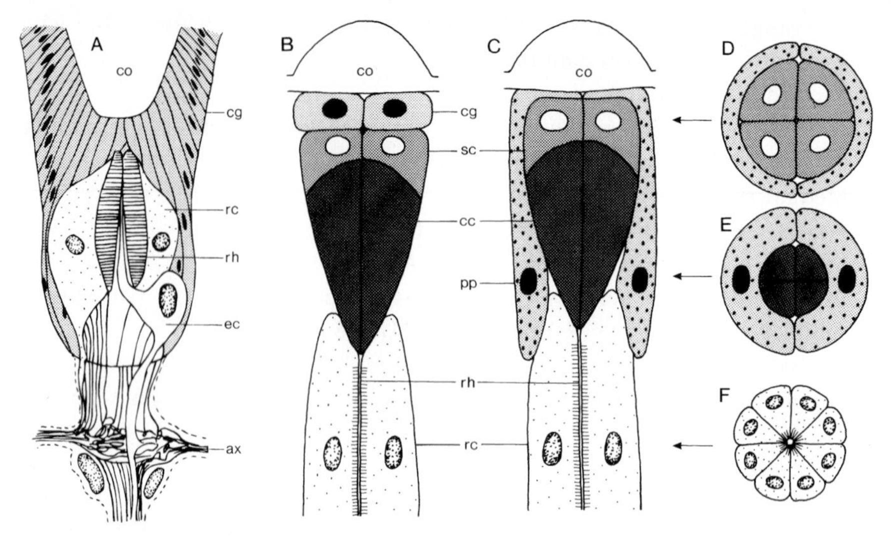

Fig. 39. Construction of the ommatidium in the Euarthropoda compound eye. **A–C** Longitudinal sections. **D–F** Cross sections. **A** *Limulus polyphemus* (Xiphosura, Chelicerata). The circular corneal lens with a deep inner cone is produced from numerous corneagenous cells. A crystalline cone is absent. A crown of 4–20 retinula cells form the rhabdome with a star-shaped cross section. **B–F** Mandibulata. **B** Ground pattern of the Crustacea (taken over from the stem species of the Mandibulata). The ommatidium consists of (1) a cuticular lens: produced from two corneagenous cells, (2) a four-part eucone crystalline cone: produced from four semper cells (crystalline cone cells), and (3) a retinula of eight retinula cells with rhabdome in the center. **C** Autapomorphy in the ground pattern of the Insecta. After secretion of the lens the two corneagenous cells become primary pigment cells of the compound eye; they sink downwards and cover the crystalline cone. **D–F** Cross sections through diagram **C**: **D** at the level of the semper cells, **E** at the level of the primary pigment cells, **F** at the level of the retinula. *ax* Axon; *cc* crystalline cone; *cg* corneagenous cell; *co* cornea; *ec* eccentric cell; *pp* primary pigment cell; *rc* retinula cell; *rh* rhabdome; *sc* semper cell (crystalline cone formation cell). (**A** Weygoldt 1996; **B–F** outline after Paulus 1979)

planation is somewhat circuitous. In the Cephalocarida (Crustacea), which are cited as "proof", the second maxillae are identical to the following trunk appendages. This could, however, be the result of a derived persistence of an ontogenetic state or, just as likely, a secondary extension of the genetic information for formation of trunk appendages. The undisputed hypothesis of a reduction of the second antennae in the stem lineage of the Tracheata is untroubled by this question.

Compound eyes as realized in the Xiphosura (Chelicerata) are placed in the ground pattern of the Euarthropoda. The corneal lens of the individual ommatidium is produced from numerous corneagenous epidermis cells. There is no crystalline cone. The rhabdome consists of a variable number of retinula cells (4–20) (Fig. 39 A).

The apomorphous state in the ground pattern of the Mandibulata includes the following elements. The corneal lens of the ommatidium is secreted by two corneagenous cells. The crystalline cone is an evolutionary novelty and is formed from four cells. The rhabdome consists of eight retinula cells (Fig. 39 B).

Four median eyes (ocelli) which are separated from one another belong in the ground pattern of the Euarthropoda. This is the case seen in the Pantopoda (Chelicerata) as well as in various taxa of the Crustacea and Insecta. Independent transformations of the ground pattern led to two median eyes in the Euchelicerata (Xiphosura + Arachnida), to three ocelli within the Crustacea and Insecta and to a complete reduction in the stem lineage of the Myriapoda. The grouping to form a nauplius eye is interpreted as an autapomorphy of the Crustacea.

Six pairs of nephridia in segments 2–7 in the Xiphosura are carried over from the ground pattern of the Euarthropoda. Although in the adult *Limulus*, from the embryonic disposition, only the middle four pairs (segments 3–6) are fully developed. In the stem lineages of various subtaxa of the Euarthropoda an independent reduction to 1–2 pairs of segmental organs has taken place. Two pairs of nephridia belong in the ground pattern of the Arachnida (segments 4 and 6), in the ground pattern of the Crustacea (segments 2 and 5) and in the ground pattern of the Tracheata (segments 4 and 5); the latter is expressed in the Myriapoda (Chilopoda, Symphyla). In the Insecta only one pair of nephridia remains on segment 5; they are here transformed into labial glands.

Chelicerata – Mandibulata

The stem species of the Euarthropoda was an aquatic organism. The Chelicerata and the Mandibulata as the highest-ranking adelphotaxa comprise unities having primary water-living species; however, both also include monophyla in the stem lineages by which the land was conquered.

Original marine unities of the Chelicerata are the Pantopoda and the Xiphosura; terrestrial organisms are the Arachnida. In the Mandibulata the Crustacea are primarily aquatic and the Tracheata are terrestrial.

Chelicerata

With the **fusion of three trunk segments to the head of the Euarthropoda** (p. 81), a division of the body into two large tagmata – **prosoma** and **opisthosoma** – took place in the stem lineage of the Chelicerata. This process gave rise to the first fundamental feature of the Chelicerata.

Autapomorphies (Figs. 36 → 4; 40 → 1)

- Prosoma – Opisthosoma + caudal spine.
  The prosoma consists of the acron and seven segments; it has six pairs of rod-shaped appendages on segments 2–7. The opisthosoma is comprised of the following 12 trunk segments and the caudal spine – a differentiation product of the 13th segment.
- Opisthosoma: Mesosoma – Metasoma.
  Subdivision of the Opisthosoma into a mesosoma of seven segments with appendages (1–7) and the metasoma from five metameres (8–12) without appendages. The mesosoma appendages 2–7 are plate-shaped (2=genital opercula, 3–7 with book gills). In the mesosoma the body covering of the segments is differentiated into tergite and sternite, in the metasoma segments it is annular.
- Absence of appendages on prosoma segment 1 (Fig. 38).
  The lack of the pair of appendages, which are designated antennae 1 in the Crustacea, is postulated to be a product of reduction.
- Absence of deutocerebrum in the brain.
  The reduction of appendages on prosoma segment 1 is correlated to the reduction of the corresponding brain section.
- Evolution of chelicerae.
  In the stem lineage of the Chelicerata the appendages on prosoma segment 2 (Fig. 38) (=antennae 2 of the Crustacea) evolved into three-part grasping organs. The two distal articles of the chelicerae form a pincer-like structure.
- Genital pores in opisthosoma segment 2.
  The primary paired genital openings (plesiomorphy) are shifted from the posterior part of the body (ground pattern Arthropoda) to the anterior part of the opisthosoma (Fig. 38).

Systematization

Chelicerata
- **Pantopoda**
- **Euchelicerata**
  - **Xiphosura**
  - **Arachnida**

Although omitting the Pantopoda, the computer cladogram from DUNLOP & SELDEN (1997) shows for the recent Chelicerata a sister group kinship between the Xiphosura and the Arachnida as well as one within the Arachnida between the Scorpiones and the Lipoctena (p. 97).

Pantopoda – Euchelicerata

From the ground pattern of the Euarthropoda paired lateral compound eyes and four median eyes were carried over into the ground pattern of the Chelicerata. Their alternative continuance within the Chelicerata can be used as the first justification of the systematization into the adelphotaxa Pantopoda and Euchelicerata.

The reduction of the dorsal light-sensory organ to two median eyes is an incontestable autapomorphy of the Euchelicerata. Otherwise the ground pattern of the Chelicerata is largely retained in the marine Xiphosura.

The original four median eyes are only found in the Pantopoda. Apart from the loss of the compound eyes the reduction of the opisthosoma to a peg must be considered an autapomorphy of the Pantopoda. Because recent Pantopoda possess only a tiny rudiment of the opisthosoma, the interpretation of certain opisthosoma features – number of segments, place of the genital pores, genital opercula – as ground pattern autapomorphies of the Chelicerata can only be given with reservation. The few fossil Pantopoda from the Devonian with a segmented opisthosoma (†*Palaeopantopus*, †*Palaeoisopus*) give no clue as to how far these features were realized in the stem species of the Chelicerata, or if they first evolved after separation of the stem lineages of the Euchelicerata and Pantopoda.

Pantopoda

The Pantopoda (Pycnogonida) are bizarre marine organisms with a tiny body stem and very long appendages. Their peculiar proportions "force" the midgut and the gonads to develop diverticula which extend into the legs. A rostral proboscis with muscular pharynx serves as a pump and grinding organ for carnivorous feeding. Many "sea spiders" graze on colonies of hydropolyps. *Pycnogonum litorale* punctures the walls of sea anemones and sucks up their tissues.

Autapomorphies (Fig. 40 → 2)

- Habitus.
  Tube-like narrowing of the central body in combination with the differentiation of very thin walking legs (Fig. 41 A, B).
- Proboscis.
  Piercing and sucking organ terminal on the prosoma (Fig. 41 C, D). Mouth at the tip of the proboscis.
- Parapalps.
  A pair of thin feeler-like appendages following the chelicerae. No locomotory function.

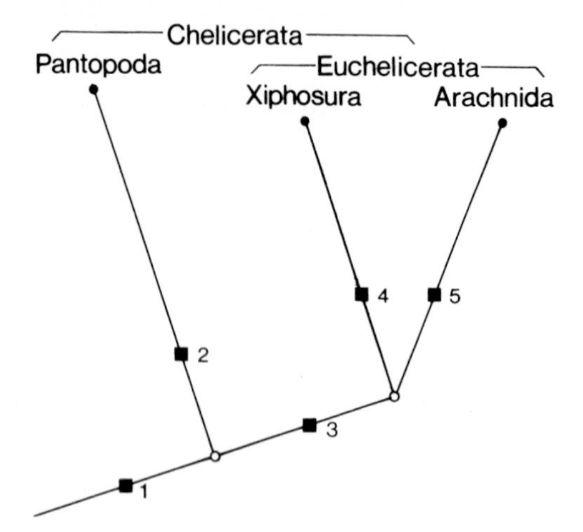

Fig. 40. Diagram of the phylogenetic kinship between the highest-ranking subtaxa of the Chelicerata.

- Ovigera.
  A further pair of thin appendages just in front of the first walking leg pair – likewise without a locomotory function (Fig. 41 D). In the male the eggs are attached to the ovigera in a ball; brooding follows at least until the hatching of the protonymph larvae. The ovigera are always more weakly developed in the female.
- Prosoma formed of two sections.
  1. Undivided anterior section with four appendage pairs: chelicerae, parapalps, ovigera, walking leg pair 1.
  2. Divided posterior section with 3 appendage pairs: walking legs 2, 3, 4.
- Reduction of the opisthosoma.
  A tiny appendage on the prosoma between the insertion of the last walking legs (Fig. 41 C).
- Branched gonads.
  From the primary paired testes and ovaries in the trunk, diverticula extend into the four walking leg pairs.
- Location of the genital pores on the walking legs.
  In both sexes genital openings are found on the second segment of the locomotory appendages. The ground pattern number has not yet been determined. Variation between one pair and four pairs on all walking legs.
- Reduction of the compound eye.

I denote as **parapalps** the second appendage pair of the Pantopoda – i.e., not as pedipalps as is usually the case. This has the following grounds.

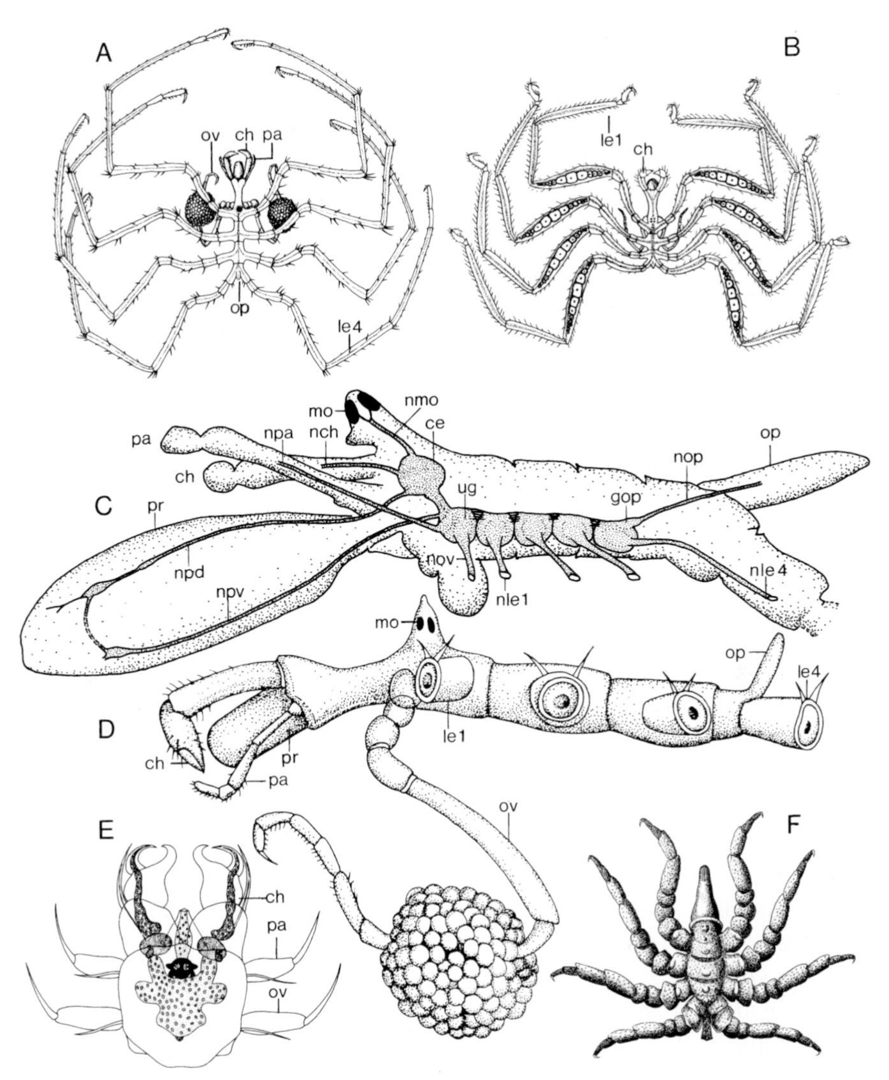

Fig. 41. Pantopoda. **A** *Nymphon rubrum*. Male in dorsal view. Ovigera with ball of eggs. **B** *Pallene brevirostris*. Female. Without parapalps. Weakly formed ovigera. Development of eggs in the walking legs. **C** *Achelia echinata*. Central nervous system in side view. Innervation: chelicerae and median eyes from the brain (supraesophageal ganglion); parapalps and ovigera from the subesophageal ganglion; walking legs 1–4 from separate ganglia. Proboscis with a dorsal unpaired nerve from the brain and ventral paired nerves from the subesophageal ganglion. **D** *Nymphon rubrum*. Male. Body stem with chelicerae, parapalps and ovigera in side view, walking legs removed. **E** *Achelia echinata*. Protonymph larva with three appendage pairs. **F** *Pycnogonum litorale*. Female. Dorsal view. *ce* Cerebrum (supraesophageal ganglion); *ch* chelicera; *gop* opisthosoma ganglion; *le* walking leg; *mo* median ocellus (median eye); *nch* chelicera nerve; *nle* walking leg nerve; *nmo* median eye nerve; *nop* opisthosoma nerve; *nov* ovigera nerve; *npa* parapalp nerve; *npd* dorsal proboscis nerve; *npv* ventral proboscis nerve; *op* opisthosoma; *ov* ovigera; *pa* parapalp; *pr* proboscis; *ug* subesophageal ganglion. (**A, D, E, F** Moritz 1993b; **B** Fage 1949a; **C** Winter 1980)

The pedipalps of the Arachnida and the parapalps of the Pantopoda are in the first instance comparable feeler-like appendages (without a locomotory function) having an identical location behind the chelicerae. The widespread use of the same term implies a homology between the appendages of the Arachnida and the Pantopoda in their construction as feeler-like organs. However, in an adelphotaxa relationship between the Pantopoda and the Euchelicerata, this is not possible. According to this hypothesis the parapalps of the Pantopoda and the pedipalps of the Arachnida must have developed independently of one another from the second appendage pair of the chelicerate prosoma – that is, from appendages which in the ground pattern of the Chelicerata were developed as walking legs, as is the case seen in the Xiphosura today.

In their ground pattern the **Pantopoda** possess **seven pairs of appendages on the prosoma,** one pair more than the Euchelicerata. These are the chelicerae with innervation from the brain (tritocerebrum), the parapalps and ovigera with innervation from the subesophageal ganglion, as well as four pairs of walking legs with four pairs of separate ganglia. The increase of one appendage pair is an apomorphy of the Pantopoda – no matter if this came about due to addition or a splitting of an appendage. Having seven appendages is also not constant. In some cases extra walking legs evolved with corresponding ganglia (*Pentanymphon, Pentapycnon, Decolopoda, Dodecolopoda*) and in others chelicerae, parapalps and ovigera are completely reduced (♀ *Pycnogonum*).

	Chelicera	**Parapalps**	**Ovigera**	**Walking legs**
Nymphon	+	+	+	4
Pentanymphon	+	+	+	5
Pentapycnon	+	+	+	5
Decolopoda	+	+	+	5
Dodecolopoda	+	+	+	6
Pycnogonum ♂	–	–	+	4
Pycnogonum ♀	–	–	–	4

In comparison with their adelphotaxon Euchelicerata the Pantopoda have only a few original features.

A basic plesiomorphy is the above-mentioned existence of four median eyes, whereby their elevation onto a tubercle on the prosoma must certainly be regarded as an apomorphy. The free ganglion pairs in the segments of the walking legs may also be an original feature.

The common protonymph larva with three appendage pairs (chelicerae, parapalps, ovigera) is possibly another plesiomorphy (Fig. 41 E). In the Xiphosura, at any rate, the "trilobite larva" already has nine pairs of appendages and the full number of segments on hatching (Fig. 42 I).

Euchelicerata

Autapomorphies (Fig. 40 → 3)

- Two median eyes.
  In comparison with the existence of four median eyes in the ground pattern of the Euarthropoda and a corresponding state in the Pantopoda, a reduction to two median eyes in the stem lineage of the Euchelicerata must be postulated.
- Concentration of the central nervous system.
  Formation of a buccal nerve ring: the protocerebrum and tritocerebrum (innervation of the chelicerae) lie at the anterior; at the posterior of the ring the ganglia of prosoma appendages 2–6 and the ganglia of opisthosoma appendages 1+2 are united. Only after this do free ganglia pairs appear.
  The state just outlined is realized in the Xiphosura. This is an apomorphy compared with the chain of separate trunk ganglia in the ground pattern of the Mandibulata. However, it did not evolve in the stem lineage of the Chelicerata. The Pantopoda have free ganglia in the segments with walking legs.
  We postulate that the concentration process first occurred in the stem lineage of the Euchelicerata, and is therefore a derived feature of this unity.

Xiphosura – Arachnida

Xiphosura

Limulus polyphemus from the Atlantic coast of the USA, as well as *Tachypleus* and *Carcinoscorpius* with three species in southeast Asia, form the remnants of a unity of marine spiders, which first appeared in the Silurian and which once had a worldwide distribution. With a length of up to three fourths of a meter the horseshoe crabs are the giants of the recent Chelicerata.

We place particular emphasis on some of the basic organizational traits of the Xiphosura as they are important for the goals of phylogenetic systematics.

The **horseshoe-shaped anterior body** and the **box-like posterior body** are connected by a hinge. This conspicuous two-part division does not correspond exactly to the morphological division of the chelicerate body into prosoma and opisthosoma. On the contrary, the last prosoma segment 7 broadly overlaps onto the posterior part of the body laterally, whereas opisthosoma segment 1 and part of the second segment extend over the joint region and into the anterior part of the body (Fig. 42 F).

Following the chelicerae are five rod-shaped prosoma appendages; they are all designed for locomotion. All walking legs end distally in small chelae. Additionally, the last leg carries four spoon-shaped setae that splay out when placed on sandy substrates (Fig. 42E). The coxae of the chelicerae and the walking legs surround the elongated mouth to the anterior and the sides.

Let us now consider the opisthosoma. Leaving aside the overlap of the prosoma, the mesosoma alone provides the box-shaped posterior body with seven appendage pairs. In contrast, the metasoma in the Xiphosura is much reduced. It becomes a short rod formed of three fused segments to which the caudal spine is attached (Fig. 42G).

The first opisthosoma segment, which penetrates into the anterior body, marks with its appendages the caudal limit of the mouth; these are the chilaria – short pistil-like structures. The following second opisthosoma segment, which is also pushed to the anterior, bears the genital operculum – a single plate formed from fused leaf-like appendages which has genital pores on its underside. The genital operculum covers the book gill bearing appendages of opisthosoma segments 3–7, which more strongly preserve the character of paired segmental appendages. Their small three-part inner branches are interpreted as telopodites (endopodites); the flat outer branches (exites) bear on their posterior the familiar book gills of the horseshoe crabs in the form of delicate, multi-folded lamellae (Fig. 42H).

The Xiphosura are generally regarded as representing the "most original" Chelicerata. Indeed, in comparison with their sister group Arachnida, the Xiphosura have a broad spectrum of plesiomorphic features that can

Fig. 42. Xiphosura. *Limulus polyphemus.* **A–B** Living organism in dorsal and ventral view (North American Atlantic Coast). Genital operculum covers about half of the region of the gill bearing leaf-like appendages. **C** Prosoma appendage 2. Endite of the coxa equipped with spines. Tarsus forms chela with a movable finger. **D** Prosoma appendage 6 with flagellum and spoon-shaped attachments. **E** Distal end of prosoma appendage 6. Attachments splayed out. End article chelate. **F** Segmentation in the boundary area between the anterior and posterior body. Prosoma segment 7 broadly overlaps the joint region onto the posterior part of the body. Opisthosoma segment 1 and part of the second segment extend over the joint region and into the anterior part of the body. **G** Juvenile in ventral view. Genital operculum (*black border*) covers the five pairs of flat appendages with gills. **H** Leaf-like appendage of the opisthosoma. Inner side. Gill lamellae in two lateral rows. **I** First larval stage ("trilobite larva") with complete segment number of the adult. The segment structures of the opisthosoma emerge from underneath the cuticle. **J** Locomotion on firm substrate by pushing the body with the splayed appendages of the last walking legs (Prosoma appendages 6). *ch* chelicera; *ci* chilaria; *oe* opisthosoma appendage; *og* genital operculum; *os* opisthosoma segment; *pe* prosoma appendage; *ps* prosoma segment. (**A–B** Originals; **C** Fage 1949b; **D–G,J** Ankel 1958; **H** Moritz 1993b; **I** Gerhardt 1941) ▶

mostly be correlated with their aquatic mode of life. The plesiomorphies range from carnivory and the muscular gizzard through the five pairs of locomotory prosoma appendages and the gills on the opisthosoma to the free release of gametes and external fertilization of eggs.

On the other hand, from a comparison with recent Arachnida as well as with extinct members from the stem lineages of the Arachnida and Xiphosura, a number of derived characteristics can be highlighted by means of which the recent Xiphosura can undoubtedly be designated a monophylum.

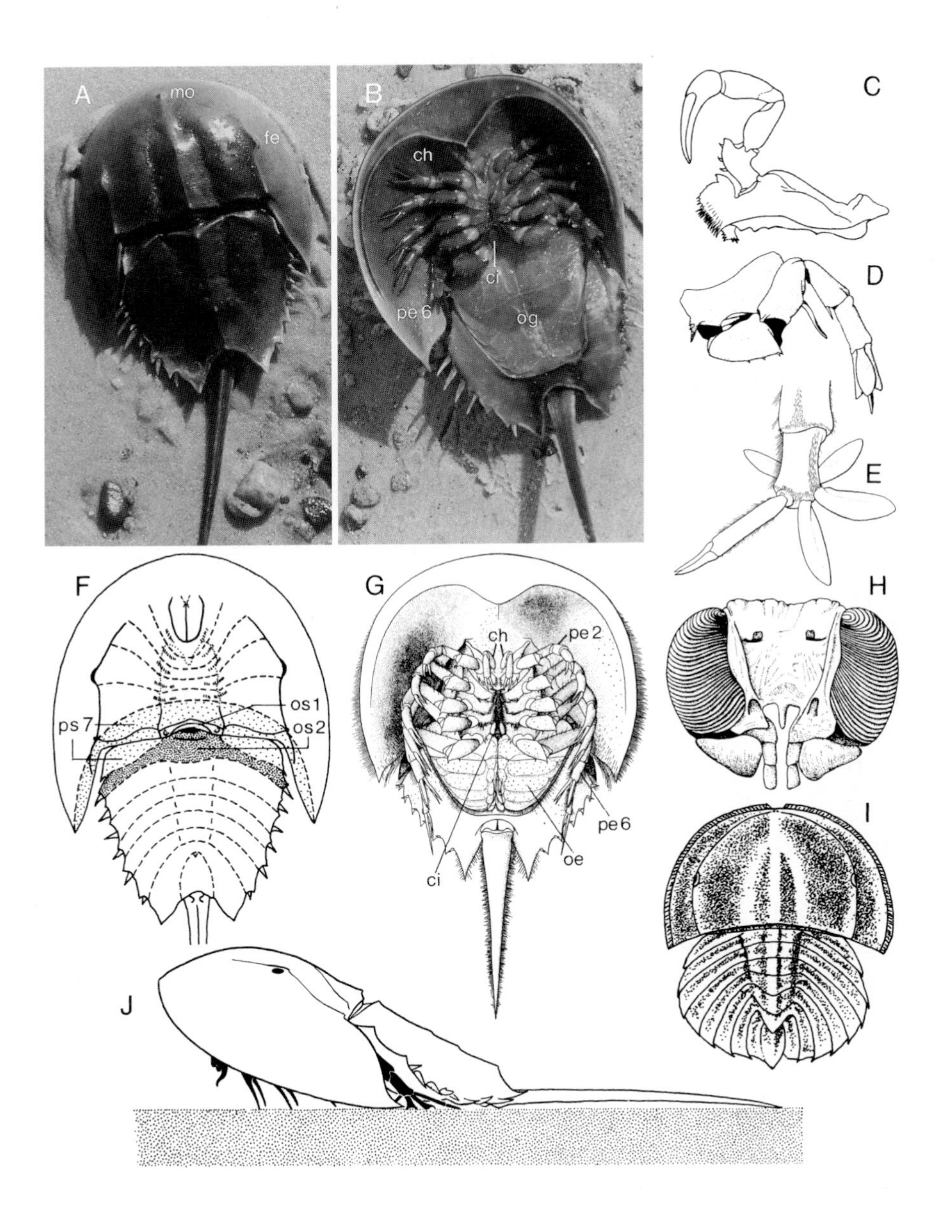

Autapomorphies (Fig. 40 → 4)

- Fusion of the segments of the mesosoma.
  Postulated for the last common stem species of the recent Xiphosura. The mesosoma is segmented in stem lineage members of the Xiphosura, in the fossil Eurypterida from the stem lineage of the Arachnida and in the ground pattern of the Arachnida – as in the Scorpiones.
- Reduction of the metasoma to three segments and their fusion.
  The plesiomorphic state of five separate metasoma segments is realized in the Scorpiones from the adelphotaxon Arachnida as well as in the Eurypterida from the stem lineage of the Arachnida.
- Chelae on the rod-shaped locomotory leg pairs of the prosoma.
  Since a comparable arrangement in the stem lineage of the sister group Arachnida (†Eurypterida) does not exist, it can be postulated that the evolution of small chelae on the appendages of prosoma segments 3–7 must have first occurred in the stem lineage of the Xiphosura.
- Extra spoon-shaped setae on the last prosoma leg pair.
  Likewise only realized in the Xiphosura. The broad supports are spread out on the sediment to push the body forward.
- One pair of excretory pores for the coxal nephridia.
  Nephridia are developed in ontogeny in all six prosoma segments with appendages (plesiomorphy), but are not present in segments 2 and 7 of the adult. The remaining four organs of segments 3–6 unite on each side to form an excretory tube; this opens in a pore between the sixth and seventh prosoma segments (Fig. 38).

The construction of the chilaria as small pistils on the first opisthosoma segment may be a further characteristic feature of the Xiphosura. Their manifestation as paired appendages is, however, doubtless a plesiomorphy. In the stem lineage of the Arachnida, they fuse to form a single plate – respectively, the metastoma of the Eurypterida and the metasternum of the Scorpiones.

Arachnida

Unbelievably, it is still argued as to whether the terrestrial Arachnida form a descent community, although they are easily established as a monophylum through a large number of derived characteristics. Understandably, however, comparable requirements necessary for colonizing the land in the Arachnida and the Tracheata have led to very similar, yet independent, answers – and to homonymous terms such as tracheae, Malpighian tubules and trichobothria in the two large unities of terrestrial Euarthropoda.

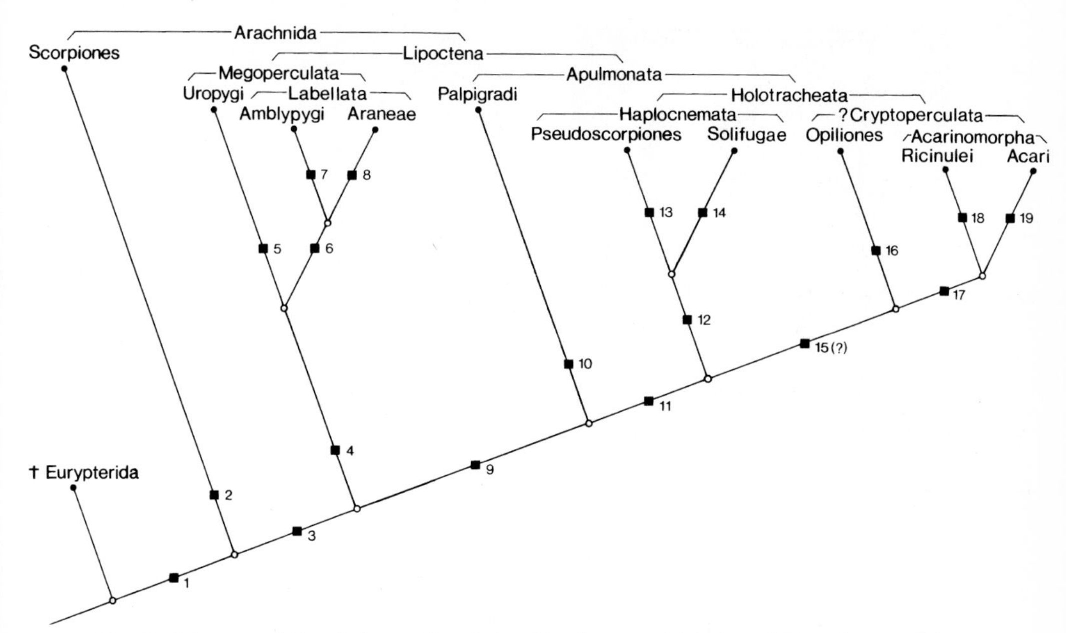

Fig. 43. Diagram of the phylogenetic relationships between highest-ranking subtaxa of the Arachnida. The †Eurypterida are placed in the stem lineage of the Arachnida

Autapomorphies (Figs. 40 → 5; 43 → 1)

- Terrestrial air-breathing stem species.
- Pedipalp.
  Following the chelicerae the first locomotory leg pair is transformed into a feeler organ. Since the homologous appendage in the †Eurypterida from the stem lineage of the Arachnida bore no chelae (Fig. 45 A), a non-chelate pedipalp is probably part of the ground pattern of the recent Arachnida.
- Locomotion with four pairs of prosoma appendages.
  Due to the evolution of the pedipalp only prosoma appendages 3–6 remain for locomotion. The absence of chelae in the fossil Eurypterida and recent Arachnida is a plesiomorphy in comparison with the Xiphosura.
- Fusion of the appendages of the first opisthosoma segment.
  In the Xiphosura there are small pistils – the chilaria – on the first opisthosoma segment. In the same position in the Scorpiones there is an unpaired angled plate – the metasternum (Fig. 45 C, D). This is interpreted as a fusion product of the appendages. From the existence of a corresponding plate with an identical position in the †Eurypterida – the metastoma (Fig. 45 A) – an important argument results for placing the †Eurypterida in the stem lineage of the Arachnida.

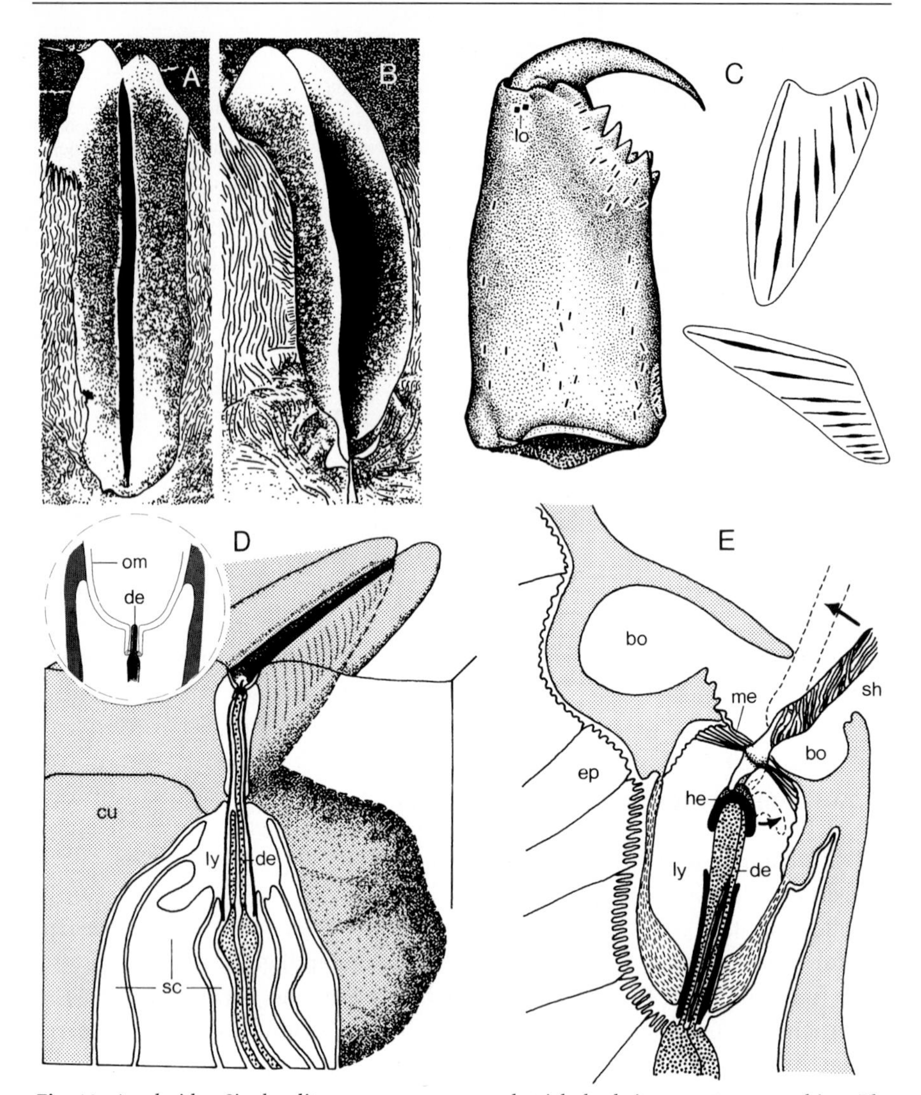

Fig. 44. Arachnida. Single slit sensory organs and trichobothria as autapomorphies. The compound lyriform slit sensory organs form an autapomorphy of the subtaxon Lipoctena. **A–B** Single slit of *Cupiennius salei* (Araneae) on the front of a walking leg tarsus; aerial and oblique side view. **C** Lyriform slit sensory organs of *Cupiennius salei* (Araneae). *Left*: Inner side of a chelicera. Two organs are found (*black squares*) in the region of the poison claw joint on both sides. *Right*: The longitudinal axis of the slits in the two organs are at right angles to one another. **D** Cross-sectional diagram through a single tarsal slit of *Cupiennius salei* (Araneae). The slit between the cuticular lips is covered by an outer membrane in which a dendrite of the organ is firmly anchored (insertion). Narrowing of the slit leads to release of nerve impulses. **E** Trichobothrium of *Tegenaria derhami* (Araneae). The cylindrical organ follows a beaker-shaped depression in the cuticle. An easily movable sensory hair is suspended in a thin membrane. When moved a rudder-like structure presses against the end of the dendrites. *bo* Bothridium; *cu* cuticle; *de* dendrite; *ep* epidermis; *he* helmet; *lo* lyriform organ; *ly* lymph space; *me* membrane; *om* outer membrane; *sc* sheath cell; *sh* sensory hair. (**A-C** Barth & Libera 1970; **D** Barth 1971, 1972; **E** Christian 1973)

- Booklungs from plate-shaped appendages of opisthosoma segments 3–6.
  Evolution from the book gills of aquatic Chelicerata – as realized in the Xiphosura. The gill lamellae were embedded ventrally in lung cavities in the body and transformed into lamellae for air-breathing. The plate-shaped appendages are themselves fused with one another to form lung covers. Slit-like openings lead to the lung cavities.
  Tracheae in the form of air-conducting tubes do not belong in the ground pattern of the Arachnida. They developed first within the Arachnida in connection with the reduction of the booklungs.
- Extraintestinal digestion – fluid feeding – precerebral sucking pump.
  Due to intense narrowing of the mouth the uptake of solid food particles – in contrast to the Xiphosura – is no longer possible. Digestive enzymes from the midgut are secreted externally onto the prey. In the foregut a suction pump is used for the uptake of the liquid product.
- Two pairs of coxal nephridia (coxal glands) on the prosoma.
  In contrast to the development of six pairs of coxal nephridia on all six appendage-carrying prosoma segments of the Xiphosura, the existence of two pairs of coxal nephridia is an apomorphy in the ground pattern of the Arachnida. These coxal nephridia lie on prosoma segments 4 and 6; they open on the first and third walking leg pair.
  The ground pattern state of the Arachnida is only realized in the Megoperculata (Uropygi, Amblypygi, Araneae). Further reductions within the Arachnida affect the coxal nephridia on the first walking leg pair segment (Scorpiones, Pseudoscorpiones, Ricinulei, Opiliones) and on the third walking leg pair segment (Solifugae, Palpigradi, Acari) (WEYGOLDT & PAULUS 1979).
- Malpighian tubules.
  New excretory arrangement of entodermal origin. Outgrowths of the midgut. Convergence in the evolution to a functionally comparable, but ectodermal organ in the stem lineage of the Tracheata.
- Further concentration of the central nervous system.
  The ground pattern state of the nervous system of the Euchelicerata is modified through the union of further opisthosoma ganglia pairs into the buccal nerve ring. In the ground pattern of the Arachnida there is a "subesophageal ganglion" in which ganglia 3–7 of the prosoma and the four anterior ganglia of the opisthosoma are united. Fusion of further opisthosoma ganglia takes place within the Arachnida.
- Five pairs of lateral eyes.
  The compound eyes of the Chelicerata are present in the †Eurypterida as stem lineage members of the Arachnida. After the Eurypterida branched off, the compound eyes became divided up into several separate single eyes. Each single eye is primarily formed from a number of ommatidia beneath a common cornea.

Five pairs of lateral eyes – as seen in a number of Scorpiones and Uropygi – can be placed in the ground pattern of the Arachnida. Multiple reductions to total loss occur within the unity.

- Trichobothrium (Fig. 44 E).
  A sensory hair inserted in a pit of the cuticle. Deflection by air currents. Convergence with the trichobothrium of the Tracheata.
- Slit sense organs.
  Long slits in the cuticle. Surrounded by thick lips and covered by a thin membrane. In every slit a dendrite of a sensory cell terminates at the membrane. Primary proprioreceptors (Fig. 44 A, B, D).
  Lyriform organs which are groups of single slit sense organs do not belong in the ground pattern of the Arachnida. They are lacking in the Scorpiones; they evolved in the stem lineage of the Lipoctena.
- Unpaired genital openings.
  The paired genital openings in opisthosoma segment 2 (Xiphosura) are united to form an unpaired pore in the stem lineage of the Arachnida.
- Indirect spermatophore transfer.
  The change to land dwelling is, as in the Tracheata, connected with the insemination of eggs within the body of the female. Again, as in the Tracheata, this process begins with the production of spermatophores that are deposited on the ground by the male and from which the female takes the sperm.

Systematization

For the phylogenetic systematization of the Arachnida two partly divergent assessments from WEYGOLDT & PAULUS (1979) and SHULTZ (1989, 1990) are available. In the following hierarchical tabulation they are placed opposite one another.

To begin with some agreements can be emphasized. In both relationship hypotheses the Pseudoscorpiones and the Solifugae form the sister groups of a unity Haplocnemata. Furthermore the Ricinulei and the Acari are united as the Acarinomorpha (or Acaromorpha). In both concepts the Uropygi, Amblypygi and Araneae form a monophylum, albeit with different adelphotaxa hypotheses.

WEYGOLDT & PAULUS (1979)

- **Arachnida**
  - **Scorpiones**
  - **Lipoctena**
    - **Megoperculata**
      - **Uropygi**
      - **Labellata**
        - **Amblypygi**
        - **Araneae**
    - **Apulmonata**
      - **Palpigradi**
      - **Holotracheata**
        - **Haplocnemata**
          - **Pseudoscorpiones**
          - **Solifugae**
        - **Cryptoperculata**
          - **Opiliones**
          - **Acarinomorpha**
            - **Ricinulei**
            - **Acari**

SHULTZ (1990)

- **Arachnida**
  - **Micrura**
    - **Megoperculata**
      - **Palpigradi**
      - **Tetrapulmonata**
        - **Araneae**
        - **Pedipalpi**
          - **Amblypygi**
          - **Uropygi**
    - **Acaromorpha**
      - **Ricinulei**
      - **Acari**
  - **Dromodopoda**
    - **Opiliones**
    - **Novogenuata**
      - **Scorpiones**
      - **Haplocnemata**
        - **Pseudoscorpiones**
        - **Solifugae**

A fundamental difference occurs in the estimation of the position of the Scorpiones in the phylogenetic system of the Arachnida. According to WEYGOLDT & PAULUS (1979), the Scorpiones form the sister group of all other taxa – grouped together under the name Lipoctena. In contrast, SHULTZ (1990) places the Scorpiones as a strongly subordinate subtaxon of the Arachnida; they appear here as the sister group of the Haplocnemata (Pseudoscorpiones + Solifugae)[10]. This hypothesis forces the conclusion that convergence occurred three times in the evolution of those apomorphic features which the Lipoctena have in common (p. 100) – i.e., in the stem lineages of the Haplocnemata, of the Opiliones, and of the Micrura as the rest of the Arachnida. This is as unlikely as the hypothesis put forward by SHULTZ (1990) of an independent evolution of the booklungs in the Scorpiones and the "Tetrapulmonata" (Araneae, Amblypygi, Uropygi).

As a consequence of the weaknesses mentioned, in the following overview we orient ourselves using the phylogenetic system for the Arachnida developed by WEYGOLDT & PAULUS (1979).

---

[10] Addition during translation: The Scorpiones occupy an identical, subordinate position in the latest kinship diagram of the Arachnida from W.C. WHEELER & C.Y. HAYAHI (Cladistics 14, 173–192, 1998). On the grounds already mentioned, I reject this hypothesis and still favor a systematization of the Arachnida into the adelphotaxa Scorpiones and Lipoctena.

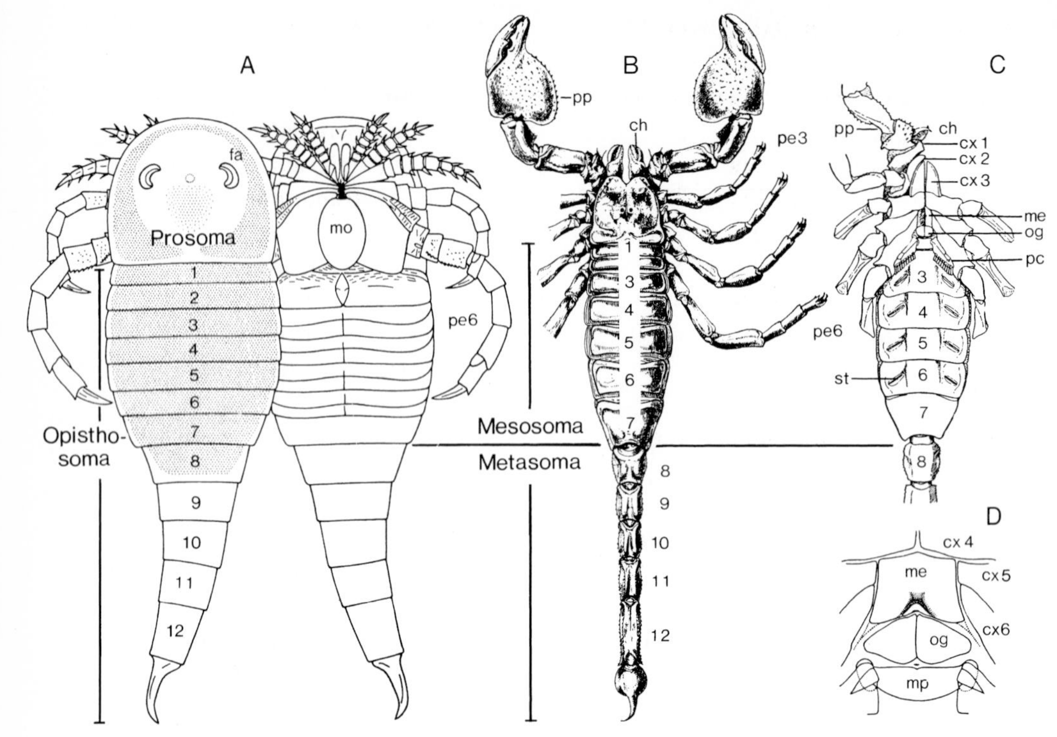

Fig. 45. †Eurypterida (stem lineage representatives of the Arachnida) and Scorpiones. **A** Eurypterida. *Moselopterus ancylotelson* (Devonian). Reconstruction in dorsal and ventral view. With compound eyes more primitive than recent Arachnida. **B–D** Scorpiones. **B–C** *Pandinus* sp. in dorsal and ventral view. **D** *Belisarius xambeui*. Appendage derivatives of opisthosoma segments 1 (metasternum) and 2 (genital opercula, pectines). Ventral view. *ch* Chelicera; *cx* coxa of the pedipalp (1) and of the following walking leg pairs (2–6); *fa* compound eye; *me* metasternum; *mo* metastoma; *mp* median plate between the combs; *og* genital operculum; *pc* pectines (comb); *pe* prosoma appendage; *pp* pedipalp; *st* spiracle of the booklung. (**A** Müller 1994; **B** Kästner 1941; **C** Snodgrass 1965; **D** Millot & Vachon 1949 a)

Scorpiones – Lipoctena

Scorpiones

The Scorpiones are in many respects more primitive than the other terrestrial Arachnida.

Plesiomorphies

- Opisthosoma of 12 segments.

  A long opisthosoma with seven mesosoma segments and five metasoma segments is carried over from the ground pattern of the Chelicerata into the stem lineage of the Arachnida. In this feature the stem species of recent Arachnida is more primitive than the stem species of the recent Xiphosura.

  An opisthosoma with 12 segments is realized in the fossil †Eurypterida as stem lineage representatives of the Arachnida.

- Four pairs of booklungs.
  Development of booklungs in opisthosoma segments 3–6. In the ground pattern of the sister group Lipoctena only two pairs of booklungs are found. The number of gas-exchange organs in the stem lineage of the Arachnida is reduced by one pair. The Xiphosura have five pairs of book gills on opisthosoma appendages 3–7.
- Long chain of free ganglia in the opisthosoma.
  The opisthosoma contains seven pairs of separate ganglia: the last three ganglia pairs of the mesosoma + (only) four pairs in the metasoma, since the ganglia of opisthosoma segments 11 and 12 have fused together.
- Caudal spine.

Autapomorphies (Fig. 43 → 2)

- Chelate pedipalps.
  The six-part palps bear large chelae at their ends (Fig. 45B). As well as functioning as feeler organs (equipped with trichobothria), they serve to seize mobile prey.
  Similar chelae are found on the pedipalps of the Uropygi, Pseudoscorpiones and Ricinulei. Established kinship hypotheses lead to the assumption of convergent evolution.
- Opisthosoma segment 2 with two pairs of appendage derivatives.
  According to the current commonly held view, the appendages of segment 2 are divided and differentiated into two very different elements:
  1. The small genital operculum as an opening place for the genital aperture develops from the anterior section.
  2. The pectines are very conspicuous differentiation products of the posterior section (Fig. 45C, D). Two laterally projecting combs are densely equipped with cuticular tubes. The function of the mechanoreceptors is unclear. Similar structures are not found in any other Arachnida group.
- Caudal spine with poison glands.
  Two poison glands in the last body section give the caudal spine its characteristic form as a stinging apparatus. The globular, poison gland receptacle tapers distally to a crooked tip under which the glands open.
  To sting prey and for defense the opisthosoma with stinging apparatus is raised bow fashion over the front of the body.
- A pair of coxal nephridia.
  In contrast to the ground pattern of the Arachnida with two pairs of coxal nephridia, the formation of only one pair of nephridia in the segment of the third walking legs is an apomorphy of the Scorpiones.
  Convergent reduction to one pair of coxal nephridia in the Palpigradi and within the Holotracheata.

Scorpions are widespread in the tropics and subtropics. *Buthus occitanus* and several *Euscorpius* species are found in southern Europe.

Paleozoic "Scorpiones" should have been aquatic organisms with gills (STØRMER 1970; KJELLESVIG-WAERING 1985). The change from gills to lungs is hypothesized to have occurred during the Late Devonian; Scorpiones with lungs existed in the Late Carboniferous. In the argument of KJELLESVIG-WAERING (1985) for a separate transfer from water to land within the unity Scorpiones, the rest of the Arachnida and the common apomorphic features between these and the Scorpiones remain totally unconsidered. The numerous autapomorphies listed above, which are interpreted as belonging to the monophylum Arachnida, must point for point be proven to be the result of convergent evolution in the Scorpiones and the Lipoctena. There are no grounds for this.

If the interpretation of water-living "Scorpiones" with gills is correct, it remains to be shown how far Paleozoic "Scorpiones" can be considered to be members of the stem lineage of the whole Arachnida.

Lipoctena

Autapomorphies (Fig. 43 → 3)

- Lyriform organ.
  Scorpiones have slit sense organs in the form of individual slits. The close proximity of single slits of increasing length with a parallel orientation must have evolved in the stem lineage of the Lipoctena (Fig. 44 C). Lyriform sense organs are found exclusively on the appendages, mostly in the vicinity of joints.
- Network of connected microvilli fringes of the light sensory cells in the lateral eyes.
  The Scorpiones have taken over ommatidia from the compound eye of the Chelicerata ground pattern into their lateral eyes. The individual retinulae consist of 5–10 crown-shaped sensory cells with a star-shaped rhabdome inside. (Fig. 46 A). In the Lipoctena there are, however, no ommatidia; the retinulae are reduced. The tightly packed cells of the uniform retina carry peripheral microvilli that form a continuous network with one another (Fig. 46 B; WEYGOLDT & PAULUS 1979; WEYGOLDT 1996).
- Two pairs of booklungs.
  In contrast to the four pairs of booklungs in the Scorpiones, two pairs of booklungs can be placed in the ground pattern of the Lipoctena as a derived feature. These, however, are only realized in the Megoperculata – on opisthosoma segments 2 and 3. In the stem lineage of the adelphotaxon Apulmonata booklungs have been totally eliminated.

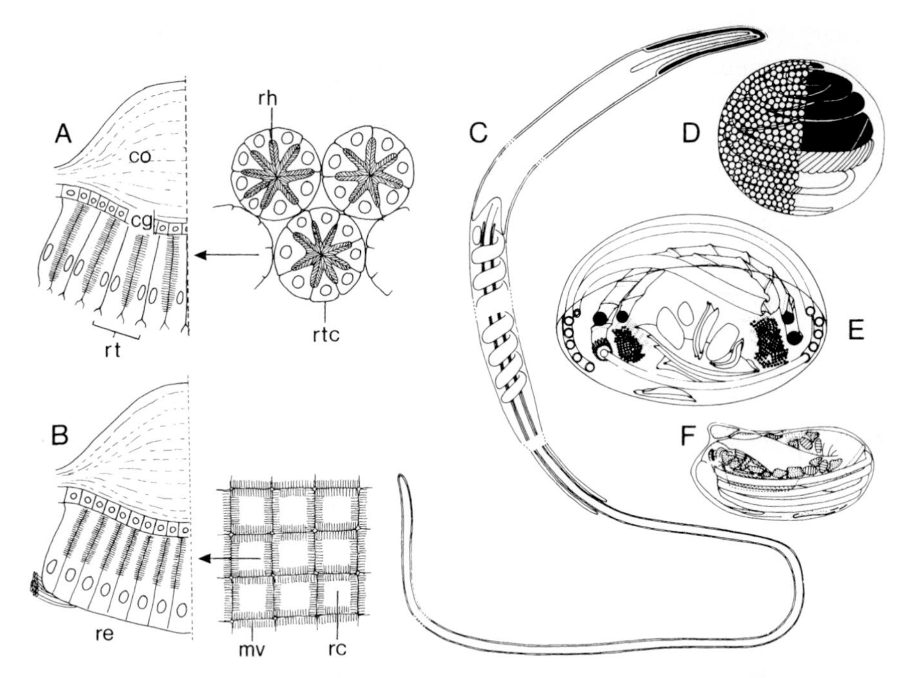

Fig. 46. Lateral eyes and sperm of the Arachnida. Comparison of the plesiomorphous manifestation in the Scorpiones with the autapomorphies of the Lipoctena. **A–B** Lateral eyes. Longitudinal section through a halved eye (*left*) and cross section at the level of the light sensory cells (*right*). **A** Scorpiones. The light-sensory cells form enclosed crowns (retinulae) with a central star-shaped rhabdome. **B** Lipoctena. Lateral eyes with single eversed retina (ground pattern state). The light-sensory cells possess lateral microvilli fringes which are connected to one another in a network. **C–F** Sperm. **C** Scorpiones (*Hadrurus*). Thread-like sperm with tail. **D–F** Lipoctena. Coiling in the spermatid stage. **D** Amblypygi (*Tarantula*). **E** Pseudoscorpiones (*Cheiridium*). **F** Araneae (*Eurypelma*). *cg* Corneagenous cell; *co* cornea; *mv* microvilli; *rc* retina cell; *re* retina; *rh* rhabdome; *rt* retinula; *rtc* retinula cell. (**A–B** Paulus 1979; **C–F** Alberti 1995)

- Coiling of the sperm during spermiogenesis.
  The Scorpiones have an elongated sperm with a head, midsection and tail (Fig. 46C). In the stem lineage of the Lipoctena a strange process of coiling in the spermatid takes place. The nucleus, midsection with mitochondria and the cilium are coiled up in the cytoplasma of the last spermatid (WEYGOLDT & PAULUS 1979). The resulting round and immobile sperm (Fig. 46D–F) are realized in all Megoperculata and Pseudoscorpiones. Further changes to aciliar sperm occur within the Lipoctena (Palpigradi, Solifugae, Acari).
- Flagellum on the opisthosoma.
  The Lipoctena do not possess the uniform caudal spine of the Xiphosura and the Scorpiones. In its place in the Uropygi and the Palpigradi there is a long annulated appendage – the flagellum (Figs. 47, 54). In general, this is

interpreted as being a modified caudal spine. The flagellum can be viewed as an autapomorphy in the ground pattern of the Lipoctena. It must have then been independently reduced at least twice – in the stem lineage of the Labellata and in the stem lineage of the Holotracheata.

The name Lipoctena (without combs) for all terrestrial Arachnida with the exception of the Scorpiones refers to a plesiomorphic feature and is therefore flawed from the point of view of phylogenetic systematics. Like many other common names it originates from a time in classification when the establishment of taxa and their naming due to autapomorphies was not yet seen as a central mission.

Megoperculata – Apulmonata

Within the Lipoctena the Megoperculata and the Apulmonata can be regarded as sister groups; the justifications for their being designated as monophyla is of varying quality. For the Megoperculata with a number of original features clear autapomorphies result from the construction of the two-part chelicerae and the structure of the sperm cilia. For the Apulmonata a narrowing of the body and a correlated total loss of booklungs have been postulated to be autapomorphies.

Megoperculata

Autapomorphies (Fig. 43 → 4)

- Subchelate chelicerae of two articles.
  The Scorpiones and various Apulmonata have the original chelicerae of three articles; the moveable end article works against the fixed projection of the second article (chelate condition). In the stem lineage of the Megoperculata, a new mechanism evolved through reduction of one article. A construction using the same principle as a jack-knife developed, in which the terminal article strikes as a claw against the immobile basal segment (Fig. 44C).
- Sperm cilia with $9 \times 2 + 3$.
  Three microtubules in the axis of the cilium are found in the Uropygi, Amblypygi and Araneae (ALBERTI 1995). These apomorphic structural elements belong, without doubt, in the ground pattern of the Megoperculata.

For the **systematization** of the three monophyletic subtaxa Uropygi, Amblypygi and Araneae two hypotheses exist. In the first the Uropygi and Amblypygi are united as the Pedipalpi. In the other competing view the Amblypygi and Araneae form a monophylum Labellata.

We follow here the Labellata hypothesis, this will be justified later (p. 105).

Uropygi – Labellata

Uropygi

Autapomorphies (Fig. 43 → 5)

- Powerful pedipalps with chelae.
  The pedipalps function as capturing legs; they work horizontally (Fig. 47 B, E).
- Camerostom.
  Special pre-buccal chamber with a roof formed from the upper lip, and a floor from the large coxae of the pedipalps that have medially grown together (Fig. 47 C).
- Prosoma appendages 3 evolved into tactile organs.
  Tarsi weakly subdivided.
  The third prosoma appendages lose their original function as walking legs; they are differentiated into tactile organs. The appendages are clearly lengthened, the tarsi are subdivided into eight or nine sections (Fig. 47 F). The coxae remain in the embryonic lateral position. For walking, appendages 4–6 are used and the tactile legs are carried raised.
- Anal glands (defensive glands).
  In the opisthosoma there are pairs of large glands that open in the 12th segment adjacent to the anus. The secretion, which has acetic acid as its main component, is sprayed at attackers.

The large, mostly several cm long, whip scorpions of the tropics and subtropics are grouped under the name **"Thelyphonida"** (Fig. 47 B). They represent the original niveau of the Uropygi, but are, however, not established as a monophylum. The opisthosoma consists of 12 segments; the tapered waist-like first opisthosoma segment is perhaps a forerunner of the petiole of the Labellata. Only the last three segments form closed rings like the metasoma segments of the Scorpiones. Five pairs of lateral eyes (WEYGOLDT 1996); two pairs of booklungs in opisthosoma segments 2 and 3. *Thelyphonus, Mastigoproctus.*

The small **Schizomida** (Fig. 47 A) can be established as a monophylum due to the numerous **autapomorphies** which are connected with their mode of life as interstitial soil inhabitants: Shortening of the body length (ca. 5–13 mm), eyeless, only one pair of booklungs (in the second opisthosoma segment), pedipalps without chelae. Flagellum of three segments in the ♀ fused to form a stalked plate in the ♂.
Schizomus.

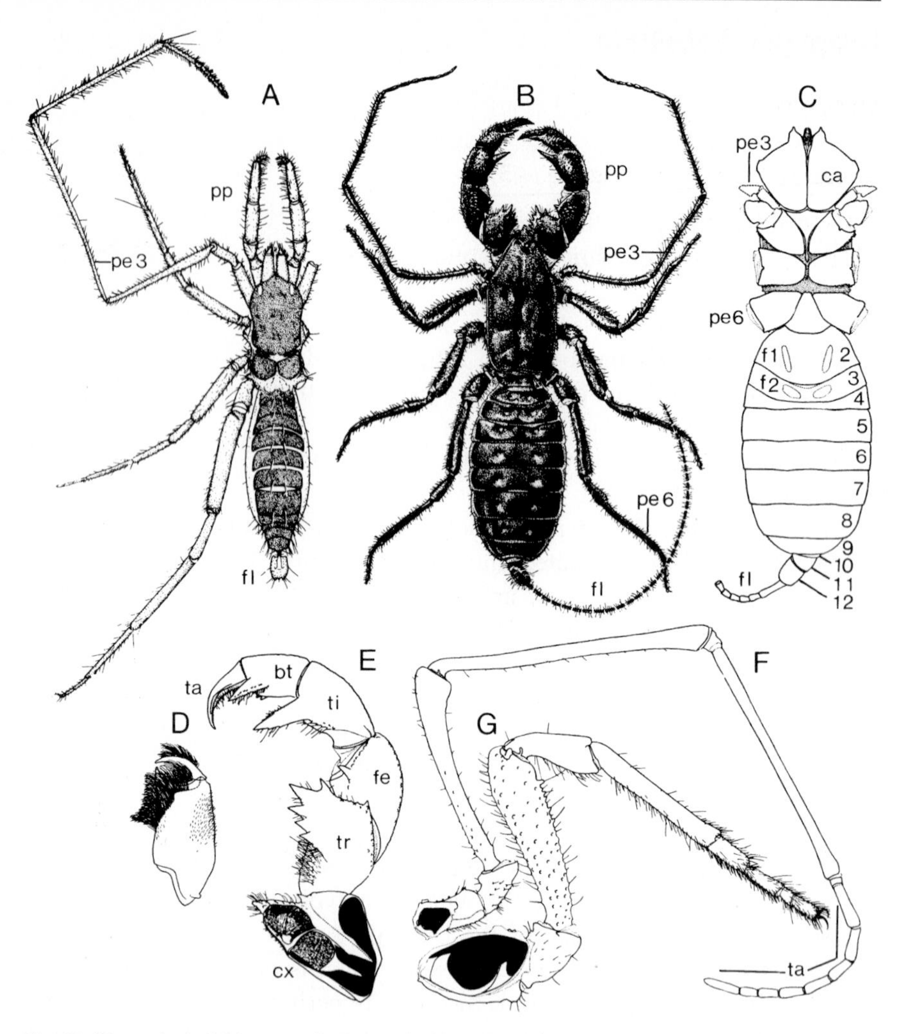

Fig. 47. Uropygi. **A** *Schizomus siamensis* (Schizomida). Male in which the flagellum forms a stalked plate (anchor organ for the female in indirect sperm transfer through spermatophores). **B–G** *Mastigoproctus giganteus* ("Thelyphonida"). **B** Dorsal view with the large pedipalps and the third prosoma appendage as sensory organs. Opisthosoma segment 1 with a waist-like constriction, but with tergite. **C** Ventral view. Coxae of the pedipalps form the floor of the camerostom. Opisthosoma segment 1 not visible. Paired booklungs in opisthosoma segments 2 and 3 (plesiomorphy). Opisthosoma segments 10–12 form small closed rings. **D** Chelicera in the form of a jack-knife. **E** Chelate pedipalp. Tarsus works against a projection of the basitarsus. **F** Prosoma appendage 3 as a sensory organ with a tarsus of nine articles. **G** Walking leg. Pretarsus with two claws. *bt* Basitarsus; *ca* camerostom; *cx* coxa; *f* booklung; *fe* femur; *fl* flagellum; *pe* prosoma appendage; *pp* pedipalp; *ta* tarsus; *ti* tibia; *tr* trochanter. (**A** Moritz 1993b; **B–G** Millot 1949a)

Labellata

Autapomorphies (Fig. 43 → 6)

- Petiole.
  Restriction of the first opisthosoma segment to form a narrow pedicellus that connects prosoma and opisthosoma (Fig. 48 B).
- Post-cerebral sucking stomach.
  Following the transformation of the pharynx into a suction pump, a sucking stomach is differentiated behind the brain, which becomes the main pump for nutrient uptake. The sucking stomach is equipped with dorsal and lateral radiating dilators.
- Sternum and foremost prosoma sternite.
  The sternites of the four walking leg segments are fused to form a large undivided sternal plate (Figs. 48 B; 51 B). According to WEYGOLDT (1972), the prosoma sternite directly in front of the sternum (lower lip, labium) is also a synapomorphy of the Amblypygi and Araneae.
- Loss of the subdivision of opisthosoma into mesosoma and metasoma.
  Amblypygi and Araneae have the original 12 opisthosoma segments. Compared with the Scorpiones and Uropygi, however, there are no circular differentiated segments at the posterior. In the stem lineage of the Labellata the opisthosoma evolved to form a single sac-shaped structure.
- Reduction of the flagellum.
  The flagellum realized in the Uropygi is missing in the Amblypygi and Araneae. Using the principle of parsimony we postulate a single reduction in a common stem lineage of these two unities.
- All opisthosoma neuromeres in the subesophageal ganglion.
  In the Uropygi the last five neuromeres of the body remain as a united ganglion in opisthosoma segment 8. In contrast, in the Labellata, all 12 opisthosoma neuromeres are found in the subesophageal ganglion of the prosoma.
- Sperm nucleus with post-centriol lengthening.
  The Labellata hypothesis is supported by the ultrastructure of the sperm. In the Amblypygi and Araneae a projection of the nucleus extends over the base of the axoneme to the posterior; the cell is asymmetrical. In the Uropygi this is not the case (ALBERTI 1995).

Amblypygi – Araneae

Amblypygi

Autapomorphies (Fig. 43 → 7)

- Prosoma appendage 3 an antenna-like tactile organ with subdivision of tibia and tarsus.

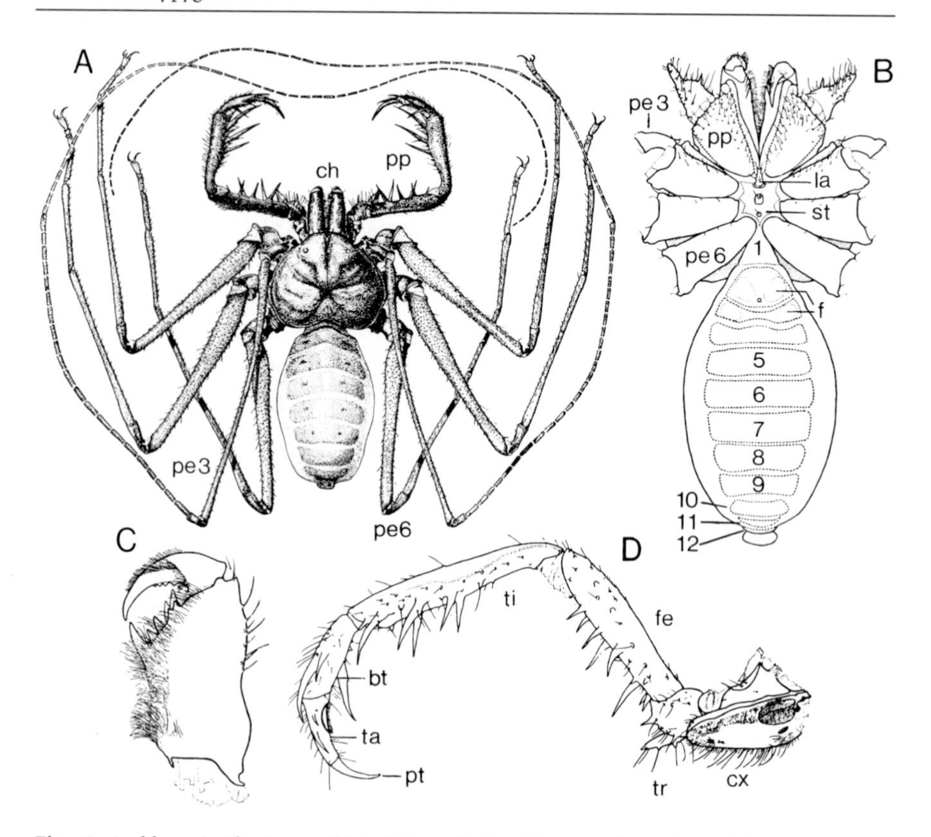

Fig. 48. Amblypygi. *Charinus milloti.* **A** Dorsal view. Extreme elongation of the prosoma appendages 3. Pedipalps large raptorial legs. Flat opisthosoma. **B** Ventral view. Opisthosoma of 12 segments (plesiomorphy). Segment 1 tapers to the petiole (synapomorphy with the Araneae). Booklungs in segments 2 and 3 (plesiomorphy). **C** Chelicera like a jack-knife. **D** Pedipalp a raptorial leg. The tarsal section of three segments can be struck against the tibia. *bt* Basitarsus; *ch* chelicerae; *cx* coxa; *f* booklung; *fe* femur; *la* labium; *pe* prosoma appendage; *pp* pedipalp; *pt* posttarsus; *st* sternum; *ta* tarsus; *ti* tibia; *tr* trochanter. (Millot 1949b)

The long, thin and extremely mobile tactile leg is inserted above the coxae of the other legs. Tibia and tarsus consist of numerous segments (Fig. 48 A). Example: the ♂ of *Damon medius* (length 22 cm) has a tibia (length 8.8 cm) of 37 segments; the tarsus (length 8 cm) has 92 segments.

- Pedipalps form large predatory legs with folding tarsal segments.

  The inner sides of the tibia and tarsus are equipped with prominent spines. The pedipalps can be deployed either together as a capturing basket or singly as grasping organs (Fig. 48 D).

After the Uropygi, the Amblypygi form the second monophylum of the Arachnida in which the first walking leg pair is transformed into tactile legs. The existence of tactile legs has been seen as a synapomorphy leading to the grouping of the Uropygi and Amblypygi into a unity Pedipalpi. This interpretation, however, is in conflict with several presumed autapomorphies of a unity Labellata consisting of the Amblypygi and Arachnida. Consequently, a convergent evolution of the tactile legs in the Uropygi and Amblypygi must be considered. There are, however, considerable differences in length, subdivision, and insertion. I therefore interpret, with some reserve, the tactile legs of the Uropygi and Amblypygi as an autapomorphy for each group.

The tail-less whip scorpions are tropical organisms in biotopes with high humidity. To secrete themselves in suitable crevices in the soil they have evolved a flat body with a dorsoventrally compressed opisthosoma.

Phrynus, Damon, Charon.

Araneae

The "evolutionary success" of the spiders with the genesis of 35,000 species is connected with the evolution of a unique feature in the stem lineage of the Araneae – **the ability to spin a web**. Silk glands in the opisthosoma produce a silk (fibroin), which is emitted from spinnerets through special spigots. The **spinning apparatus** can be used for different functions. The spiders' silk serves for the production of egg cocoons and molt cocoons, the lining of burrows, the formation of walking threads, and naturally also for the production of threads and complex webs for catching prey. An example of a singular specialization is the "extraintestinal stomach" of the Uloboridae (Araneomorphae). In an envelope spun around the prey the food is liquefied before uptake in the stomach of the spider (PETERS 1982).

Autapomorphies (Figs. 43 → 8; 50 → 1)

- Spinning apparatus in the opisthosoma.
  1. Silk glands.
  In the taxon Mygalomorphae a single type of gland is found (PALMER 1985). Such uniform silk glands are a part of the ground pattern of the Araneae. Within the unity up to eight different silk glands each with a specific function then evolved – as in the Araneidae (orb weaving spiders).
  2. Spinnerets.
  Short, articulated spinnerets are morphologically appendages of opisthosoma segments 4 and 5. During development the single appendages divide resulting in 4 spinneret pairs in the ground pattern of the Araneae; originally they lie ventrally in the middle of the opisthosoma (Fig. 51). Eight functioning spinnerets are only found, however, in developmental

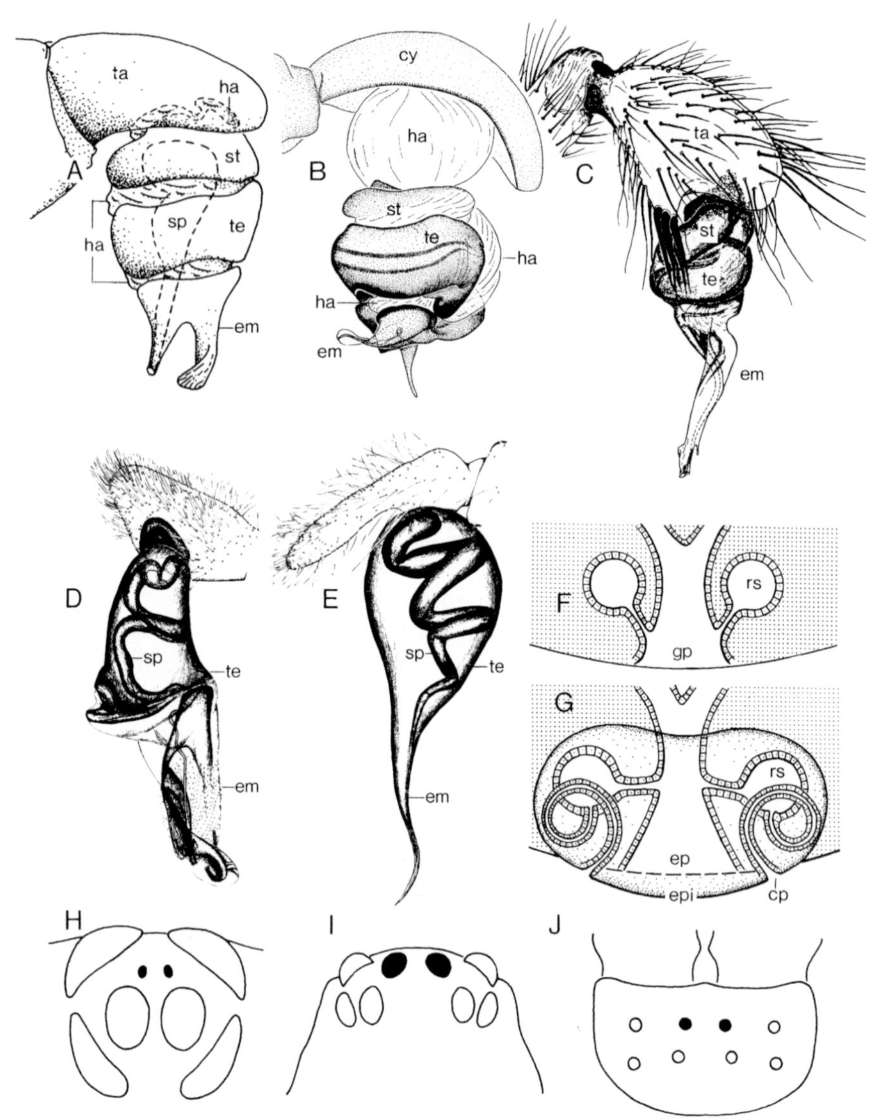

Fig. 49. Araneae. Male palp organs, female genital system and arrangement of the eight eyes. **A** Three-part palp organ in the ground pattern of the Araneae. **B** Three-part palp organ from *Uroctea durandi* (Entelegynae). **C** Three-part palp organ of *Hypochilus gertschi* (Palaeocribellatae). **D** Two-part bulb of *Dysdera erythrina* (Haplogynae). **E** Undivided pear-shaped bulb of *Segestria senoculata* (Haplogynae). **F** Primary state of the female genital organs with one genital pore. **G** Apomorphous entelegynous state with epigynum, one egg-laying opening and two copulatory pores. **H–J** Eye arrangements. Median eyes in black. **H** *Liphistius* (Mesothelae). Lateral eyes in triads. **I** *Hypochilus gertschi* (Palaeocribellatae). Lateral eyes in triads. **J** *Tetragnatha* (Entelegynae). Arrangement of the eyes in two rows. *cp* Copulatory pore; *cy* cymbium; *em* embolus; *ep* egg-laying pore; *epi* epigynum; *gp* genital pore; *ha* hematodocha; *rs* receptaculum seminis; *sp* spermophore; *st* subtegulum; *ta* tarsus; *te* tegulum. (**A** Kraus 1984; **B** Baum 1973; **C** Kraus 1978; **D–E** Schult 1983; **F–G** Weygoldt 1996; **H–I** Kraus & Kraus 1993 a; **J** Foelix 1992)

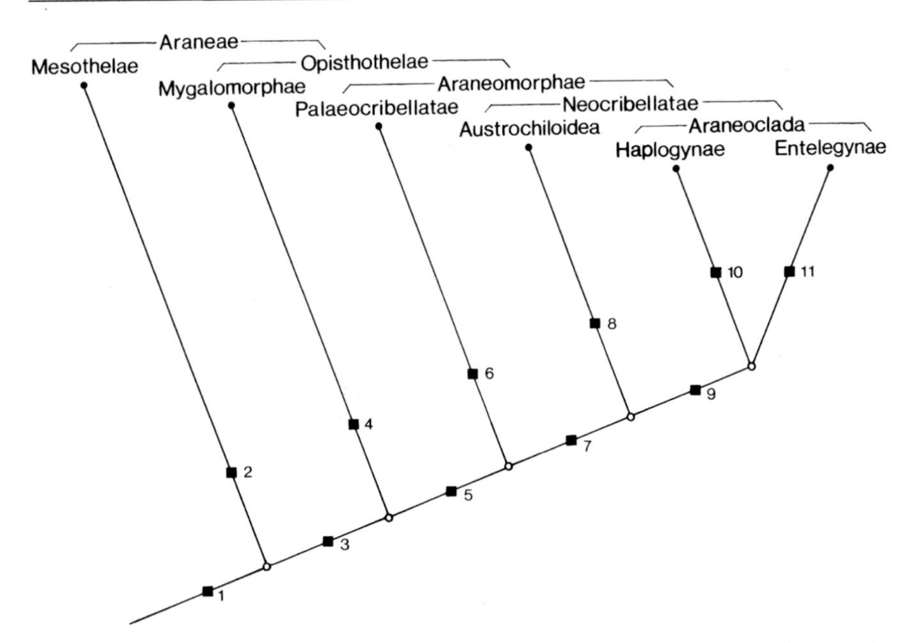

Fig. 50. Diagram of the phylogenetic relationships between the high-ranking subtaxa of the Araneae

stages of the Mesothelae. Important modifications exist in the adult Mesothelae as well as in other subtaxa; in some cases these can be employed as autapomorphies to establish various monophyla within the Araneae.

- Pedipalp used for sperm transfer.

  In the stem lineage of the Araneae, the indirect transfer of sperm using spermatophores – as still realized in the adelphotaxon Amblypygi – is replaced by direct sperm transfer. This change takes place through the deployment of the male pedipalps.

  The palp organ (Fig. 49 A–E) inserts at the cymbium which is usually a shuttle-shaped extension of the tarsus. For the ground pattern of the Araneae a three-part organ with subtegulum, tegulum and embolus has been hypothesized (KRAUS 1984).

  Pliable zones (hematodocha) connect the sclerites. A spermaphore in the interior serves in the uptake of sperm. To fill the copulation organ the male spins a tiny sperm net. A drop of sperm is then deposited here from its own genital opening and taken up in the palp organs. During copulation, the palp organs are inserted into the female genital opening. The pliable expansive sections extend the organ, sclerites help to anchor it in position. The sperm is then passed into the seminal receptacles of the female.

- Chelicerae with poison glands.
  Small poison glands limited to the chelicerae basal segments (Mygalomorphae: Theraphosidae) can be placed in the ground pattern of the Araneae. Major enlargements of the glands then take place within the Araneae. In the Neocribellatae the glands extend far into the prosoma (Fig. 53 G). The cylindrical gland bodies are spirally surrounded by cross-striated muscles which press out the poison by contraction. The glands open under the tip of the chelicerae claws.

The **two-part subchelate chelicerae** appear in the Araneae in **three different forms.**

Orthognathy. The basal segments sit horizontally on the prosoma. The terminal segments (claws) fold in parallel to one another ventrally. Captured prey is clamped between the basal and terminal segments. The majority of the Mygalomorphae (Fig. 52 A, B).

Plagiognathy. The basal segments are splayed out and directed diagonally downwards. The claws work diagonally against one another. Mesothelae, subtaxa of the Mygalomorphae, Austrochiloidea, Palaeocribellatae (Figs. 51 B; 53 A).

Labidognathy. The vertically oriented basal segments sit under the prosoma. The claws are turned 90° inwards and stand at right angles to the long axis of the body. Captured prey is clamped between the right and left claw. Araneomorphae (Fig. 53 B, H).

Recently, it has been discussed that plagiognathy may be the plesiomorphic state of spiders (KRAUS & KRAUS 1993 a). However, orthognathy in other Megoperculata (Uropygi, Amblypygi) points to the adoption of a corresponding orthognathic state in the ground pattern of the Araneae. From this plagiognathy and labidognathy then evolved in different subtaxa of the Araneae.

Spiders have as a plesiomorphy **eight eyes** in their **ground pattern** – two eversed main eyes (= median eyes from the ground pattern of the Euchelicerata) and six inverted lateral eyes (= lateral eyes from the ground pattern of the Arachnida; Fig. 49 H–J).

The common ordering of the eight eyes into two transverse rows is probably a derived state within the Araneae. The arrangement of the lateral eyes in triads appears to be the plesiomorphic state, as demonstrated by their distribution within the Araneae (Mesothelae and Palaeocribellatae amongst others) and their manifestation in the adelphotaxon Amblypygi (KRAUS & KRAUS 1993 a).

Two pairs of booklungs and an opisthosoma with 12 segments are original features from the ground pattern of the Megoperculata. Compared with the derived states in the Uropygi and Amblypygi, undifferentiated leg-like pedipalps in the female and prosoma appendages 3 as a normal walking leg pair are plesiomorphies.

In the female the following states belong in the ground pattern of the Araneae. From each of the paired ovaries oviducts arise that unite to form an unpaired terminal section (uterus, bursa copulatrix). The latter opens ventrally with a genital pore at the posterior edge of the second opisthosoma segment. Paired seminal receptacles for the storage of foreign sperm are connected with the vagina (Fig. 49 F). This primary state is realized from the Mesothelae to the haplogynic spiders within the Opisthothelae. In the stem lineage of the Entelegynae the evolution of entelegynous female genitalia took place with three openings and the epigynum (Fig. 49 G).

Systematization

Phylogenetic systematics has produced well-grounded hypotheses concerning the high-ranking adelphotaxa kinships (PLATNICK & GERTSCH 1976; PLATNICK 1977; HAUPT 1983; RAVEN 1985; PLATNICK et al. 1987; CODDINGTON 1990; CODDINGTON & LEVI 1991; PLATNICK et al. 1991; FOELIX 1992 amongst others). The resulting system is represented in the following hierarchical tabulation and the equivalent relationship diagram. The evidence is discussed more fully below.

Araneae
- **Mesothelae**
- **Opisthothelae**
  - **Mygalomorphae**
  - **Araneomorphae**
    - **Palaeocribellatae**
    - **Neocribellatae**
      - **Austrochiloidea**
      - **Araneoclada**
        - **Haplogynae**
        - **Entelegynae**

Mesothelae – Opisthothelae

Mesothelae

Around 40 species of the taxa *Liphistius* and *Heptathela* from Asia form the sister group of all other spiders; they are distinguished by several ple-

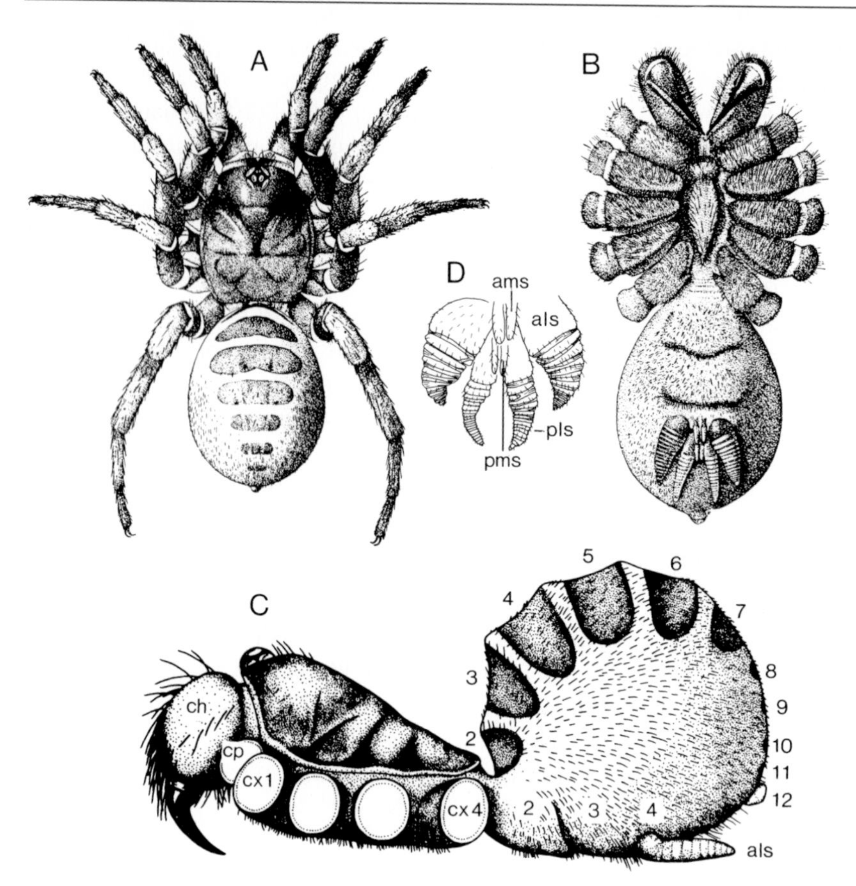

Fig. 51. Mesothelae. **A–C** *Liphistius malayanus*. **A** Dorsal view with tergites on the opisthosoma. **B** Ventral view. Plagiognathous position of the chelicerae claws. Eight spinnerets ventrally on the opisthosoma. **C** Lateral view. Narrowing between the prosoma and opisthosoma through a stalk-like first opisthosoma segment (petiole). **D** *Liphistius desultor*. Spinnerets in ventral view. *als* anterior lateral spinneret; *ams* anterior median spinneret; *ch* chelicera; *cp* coxa of the pedipalp; *cx* coxae of walking legs 1–4; *pls* posterior lateral spinneret; *pms* posterior median spinneret. 2–12=opisthosoma segments (2–11 with tergites, 12 anal tubercle). (**A, B** Millot 1949c; **C** Moritz 1993b; **D** Raven 1985)

siomorphic organizational traits. Only in the Mesothelae are signs of opisthosoma segmentation externally visible through the development of ten tergites. Also only seen in the Mesothelae are eight ventral spinnerets in the middle of the opisthosoma as appendages of segments 4 and 5.

The question of the monophyly of the Mesothelae is discussed by PLATNICK & GERTSCH (1976), HAUPT (1983) and PLATNICK & GOLOBOFF (1995). From the possible apomorphic peculiarities the following appear to be sufficiently well established to be considered as autapomorphies.

Autapomorphies (Fig. 50 → 2)

- Reduction of the middle spinnerets.
  The middle spinnerets of opisthosoma segments 4 and 5 are small, mainly or fully non-functional pegs in the adult (Fig. 51 D).
- Spurs on the tibia of walking legs 1, 2 and 3.
  A pair of large, flat spurs sits distally on the sides of tibiae 1–3; they are fused to the tibia without joints (present in the female juvenile and adult; in the male only in the juvenile). The tips of the spurs project above oval, non-sclerotized fields on the sides of metatarsals 1–3. The structures may be proprioreceptors that register lateral deflections of the leg (PLATNICK & GOLOBOFF 1985).
- Concavity at the anterior inner corner of the coxae of the fourth walking leg pair (Fig. 51 B).
- Plagiognathy.
  Inclined position of the chelicerae in comparison with the original orthognathy (Fig. 51 B).

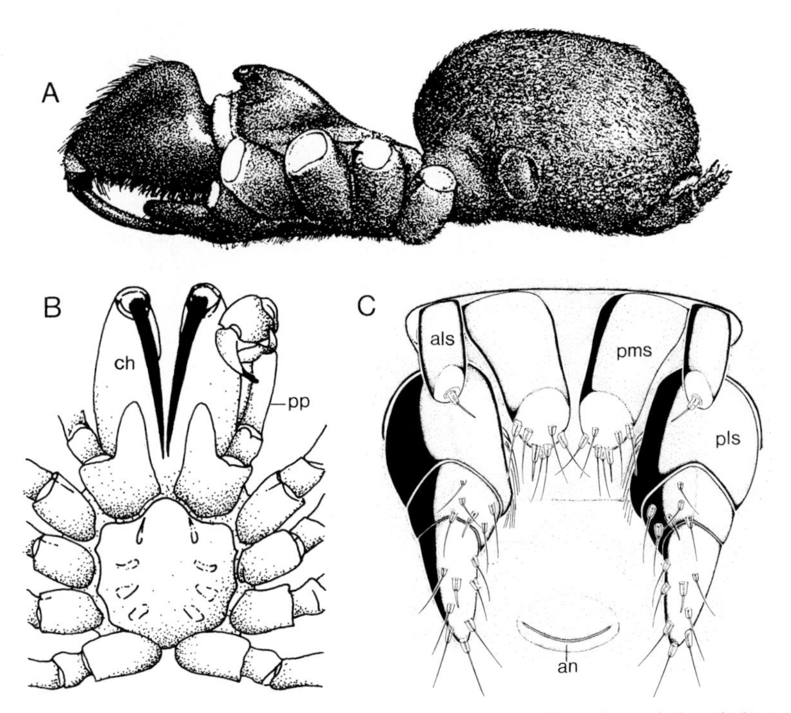

Fig. 52. Mygalomorphae. **A** *Atypus affinis*. Side view. Orthognathy. Claw of the chelicera folded back ventralwards. Spinnerets at the posterior of the opisthosoma. **B** *Atypus bicolor*. Prosoma from underneath. Ventral view of the orthognathous chelicerae. **C** *Atypus affinis*. Spinning apparatus in ventral view. Ground pattern of the Mygalomorphae with six spinnerets; the front middle spinnerets are missing. *als* Anterior lateral spinneret; *an* anus; *ch* chelicera; *pls* posterior lateral spinneret; *pms* posterior median spinneret; *pp* pedipalp. (**A** Millot 1949 c; **B** Snodgrass 1965; **C** Glatz 1973)

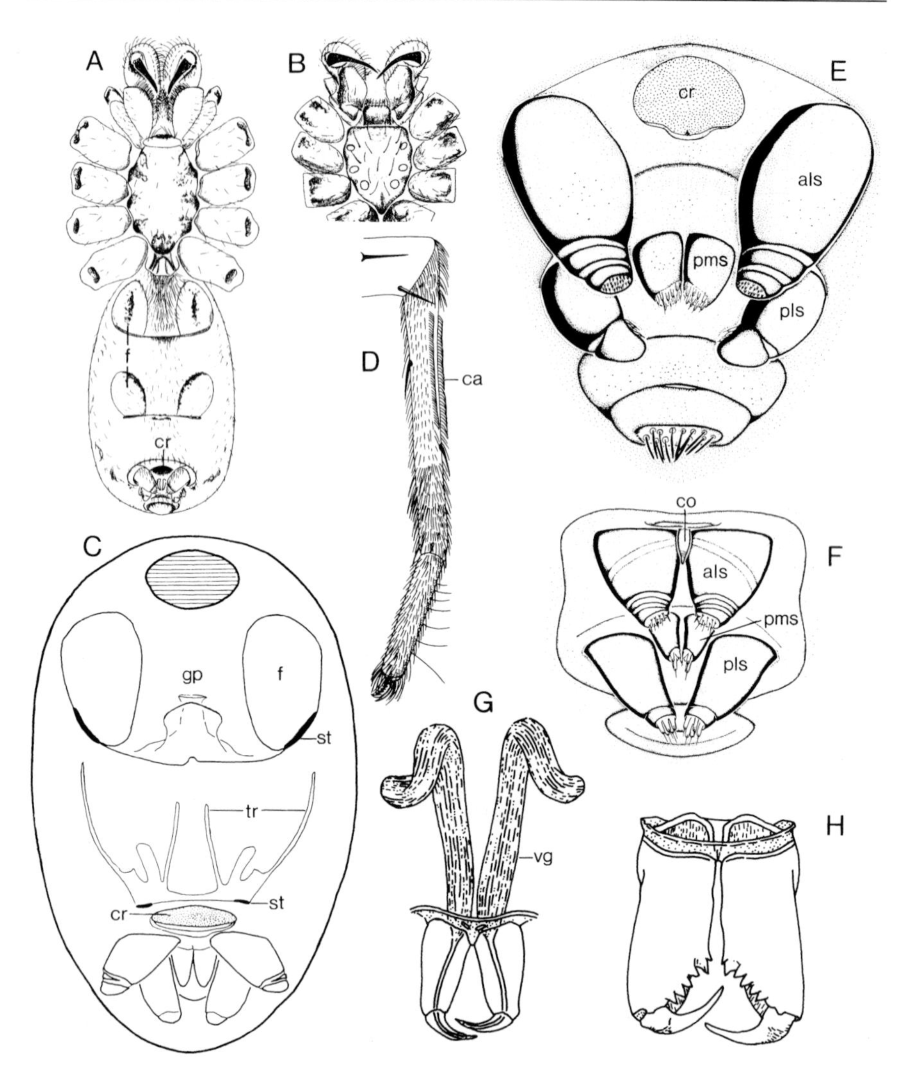

Fig. 53. Araneomorphae. **A** *Hypochilus thorelli* (Palaeocribellatae). Female in ventral view. Plagiognathous position of the chelicerae. **B** *Ectatosticta davidi* (Palaeocribellatae). Female. Ventral view of the prosoma. Labidognathous position of the chelicerae. **C** *Austrochilus* (Austrochiloidea). Diagram of the ventral view of the opisthosoma. One pair of booklungs, one pair of tracheal tubes and the cribbellate spinning apparatus. **D** *Amaurobius ferox* (Entelegynae). Calamistrum (silk comb) on the metatarsus of the fourth walking legs. **E** *Hypochilus gertschi* (Palaeocribellatae). Spinning apparatus with cribellum. Ventral view. **F** *Segestria senoculata* (Haplogynae). Spinning apparatus with colulus between the anterior spinnerets. Ventral view. **G** *Latrodectes mactans* (Entelegynae). Labidognathous chelicerae. Poison glands inserted deep in the prosoma. **H** *Ancylometes* (Entelegynae). Labidognathous chelicerae. Frontal view. *als* Anterior lateral spinneret; *ca* calamistrum; *co* colulus; *cr* cribellum; *f* booklungs; *gp* genital pore; *pls* posterior lateral spinneret; *pms* posterior median spinneret; *st* spiracle; *tr* tracheal tube; *vg* poison gland. (**A, B** Forster, Platnick & Gray 1987; **C** Marples 1968; **D** Moritz 1993b; **E, F** Glatz 1972; **G, H** Snodgrass 1965)

Opisthothelae

Autapomorphies (Fig. 50 → 3)

- Sac-shaped opisthosoma.
  In the stem lineage of the Opisthothelae a uniform opisthosoma evolved without external segmentation. The tergites and sternites realized in the Mesothelae are lost within the Opisthothelae. In the Palaeocribellatae five tergites are still seen.
- Spinnerets at the posterior.
  The spinneret complex has migrated from its primary mid-position to the posterior end of the opisthosoma (Fig. 52 A).

Mygalomorphae – Araneomorphae

Mygalomorphae

Autapomorphy (Fig. 50 → 4)

- Six spinnerets.
  The anterior mid spinnerets are completely missing. Six spinnerets form the derived ground pattern state of the Mygalomorphae (Fig. 52 C). Additionally, the anterior lateral spinnerets are greatly reduced (RAVEN 1985).

Two pairs of booklungs are a plesiomorphy in the ground pattern of the Mygalomorphae, likewise the already discussed orthognathy. Since phylogenetic systematics tries, wherever possible, to avoid using names which are derived from plesiomorphies the name Orthognatha will no longer be used.

Well-known members of the Mygalomorphae form the tropical Theraphosidae (bird spiders and tarantulas); these include the largest representatives of the Araneae. In Europe, Ctenizidae (trap-door spiders) and Atypidae (purse-web spiders) are common.

Araneomorphae

Autapomorphies (Fig. 50 → 5)

- Cribellum.
  Transformation of the foremost middle spinnerets to form a flat, oval plate, over which strong sticky silk is emitted (Fig. 53 C, E).
- Calamistrum.
  Connected with the cribellum a special comb of regular, longitudinal rows of setae is developed on the metatarsus of the fourth walking leg

pair (Fig. 53 D). The calamistrum combs out the silk from the spigots of the cribellum.
- One pair of coxal nephridia.
  The Mesothelae and the Mygalomorphae have each carried over from the ground pattern of the Arachnida two pairs of coxal nephridia; the openings are on the coxae of walking leg pairs 1 and 3. In the Araneomorphae only the anterior pair is present.

The postulated complex of cribellum and calamistrum is only found in some 100 "real spiders". In the majority of the 32,000 species of Araneomorphae in the place of the cribellum either a small (?) non-functional cone (colulus) is found (Fig. 53 F), or a comparable element is totally absent.

These purely quantitative distributions appear, however, to be of lesser importance. Much more important is the fact that in diverse close-kinship groups of the Araneomorphae, species exist with and without a cribellum. For this distribution pattern there are principally two possible explanations: (1) Cribellum and calamistrum evolved once in the stem lineage of the Araneomorphae and were later lost convergently. (2) The stem species of the Araneomorphae did not possess a cribellum; cribellum and calamistrum evolved several times independently in different lineages within the Araneomorphae. Using the principle of parsimony the first hypothesis with a single evolution of cribellum and calamistrum is to be favored.

Palaeocribellatae – Neocribellatae

Palaeocribellatae

With the two taxa *Hypochilus* (North America) and *Ectatosticta* (China) the Palaeocribellatae form the species-poor sister group of all other Araneomorphae which are combined under the name Neocribellatae.

The Palaeocribellatae are characterized by a mass of original features – two pairs of booklungs, confinement of the poison glands to the chelicerae, five tergites on the opisthosoma.

However, PLATNICK (1977) and FORSTER et al. (1987) interpret some apomorphic peculiarities as synapomorphies of *Hypochilus* and *Ectatosticta*; using these they established a monophylum Palaeocribellatae.

Autapomorphies (Fig. 50 → 6)

- Extension of midgut diverticula into the chelicerae.
- Notch on the surface of the chelicerae to accommodate the tips of the long claws.

Hypochilus has a plagiognathic position in the chelicerae; in contrast, in *Ectatosticta* labidognathy is almost fully realized (Fig. 53 A, B; GERTSCH 1958).

Neocribellatae

The feature of labidognathy, which is usually used to establish a monophylum Neocribellatae, appears to be somewhat problematic. Firstly, *Ectatosticta* (Palaeocribellatae) has labidognathous chelicerae – and, secondly, *Hickmania* and *Austrochilus* (Austrochiloidea) within the Neocribellatae have plagiognathous chelicerae (GERTSCH 1958).

According to PLATNICK (1977) and FORSTER et al. (1987), the following features can be considered as autapomorphies of the Neocribellatae.

Autapomorphies (Fig. 50 → 7)

- Extension of the poison glands into the prosoma (Fig. 53 G).
- Simple U-shaped passageway of the coxal nephridia.
  The Mesothelae, Mygalomorphae and Palaeocribellatae have a strongly coiled passageway, which is interpreted as a plesiomorphic state.
- Absence of the fifth ventral endosternite on the opisthosoma.
  Present in *Hypochilus* and *Ectatosticta*.

Austrochiloidea – Araneoclada

Austrochiloidea

Austrochilidae (*Austrochilus* and *Thaida* – Chile; *Hickmania* – Tasmania) as well as the Gradungulidae (*Gradungula* and other "genera" – New Zealand, Australia) form "basal" taxa of the Neocribellatae. Their grouping into a monophylum is based on one feature (FORSTER et al. 1987).

Autapomorphy (Fig. 50 → 8)

- The anterior edge of the prosoma forms a triangular projection (clypeus) being a short cap-like area over the base of the chelicerae.

This rather weak justification brings us to a further problem. The majority of the Austrochiloidea possess the plesiomorphic state of two booklungs; in *Austrochilus* and *Thaida*, however, "tracheal tubes" replace the second pair of booklungs (Fig. 53 C). The cuticle of these tubes is covered with tiny projections. "The same appearance is seen on the lamellae of the lungs" (MARPLE 1968, p. 17). "Despite the tracheate appearance of the structures found in *Austrochilus* and *Thaida*, the organs are perhaps more appropriately described as modified booklungs than as tracheae" (FORSTER et al. 1987, p. 92).

Araneoclada

Autapomorphies (Fig. 50 → 9)

- Three pairs of heart ostia.
  A reduction in comparison to the four pairs of ostia in the Austrochiloidea. Further reductions within the Araneoclada.
- Extended arrangement of the intestine in the opisthosoma.
  Mesothelae, Mygalomorphae, Palaeocribellatae and Austrochiloidea have in common an M-shaped coiled intestine in the opisthosoma. Within the Araneae this state is considered as original (FORSTER et al. 1987).
- (?) Tracheae.
  "The most salient may be the transformation of the posterior booklungs into tracheae, the first appearance of a tracheal system in spiders" (CODDINGTON & LEVI 1991, p. 576).

The passage just cited concerning the tracheal system of the Araneoclada must be regarded with reservation. Thorough proof is still needed of how far the above-mentioned "tracheal tubes" in *Austrochilus* and *Thaida* evolved convergently to the tubular tracheae of the Araneoclada or whether they are homologous with them.

In the ground pattern of the Araneoclada belong the anterior pair of booklungs as a plesiomorphy; within the Araneoclada, however, they too are replaced by tracheae.

The tubular tracheae in opisthosoma segment 3 have primary paired spiracles. Their uniting to form an unpaired opening, and their transfer to a position in front of the anus, belong to the manifold modifications of the tracheal system within the Araneoclada that we cannot follow further here.

Haplogynae – Entelegynae

Haplogynae

The common name Haplogynae is based on a plesiomorphy of the Araneae – a female genital opening without an epigynum for copulation and egg laying (Fig. 49 F). However, the haplogynous spiders can be established as a monophylum through several other characteristic features.

Autapomorphies (Fig. 50 → 10)

- Basal fusion of the chelicerae.
- Chelicerae with a lamina along the claw groove in place of teeth.
- Synspermia (?).
- Secondary simple palp organ (?).

Within the Arachnida the fusion of two to four sperm (or incomplete separation of spermatids) in the testes is only known from the haplogynous spiders, and this in only a few taxa (*Segestriidae, Dysderidae, Scytodidae, Sicariidae)* (ALBERTI & WEINMANN 1985; ALBERTI 1995).

For the ground pattern of the Araneae a three-part palp organ has been suggested (p. 109) – as is commonly seen in the Mesothelae, Mygalomorphae (Atypidae), Palaeocribellatae (*Hypochilus*), Austrochiloidea and Entelegynae. Therefore the simple palp organ of haplogynous spiders (Fig. 49 D, E) must be interpreted as a derived state with secondary fusions (KRAUS 1978, 1984; SCHULT 1983). Details of the ground pattern state for the unity Haplogynae are not known at present.

With cribellum: *Filistata.* Without cribellum: *Dysdera, Pholcus, Scytodes, Sicaria, Segestria.*

Entelegynae

Autapomorphy (Fig. 50 → 11)

- Entelegynous genitalia with epigynum.

Paired copulation pores open in the female in the epigynum – a sclerotized plate on the opisthosoma with pits and protuberances to anchor the male palp organs during copulation. Sperm ducts lead from the copulation pores into the seminal receptacles – storage organs for foreign sperm. Fertilization ducts connect the seminal receptacles with the genital atrium. The opening for egg laying is located behind the epigynum; it corresponds to the single genital opening in the ground pattern of the Araneae (Fig. 49 G).

The majority of the Araneae belong to the entelegynous spiders. I will mention here only a few well-known taxa in alphabetical order.

With cribellum: *Amaurobius, Dictyna, Eresus, Oecobius, Uloborus.* Without cribellum: *Agelena, Araneus, Argyroneta* (aquatic), *Clubiona, Linyphia, Pardosa, Pholcus, Pisaura, Salticus, Tetragnatha, Theridion, Thomisus, Uroctea.*

Apulmonata

After the presentation of the Megoperculata we come to the second highest-ranking and at the same time species-rich unity of the Lipoctena. The Apulmonata have been interpreted as a possible sister group of the Megoperculata (WEYGOLDT & PAULUS 1979). In comparison with the Megoperculata, however, the monophyly of the Apulmonata appears less well grounded.

Autapomorphies (Fig. 43 → 9)

- Smaller body size.
- Complete absence of booklungs.

The smaller body size of prosoma + opisthosoma to near or less than 1 cm is strongly correlated with the absence of booklungs. The bodily dimensions of the Apulmonata must be interpreted as a result of a secondary decrease in size in contrast to the Megoperculata, which are mostly large. The loss of the primary booklungs is connected with this decrease. Within the Apulmonata, only in the Solifugae do clearly larger species appear again; they are equipped with a new gaseous exchange system of connected tubular tracheae.

An opisthosoma of 12 segments (Pseudoscorpiones, original number in the Acari) with a flagellum (Palpigradi) belongs in the ground pattern of the Apulmonata and its sister group the Megoperculata. The ground pattern of the Apulmonata further includes three-part chelate chelicerae – carried over from the stem species of the Arachnida. In this feature the Apulmonata are more original then the Megoperculata which have two-part chelicerae (subchela).

Palpigradi – Holotracheata

Palpigradi

Species-poor group of very small Arachnida on average 1–1.5 mm in length. *Prokoenenia*, *Eukoenenia*.

Hygrophilous, interstitial organisms in tropical and subtropical terrestrial soils. Occasionally seen in southern Europe; *Eukoenenia austriaca* has been found in Austria.

Autapomorphies (Fig. 43 → 10)

- Opisthosoma of 11 segments.
- Subterminal position of the mouth.
  Migration of the oral slit to the tip of a small proboscis with upper and lower lips (Fig. 54C).

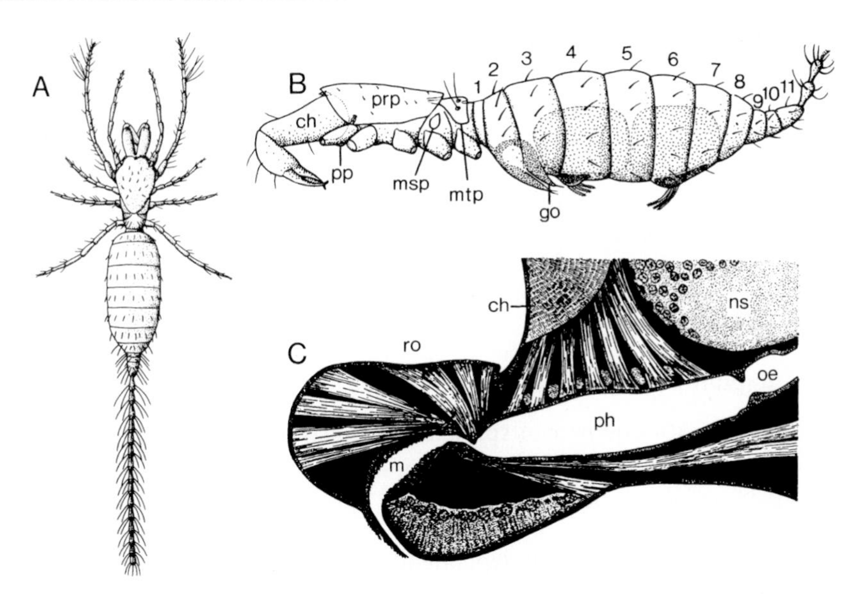

Fig. 54. Palpigradi. *Eukoenenia mirabilis.* **A** Female in dorsal view. **B** Female in lateral view. Division of the prosoma with pro-, meso- and metapeltidium. Opisthosoma of 11 segments. Only the two basal articles of the flagellum figured. **C** Rostrum (proboscis) in sagittal section. Subterminal position of the oral opening. *ch* Chelicera; *go* genital opening; *m* mouth; *msp* mesopeltidium; *mtp* metapeltidium; *ns* nervous system; *oe* esophagus; *ph* pharynx; *pp* pedipalp; *prp* propeltidium; *ro* rostrum. (**A, B** Moritz 1993b; **C** Millot 1949d)

- Tactile leg.
  The third prosoma appendage is lengthened; the number of articles increases distally (Fig. 54 A).
- Aciliar sperm.
  Coiled sperm are carried over from the ground pattern of the Lipoctena. The reduction of the cilium is new (ALBERTI 1979).
- Reductions due to miniaturization.
  As well as the absence of booklungs (ground pattern of the Apulmonata) endites on the prosoma appendages, eyes and Malpighian tubules are missing. Remnants of the circulatory system exist as a tubular heart.
- Coxal nephridia in the third walking leg segment reduced.

Despite far-reaching reductions a few original features in the organization of the Palpigradi remain. The following can be counted as **plesiomorphies:**

- Three-part chelate chelicerae.
- Subdivision of the opisthosoma into a large mesosoma and a metasoma of three segments.
- Multiarticulate flagellum as an appendage on the opisthosoma (ground pattern feature of the Lipoctena); within the Apulmonata only seen in the Palpigradi (Fig. 54 A).

The prosoma is subdivided into an anterior section of four segments with a uniform dorsal propeltidium and two free, mobile segments with the separated sclerites mesopeltidium and metapeltidium (Fig. 54 B). The interpretation of these structures is disputed. WEYGOLDT & PAULUS (1979) interpret the two free segments as the result of secondary division to optimize flexible locomotion in soil cracks and crevices. A corresponding behavior in the tiny Schizomida (Uropygi) must be seen as convergence.

Holotracheata

Autapomorphies (Fig. 43 → 11)

- Tracheae with one pair of spiracles on opisthosoma segment 3.
- Reduction of the flagellum on the opisthosoma.
- Maximum of two pairs of lateral eyes.

Tracheae evolved twice independently in the Arachnida, once within the Araneae in the stem lineage of the Araneoclada and once in the Holotracheata. This argument is the conclusive result of already established kinship hypotheses. Neither the Holotracheata as a whole nor any of its subtaxa can be interpreted as the sister group of an Araneae unity with tracheae.

For the interpretation of the Holotracheata tracheae we must turn again to the principle of parsimony. We postulate a single evolution of tracheae from opisthosoma appendages in the stem lineage of a unity Holotracheata. In addition, we place the simplest state in the **ground pattern of the Holotracheata.** This is **tracheae with one pair of spiracles**, as realized in the Opiliones, the Ricinulei and parts of the Acari. The spiracles probably belong primarily in the third opisthosoma segment. Increases and changes in position have led to far-ranging modifications of this pattern.

Haplocnemata – ? Cryptoperculata

Haplocnemata

Due to the commonly derived construction of their chelicerae Pseudoscorpiones and Solifugae are established as sister groups.

Autapomorphies (Fig. 43 → 12)

- Chelate chelicerae with two articles. The movable finger is inserted ventrally.
- Paired spiracles on opisthosoma segments 3 and 4 (Fig. 55 G). A doubling in comparison with the ground pattern of the Holotracheata.

Pseudoscorpiones – Solifugae

Pseudoscorpiones

With a body size decreased to less than 1 cm the Pseudoscorpiones are basically limited to crevices in soil with high humidity. However, a wide range of varying land biotopes have been colonized down to the tidal zones of marine beaches. Among the ca. 30 species found in Germany, there are a number of so-called synanthropic species. As predators the Pseudoscorpiones can even find suitable prey in dusty libraries, which has given the cosmopolitan species *Chelifer cancroides* the charming name of the book scorpion.

Restructuring of the pedipalps to a pincer-like grasping organ in the stem lineage of the Pseudoscorpiones has led to perfect copies of the pedipalps of the Scorpiones. From these the name of the taxa is derived. This is a classic example of a far-reaching congruence between two taxa, in which we have to presume, a priori, a synapomorphy; a posteriori, however, we are forced to the assumption of convergence with regard to the kinship ties of the Pseudoscorpiones which place them in a subgroup of the Apulmonata.

Autapomorphies (Fig. 43 → 13)

- Chelate pedipalp with poison glands.
  In comparison with the Scorpiones the Pseudoscorpiones have gone a step further in the evolution of the pedipalp (Fig. 55 A, C). They have evolved poison glands in the terminal part of the pedipalp. The glands open primarily on the mobile finger, less often additionally (or only) on the fixed finger process.
- Chelicerae with silk glands and serrulae (Fig. 55 E, F).
  On the ventral, mobile finger of the two-part chelicerae silk glands open on a special silk protuberance (galea). The glands lie in the prosoma, but can extend far into the opisthosoma. The silk is used to line small nests in the ground (molting, brooding, hibernation).
  The fingers of the chelicerae possess comb-like cleaning structures (exterior and interior serrula).
- Pretarsus of the walking leg with arolium.
  As well as two claws a tiny pretarsus carries a special clasper (arolium). Using this the Pseudoscorpiones can climb on smooth surfaces (Fig. 55 D).
- Absence of median eyes.
  The assumption of a single reduction in the stem lineage of the Pseudoscorpiones offers the simplest explanation.
- Absence of Malpighian tubules.

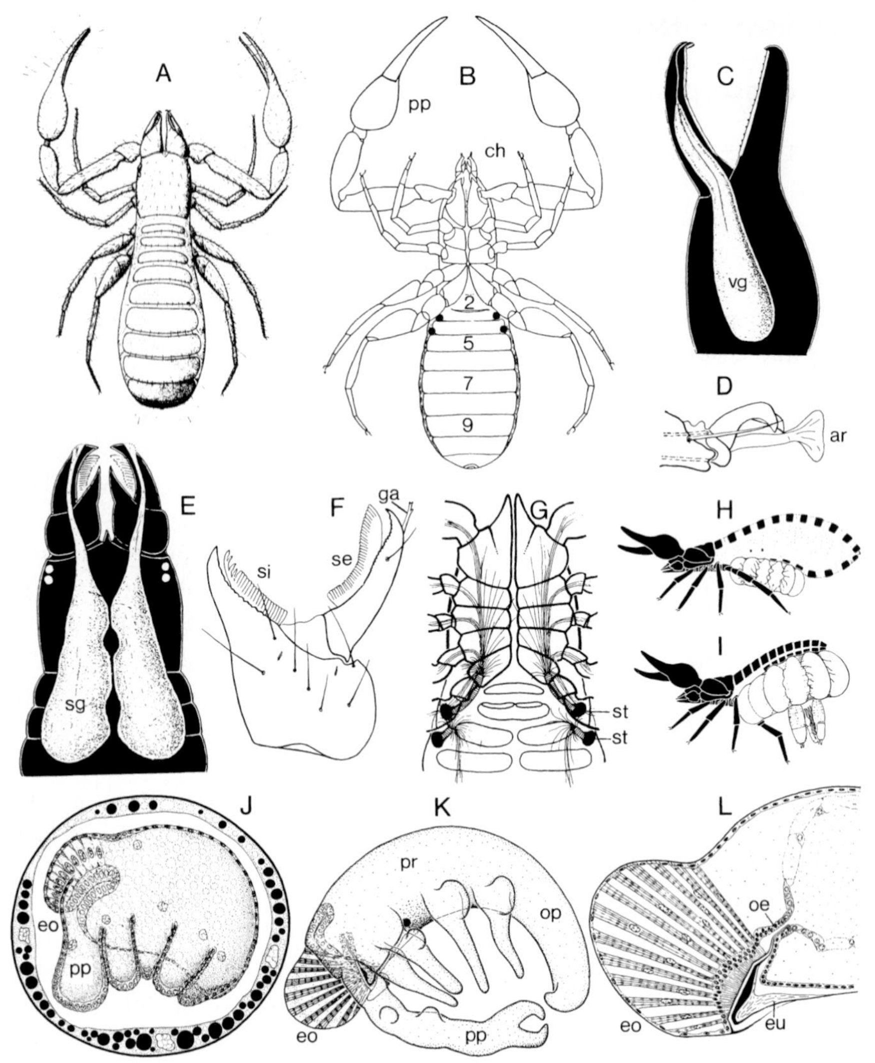

Fig. 55. Pseudoscorpiones. **A** *Neobisium muscorum*. Dorsal view. **B** *Hesperolpium sleveni*. Ventral view. Spiracles black points. **C** *Cordylochernes macrochelatus*. Poison gland with opening on the mobile finger of the pedipalp. **D** *Hesperolpium sleveni*. End of a walking leg with arolium. **E** *Neobisium simoni*. Silk glands in the prosoma and foremost part of the opisthosoma. Opening in the chelicerae. **F** *Microcreagris sequoiae*. Two-part chelicera from the exterior. The silk glands open at the galea on the mobile finger. **G** *Chernes cimicoides*. Tracheal system with two pairs of spiracles on opisthosoma segments 3 and 4. **H–I** *Pselaphochernes scorpioides*. Brood care. Lateral view of female. **H** Brood-sac with newly molted embryos. **I** Protonymph hatches out of the brood-sac. **J–L** Development. **J** *Pselaphochernes scorpioides*. Early embryo in embryonic envelope. **K** *Serianus carolinensis*. Second embryonic stadium with pump organ. **L** *Neobisium muscorum*. Embryonic pump organ in sagittal section. *ar* Arolium; *ch* chelicera; *eo* embryonic upper lip of the pump organ; *eu* embryonic lower lip; *ga* galea; *oe* esophagus; *op* opisthosoma; *pp* pedipalp; *pr* prosoma; *se* serrula exterior; *sg* silk gland; *si* serrula interior; *st* spiracle; *vg* poison gland. (**A, G** Weygoldt 1966; **B–F** Vachon 1949; **H–J** Weygoldt 1964; **K** Weygoldt 1971; **L** Weygoldt 1965)

- Brooding, embryonic envelope and larval pump organ (Fig. 55 H–L).
  The female does not lay her eggs but carries them on the ventral side of the opisthosoma. The eggs are either deposited in a liquid drop (? primary behavior in the Chthoniidae) or distributed in a brood-sac formed from gland secretions of the genital atrium.
  The female secretes nutritional fluid into the brood-sac, which the embryo takes up in two phases. First of all an embryonic envelope of blastomeres collects the fluid; the envelope is later sucked up by the embryo. In the first embryonic stage a special larval organ forms – a pump organ with a strong muscular upper lip for the uptake of nutrients. When the protonymph is ready to hatch the larval organ is broken down.
- Coxal nephridia in the first walking leg pair segment reduced.

The Pseudoscorpiones carry over the original number of 12 opisthosoma segments (excluding the caudal spine) from the ground pattern of the Chelicerata. The 12th segment is, however, only a tiny anal cone.

The Pseudoscorpiones are more original than the adelphotaxon Solifugae in the deposition of spermatophores and their indirect uptake by the female.

The paired spiracles in opisthosoma segments 3 and 4 come from the ground pattern of the Haplocnemata. Short, sturdy tracheal stems are attached to the spiracles. These then split up into numerous, unbranched capillary tubes (Fig. 55 G).

Two pairs of lateral eyes come from the ground pattern of the Holotracheata. Reduction within the Pseudoscorpiones has led to species with only one pair of eyes and to eyeless species.

Solifugae

Striking aspects of the Solifugae are the large chelicerae, the barrel-shaped opisthosoma, and prosoma appendages thickly covered with long tactile setae and spines. With an average body size of more than several cm the Solifugae are by far the largest of the Apulmonata.

The Solifugae are predominantly nocturnal animals of the steppes and deserts of warm climates; mostly they "flee from the sun". In southern Europe several wind scorpions are to be found from, amongst others, the taxa *Galeodes* and *Gluviana*.

Autapomorphies (Fig. 43 → 14)

- Mobile prosoma.
  The division of the prosoma into an anterior part with a uniform propeltidium and two free segments is the same as that of the Palpigradi. Since the sister group of the Solifugae, the Pseudoscorpiones, have an

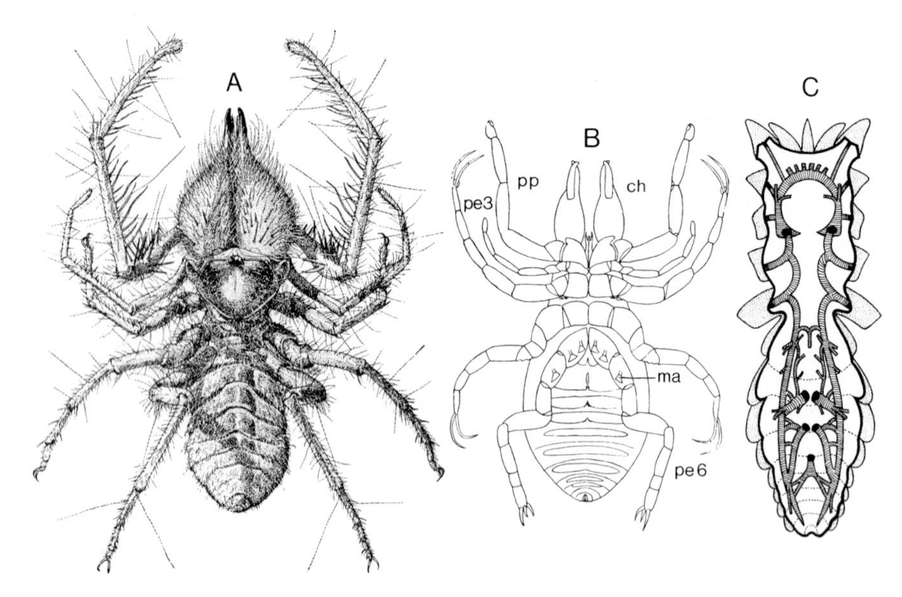

Fig. 56. Solifugae. **A** *Galeodes arabs*. Habitus in dorsal view. **B** *Mossamedessa abnormis*. Ventral view of female. Malleoli as sensory plates on the fourth walking leg pair. **C** *Galeodes*. Tracheal system with longitudinal stems. Spiracles (*black*): Prosoma = one pair between second and third walking leg pairs. Opisthosoma: One pair in the third and another in the fourth segment; one unpaired spiracle in the fifth segment. *ch* Chelicera; *ma* malleolus; *pe* prosoma appendage; *pp* pedipalp. (**A, B** Millot & Vachon 1949b; **C** Hennig 1994)

undivided prosoma an independent development of the phenomenon in the Palpigradi and the Solifugae must be postulated.

- Opisthosoma of 11 segments.
- Pedipalp with clasper (Fig. 56 A, B).

  The original leg-like pedipalp has on its tip an adhesive organ with a dual function – for climbing and sticking to smooth surfaces, and for capturing small prey (insects).
- Tracheal system with longitudinal stems and additional spiracles.

  Two tracheal longitudinal stems that are connected to each other traverse the body. The longitudinal stems have numerous branches. These can fan out and penetrate into the appendages. To the primary paired spiracles in opisthosoma segments 3 and 4 (ground pattern Haplocnemata) the following new structures are added: an unpaired spiracle in the fifth opisthosoma segment and a pair of spiracles in the prosoma between the second and third walking leg pair (Fig. 56 C).
- Malleolus (racquet organ).

  Stalked, racquet-shaped sensory plates ventral on the proximal segment of the last walking leg. Probably a chemoreceptor (Fig. 56 B).

- Rostrum.
  Mouth at the tip of a projection under the chelicerae. Convergence to the Palpigradi.
- Direct transfer of spermatophores.
  1. Directly from the male into the female genital opening or
  2. Picked by the chelicerae of the male and inserted by them into the female pore. It is unclear which belongs in the ground pattern of the Solifugae.
- Sperm without cilium.
- Coxal nephridia in the third walking leg segment reduced.

In comparison with the adelphotaxon Pseudoscorpiones, the Solifugae remain more original in the manifestation of simple leg-like pedipalps and in possessing median eyes.

The decrease in the number of lateral eyes (anyway only weakly developed) to a maximum of two pairs corresponds to the sister group Pseudoscorpiones. This state could have been taken over from the stem species of the Holotracheata.

? Cryptoperculata

The monophyly of a possible kinship group consisting of the taxa Opiliones, Ricinulei and Acari is not yet proven (? ■ 15 in Fig. 43). In the first place WEYGOLDT & PAULUS (1979) put forward the evolution of the second walking leg pair into a tactile leg as an autapomorphy for a unity Cryptoperculata.

In the Ricinulei the facts are clear. The second walking leg pair is elongated and furnished with a five-part tarsus. The Ricinulei feel with this leg pair, while they move with appendages 1, 3 and 4.

Similarly, in the Opiliones, the second leg pair is generally the longest appendage. It remains questionable, however, if this state can be postulated for the ground pattern of the Opiliones. In the short-legged mite-like Sironidae (*Siro*) the second leg pair is not prominent; on the contrary, the longest here is the first leg pair. The tarsi of the walking legs are usually undivided (MARTENS 1978), and this is probably a plesiomorphic state (see below).

Likewise, I do not find any convincing argument for interpreting a tactile organ developed from the second walking leg pair to be a feature of the ground pattern of the Acari.

The previously legitimate suspicion that aciliar sperm is a further autapomorphy of the Cryptoperculata has become untenable. The Ricinulei have sperm with a cilium (ALBERTI 1995).

Until the unanswered problems have been resolved, the systematization of WEYGOLDT & PAULUS (1979) will be used.

Opiliones – Acarinomorpha

Opiliones

Long-legged creatures with a cone-like body a few mm in length that climb elegantly about the tips of vegetation or in the meshwork of bushes. Daddy-long-legs or harvestmen are the embodiment of the Opiliones. A unique grasping organ is formed from the division of the tarsus (Fig. 57 A). Members of the Phalangiidae such as *Phalangium opilio* or *Opilio parietinus* have up to 50 tarsal segments. A continuous claw tendon is attached to a claw flexor in the tibia. By contraction of the flexor the tarsus can wind lasso-like around thin objects such as grass or small twigs.

Long legs with a multiarticulate tarsus do not, however, belong in the ground pattern of the Opiliones; they evolved first within the Opiliones. In the primary short-legged, soil-dwelling *Siro* species that resemble mites, tarsus and metatarsus each have one article (Fig. 57 D). The taxon *Trogulus* with flat species that are also soil dwelling, has two tarsal articles on both front walking leg pairs, and three articles on the rear walking legs (Fig. 57 C).

Autapomorphies (Fig. 43 → 16)

- Prosoma and opisthosoma broadly fused. Opisthosoma of ten segments.
  The result of fusion of the prosoma and opisthosoma is an egg-shaped to round body. Of the ten embryonic opisthosoma segments the pregenital segment disappears during development.
- Displacement of the genital opening rostralwards to a position between the coxae of the prosoma appendages (Fig. 57 E).
- Penis and ovipositor (Fig. 57 F, G).
  The genital copulatory organ and the oviposition tube are likewise formations of the ectodermal genital atrium penetrating deep into the body. In both sexes a distally directed cone grows from the basal bulge of the atrium.
  The penis – a chitin tube with distal glans – is inserted during copulation through the female genital opening into the oviposition tube; the sperm reach the seminal receptacle through the tube.
  The ovipositor, of variable construction (MARTENS et al. 1981), remains in the body interior during copulation. To lay eggs it is extruded by an increase in hemolymph pressure. The eggs are deposited by the tube into crevices in the ground.
- Median eyes on a common eye tubercle.
  The possession of paired median eyes is original; their position on a tuber oculorum medially on the prosoma is, in contrast, a derived feature (Fig. 57 B).

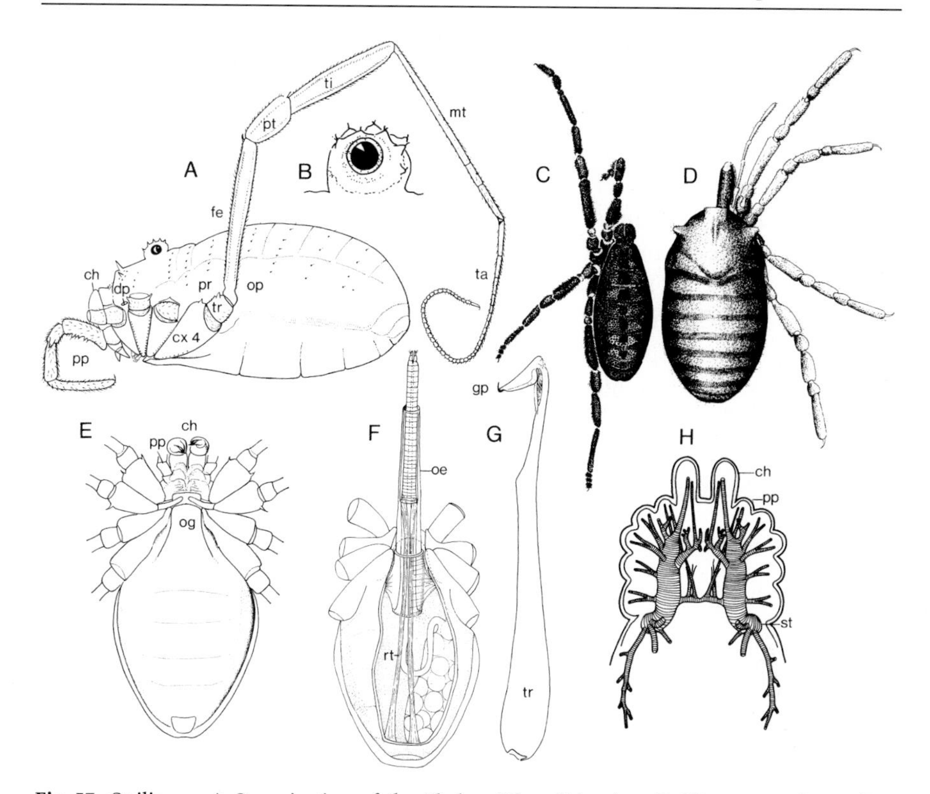

Fig. 57. Opiliones. **A** Organization of the Phalangiidae. Side view. **B** *Mitopus morio*. Median eye on eye tubercle. **C** *Trogulus nepaeformis*. **D** *Siro duricorius*. **E** Organization of the Phalangiidae. Ventral view. A prominent genital operculum (new formation) covers the genital opening, but does not belong, however, in the ground pattern of the Opiliones. **F** *Phalangium opilio*. Ovipositor protruded. **G** *Phalangium opilio*. Lateral view of penis. Division into truncus and glans. **H** *Opilio parietinus*. Main branch of the tracheal system with paired spiracles in the third opisthosoma segment. *ch* Chelicera; *cx* coxa; *dp* pore of the defense gland; *fe* femur; *gp* glans penis; *mt* metatarsus; *oe* ovipositor envelope; *og* operculum genitale; *op* opisthosoma; *pp* pedipalp; *pr* prosoma; *pt* patella; *rt* retractor muscle; *st* spiracle; *ta* tarsus; *tc* truncus; *ti* tibia; *tr* trochanter. (**A–E, G** Martens 1978; **F** Martens et al. 1981; **H** Moritz 1993b)

- Lateral eyes absent.

  Using the principle of parsimony, a single reduction in the stem lineage of the Opiliones is postulated.
- Repugnatorial glands in the prosoma.

  Paired glands which produce a defense secretion are developed in all Opiliones. The openings of the defense glands lie mainly dorsalwards on the prosoma. They can, however be shifted laterally to under the edge of the prosoma.
- Absence of Malpighian tubules.
- Absence of coxal nephridia in the first walking leg segment.

Two median eyes and three-part chelicerae with chelae are original features of the Opiliones. Simple, leg-like pedipalps with a tactile function also belong in the ground pattern. In the Laniatores the pedipalps evolved into predatory legs with which prey is held fast.

Finally, the existence of a pair of spiracles in the morphological third opisthosoma segment must be seen as a plesiomorphy, likewise the structure of gaseous exchange organs as tubular tracheae (Fig. 57H).

According to MARTENS et al. (1981), the **Cyphopalpatores** (here, all the taxa mentioned in the above text) and the **Laniatores**, distributed mainly in the tropics and subtropics, form the highest-ranking sister groups of the Opiliones. The argument based on the structure of the ovipositor cannot be followed further here.

Acarinomorpha

Autapomorphies (Fig. 43 → 17)

- Larva with six legs.
- Three nymphs.

The first freely mobile stage is a **larva with three leg pairs**. Three nymphs with four leg pairs follow – the proto-, deuto- and tritonymph. The molting of the tritonymph results in the adult.

Ricinulei – Acari

Ricinulei

Species-poor unity of small, strongly sclerotized soil dwellers in the tropics. Body length at most 6 mm. *Ricinoides* in Africa, *Cryptocellus* and *Pseudocellus* in America.

The tick spiders are characterized by a series of marked features. We place the apomorphy associated with their name first.

Autapomorphies (Fig. 43 → 18)

- Cucullus (hood).
  Division of a hinged flap at the anterior of the prosoma that covers the chelicerae when lowered.
- Opisthosoma of ten segments with four large elements in the middle.
  Only four opisthosoma segments (4–7) with tergite and sternite are visible at the exterior. Segments 1–3 are underneath the prosoma and hidden between the coxae of the fourth walking leg pair. Segments 8–10 are telescoped into segment 7 (Fig. 58A, E). A broad coupling binds the prosoma with the fourth opisthosoma segment.

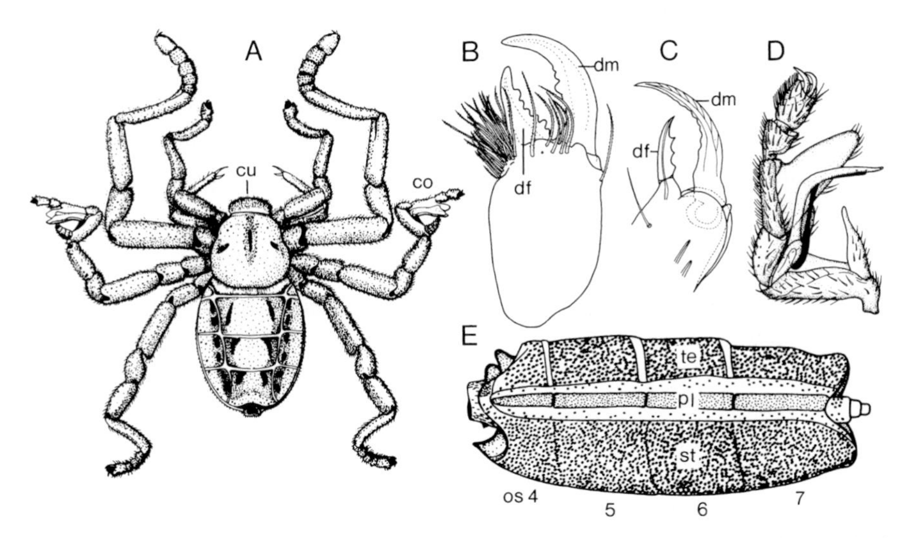

Fig. 58. Ricinulei. **A** *Cryptocellus isthmus.* Male in dorsal view. Cucullus on the front end of the prosoma. Copulatory organs on the third walking leg pair. **B** *Pseudocellus.* Two-part chelicera (pincers) with a mobile finger. **C** *Ricinoides hanseni.* Tip of the chelate pedipalp. **D** *Pseudocellus paradoxus.* Copulatory organ on the metatarsus and tarsi 1 and 2 of the third walking leg. **E** *Ricinoides hanseni.* Male. Opisthosoma in side view. Four large segments with tergites, pleurites and sternites. *co* Copulatory organ; *cu* cucullus; *df* digitus fixus; *dm* digitus mobilis; *os* opisthosoma segment; *pl* pleurite; *st* sternite; *te* tergite. (Moritz 1993b)

- Two-part chelate chelicerae.
  Established as a derived state in the tick spiders in comparison with the three-part chelicerae in the ground pattern of the sister group Acari (Fig. 58B).
- Chelate pedipalp.
  The tarsus works as a movable finger against a fixed process of the tibia (Fig. 58C).
- Second walking leg pair lengthened to form a tactile organ.
  In as much as this feature does not belong in the ground pattern of the Cryptoperculata (p. 127), it must be seen as an autapomorphy of the Ricinulei.
- Copulatory organ on the third walking leg pair.
  Through a complicated sperm storage and transfer apparatus on the metatarsus and two subsequent tarsal segments, the third walking legs of the male are transformed into gonopods (Fig. 58A, D). Sperm is taken from the genital opening and transferred directly into the female.
- Sieve tracheae with spiracles on the prosoma.
  A tracheal system with one pair of spiracles is probably carried over from the ground pattern of the Holotracheata. Apomorphous states, however, are (1) the position of the spiracles on the posterior of the pro-

soma, (2) their opening into an atrium of the prosoma, and (3) tracheal capillaries that originate in great numbers from the atrium.
- Absence of eyes.
- Absence of coxal nephridia on the first walking leg segment.

In having ciliar sperm the Ricinulei remain more original than the Acari.

Acari

With 35,000 known species only a few mm in length the Acari form the second largest unity of the Arachnida besides the Araneae; however, the spectrum of biotopes colonized by the Acari is much larger than that of the spiders.

The mites are primary free-living inhabitants of terrestrial soils. From the land various evolutionary lineages have led to fresh water ("Hydrachnellida") and to the sea (Halacarida). Independently repeated plant and animal parasites evolved – in the latter also regular inhabitants on and in man. The harmless hair follicle mite *Demodex folliculorum*, the annoying scabies mite *Sarcoptes scabiei* in the skin, and the blood sucker *Ixodes ricinus*, carrier of diseases (lyme disease, encephalitis), are just a few.

In contrast to the Araneae no phylogenetic system for the Acari exists as yet. I therefore limit myself to the justification of the Acari as a monophylum by means of their autapomorphies and the listing of some of the plesiomorphies in the ground pattern.

Autapomorphies (Fig. 43 → 19)

- Broad fusion of prosoma and opisthosoma.
- Body division into gnathosoma and idiosoma.
  The basic characteristic feature of the Acari is the particular separation of a rostral gnathosoma with chelicerae and pedipalps (Fig. 59 A).
  The tectum pictured in the schemes as a dorsal covering of the gnathosoma obviously does not yet belong in the ground pattern of the Acari. In the Opilioacarida the chelicerae without any dorsal covering are visible (Fig. 59 B).
- Aciliar sperm.
  A derived peculiarity in comparison with the sister group Ricinulei, for which sperm with cilia have recently been proven.
- Coxal nephridia in the third walking leg segment reduced.

The following features can be placed as **plesiomorphies** in the ground pattern of the Acari.
- Opisthosoma of 12 segments.
  This state is only realized in the Opilioacarida (Notostigmata). A decrease in the number of segments followed within the Acari.

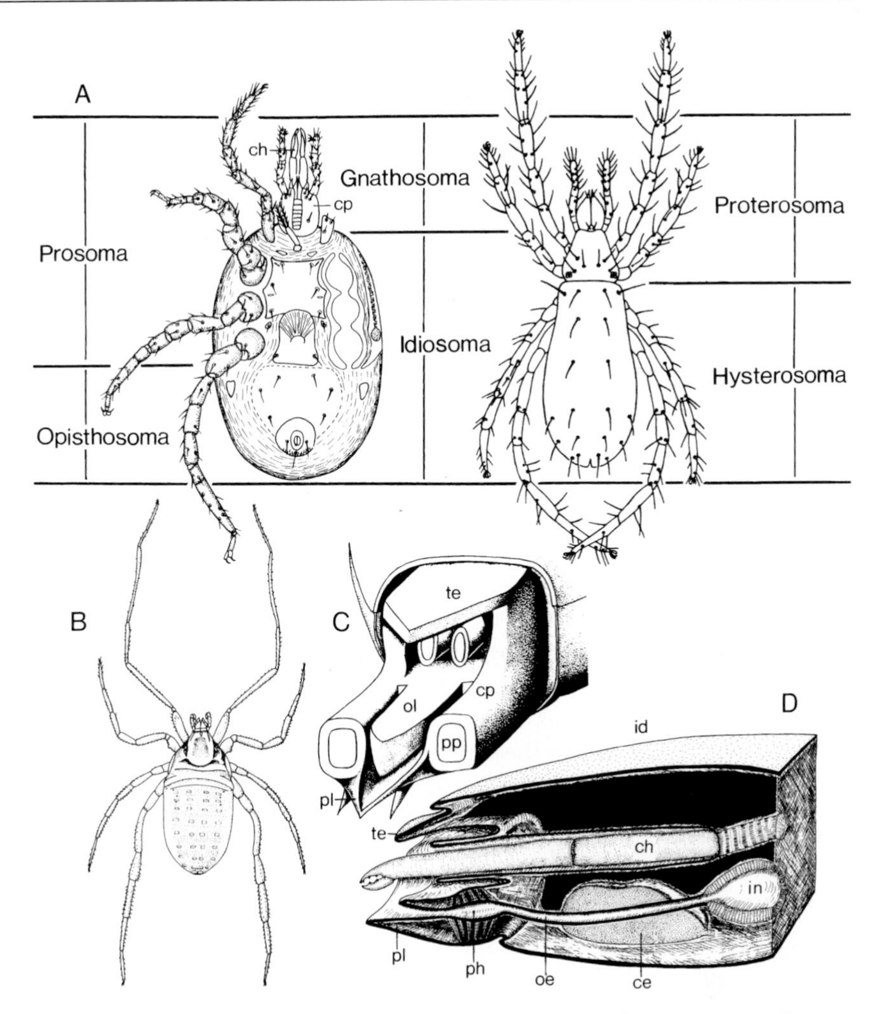

Fig. 59. Acari. **A** Two examples of the division of the mite body. *Left* Gamasina. Female, ventral view (Anactinotrichida). *Right Rhagidia gelida* (Actinedida, Actinotrichida). The ground pattern of the Chelicerata with prosoma and opisthosoma is overlain by a subdivision of the anterior section with chelicerae and pedipalps as a gnathosoma from the rest of the body, which is then named the idiosoma. In the Actinotrichida a transverse furrow is also found between the second and third walking leg pairs, which leads to a division of the body into a proterosoma and hysterosoma. As a result the insertion of the walking leg pairs can move apart from one another so that a broad gap between the two anterior and the two posterior walking leg pairs can occur. **B** *Opilioacarus segmentatus* (Opilioacarida, Anactinotrichida). Original formation of the opisthosoma of 12 segments and a gnathosoma, in which the chelicerae are not covered dorsally. **C-D** Gnathosoma as an autapomorphy of the Acari. **C** Diagram in oblique view from the front. **D** Sagittal section (Anactinotrichida). A dorsal tectum (subdivision of the prosoma) fuses laterally with the coxae of the pedipalps. These fuse ventrally with each other through blade-like offshoots. A closed futteral results in which the chelicerae lie. *ce* Cerebrum; *ch* chelicera; *cp* coxa of the pedipalp; *id* idiosoma; *in* intestine; *oe* esophagus; *ol* upper lip; *ph* pharynx; *pl* coxal offshoot of the pedipalp; *pp* pedipalp; *te* tectum. (**A** Karg 1971, Moritz 1993b; **B,C** Moritz 1993b; **D** Eisenbeis & Wichard 1985)

- One pair of median eyes and three lateral eyes.
  Light-sensory organs are discontinuously distributed within the Acari. *Paracarus hexophthalmus* (Opilioacarida) is the only mite species with three pairs of lateral eyes (WEYGOLDT 1996). In the rest one or two pairs are common.
- One pair of spiracles.
  Common within the Acari. However, their position in the region of the coxae of the third and fourth walking leg pairs is apomorphous in contrast to the opisthosomal position in the ground pattern of the Holotracheata (p. 122).
  In this interpretation the four pairs of spiracles in the Opilioacarida (opisthosoma segments 9–12) must be seen as the result of a secondary increase.
- Indirect spermatophore transfer (?).
  As well as direct transfer of spermatophores using a genital copulatory organ, or through use of the chelicerae, an indirect transfer of spermatophores exists in mites. This can take place with or without partner contact as in the fresh-water *Arrenurus globator* or in the Oribatida (beetle mites). So long as there is no phylogenetic system for the mites it must remain open as to whether an indirect transfer of spermatophores in some of the Acari is carried over from the ground pattern of the Arachnida or has evolved anew within the Acari.

Mandibulata

Autapomorphies (Fig. 36 → 5)

- Second antennae.
  Sensory organs, differentiation products of the second head segment appendages.
- Mandibles.
  Mouthparts from the third head segment appendages. The elements from which the Mandibulata derive their name.
  The mandibles bear a sturdy endite on the protopodites; these blades serve to manipulate food.
- First maxillae.
  Mouthparts from the fourth head segment appendages.
- Second maxillae (?).
  Mouthparts from the fifth head segment appendages. Whether these originated once in the stem lineage of the Mandibulata or convergently in the Crustacea and Tracheata is disputed.

- Crystalline cone.
  New element in the ommatidia. Formed from four crystalline cone cells (Fig. 39 B).
- Cornea and rhabdome.
  Corneal lens secreted by two corneagenous cells. Rhabdome formed from relatively few retinula cells (ground pattern probably eight) (Fig. 39 B).
- Molt gland.
  Production of a molt hormone in a gland which develops embryonically from ectodermal cells of the second maxillary segment. In the Crustacea the molt gland is known as the Y-organ and in the Tracheata as the prothorax gland (WÄGELE 1993).

Crustacea – Tracheata

Crustacea and Tracheata are adelphotaxa. This is a common hypothesis that I accept and will explain in detail. The hypothesis has been questioned due to ribosomal DNA analysis (FRIEDRICH & TAUTZ 1995) and special congruencies in the compound eyes and nervous system of the Crustacea and Insecta (NILSSON & OSARIO 1997; WHITINGTON & BACON 1997; DOHLE 1997).

Crustacea

Crustacea are primary aquatic organisms, i.e., they have retained the aquatic habitat of the stem species of the Mandibulata. It is understandable that in correlation with their aquatic mode of life many features of Crustacea organization were carried over as plesiomorphies from the ground pattern of the Mandibulata. Possibly the Crustacea form a paraphyletic collection of relatively plesiomorphic Mandibulata compared with the terrestrial Tracheata (MOURA & CHRISTOFFERSEN 1996).

However, the following are two features that can undoubtedly be interpreted as derived characteristics of the Crustacea (LAUTERBACH 1983 a; WÄGELE 1993).

As a further autapomorphy WALOSSEK & MÜLLER (1997) suggest the nauplius larva with three pairs of appendages that may have originated in the stem lineage of the Crustacea from a head larva with four appendage pairs.

Autapomorphies (Figs. 36→6; 60 → 1)

- Nauplius eye.
  For the ground pattern of the Euarthropoda four isolated median eyes on the head were hypothesized. In the Crustacea the four pigment-cup

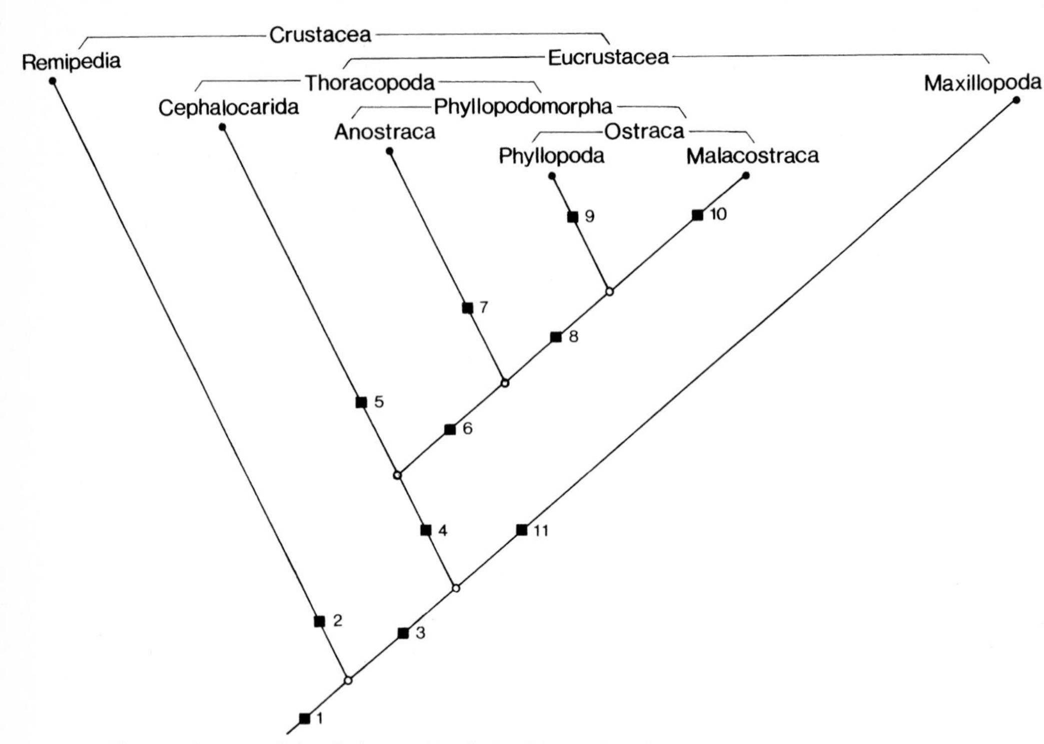

Fig. 60. Diagram of the phylogenetic relationships within the Crustacea

ocelli have moved together. They form a single, external uniform nauplius eye.

- Antennal and maxillary nephridia.
  Of the six pairs of segmental nephridial organs in the ground pattern of the Euarthropoda (p. 81), only two pairs exist in the Crustacea: on the second antennal and second maxillary segments. The process of reduction of these excretory structures to one pair of antennal nephridia and one pair of maxillary nephridia (formerly known as antennae and maxillae glands) is hypothesized to have occurred in the stem lineage of the Crustacea.
- (?) Nauplius larva with antenna 1, antenna 2 and mandible. Mouth within an atrium oris covered by the labrum.

Further ground pattern features that are useful diagnostically in making a distinction from the sister group Tracheata are not of any use in defining the Crustacea as a monophylum since they were already present in the stem species of the Mandibulata and could have originated even further back in the stem lineage of the Euarthropoda.

Such plesiomorphies are the existence of the second antennae, the branched appendages with endo- and exopodites, the manifestation of midgut glands, and the telson with posterior furca. They all undergo modifications in the stem lineage of the Tracheata and in some cases can be totally reduced.

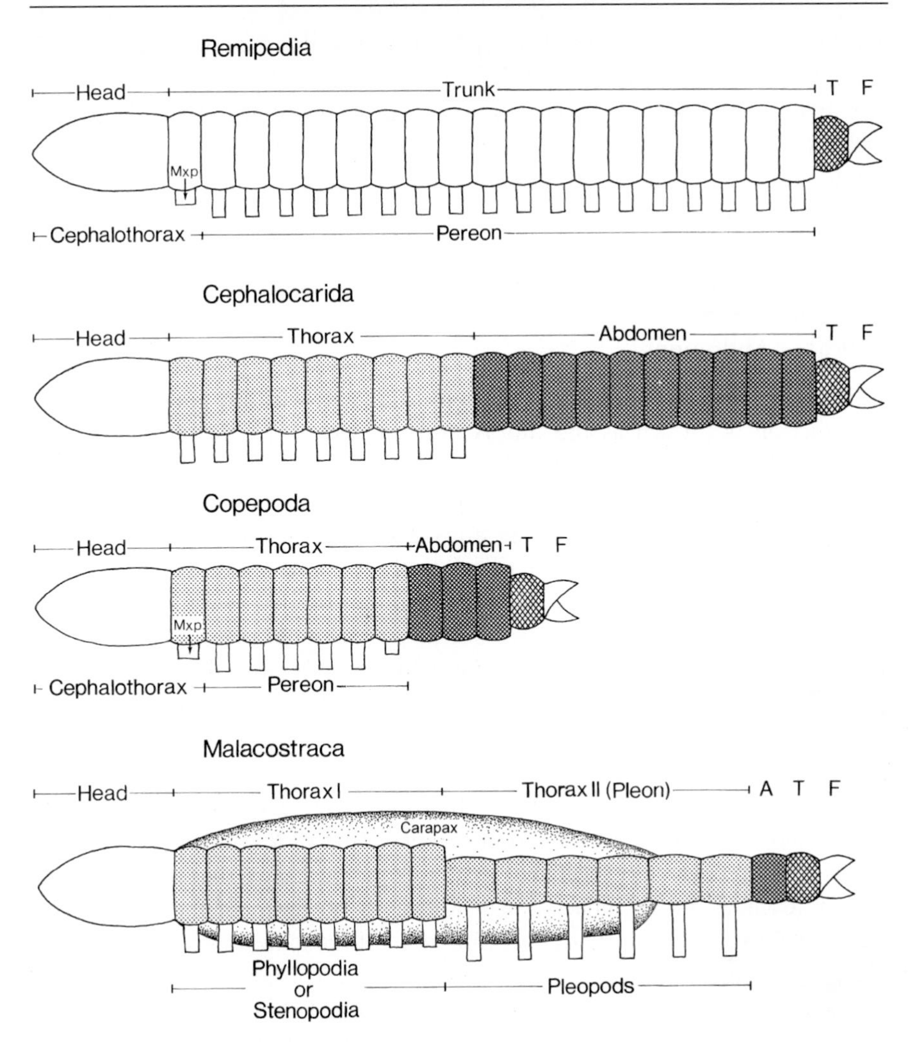

Fig. 61. Body division in some taxa of the Crustacea. Explanations in the text. *A* abdomen; *F* furca; *Mxp* maxilliped; *T* telson

Evolutionary Interpretation and Terminology of the Body Division

The terminology concerning the division of the Crustacea body is not uniform and can lead to misunderstandings about the evolutionary worth of single sections when comparing different taxa. We use the major terms with the following meanings (Fig. 61).

Trunk

When all postcephalic segments are uniformly equipped with appendages, then we have a trunk without subdivision. There is always a terminal telson, without the character of a segment and without appendages, which bears the furca in the form of a pair of processes.

A trunk without any differentiation – as is postulated for the ground pattern – is not seen in any recent Crustacea, although the Remipedia come closest to this state. Here only the first trunk leg pair is transformed into maxillipeds (see below) and is functionally connected with the mouthparts. All other trunk segments bear the same kind of swimming legs with endo- and exopodites; we judge this to be a plesiomorphic state.

Thorax – Abdomen

When the trunk is divided into sections with and without appendages, the first section is known as the thorax and the second as the abdomen. The term trunk remains to describe the two sections together.

The division thorax – abdomen applies to the majority of the Crustacea, which are grouped together in the unity Eucrustacea. We postulate a single evolution of the basic differentiation into thorax and abdomen in the stem lineage of the Eucrustacea. We are then confronted with considerable changes in proportions of the two sections to one another in certain taxa. Despite this, in many unities, determinable homologous relations exist to precise numbers of segments. An example is the large monophylum Maxillopoda with seven thoracic and three abdominal segments in their ground pattern.

The Malacostraca merit special attention. They have 14 thoracic segments and an appendageless abdomen consisting of a single segment. The latter is moreover only realized in the Leptostraca; it disappears in the stem lineage of the Eumalacostraca. Above all however, in the Malacostraca an evolutionary differentiation into two different thoracic sections took place – a unique process within the Crustacea. Thorax section 1 of eight segments bears primary leaf-shaped phyllopodia (Leptostraca) and secondary rod-shaped stenopodia (Eumalacostraca). In the ground pattern, thorax section 2 has six pairs of plate-shaped swimming legs; the section itself is called the pleon and the appendages are known as pleopods. It is morphologically false, and correspondingly terminologically inconsistent, to call only the foremost eight segments a thorax and to contrast this with the pleon (sometimes even called the abdomen), i.e. to use the formulation: thorax – pleon (abdomen).

On the contrary, both sections are essential components of the Malacostraca thorax (p. 159). This interpretation corresponds to the concept of WALOSSEK & MÜLLER (1997).

Cephalothorax – Pereon

Occasionally, a trunk segment or a few trunk metameres become fused with the head. A cephalothorax results. In contrast, the rest of the trunk with appendages is terminologically known as the pereon.

Maxillipeds – Pereopods

The cephalothorax formation is a consequence of the transformation of anterior trunk legs into maxilla-like structures that together with the mouthparts are used in feeding. These structures are named maxillipeds, all subsequent unchanged appendages pereopods.

The process of the evolution of maxillipeds and the shaping of the anterior body to a cephalothorax is repeated convergently within the Crustacea. A pair of maxillipeds developed in the stem lineages of the Remipedia and the Copepoda; up to five maxillipeds evolved in the Malacostraca (Stomatopoda).

Carapace

We now come to the carapace of the Crustacea. A plethora of meanings hidden in this one term caused WALOSSEK (1993) to conclude with resignation that the use of the term carapace should be discontinued. This could, however, lead to a loss of possibilities in interpretation and communication – also, or especially, when it is clear that the structures defined as a carapace do not necessarily have to be homologous within the Crustacea.

Together with GRUNER (1993), I define under the term carapace a shield-like skeletal plate that originates from the posterior part of the head or the foremost section of the trunk and then more or less fully covers the trunk or, in the form of a bivalved shell, encloses the body and sometimes also the head.

Established kinship hypotheses force us to the assumption that a carapace so defined has evolved twice independently within the Crustacea as a bivalved shell – once within the Thoracopoda in the stem lineage of the Ostraca (Phyllopoda + Malacostraca) and once within the Maxillopoda in the stem lineage of the Thecostracomorpha (Ascothoracida + Cirripedia and Ostracoda). I will discuss the justification for this case by case.

Systematization

The monophyly of numerous subtaxa of the Crustacea to which the categories order, class, etc., are ascribed in traditional classifications is well established. We will demonstrate this in detail for the Remipedia, Cephalocarida, Anostraca, Phyllopoda, Malacostraca, Copepoda, Mystacocarida, Tantulocarida, Ascothoracida, Cirripedia, Ostracoda and Branchiura.

In contrast, there is fierce debate over the kinship relations of these unities to one another with a multitude of differing opinions. Data from 18S rDNA analysis "represent but a relic, and not the Holy Grail itself" (SPEARS & ABELE 1997). I see the possibility through strict usage of the principle of parsimony to specify a hierarchy of high-ranking Crustacea adelphotaxa kinships and lay them open to intersubjective examination (Figs. 60, 71, 80).

- **Crustacea**
  - **Remipedia**
  - **Eucrustacea**
    - **Thoracopoda**
      - **Cephalocarida**
      - **Phyllopodomorpha**
        - **Anostraca**
        - **Ostraca**
          - **Phyllopoda**
          - **Malacostraca**
            - **Leptostraca**
            - **Eumalacostraca**
              - **Stomatopoda**
              - **Caridoida**
                - **Decapoda**
                - **Xenommacarida**
    - **Maxillopoda**
      - **Copepodomorpha**
        - **Copepoda**
        - **Mystacocarida**
      - **Progonomorpha**
        - **Tantulocarida**
        - **Thecostracomorpha**
          - **Thecostraca**
            - **Ascothoracida**
            - **Cirripedia**
          - **Ostracoda**
          - **? Branchiura**
          - **(+ Pentastomida)**

Let us begin at the highest level with a presumed sister group relationship between the Remipedia and the Eucrustacea – a grouping of all other Crustacea (SCHRAM 1986). The **Remipedia** represent the only unity of the Crustacea that has a uniform trunk composed of numerous segments, which are regularly equipped from anterior to posterior with biramous appendages (excepted are the maxillipeds of the first trunk segment). In the uniform provision of the trunk segments with locomotory appendages there is agreement with the ground pattern of the sister group Tracheata (realized in the Myriapoda) as well as with the ground pattern of the Arthropoda in general (realized in the Onychophora). Consequently, a strict parsimonious assessment must place the state seen in the Remipedia as a plesiomorphy in the ground pattern of the Crustacea – in other words it must postulate not only appendages on all trunk segments, but has no other choice than to postulate that these locomotory appendages in the form of swimming legs belong in the ground pattern. This argument compels the hypothesis of primary nutrient uptake in the Crustacea using only the mouthparts – independent of the mode of feeding which is assumed in detail (SCHRAM 1986).

Please note that we are dealing here with the interpretation of one particular characteristic feature. Obviously, and as it should be, in the case of the Remipedia as well as in general it must be emphasized that the group does not thus automatically become in their total organization the "most primordial one of the Crustacea" (WALOSSEK 1993, p. 65). The group may be derived in many other respects – in their feeding as carnivores, in the formation of a cephalothorax, in the absence of eyes or hermaphroditism (p. 144). These do not affect the evidence concerning the primary state of the trunk at all.

We now come to the apomorphic alternative – the limiting of appendages to the anterior part of the trunk. The process of the division of the Crustacea trunk into a thorax with appendages and an abdomen without legs can be postulated for the stem lineage of the **Eucrustacea** and as a result becomes an autapomorphy of this unity.

Within the Eucrustacea the Thoracopoda (HESSLER & NEWMAN 1975; HESSLER 1992) and the Maxillopoda (DAHL 1956) form the highest-ranking adelphotaxa.

In the stem lineage of the **Thoracopoda** the swimming legs were transformed into broad turgor appendages. "Polyramous limbs" used for microphagous feeding with filter combs on the interior and epipodites on the exterior developed from "biramous limbs" (SCHRAM 1986). The broadening of the function of the primary locomotory appendages leads to "thoracophagy" that is basically identical with the formation of suction pressure-chambers between successive legs in the Cephalocarida, Anostraca, Phyllopoda and Malacostraca (Leptostraca). These filter small particles from the water flowing through them; the filtrate is then transported in a ventral furrow towards the head. The principle of parsimony leads to the hypothesis of a single evolution of a postcephalic filter-feeding apparatus in a stem lineage common to these unities. As a result the stem species of the Thoracopoda retained a comparatively high number of trunk segments

– at least as many as the 19 segments realized in the Cephalocarida and Anostraca.

For the two features discussed, the sister group **Maxillopoda** have the following alternatives. The Maxillopoda retain the biramous swimming legs as a plesiomorphy from the ground pattern of the Crustacea. In their stem lineage there is then an extreme shortening of the trunk to a constant ten segments + telson with furca. The thorax consists of seven segments in the ground pattern of the Maxillopoda of which six are equipped with locomotory thoracopods. In contrast, the legs of the seventh thoracic metamere are transformed into copulatory organs. The abdomen has three segments without appendages. This segmentation pattern forms the essential autapomorphy of the monophylum Maxillopoda, in which the Copepoda, Mystacocarida, Tantulocarida, Ascothoracida, Cirripedia and Ostracoda can assuredly be grouped; uncertainty still surrounds the position of the Branchiura.

The resurrection of the Entomostraca as sister group of the Malacostraca (WALOSSEK & MÜLLER 1997) is not compatible with the systematization of the Crustacea established here.

Remipedia – Eucrustacea

Remipedia

The Remipedia are stygian organisms that inhabit underground cave passageways connected to the sea.

Speleonectes, Lasionectes.

Only a few species are known from the Bahamas, the Yucatan Peninsula and the Canary Islands. The body can reach a length of 4.5 cm and the trunk can have up to 38 segments. The first segment is fused with the head to form a cephalothorax; its appendages are transformed into uniramous maxillipeds. The cephalothorax is covered by a flat cephalic shield. The other trunk segments have pleurotergites and bear uniform lateral swimming legs with four-part endopodites and three-part exopodites (Fig. 62 A, C). To the anterior and posterior the appendages are smaller. The telson with furca follows the trunk (Fig. 62 D).

We interpret the uniform trunk with appendages on all segments as a plesiomorphous state. Within the Crustacea the construction of the first antennae in the form of two multiarticulate branches is unique (Fig. 62 F). The interpretation of these is disputed. Otherwise the Remipedia are distinguished by numerous derived characteristic features which we list in the following overview.

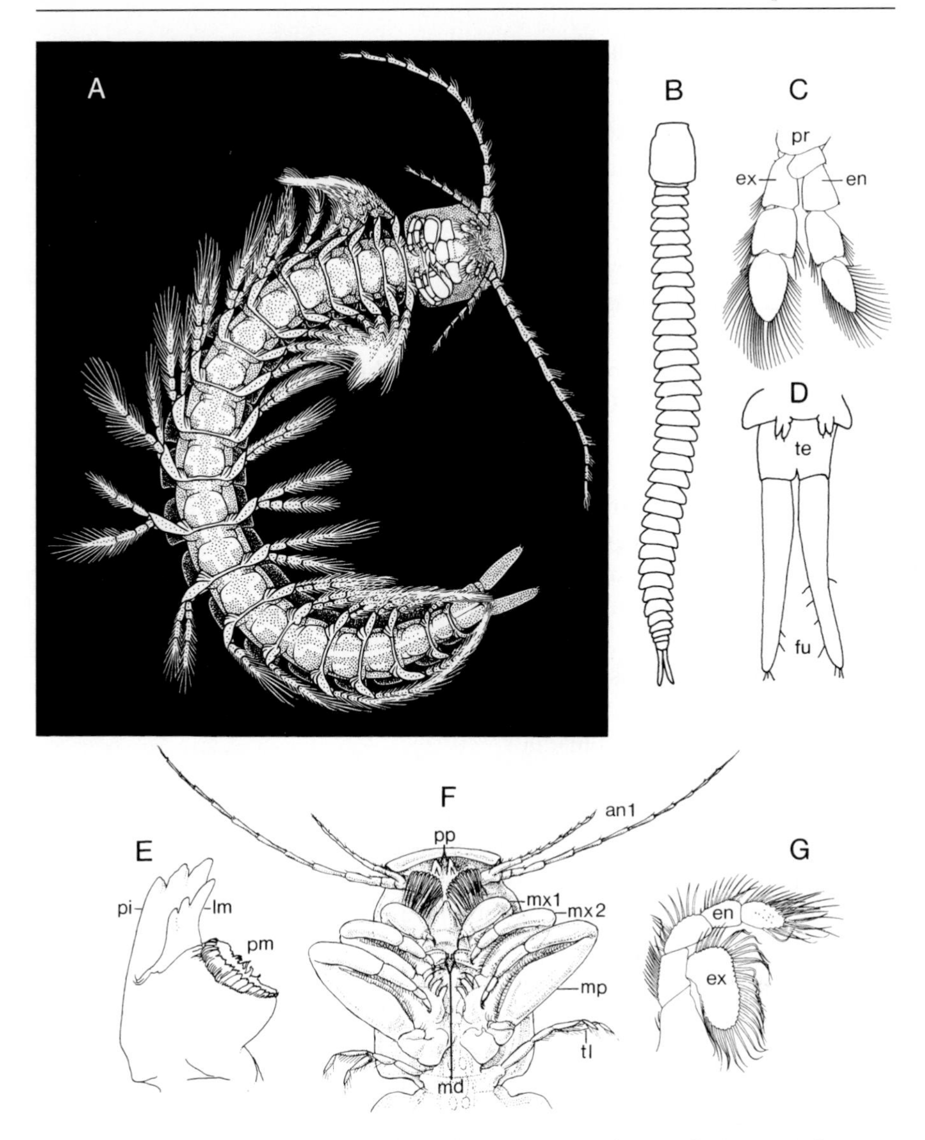

Fig. 62. Remipedia. **A** *Speleonectes ondinae*. Swimming animal. Ventral view. **B** *Speleonectes tulumensis*. Body with headshield and pleurotergites on the trunk segments. **C** *Lasionectes entrichoma*. Appendage with endo- and exopodite (trunk segment 10). **D** *Speleonectes tulumensis*. Posterior end. Last trunk segment before the telson with small appendages. **E** *Lasionectes entrichoma*. Right mandible. **F** *Lasionectes entrichoma*. Head from underneath. Second antennae covered by a row of long esthetascs on the base of the first antennae. **G** *Lasionectes entrichoma*. Small second antenna. *an* Antenna; *en* endopodite; *ex* exopodite; *fu* furca; *lm* lacinia mobilis; *md* mandible; *mp* maxilliped; *mx* maxilla; *pi* pars incisiva; *pm* pars molaris; *pr* protopodite; *pp* preantennal process; *te* telson; *tl* trunk leg. (**A** Schram, Yager & Emerson 1986; **B, D** Yager 1987; **C, E, G** Yager & Schram 1986; **F** Yager 1991)

Autapomorphies (Fig. 60→2)

- Cephalothorax.
  Attachment of the first trunk segment to the head and differentiation of the appendages to uniramous maxillipeds (Fig. 62 F).
- Second antenna short, with endopodite of three articles and exopodite of one article (Fig. 62 G).
- Asymmetric mandibles without palps.
- First and second maxillae uniramous. Grasping organs for carnivorous feeding (Fig. 62 F).
- Large glands of unknown function in the head that extend into the second trunk segment. Opening of these "maxillary glands" in the first maxillary segment.
- Hermaphroditism.
  An unpaired ovary in the second trunk segment; separate openings of paired oviducts in segment 8. Two testes in trunk segments 8–11; opening of the vasa deferentia in segment 15.
- Preantennal processes of unknown function, ventral before the first antennae (Fig. 62 F).
- Absence of a nauplius eye and compound eyes correlated to their stygian mode of life.
- Absence of antennal nephridia.

Since the ontogeny of the Remipedia is unknown, we do not know anything about the development of a nauplius eye. Nephridia only exist on the second maxillary segment. This is consistent with the placing of the Remipedia in the Crustacea; the basis for this argument, however, still remains weak.

The hypothesis of an adelphotaxa relationship between the Remipedia and the Tracheata (MOURA & CHRISTOFFERSEN 1996) is not valid. It is not possible to interpret the existence of a second antenna in the Remipedia (even when it is only small) and the lack of a second antenna in the Tracheata as a synapomorphy of the two unities. Also the statement of a common position of the gonopores on the last body segment remains unintelligible.

Eucrustacea

Autapomorphy (Fig. 60 → 3)

- Thorax - Abdomen.
  Division of the trunk into a thorax with swimming legs and an abdomen without appendages.

Thoracopoda – Maxillopoda

Thoracopoda

Autapomorphy (Fig. 60 → 4)

- Turgor appendages with epipodites as elements of a thoracic filter apparatus for microphagous feeding.

HESSLER (1992) interpreted the **existence of epipodites** as a **common apomorphy** of the Cephalocarida, Anostraca, Phyllopoda and Malacostraca. Epipodites in the form of an attachment on the outer side of the appendages (exites) are only found in this unity of the Crustacea. The hypothesis of a single evolution of these connected with the development of turgor appendages is the most parsimonious explanation. I find that the different insertions of the epipodites – onto the exopodites in the Cephalocarida and onto the protopodites in the Phyllopodomorpha – can best be explained as being due to a shift in position rather than by being due to independent origin (WALOSSEK 1993).

With this we come to the highest-ranking **adelphotaxa relationship** within the Thoracopoda. In the turgor appendages of the **Cephalocarida** the endopodite is round and formed from five to six elements; the exopodite is flat and formed of three elements. In contrast, in the **Phyllopodomorpha** (Anostraca, Phyllopoda, Malacostraca: Leptostraca), due to total loss of articulation, both appendage branches are differentiated into leaf-like structures. I interpret these phyllopodia as an evolutionary novelty that originated only once – in a stem lineage common to the Anostraca, Phyllopoda and Malacostraca. The structurally different manifestations of the turgor appendages are obviously connected with differences in the uptake of fine particles. The original state is represented by the benthic Cephalocarida which are suspension feeders; mud particles are stirred up from the loose sediment and sucked in between the thoracopods. In the Phyllopodomorpha, however, the filtration of fine particles from open water is a derived state in their ground pattern.

Cephalocarida – Phyllopodomorpha

Cephalocarida

Inhabitants of muddy substrates, mainly in the littoral areas of sea coasts.

Hutchinsoniella, Lightiella.

Only a few species are known from North and South America, South Africa, New Caledonia, New Zealand and Japan.

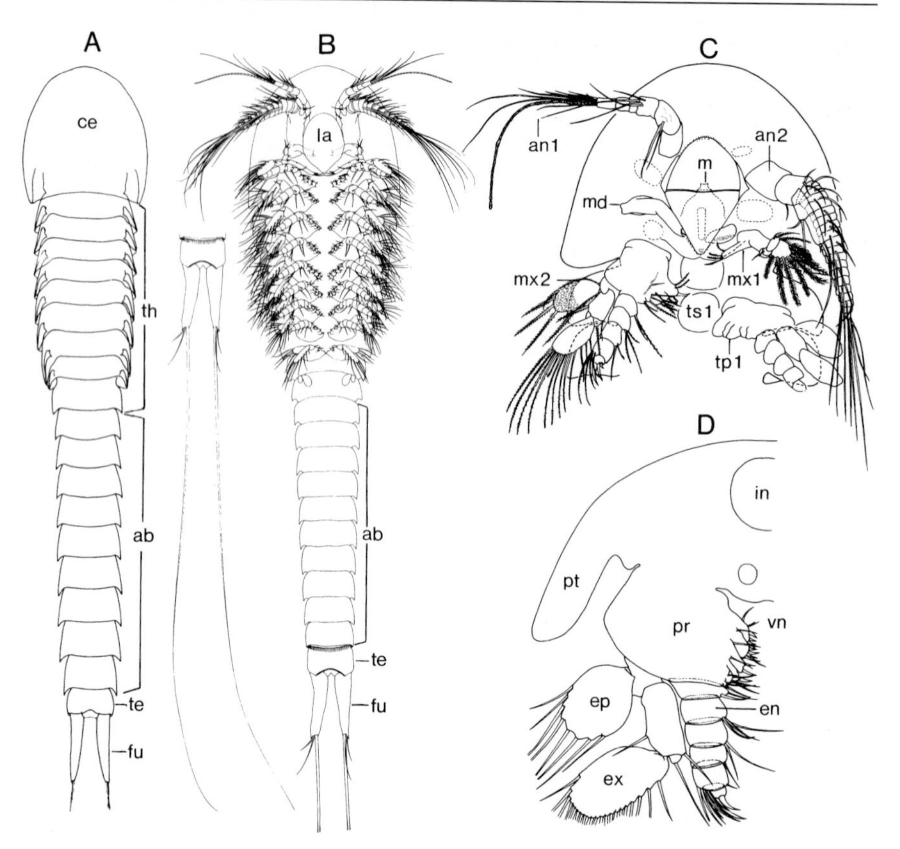

Fig. 63. Cephalocarida. **A–B** *Hutchinsoniella macracantha.* **A** Dorsal view. **B** Ventral view. *Left*: telson and furca with two extremely long setae. **C** *Lightiella incisa.* Head and thoracopod 1. Ventral view. Appendages only shown alternatively on one side of the body. Maxilla 2 structurally identical to the thoracopods. **D** *Hutchinsoniella macracantha.* Cross section through the thorax. Ventral food groove between the appendages. *ab* Abdomen; *an* antenna; *ce* cephalon; *en* endopodite; *ep* epipodite; *ex* exopodite; *fu* furca; *in* intestine; *la* labrum; *m* mouth; *md* mandible; *mx* maxilla; *pr* protopodite; *pt* pleurotergite; *te* telson; *th* thorax; *tp* thoracopod; *ts* thorax segment; *vn* food groove. (**A** Sanders 1955; **B** Sanders 1963; **C, D** Gruner 1993)

The Cephalocarida are small organisms between 2 and 3.7 mm in length; additionally, they have long furcal setae.

The head is horseshoe shaped and covered by a cephalic shield. The trunk consists of 19 segments – nine thoracic segments with appendages and ten abdominal segments without appendages. The trunk segments bear pleurotergites as lateral folds.

The second maxillae and the first seven thoracopods are identical turgor appendages with a leaf-like protopodite, a round, multiarticulate endopodite and a flat exopodite of three articles; on the latter sits a lobe-shaped epipodite (Fig. 63 C, D). Thoracopods 8 and 9 are greatly reduced. During ontogeny the antennal nephridia appear first; maxillary nephridia develop in the adult.

In the anameric development the nauplius larva is absent. The first stage is a metanauplius with three (*Hutchinsoniella macracantha*) or seven (*Lightiella incisa*) postcephalic segments. In *H. macracantha* five juvenile stages follow 13 metanauplius stages before the adult is formed. The molting of the last metanauplius to the first juvenile stage marks the change from larval feeding using the mouthparts to adult feeding in the form of filtration with the thoracopods.

Autapomorphies (Fig. 60 → 5)

- Hermaphroditism.
  Ovaries in the head; testes in thoracic segments 7–12. Oviducts and vasa deferentia unite on each side of the sixth thoracomere. Opening of the gonoducts on the protopodite of the sixth thoracopod.
- No nauplius larva.
- Absence of nauplius eye and compound eyes.
- Mandibles of the adult without palps (Fig. 63 C).

Phyllopodomorpha[11]

Autapomorphies (Fig. 60 → 6)

- Formation of the turgor appendages to phyllopodia through loss of all articles in the endo- and exopodites.
- Filtration of fine particles from open water.
- Compound eyes on stalks.

Within the Phyllopodomorpha the non-carapaced **Anostraca** occupy the same hierarchic level as the carapaced Phyllopoda + Malacostraca. As in the Cephalocarida the lack of a carapace is judged to be a plesiomorphy since there is no indication of reduction. On the other hand, the hypothesis of a single evolution of a bivalved shell in a common stem lineage of the Phyllopoda and Malacostraca offers the simplest explanation for this feature. Within the Phyllopodomorpha I assess the formation of a carapace with adductor muscle which largely envelops the body as the product of a single evolution and accordingly unite the Phyllopoda and Malacostraca to a monophylum **Ostraca**.

The Leptostraca (= Phyllocarida) are the only Malacostraca with turgor appendages and a bivalved carapace; both features are modified within the Malacostraca.

[11] tax. nov.

WALOSSEK (1993, 1995) refers to differences in details between the Leptostraca and "Branchiopoda" (Anostraca and Phyllopoda) and denies that there is a homology of the turgor appendages, but offers no explanation for their separate evolution. At the same time, within the monophyletic Phyllopoda, a serious modification of the turgor appendages of the Notostraca together with a relinquishing of their filter function is conceded (p. 158). I therefore reiterate that it is much simpler to hypothesize **a single evolution of the turgor appendages** of the Crustacea with subsequent alterations in different stem lineages, than two separate evolutions leading to basically identical results in the Anostraca and Phyllopoda on the one hand and in the Malacostraca on the other.

With the "**stalked compound eye**" we have another apomorphic feature for the monophyly of the unity Phyllopodomorpha. Within the Crustacea only the Anostraca and the Malacostraca have their compound eyes on the tip of a mobile stalk. We turn again to the principle of parsimony, and postulate a single evolution of the identical phenomenon in a common stem lineage of the Anostraca and Ostraca. Of course, stalked compound eyes may have already developed in the stem lineage of the Thoracopoda. Since eyes are reduced in the Cephalocarida, this remains an unprovable possibility.

The hypothesis provokes an explanation for the "**sessile compound eye**" as an autapomorphy of the Phyllopoda (p. 150). The evolutionary process of displacement under the skin appears more understandable to me if we proceed from an already mobile, stalked eye rather than from the primary state of a compound eye fixed in the cuticle (ground pattern of the Crustacea).

In the proposed systematization the traditional "Branchiopoda" become a paraphyletic collection of the Anostraca and Phyllopoda within the Phyllopodomorpha. The group "Branchiopoda" should therefore be removed from the systematics of the Crustacea.

Anostraca – Ostraca

Anostraca

The Anostraca without carapace live in stagnant, mostly ephemeral inland waters. Common fresh-water species with a life cycle restricted to a few months of the year are the "Spring form" *Siphonophanes* (*Chirocephalus*) *grubei* and the "Summer form" *Branchipus schaefferi*. A species complex under the name *Artemia salina* lives worldwide in saline-rich lakes and ponds up to the salt lagoons of sea coasts.

Compared with the Cephalocarida, the Anostraca are not bottom dwellers; rather they swim permanently on their backs. Food particles are filtered from the open water.

The following body division belongs in the ground pattern of the fairy shrimps: Cephalon, 13 thoracic segments with 12 (or 13) appendages or appendage derivatives, six abdominal segments without appendages, telson and furca with flat branches.

Thoracic segments 1–11 bear phyllopodia. The following segments 12 and 13 carry the genital apparatus; the male has paired penes on segment 12, in the female an egg-sac develops from the area of segments 12 and 13.

In having 19 trunk segments congruence with the Cephalocarida can be seen. This may be the ground pattern number for the Eucrustacea trunk (p. 142).

Autapomorphies (Fig. 60 → 7)

- First antenna small, thread-like.
- Second male antenna a large grasping organ (copula) (Fig. 64 A, B).
- Mandible without palp.
- Second maxilla greatly reduced.
- Genital segments.
  Thoracic segments 12 and 13 are incorporated into the reproductive apparatus (Fig. 64 C, D).
- Nauplius eye with three ocelli.

Plesiomorphic ground pattern features: compound eyes, antennal and maxillary nephridia, gonochorism.

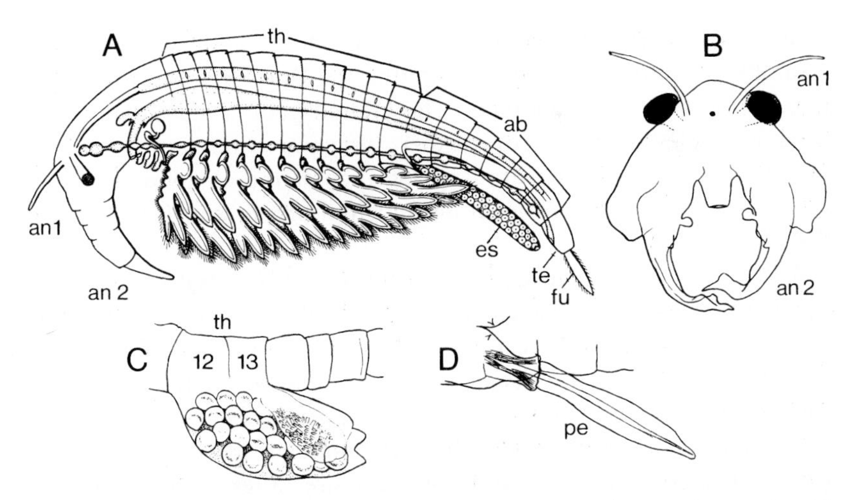

Fig. 64. Anostraca. **A** Diagram of organization. **B** *Siphonophanes grubei*. Head of a male with large second antennae as grasping organs. **C** *Chirocephalus josephinae*. Female. Genital diplo-segment with brood-sac. Side view. **D** *Artemia salina*. Male. Penis on the 12th thoracic segment. Side view. *ab* Abdomen; *an* antenna; *es* egg-sac; *fu* furca; *pe* penis; *te* telson; *th* thorax. (**A** Remane 1957; **B–D** Flössner 1972)

Ostraca [12]

Autapomorphy (Fig. 60 → 8)

- Carapace in the form of a bivalved shell with adductor muscle. The shell largely envelops the body.

This interpretation is valid within the monophylum Phyllopodomorpha. A convergent evolution to a similar carapace with adductor muscle took place within the Maxillopoda (p. 191).

Phyllopoda – Malacostraca

Phyllopoda

Autapomorphies (Figs. 60 → 9; 65 → 1)

- Sessile compound eye.
  With the following subfeatures: (1) approximation of the eyes to the middle of the head. (2) Sinking of the eyes to below the surface of the head. A fold of skin grows over the compound eyes and encloses them in an eye chamber with ectodermal lining. The eyes are suspended in the chamber and are mobile. There is a primary connection to the surrounding water through an eye pore (Fig. 66 A).
- Around 30 thoracopods (exact number not determinable).

Above all the **sessile compound eye** beneath the body surface is a significant apomorphic peculiarity for uniting the Onychura and the Notostraca into a monophylum Phyllopoda. The ground pattern state outlined is realized in the Notostraca. For further systematization useful evolutionary modifications are (1) the drawing together of the compound eyes to the center, (2) their fusion to form an unpaired organ, and (3) the sealing of the eye-chamber pore.

A body size of 1–2 cm and ca. 30 thoracopods – realized in the Spinicaudata sensu stricto ("Conchostraca") – belong in the ground pattern of the Phyllopoda. The latter can be interpreted as a further autapomorphy of the Phyllopoda compared with the lower number in the Cephalocarida, Anostraca and Malacostraca (14 thoracic appendages).

Body size and thoracopod number undergo extensive modifications within the Phyllopoda – a minimization to only a few mm and only six ap-

[12] tax. nov.

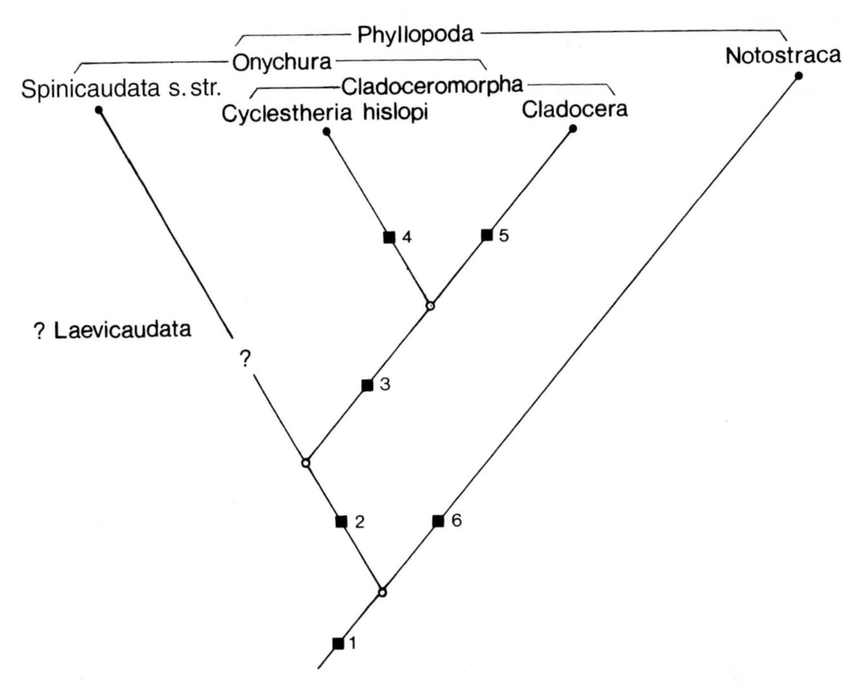

Fig. 65. Diagram of the phylogenetic relationships within the Phyllopoda

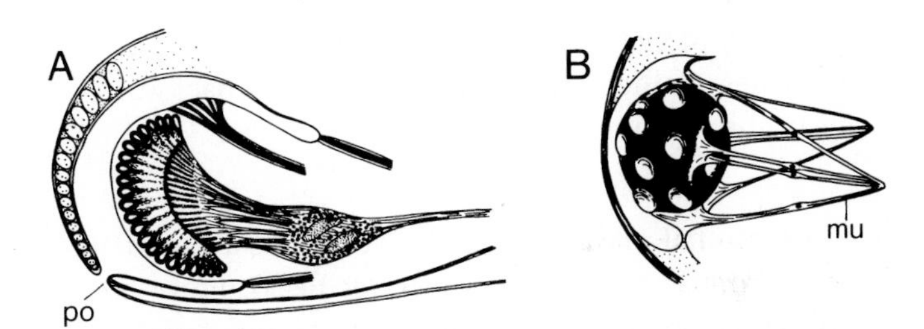

Fig. 66. Phyllopoda. Sessile compound eye as an autapomorphy of the unity. **A** *Limnadia lenticularis* (Spinicaudata sensu stricto). Eye chamber with pore. **B** *Daphnia magna* (Cladocera). Eyes completely fused. Eye chamber closed. *mu* Muscles; *po* pore. (Gruner 1993)

pendages in the Cladocera; a maximization to 7–8 cm and a correspondingly high number of about 70 appendages in the Notostraca.

A plesiomorphous feature of the Phyllopoda is the nauplius eye with four ocelli.

Onychura – Notostraca

Onychura (Diplostraca)

Autapomorphies (Fig. 65 → 2)

- Growth rings on the carapace.
  In molting the outer wall of the valve is not shed; the old wall remains on the newly enlarged cuticle. Thus growth rings are formed which document the number of molts (Fig. 67 A).
- Median placement of the compound eyes in close proximity – but not fused.
- Strongly developed biramous second antenna with a locomotory function.
- First thoracopod of the ♂♂ with hooks for copulation. A movable finger is differentiated from the endopodite.
- Deposition of the eggs in a dorsal shell space between the carapace and the back of the thorax (Fig. 67).
- Attachment of the eggs (embryos) in the ♀♀ onto dorsal structures of the exopodite. These are thread-like (or pistil-like) structures on some of the thoracopods that penetrate into the dorsal shell space.
- Short legless abdomen folded downwards ventrally.
- Dorsal telson with a pair of long setae (MARTIN & CASH-CLARK 1995; OLESEN, MARTIN & ROESSLER 1997; Fig. 67 B, C).
- Branches of the furca differentiated into backward-directed claws. They clean the interior of the carapace.

Conventional classification divides the Onychura into the Conchostraca (clam shrimps) and Cladocera (water fleas). From the above list of autapomorphies, growth rings on the carapace and dorsal attachment threads on the thoracopods are missing in the Cladocera. Until recently these have been employed as useful features for defining the unity Conchostraca. They are, however, inapplicable for this purpose since *Cyclestheria hislopi* with these two features must be taken out of the Conchostraca to become the adelphotaxon of the Cladocera. In doing so growth rings and the thoracic adhesive threads must be assessed as apomorphic ground pattern features of the Onychura – the hypothesis of their reduction in the stem lineage of the Cladocera follows automatically.

The "**Conchostraca**" cannot be established as a monophylum through any apomorphy (OLESEN 1998; OLESEN, MARTIN & ROESSLER 1997). In the state of kinship analysis they form a paraphyletic group of Onychura with plesiomorphic features. What then becomes of their two subtaxa – the Spinicaudata and the Laevicaudata (LINDER 1946; FRYER 1987b)

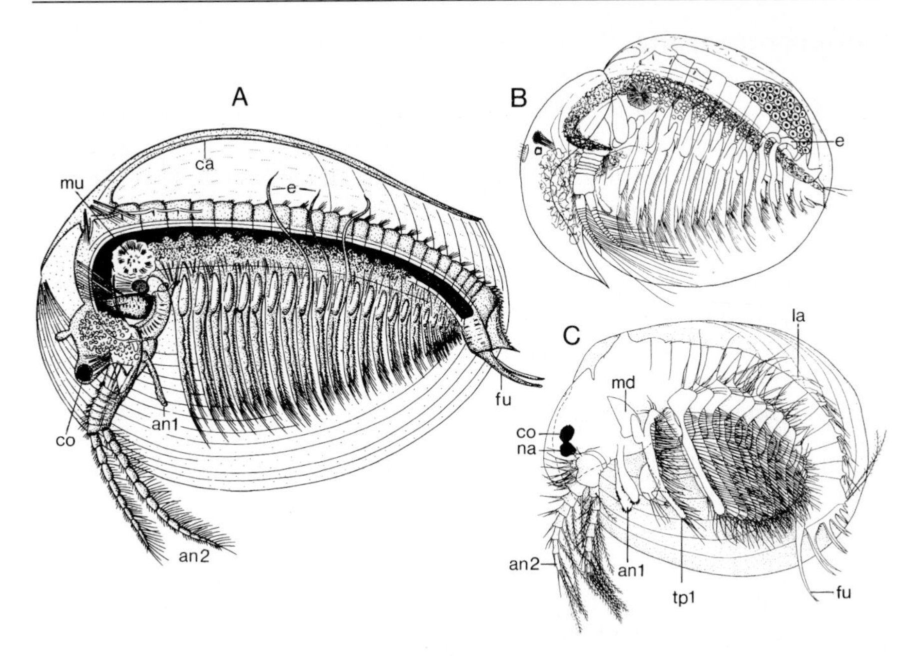

Fig. 67. Onychura. **A** *Limnadia lenticularis* (Spinicaudata sensu stricto). Female. Length 17 mm. Left half of the carapace removed. Thread-like egg carriers on thoracopods 9–11. **B** *Lynceus brachyurus* (Laevicaudata). Female. Length 6 mm. Pipe-like egg carrier on thoracopods 9 and 10. **C** *Cyclestheria hislopi* (adelphotaxon of the Cladocera). Male. *an* Antenna; *ca* carapace; *co* sessile compound eye; *e* egg carrier; *fu* furca; *la* lamellae; *md* mandible; *mu* closure muscle; *na* nauplius eye; *tp* thoracopod. (**A** Gruner 1993; **B** Pennak 1978; **C** Olesen, Martin & Roessler 1997)

whose species live mainly in small, stagnant ponds and lakes, especially in ephemeral pools and puddles.

The Spinicaudata sensu stricto, after exclusion of *Cyclestheria hislopi*, include the Leptestheriidae (*Leptestheria*), Cyzicidae (*Cyzicus*) and Limnadiidae (*Limnadia*). The **Spinicaudata in this narrow sense** are a monophylum and probably form the sister group of the Cladoceromorpha (*Cyclestheria* + Cladocera). Their basic **autapomorphy** is the formation of clasps on the first and second pair of thoracopods in the ♂♂, i.e., the additional transformation of the second pair to copulatory structures. The position of the **Laevicaudata** (? monophylum) with the Lynceidae (*Lynceus*) within the Onychura is unclear. They have no apomorphies in common with the Spinicaudata sensu stricto. Other possibilities of ordering remain, at present, vague (OLESEN et al. 1997; OLESEN 1998).

Cladoceromorpha[13]

Since the work of SARS (1887), the problem of a possible closer relationship between *Cyclestheria hislopi* and the Cladocera has been repeatedly discussed. It was a century later, however, that OLESEN et al. (1997) established precisely the hypothesis of an adelphotaxa relationship to the Cladocera as one of two possibilities of the position of the species in the system of the Onychura. Due to the following apomorphies common to both taxa, I support the interpretation of *Cyclestheria hislopi* as the sister species of the Cladocera and unite them under the name Cladoceromorpha.

Autapomorphies (Fig. 65 → 3)

- Complete fusion of the compound eyes to an unpaired organ.
- Life cycle with heterogony.
  Change between bisexual reproduction and parthenogenesis.
- Direct development without nauplius larva.
  In populations of *Cyclestheria hislopi* from Columbia dormant eggs and parthenogenetically produced eggs undergo direct development (ROESSLER 1995); in populations from Cuba metanauplius larvae hatch from dormant eggs (BOTNARIUC & VIÑA BAYES 1977).
- Ephippium.
  The well-known saddle-like capsule as a protective shell for dormant eggs may exist in its original manifestation in *Cyclestheria hislopi.* The only lightly modified carapace is used in its entirety as a protective capsule for the eggs. The deposition of the dormant eggs results in the death of the female. In contrast, in the Anomopoda (Cladocera) only part of the carapace is modified into the ephippium (Fig. 68 B) – and this without the obligatory death of the female. A homology of the ephippium in *Cyclestheria hislopi* and the Cladocera is, however, disputed; in the latter it only exists in the subtaxon Anomopoda.

Cyclestheria hislopi – Cladocera

Cyclestheria hislopi

Pantropical species found between 30 °N and 35 °S. Usually in permanent fresh-water lakes or ponds with algal mats or other water plants (ROESSLER 1995; OLESEN et al. 1997).

In the following features *Cyclestheria hislopi* is more original than the Cladocera (Fig. 67 C).

---

[13] tax. nov.

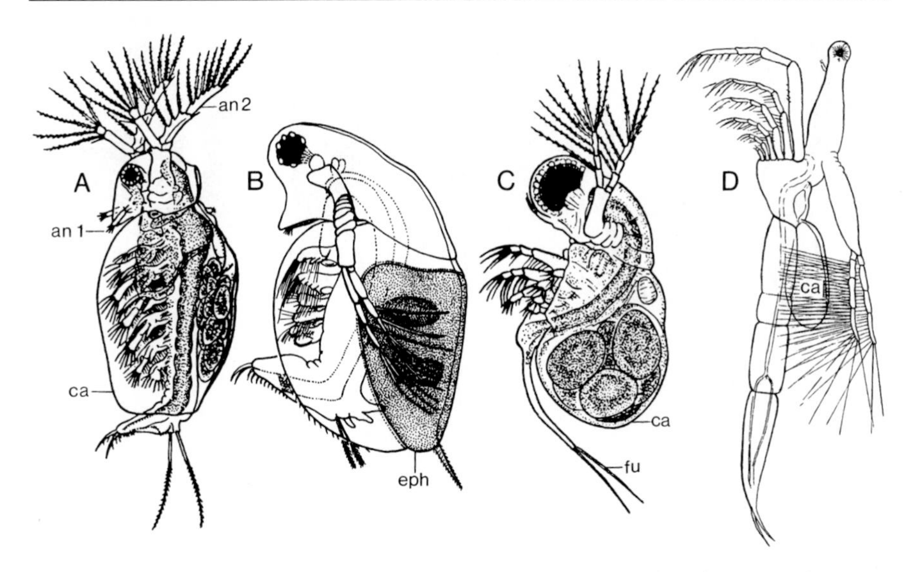

Fig. 68. Cladocera. Females. **A** *Sida crystallina* (Ctenopoda). **B** *Daphnia longispina* (Anomopoda). Female with ephippium. **C** *Polyphemus pediculus* (Onychopoda). **D** *Leptodora kindtii* (Haplopoda). *an* Antenna; *ca* carapace; *eph* ephippium; *fu* furca. (**A, C** Engelhardt 1959; **B** Gruner 1993; **D** Vollmer 1960)

- A circular carapace completely encloses the body; it has growth rings.
- Larger number of thoracopods. ♂♂ with 15, ♀♀ with 16 pairs.
- Threads for egg attachment on the exopodites of some thoracopods.
- Eye chamber with a pore (OLESEN et al. 1997). A "chitinous" cord that connects a cup-shaped depression on the head with the compound eye had already been described by SARS (1887).

Autapomorphy (Fig. 65 → 4)

- Males and females bear on their posterior thoracic segments cuticular lamellae with backward-pointing setae (Fig. 67 C). This may be a derived feature of *Cyclestheria hislopi*. Comparable formations are not known in the Cladocera.

Cladocera

Autapomorphies (Fig. 65 → 5)

- Eye chamber without connection to the exterior; pore closed (Fig. 66 B).
- Free head. As a bivalved shell the carapace only encloses the trunk (Fig. 68).
- Carapace without growth rings.

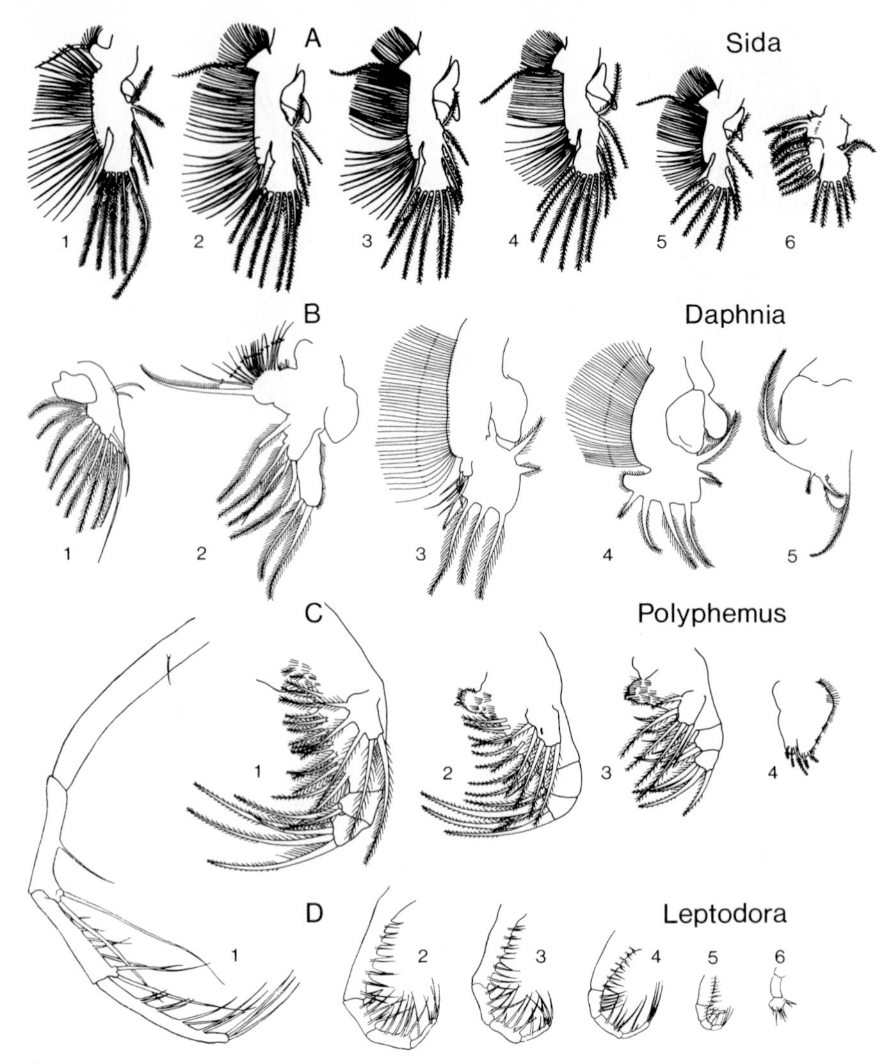

Fig. 69. Cladocera. Appendages. Females. **A** *Sida crystallina.* Six pairs of thoracopods. Appendages 1–5 are similar filter-legs; the sixth leg is clearly smaller, without filter setae. Primary leg number. Five filter-legs form the plesiomorphic state for the Cladocera. **B** *Daphnia pulex.* Five pairs of strongly differentiated thoracopods. **C** *Polyphemus pediculus.* Only four pairs of secondarily subdivided thoracopods. *Leptodora kindtii.* Six pairs of thoracopods (plesiomorphy). Long articulated rod-legs without exopodites (apomorphy) which serve in purely predacious feeding (capturing basket). (**A–C** Gruner 1993; **D** Flößner 1972)

- Six thoracopods in the ground pattern (Fig. 69), in contrast to 15–16 in *Cyclestheria hislopi.*
- Eggs in dorsal brood chamber without connection to thoracopods; reduction of the adhesive threads on the exopodites.
- Opening of the female genital pore in the brood chamber.

For some time a hypothesis of multiple independent origin of single Cladocera taxa from the "Conchostraca" has existed (FRYER 1987a,b; WALOSSEK 1993; SCHMINKE 1996) – a hypothesis according to which the Cladocera are an artificial polyphylum. This hypothesis must then (1) explain a repeated convergent evolution of all the features mentioned here as autapomorphies of the Cladoceromorpha and Cladocera, and (2) make probable adelphotaxa relationships between certain subtaxa of the "Conchostraca" and the Cladocera. For both points corresponding proof is lacking. I adhere to the **well-founded interpretation of the Cladocera as a monophylum.** The four traditional taxa of the Cladocera – Ctenopoda, Anomopoda, Onychopoda, Haplopoda (Figs. 68, 69) – are all established as monophyla. The Ctenopoda may form the sister group of the other three taxa, the Anomopoda the adelphotaxon of the Gymnomera (Onychopoda + Haplopoda). Since the kinship relations are not yet fully understood, however (MARTIN & CASH-CLARK 1995; OLESEN 1998), I limit myself to the presentation of a few, well-proven statements.

The **Ctenopoda** (*Sida, Diaphanosoma, Holopedium*) possess an original filter apparatus with six pairs of serial homonomous appendages. Striking **autapomorphies** are the modification of the first antenna to a grasping organ (♂♂) and the reduction of the second antenna to 2–3 articles per branch.

In the **Anomopoda** (*Daphnia, Bosmina, Chydorus*) two **autapomorphies** can be emphasized. With the loss of serial similarity the two foremost leg pairs are transformed into grasping structures. On the basis of the first thoracopod there are ejector hooks for expelling detritus accumulated during the filtration process.

In the stem lineage of the **Gymnomera** (Onychopoda + Haplopoda) a far-reaching **change in feeding** took place. Carnivores developed from filter feeders and rod-shaped jointed appendages evolved from phyllopodia. Furthermore the bivalved shell is transformed into a small, dorsal brood-sac and the nauplius eye disappears.

In the **Onychopoda** (*Polyphemus, Bythotrephes*) the reduction to four pairs of appendages (these still with small exopodites), and the immense enlargement of the compound eye relative to body size, are considered **autapomorphies.**

In the adelphotaxon **Leptodora kindtii** (the single species of the Haplopoda; consequently, the name Haplopoda should be dropped), a series of **unique derived features** exist. The body is elongated and cylindrical. Both maxillae are reduced while the mandibles are differentiated to rod-shaped biting structures. On the rod-legs (plesiomorphic number of six) the exopodites are totally reduced. Thoracopods 2–6 move closer together to form a 'capturing basket' open to the anterior.

Notostraca

The Notostraca (tadpole shrimps) are benthic organisms. They live mainly in small, periodic pools, but also on the bottom of large lakes.

Two central European species exist at different times of the year and can be found together with the corresponding species of Anostraca already mentioned. The "Spring form" *Lepidurus apus* appears after the thaw and colonizes in the months February to May often together with *Siphonophanes grubei*. The "Summer form" *Triops cancriformis* follows from May to October, and is often found together with *Branchipus schaefferi*.

Compared with the adelphotaxon Onychura the **change of biotope to muddy substrates** is the central autapomorphy of the Notostraca (WALOSSEK 1993, 1995) with which a whole series of further characteristics are correlated. The original filter-feeding in open water is replaced by the uptake of sediment particles with the first nine thoracopod pairs. Additionally the Notostraca can catch larger prey.

Autapomorphies (Fig. 65 → 6)

- Mode of life on muddy bottom substrates.
- Nutrient uptake from the bottom.
- Increase in body size to several cm.
- Shield-like carapace (Fig. 70 A, B).
  A dorsal keel marks the original arrangement of two valves; an adductor muscle is lacking.
- Polypody.
  Only thoracic segments 1–11 each have one pair of thoracopods. Of the following ca. 60 appendages between 2–6 pairs of legs are located on uniform body rings – these are obviously not fully segmented sections of the thorax.
- First antenna a short, thin rod on the ventral side of the head.
- Second antenna greatly reduced.
- First thoracopod with three long multiarticulate endites (sensory structures) (Fig. 70 B).
- Eleventh thoracopod in the ♀♀ with circular egg capsule: floor = disc-shaped broadening of the protopodite and exopodite; roof = epipodite (Fig. 70 D). Fertilized eggs remain only briefly in the capsule.
- Furcal branches of body length.

A plesiomorphy of the Notostraca is the separate position of the compound eyes. The sessile eyes are approximate but have not united (Fig. 70 A).

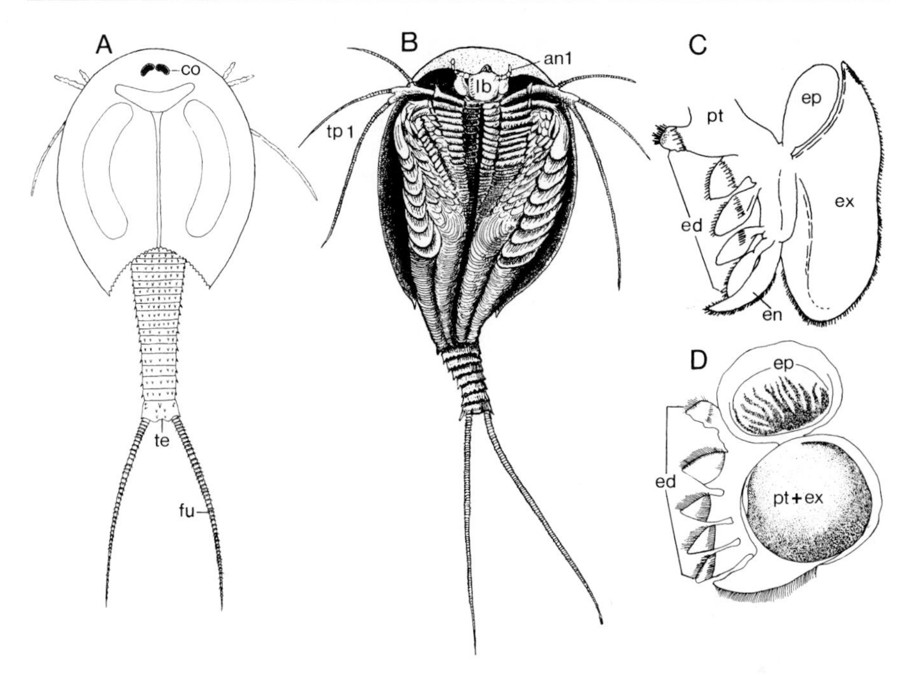

Fig. 70. Notostraca. **A** *Triops longicaudatus*. Dorsal view. Median approximation of the compound eyes, but no actual merging. **B–D** *Triops cancriformis*. **B** Ventral view. Greatly elongated first thoracopods. **C** Thoracopod from the anterior body region. Broadly spread out. **D** 11th thoracopod of the female with round egg capsule (opened). Lid of the capsule formed from protopodite and exopodite; roof originates from the epipodite. *an* Antenna; *co* compound eye; *ed* endite; *en* endopodite; *ep* epipodite; *ex* exopodite; *fu* furca; *la* labium; *pt* protopodite; *te* telson; *tp* thoracopod. (**A** Pennak 1978; **B** Vollmer 1952; **C** Gruner 1993; **D** Siewing 1985)

Malacostraca[14]

Autapomorphies (Figs. 60 → 10; 71 → 1)

- 15 trunk segments.
  14 segments with appendages form the thorax. The appendageless segment 15 represents the abdomen (Fig. 61).
- Division of the thorax into two sections with structurally and functionally different appendages (Fig. 72 B).
  Thorax I eight thoracomeres with phyllopodia (filter-legs).
  Thorax II (pleon) six thoracomeres (pleomeres) with pleopods (swimming legs).
- Constant position of the genital openings.
  Female sixth thoracic segment.
  Male eighth thoracic segment.

[14] I thank Dr. S. Richter (Berlin) for reviewing and improving the Malacostraca section.

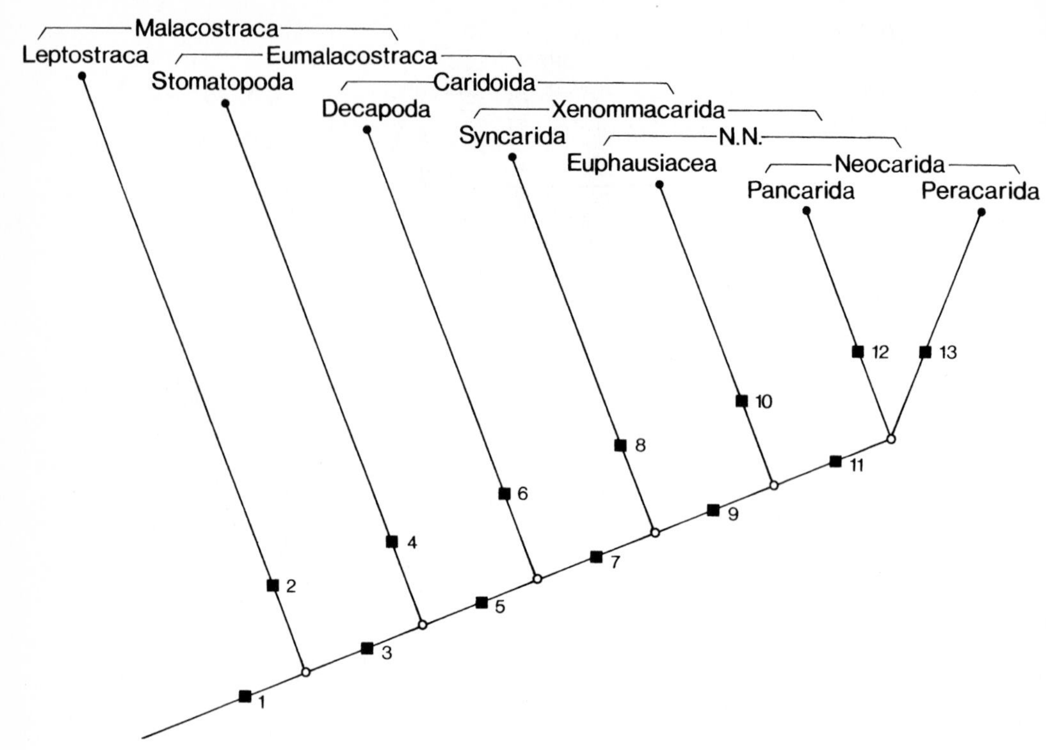

Fig. 71. Diagram of the phylogenetic relationships within the Malacostraca

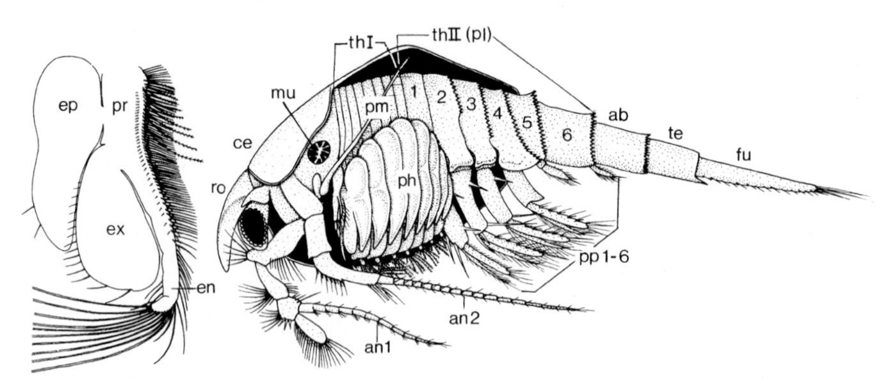

Fig. 72. Leptostraca. **A** *Nebalia bipes*. Phyllopodium (turgor appendage); spread out on one level. **B** *Nebalia*. Diagram of organization. Left valve of the carapace removed. *ab* Abdomen; *an* antenna; *ce* cephalon; *en* endopodite; *ep* epipodite; *ex* exopodite; *fu* furca; *mu* closure muscle; *ph* phyllopodia; *pl* pleon; *pm* palp of the second maxilla; *pp* pleopods; *pr* protopodite; *ro* rostrum; *te* telson; *th* thorax. (**A** Gruner 1993; **B** Gruner 1969)

- Rostrum.

  Division of a mobile plate at the anterior of the carapace. Regulates water flow to the phyllopodia with filter function (Fig. 72 B).

- Stomach with cardiac and pyloric chambers.
  Differentiation of the posterior foregut into a stomach with a double function: grinding up of food (cardiac chamber) and separation of fine and large particles (pyloric chamber).
- Ectoblast ring.
  The main part of the post-nauplial germ band originates from an ectoblast ring of exactly 19 cells (DOHLE 1972).

In the stem lineage of the Malacostraca a **differentiation of the thoracopods** took place. The eight front leg pairs retain the original structure and function as leaf-shaped filter structures, whereas the six posterior leg pairs are transformed into swimming legs. Consequently, I refer to two thoracic sections: thorax I, consisting of eight segments with eight thoracopods, and thorax II, consisting of six segments with six thoracopods. The latter is the pleon of common terminology. The terms pleon, pleomere and pleopod can then cause no confusion if they are understood to be labels for parts of the thorax.

In the ground pattern of the Malacostraca between the thorax and the telson with furca there is **a single appendageless trunk segment 15.** This is named the **abdomen**. The segment is only realized in the Leptostraca (Fig. 72 B); in the stem lineage of the Eumalacostraca it fuses with the last thoracomere (pleon segment 6).

Leptostraca – Eumalacostraca

Leptostraca

Species-poor unity of filter-feeding marine organisms. The taxon *Nebalia* with several species is found worldwide in near-coastal muddy sediments.

Autapomorphies (Fig. 71 → 2)

- First antenna with short, scale-like appendage.
- Second antenna consists only of a long multiarticulate endopodite. An exopodite is lacking.
- Second maxilla with a long palp to clean the carapace and filter-legs (the three first autapomorphies are depicted in Fig. 72 B).
- Brood care and direct development.
  On sexual maturity the female forms a brood-pouch underneath the filter-legs; this consists of fans of setae from the endopodites. In the brood-pouch direct development takes place without a nauplius larva.

Within the Malacostraca the Leptostraca represent a comparatively original niveau.

Only the Leptostraca possess a large carapace of two valves connected by an adductor muscle. To the anterior the dorsal portion of at least one (possibly two to three) thoracic segment is fused with the carapace (CASANOVA 1991; RICHTER 1994b); to the posterior the carapace encloses a considerable part of the pleon.

Only in the Leptostraca are the first eight thoracopods developed as leaf-like filter-legs (turgor appendages) (Fig. 72A). In a metachronal wave the legs form alternating sucking and pumping chambers. Assorted particles landing in the filtering setae of the endopodites arrive at a median filter passageway on the ventral side of the anterior thorax; from here they are transported to the mouthparts.

Only in the Leptostraca is the pleon with six pairs of pleopods followed by an appendageless segment that we designate the abdomen.

In addition the existence of antennal and maxillary nephridia in the Leptostraca are original features.

Eumalacostraca

Autapomorphies (Fig. 71 → 3)

- Scaphocerite.
  The exopodite of the second antenna is a large, flattened scale. A steering organ for swimming.
- Transformation of the phyllopodia into stenopodia.
  The original mode of feeding as filter feeders (Leptostraca) is given up. The leaf-shaped filter-legs on thoracic segments 1–8 are transformed into rod-shaped appendages.
- Tail fan of uropods and telson.
  The appendageless abdominal segment fuses with the sixth segment of the pleon. The furca of the telson is reduced. Pleopods 6 (uropods) are strongly developed; they are directed diagonally backwards next to the tergal plate of the telson and together with this form the tail fan (all three autapomorphies are shown in Fig. 73).

Stomatopoda – Caridoida

Stomatopoda

Large marine crustaceans with bodies that can reach up to 40 cm. The mantis shrimps live on the sea floor in cracks and crevices, in which as carnivores they lie in wait for their prey. Their main geographical range is in the tropics and subtropics.

Squilla.

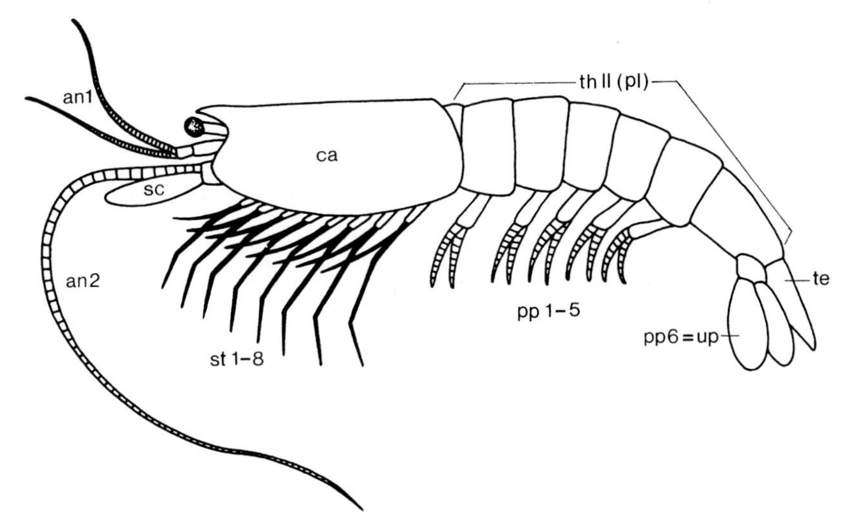

Fig. 73. Ground pattern of the Eumalacostraca. Autapomorphies: Antenna 2 with scale-like scaphocerite (exopodite). Thoracopods 1–8 developed as stenopodia; with long exopodites. Tail fan of uropods and telson. Plesiomorphies: Extensive carapace that covers thoracomeres 1–8. Five pairs of similar pleopods with endo- and exopodites. *an* Antenna; *ca* carapace; *pl* pleon; *pp* pleopods; *sc* scaphocerite; *st* stenopodia; *te* telson; *th II* second section of the thorax; *up* uropods. (Hessler 1983)

The Stomatopoda have several distinct derived peculiarities with which they can be established as a monophylum. On the other hand, in some organizational traits, they remain more original than the Caridoida. From the ground pattern of the Malacostraca they possess, like the Leptostraca, a mobile rostrum. In the pleon there are "still" a heart, midgut gland and gonads. The tail fan is bent downwards and supports the telson; the characteristic flight reaction of the Caridoida is not yet fully developed.

Autapomorphies (Fig. 71 → 4)

- First antenna triramous.
- Shorter carapace.
  A flat carapace plate (without side sections) covers the foremost four thoracic segments; however, it is apparently only fused with the first segment dorsally (CASANOVA 1993; Fig. 74A–C).
- Thoracopods 1–5 = raptorial legs.
  In the differentiation of the anterior five thoracopods to raptorial legs with subchelae the second appendages are greatly enlarged. Each of the raptorial legs is formed only from the endopodite (Fig. 74B).
- Gills on the pleopods.
  Pleopods 1–5 bear on their outer edge tube-shaped subdivided gills.

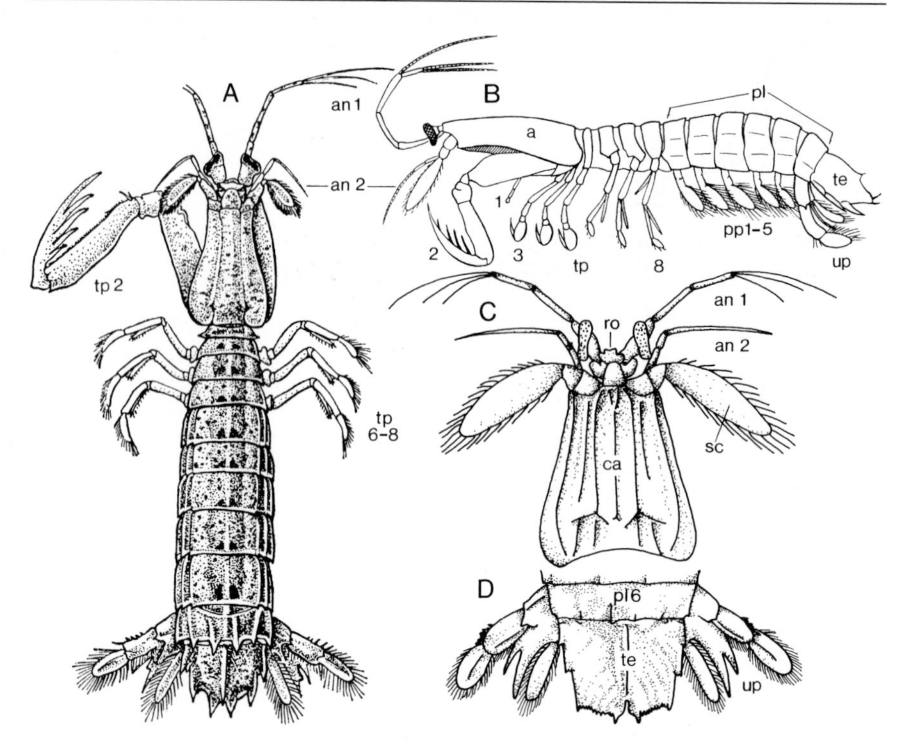

Fig. 74. Stomatopoda. **A** *Squilla desmaresti.* Dorsal view. **B** *Squilla mantis.* Organization in side view. Female. **C** Anterior end to demonstrate the scale-like exopodite (scaphocerite) of the second antenna. **D** *Squilla.* Posterior end to demonstrate the tail fan of uropods and telson. Between endo- and exopodite of the uropods there is a process of the protopodite with prongs. *an* Antenna; *ca* carapace; *pl* pleon; *pp* pleopods; *ro* rostrum; *sc* scaphocerite; *te* telson; *tp* thoracopods; *up* uropods. (**A** Riedl 1963; **B** Hennig 1994; **C** Hessler 1983; **D** Meglitsch & Schram 1991)

- Petasma.
  In the male, pleopods 1 and 2 form an organ used in sperm transfer. The copulatory organ is interpreted as a convergent structure in comparison to homonymous formations within the Caridoida.
- Only maxillary nephridia present.

Caridoida

Large sister group of the Stomatopoda. All the following taxa of the Eumalacostraca belong here.

Autapomorphies (Fig. 71 → 5)

- Anteri\timesor head with first antenna and rostrum firmly fused to the rest of the head (Fig. 75 A).

- Statocyst.
  Gravitational sense organ in the basal segment of the first antenna.
- Pleon and tail fan as escape device.
  Pleon with massive, complicated musculature through wide-ranging displacement of the inner organs. A sudden ventral strike of the pleon and tail fan leads to rapid, backward-directed flight.
- Mandibles of the larva with a row of setae between the processus incisivus and processus molaris. On the left mandible one seta is converted to a lacinia mobilis (RICHTER 1994b).

Decapoda – Xenommacarida

Clarification of the highest-ranking adelphotaxa relationships in the Caridoidea with the sister groups Decapoda and Xenommacarida was arrived at by RICHTER (1994b, 1999) through an ultrastructural analysis of the compound eyes. We outline the facts following the characterization of the Decapoda.

Decapoda

Autapomorphies (Fig. 71 → 6)

- Division into cephalothorax and pleon.
  The carapace extends over the first eight thoracomeres and fuses dorsally with them. The carapace reaches laterally down to the leg bases (branchiostegites) and there restricts a gill cavity located on the exterior of the body stem (Fig. 75 A, C). In the Decapoda the uniform body section of cephalon and thoracic segments 1–8 is generally known as the cephalothorax. There is no homology, however, with the same named body section in other taxa of the Crustacea.
- Thoracopods with four gills and an extra epipodite for gill cleaning.
- Thoracopods 1 and 2 transformed into maxillipeds used in food uptake. Whether a third maxilliped can also be placed in the ground pattern of the Decapoda is still unclear (RICHTER 1994b).
- Scaphognathite (Fig. 75 D).
  A large, lobe-shaped exopodite of the second maxilla pumps a continuous stream of water through the gill cavity. The water is sucked in from the leg pairs. It exits next to the mouthparts.

A **phylogenetic system of the Decapoda** is under development (FELGENHAUER & ABELE 1983; ABELE & FELGENHAUER 1986; CHRISTOFFERSEN 1988; RICHTER 1994b; SCHOLTZ & RICHTER 1995 amongst others). Of the classic division of the Decapoda into "Natantia" and Reptantia the former is a paraphylum, the latter can be established as a monophylum.

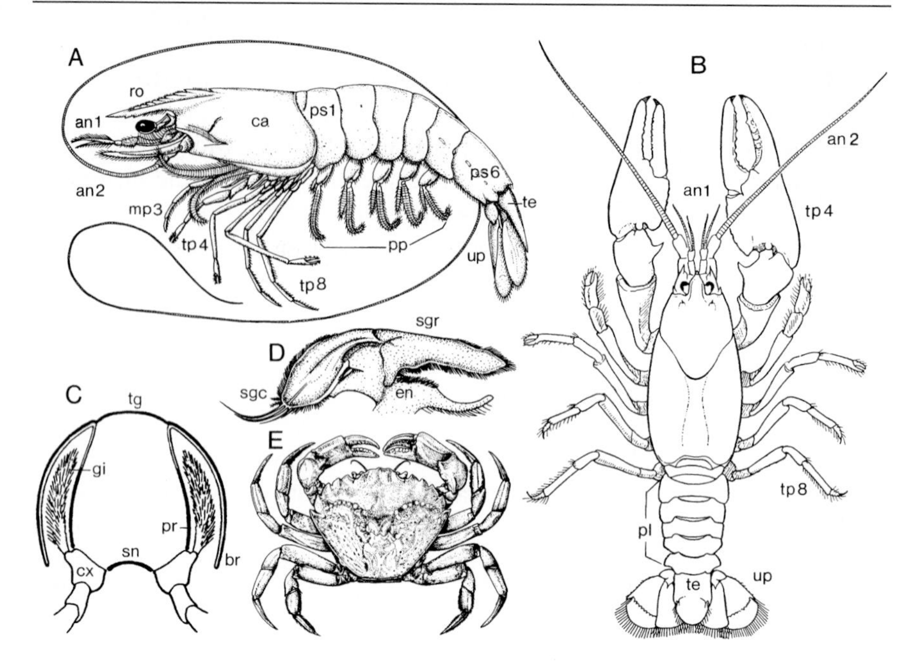

Fig. 75. Decapoda. **A** *Penaeus setiferus* (Dendrobranchiata). Original shrimp habitus of the Decapoda with a laterally compressed body. **B** *Cambarus longulus* (Reptantia, Astacida). Apomorphous, dorsoventrally flattened body. Long pleon directed backwards as a plesiomorphy within the Reptantia. **C** Cross section through the frontal part of the thorax. Diagram of the Astacida. Branchiostegites cover the gills as lateral folds of the carapace (autapomorphy of the Decapoda). **D** *Astacus astacus* (Reptantia, Astacida). Scaphognathite as a long process of the second maxilla (autapomorphy of the Decapoda). **E** *Carcinus maenas* (Reptantia, Brachyura). Apomorphous crab habitus. From above only the cephalothorax visible; the short flat pleon is tucked under the ventral side. *an* Antenna; *br* branchiostegite; *ca* carapace; *cx* coxa; *en* endopodite; *gi* gill; *mp* maxilliped; *pl* pleon; *pp* pleopods; *pr* pleura; *ps* pleon segment; *ro* rostrum; *sgc, sgr* caudal and rostral section of the scaphognathite; *sn* sternum; *te* telson; *tg* tergum; *tp* thoracopods; *up* uropods. (**A, D, E** Gruner 1993; **B, C** Snodgrass 1965)

The position of the Amphionidacea (*Amphionides reynaudii*) as a subgroup or sister group of the Decapoda is uncertain.

The highest-ranking adelphotaxa of the Decapoda today are thought to be the Dendrobranchiata and the Pleocyemata.

However, the **Dendrobranchiata** (*Penaeus, Sergestes, Lucifer*) (Fig. 75 A) may also be a paraphyletic collection of Decapoda (RICHTER 1994b) with numerous original features. Apart from the shrimp-like laterally compressed body and free-swimming habit, these are the free release of eggs into the water, a nauplius larva, leg-like thoracopods 3 ("maxillipeds 3") and a fully developed first pleon segment with swimming legs. The assessment of the dendritic, branched gills, from which the name of the unity is derived, as a plesiomorphic or apomorphic state is disputed.

In contrast, the **Pleocyemata** are a well-established monophylum – above all through brood care by the female (autapomorphy) with eggs attached to the pleopods until the larvae hatch. In the subtaxa Stenopodidea (*Stenopus*) and Caridea (*Palaemon, Crangon*) – ? monophyly justification – the shrimp habitus, epibenthic mode of life and other plesiomorphies are continued.

The **Reptantia** within the Pleocyemata with numerous apomorphic features are more strongly derived – benthic organisms with a dorsoventrally flattened body and a strongly calcified exoskeleton; fifth pereopod with chelae as a cleaning organ; first pleon segment shortened and with an articulated connection to the carapace; copula under opposition of the ventral sides of the partner.

The kinship relations of the Reptantia have been extensively clarified (SCHOLTZ & RICHTER 1995). Of the four traditional taxa, the "Palinura" (*Palinurus*), the "Astacura" (*Homarus, Astacus*) and the "Anomura" (*Galathea, Pagurus*) should be discarded as they are paraphyla. The Brachyura (*Cancer, Carcinus, Eriocheir*) form a monophylum in the system of the Reptantia.

Xenommacarida

The taxa dealt with up to now – Leptostraca, Stomatopoda and Decapoda – have ommatidia with a four-part crystalline cone, which is secreted by four cone cells. For these unities this is an old plesiomorphous feature that has been taken over from the ground pattern of the Crustacea and the Mandibulata (p. 135).

In contrast, in all the following taxa of the Malacostraca, the **crystalline cone** has only **two parts**. The crystalline cone is formed from two main cone cells, which lie on the cone. The two other cone cells are shifted distalwards as accessory cells. They do not take part, or only in a minor way, in the production of the crystalline cone.

Due to space restrictions, I concentrate on this central autapomorphy of the taxon Xenommacarida as provided by RICHTER. Two further ultrastructural features of the ommatidia are, however, integrated into the following list of apomorphic peculiarities (RICHTER 1993, 1994b, 1999).

Autapomorphies (Fig. 71 → 7)

- Two-part crystalline cone in the ommatidia of the compound eyes. Formed by two main cone cells. The two other cone cells do not take part, or only peripherally, in the production of the crystalline cone.

- Reduction of the proximal processes of the cone cells and with this the complete loss of a proximal cone stalk.
- Distal extension of the retinula cells and corresponding displacement of the cell nuclei.

Syncarida – N. N. (Euphausiacea, Pancarida + Peracarida)

Within the Xenommacarida the Syncarida form the sister group of an as yet unnamed unity of the Euphausiacea and the Pancarida + Peracarida.

Syncarida

Autapomorphies (Fig. 71 → 8)

- Absence of a carapace.
  Since a carapace covering the first eight thoracomeres belongs in the ground pattern of the Malacostraca, its absence in the Syncarida must be seen as an evolutionary loss and therefore interpreted as an autapomorphy.
- Direct development.
  Since a nauplius larva belongs in the ground pattern of the Decapoda and the Euphausiacea, there is a corresponding argument to that above for the absence of larvae.
- Fresh-water dwellers.
  The Syncarida are basically fresh-water organisms. We can, without conflict, postulate that their stem species lived in fresh waters.

The Anaspidacea and Bathynellacea are without doubt established as monophyletic subgroups of the Syncarida. Each of the apomorphies in one group is opposed by a plesiomorphous manifestation in the "partner" (NOODT 1964; SCHMINKE 1973) – with this the requirements necessary for their recognition as sister groups are fulfilled.

The **Anaspidacea** are found in Australia, New Zealand and South America. As primary surface-water dwellers (*Anaspides*) they measure several cm, in contrast the derived settlers in psammal ground waters (*Stygocaris*) only attain a few mm.

Autapomorphy: Formation of a cephalothorax through the fusion of the first thoracic segment and the head (Fig. 76 A).

Plesiomorphies: Pleon with a complete complement of six appendage pairs. Tail fan of uropods and telson.

The **Bathynellacea** (*Bathynella*) are almost all tiny inhabitants in ground water bearing gravels and sands. Single species (Taxon *Hexabathynella*)

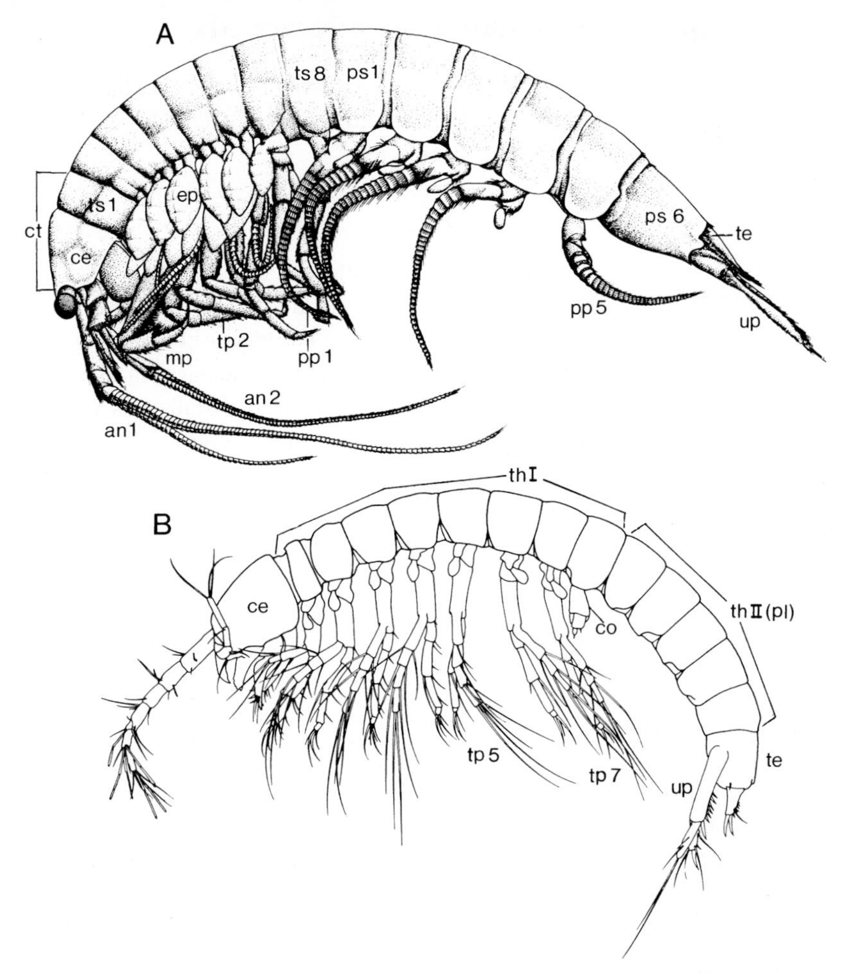

Fig. 76. Syncarida. **A** *Anaspides tasmaniae* (Anaspidacea). Female in side view. The cephalothorax "still" has a seam between the head and first thoracomere. **B** *Notobathynella longipes* (Bathynellacea). Male in side view. Thoracopod 8 a short copulatory organ. Pleopods 1–5 in this species completely reduced. *an* Antenna; *ce* cephalon; *co* copulatory organ (thoracopod 8); *ct* cephalothorax; *ep* epipodite; *mp* maxilliped; *pp* pleopods; *ps* pleon segment; *pt* pleotelson; *te* telson; *tp* thoracopods; *th I*, *th II* first and second section of the thorax; *ts* thoracic segment. (**A** Schminke 1978a; **B** Schminke 1978b)

are found on marine coasts in the boundary between limnic and marine biotopes (SCHMINKE 1973).

Autapomorphies: In the ground pattern only pleopods 1+2 and the uropods are found on the pleon. No tail fan. In the male, thoracopods 8 are a copulatory organ (Fig. 76B).

Plesiomorphy: No cephalothorax; instead eight free thoracic segments follow the head.

N. N.: Euphausiacea, Pancarida + Peracarida

An as yet unnamed monophylum with the Euphausiacea and Pancarida + Peracarida as adelphotaxa was established by RICHTER (1994b) with the following characteristic features.

Autapomorphies (Fig. 71 → 9)

- Distal displacement of the accessory cone cell nuclei.
  In the Anaspidacea (Syncarida) the four nuclei of the four crystalline cone cells (two main cone cells and two accessory cone cells) all lie in one plane on the crystalline cone. In the Euphausiacea and Peracarida the nuclei of the accessory cone cells are shifted distalwards in comparison to the main cone cells (Pancarida do not have eyes).
- Gills with a branch directed into the gill cavity.
  Euphausiacea and Lophogastrida (Peracarida) possess on their thoracopods epipodial biramous gills – one dorsolateral branch and one branch directed inwards underneath the body. The latter are not seen outside the Euphausiacea and Peracarida (Pancarida do not have any epipodites on thoracopods 2–8).

Euphausiacea

The cm-long pelagic krill are well known due to their habit of forming gigantic swarms – some species are the basic food of baleen whales.

A further impressive feature of the Euphausiacea is the light-producing organs with which they emit light (bioluminescence). The light organs appear mostly in ten-fold copies in one individual (one pair ventrally on the eye stalks; one pair on each of the coxa of thoracopods 2 and 7; four unpaired organs on the sternites of pleon segments 1–4). However, they cannot be placed in the ground pattern of the Euphausiacea without further study. In one single species with some plesiomorphic features – *Bentheuphausia amblyops* – light organs are totally missing; they may have first evolved within the Euphausiacea.

Autapomorphies (Fig. 71 → 10)

- Cephalothorax of cephalon and eight thoracic segments.
  The carapace is firmly attached dorsally to the first eight thoracomeres (Fig. 77 A, B).
- Subapical appendages on the telson.
  On each side of the last third of the tapered telson there are long, narrow processes. Similar structures are otherwise not found in the Crustacea (Fig. 77 C).

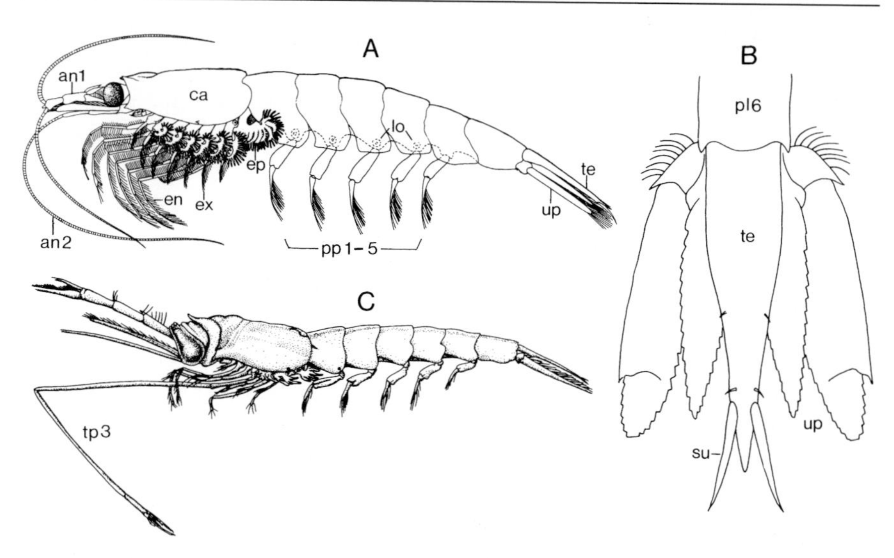

Fig. 77. Euphausiacea. **A** *Meganyctiphanes norvegica.* Filtering organism (ground pattern state). Thoracopods: Long endopodites with filter combs. Short two-part exopodites as rudder equipment. Dorsolateral branch of the gills (epipodites) projects out from underneath the carapace. **B** *Bentheuphausia amblyops.* Tail fan. Two subapical appendages on the telson (autapomorphy of the Euphausiacea). **C** *Stylocheiron suhmii.* Predator (apomorphous state). Endopodite of the third thoracopod greatly elongated. *an* Antenna; *ca* carapace; *en* endopodite; *ep* epipodite (gill); *ex* exopodite; *lo* light-emitting organ; *pl* pleon segment; *pp* pleopods; *su* subapical spine; *te* telson; *tp* thoracopod; *up* uropod. (**A, B** Hessler 1969 c; **C** Gruner 1993)

The conspicuous congruencies in the construction of the cephalothorax between the Decapoda and the Euphausiacea were, until recently, an argument for the interpretation of these two unities as adelphotaxa. Now established hypotheses postulate a closer relationship between the Euphausiacea and the Peracarida on the grounds of the special structure of the ommatidia. This forces us, a posteriori, to the assumption of a convergent evolution of the cephalothorax described above in the stem lineages of the Decapoda and the Euphausiacea. Thereby the following difference should be noted. In the Decapoda the lateral folds of the carapace overlap far down and form sealed gill cavities. In the Euphausiacea these folds are short and cover only the upper part of the gills; here, there are no gill cavities.

Neocarida [15]

The sister group relationship between the Pancarida and the Peracarida has been recognized for some time; I unite the two taxa under the name Neocarida.

Autapomorphies (Fig. 71 → 11)

- Lacinia mobilis on both mandibles of the adult.
  The lacinia mobilis is a mobile process between the pars incisiva and pars molaris of the mandibles; in the Pancarida and the Peracarida it is inserted in the vicinity of the pars incisiva (Fig. 78 D). The lacinia mobilis is found on both mandibles of adult individuals only in the Pancarida and the Peracarida (RICHTER 1994 b).
- Cephalothorax from the fusion of head and first thoracic segment.
- Transformation of the first thoracopod to a maxilliped.
- The epipodite of the maxilliped generates a ventilating current of water by oscillations in the respiratory cavity (Fig. 78 E).
- Direct development.

Pancarida – Peracarida

Pancarida

The Pancarida (Thermosbaenacea) form a species-poor unity of tiny, mm-long Malacostraca in marine and limnic subterranean waters and caves (*Monodella, Halosbaena, Tulumella*). The well-known *Thermosbaena mirabilis* has an unusual biotope – it is only known from a hot thermal spring in Tunisia.

Autapomorphies (Fig. 71 → 12)

- Short carapace.
  The first thorax section is only partly covered.
- Dorsal brood-pouch in the female.
  Development of the carapace after fertilization to a large brood-pouch in which direct development takes place (Fig. 78 C).
- On hatching thoracopods 7 and 8 are absent.
- Absence of segmental nephridial organs.
- Pleopods missing on pleon segments 3–5.
- No eyes.

---

[15] tax. nov.

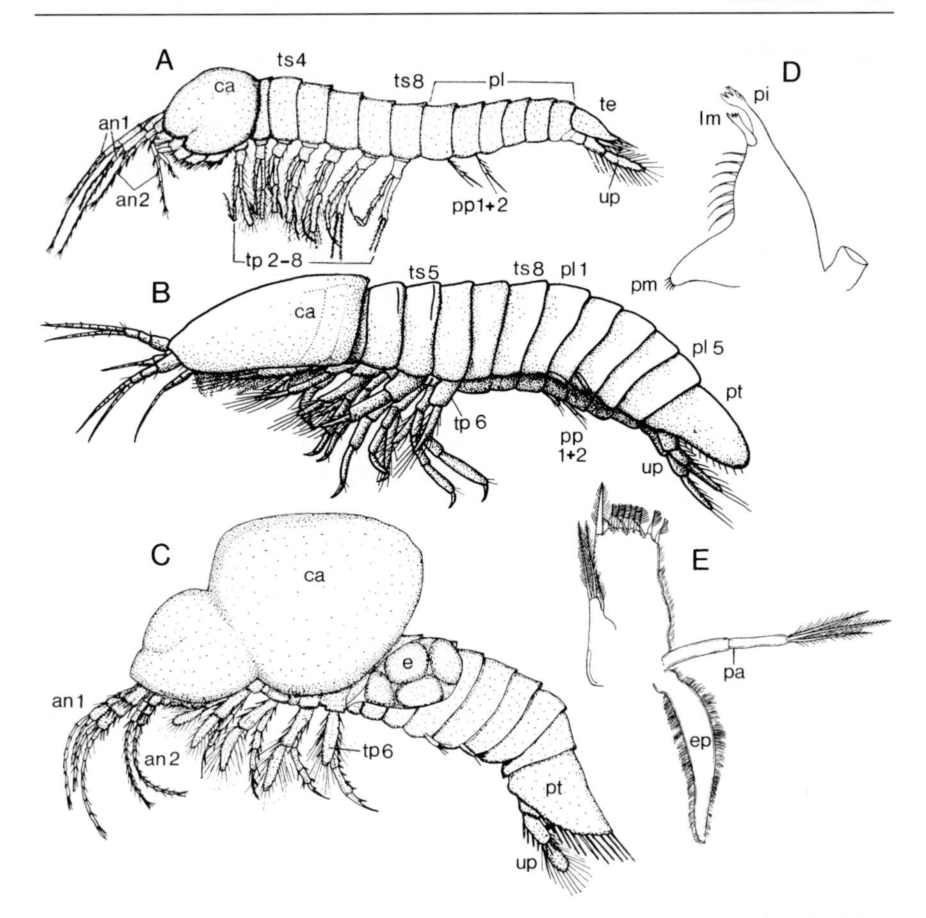

Fig. 78. Pancarida. **A** *Monodella argentarii.* Male. **B** *Thermosbaena mirabilis.* Female. **C** *Thermosbaena mirabilis.* Female in egg deposition stadium. The eggs are transported underneath the carapace, which is transformed into a large brood-pouch. **D** *Thermosbaena mirabilis.* Left mandible with pars molaris, pars incisiva and lacinia mobilis (autapomorphy of the Neocarida). **E** *Thermosbaena mirabilis.* Maxilliped with tongue-like epipodite used to produce a respiratory stream of water (autapomorphy of the Neocarida). *an* Antenna; *ca* carapace (in **C** a brood-pouch). *e* egg; *ep* epipodite; *lm* lacinia mobilis; *pa* palp; *pi* pars incisiva; *pl* pleon; *pm* pars molaris; *pp* pleopods; *pt* pleotelson; *te* telson; *tp* thoracopods; *ts* thoracic segment; *up* uropods. (**A,C** Zilch 1972; **B** Gruner 1969; **D,E** Monod 1940)

The apomorphic ground pattern features listed are realized in the organization of *Monodella* (Fig. 78 A); seven pereopods on thoracomeres 2–8 belong in the ground pattern as plesiomorphies. *Thermosbaena* is partially strongly derived – during its entire life there are no appendages on thoracomeres 7 and 8 (Fig. 78 B, C); pleotelson = a fusion of pleon segment 6 and the telson.

Peracarida

The Peracarida form the "successful" sister group of the Pancarida. They colonize in great numbers diverse marine biotopes – and have in addition conquered the fresh-water and terrestrial realms.

The outstanding characteristic feature of the Peracarida is a ventral brood-pouch formed of oostegites. A second unique feature results from the arrangement of embryonic cells on the germ band.

Autapomorphies (Fig. 71 → 13)

- Marsupium of oostegites.
  On thoracopods 2–8 the female forms basal, thin, broad lamellae that project inwards underneath the ventral side of the body where they medially overlap (Fig. 79 H, I). Thus, between the sternites and the brood plates, an enclosed brood-pouch develops in which the eggs are deposited and in which development up to juveniles takes place. The oostegites are interpreted as inwardly displaced epipodites.
- Cell pattern formation on the embryonic germ band.
  Disintegration of the caudal papilla. Ectoblasts are arranged in rows (SCHOLTZ & DOHLE 1996).

The Peracarida comprise seven unities with the traditional rank of "order", whose monophyly is established throughout.

A phylogenetic system of the Peracarida is, however, as yet not sufficiently resolved in detail (RICHTER 1994 a, b). I characterize here the Mysidacea, Amphipoda, Cumacea, Mictacea, Spelaeogriphacea, Tanaidacea and Isopoda with respect to their autapomorphies and refer to some probable kinship relations.

Mysidacea

With the monophyletic subtaxa Lophogastrida (*Lophogaster*) and Mysida (*Praunus*, *Neomysis*) the Mysidacea are relatively the most original unity of the Peracarida (Fig. 79 A).

The following plesiomorphic ground pattern features can be highlighted: Caridoid shrimp habitus. Extensive carapace that covers a large part of the first thorax section, but originally was connected to only one thoracic segment. Second antenna with scaphocerite. Six pleopods with endopodites and exopodites. Tail fan of uropods and telson. Antennal and maxillary nephridia present.

On the other hand RICHTER (1994 a, b) lists arguments for the disputed monophyly of the Mysidacea.

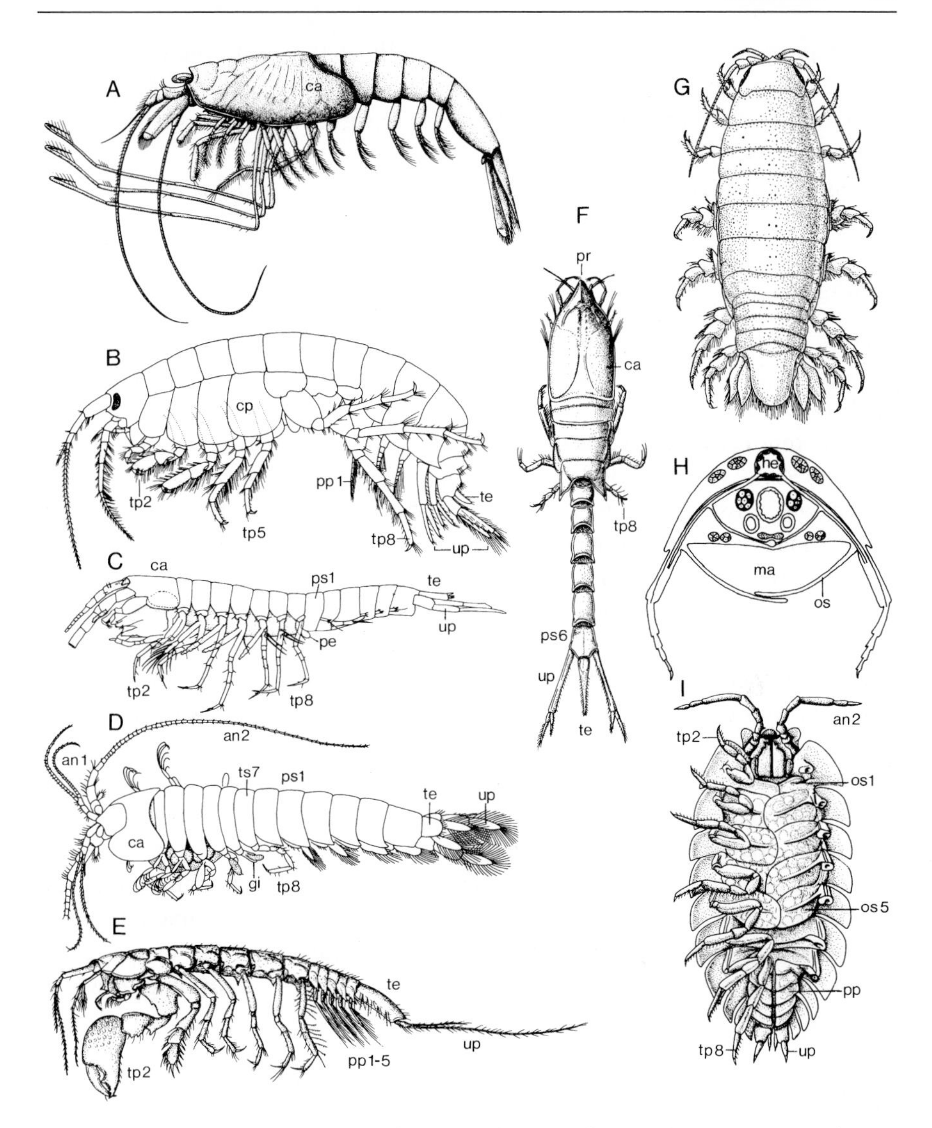

Fig. 79. Peracarida. **A** *Eucopia unguiculata* (Mysidacea, Lophogastrida). **B** *Gammarus locusta* (Amphipoda). **C** *Mictocaris halope*. Male. Flagella of the antenna truncated (Mictacea). **D** *Spelaeogriphus lepidops*. Female with marsupium. Oblique dorsal view (Spelaeogriphacea). **E** *Apseudes spinosa*. Male (Tanaidacea). **F** *Diastylis rathkei*. Female. Dorsal view (Cumacea). **G** *Cirolana borealis*. Dorsal view. With tail fan; pleotelson between the uropods (Isopoda). **H** Body cross section of the Isopoda. Diagram to demonstrate the ventral oostegites which overlap in the middle. **I** *Porcellio scaber*. Female. Ventral view. Marsupium of oostegites under thorax 1 (Isopoda). *an* Antenna; *ca* carapace; *cp* coxal plates; *gi* gill; *he* heart; *ma* marsupium; *mp* maxilliped; *os* oostegite; *pe* penis; *pp* pleopods; *pr* pseudorostrum; *ps* pleon segment; *te* telson; *tp* thoracopods; *ts* thoracic segment; *up* uropods. (**A–F, I** Gruner 1993; **G** Riedl 1963; **H** Remane 1957)

Autapomorphies: A clear protruding tooth at the posterior edge of the labrum. Specific congruencies in formation of the crystalline cone of the ommatidia. Plasmatic offshoots of the accessory cone cells surround the proximal part of the cone.

The Mysidacea are thought to be the adelphotaxon of all other Peracarida.

Amphipoda

Gammarus (marine, fresh water), *Orchestia*, *Talitrus* (semiterrestrial in tidal zones), *Talitroides* (terrestrial), *Corophium* (marine, fresh water), *Caprella*, *Cyamus* (marine).

Autapomorphies: Carapace absent. Second antenna without exopodite (Scaphocerite). Thoracopods 1–8 without exopodites. Thoracopods 2 and 3 subchelate. Oostegites only on thoracopods 3–6. Gills on thoracopods 3–8 shifted inwards; in the female they lie between ventral protuberances of the leg bases (coxal plates) and the oostegites. Maxillary nephridia absent.

In the organization of *Gammarus* probably the most ground pattern features of the Amphipoda can be found together. *Gammarus* species demonstrate the division (from which the name Amphipoda is derived) of the pereopods into four forward-directed pairs (thoracopods 2–5) and three backward-pointing pairs (thoracopods 6–8) (Fig. 79 B). They have further a characteristic differentiation of the pleopods into three anterior swimming legs (generate ventilating streams of water in the lateral resting position) and three posterior uropods as jumping legs.

The Amphipoda may form the sister group of all the following Peracarida unities which, with good justification, can be placed together in a monophylum **Mancoidea.** The essential **autapomorphy** is the manca stage in which they leave the marsupium. The manca stage is a juvenile in which the eighth thoracopod has not yet developed.

Cumacea

Diastylis, *Cumopsis*. Body length mostly between 5 and 10 mm. On marine soft substrates.

Unmistakable habitus of an egg-shaped anterior body and a long, thin pleon with pistil-like uropods (Fig. 79 F).

The Cumacea have an extensive carapace that covers several thoracic segments. Within the Mancoidea this is a plesiomorphic state. In contrast, the special arrangement of the carapace to form enclosed lateral gill chambers must be seen as a derived characteristic feature of the Cumacea.

Autapomorphies: Gill chambers with the following features. Paired lobes of the carapace combine to form a pseudorostrum to the anterior. Posterior edge of the carapace fused with the body wall, ventrally a narrow longitudinal slit. Two large epipodites from the first thoracopod pair with gills are found in the gill cavity. Their movements draw water ventrally inwards and drive it forwards under the pseudorostrum outwards. Antenna 2 without scaphocerite. As well as maxilliped 1 (ground pattern Neocarida) thoracopods 2 and 3 are also transformed into maxillipeds. Pleopods 1–5 are reduced in the female.

Mictacea

Mictocaris (Bermuda) in marine caves. *Hirsutia* (South America, Australia) in abyssal muddy substrates. Three species. Body length up to 3.5 mm.

Autapomorphies: Short carapace covers only the first thoracic segment. Strong reduction of the pleopods; they consist only of one branch and one article (male with larger, second pleopod of two articles) (Fig. 79C).

Spelaeogriphacea

Spelaeogriphus (South Africa). *Potiicoara* (Brazil). Fresh-water inhabitants in caves. Two species. Body length 7–9 mm.

Plesiomorphies: short carapace fold projects freely over the second thoracic segment; in comparison with the Tanaidacea this is plesiomorphic. Exopodites found on thoracopods 2–8.

Autapomorphy: Exopodites with one article on thoracopods 5–7 differentiated into gills (Fig. 79D).

Tanaidacea

Gigantapseudes maximus 77 mm long. Otherwise a body length of between 2 and 5 mm. *Neotanais, Tanaissus, Heterotanais.* Marine bottom dwellers in self-excavated burrows.

From the ground pattern of the Tanaidacea (SIEG 1984) two apomorphic features can quite clearly be determined that must have originated as novelties in the stem lineage of the unity as compared to the presumed sister group Isopoda.

Autapomorphies: Second thoracopod transformed into a large chela (Fig. 79E). Cephalothorax of head and two thoracic segments. Carapace fused dorsally with both of these segments. The existence of a carapace fold over thoracomeres 1 and 2 is a plesiomorphy in comparison with its absence in the Isopoda. The dorsal fusion with these thoracomeres must, however, be interpreted as an apomorphous state in comparison with the Spelaeogriphacea.

An argument for an **adelphotaxa relationship Tanaidacea – Isopoda** draws on the reduction of the exopodites on the thoracopods (present in the Spelaeogriphacea). In the ground pattern of the Tanaidacea remnants exist on thoracopods 2 and 3; in the Isopoda the exopodites are totally absent.

The pleotelson present in both taxa must be due to convergent fusion of the sixth pleomere and the telson since fossil Tanaidacea have six free pleon segments (SIEG 1984).

Isopoda

Idotea, Sphaeroma, Eurydice, Ligia (marine). *Philoscia, Oniscus, Porcellio, Armadillidium* (terrestrial). *Asellus* (fresh water).

In the ground pattern of the Isopoda (WÄGELE 1989, 1994) there are a number of apomorphic features with which the monophyly of the unity is well established.

Autapomorphies: Dorsoventrally flattened body with oval outline. Without carapace (convergence to Amphipoda). Exopodite of the second antenna vestigial. First and second maxillae without palps. Thoracopods all without exopodites (Fig. 79G,I). Heart in pleon; caudal displacement in correlation with the gill function of the pleopods. Separate molting of anterior and posterior body.

A tail fan of uropods and telson is a plesiomorphy in the ground pattern of the Isopoda. This was carried over from the ground patterns of the Eumalacostraca and Peracarida into the stem lineage of the Isopoda. In contrast, in the sister group Tanaidacea, pistil-like uropods can probably be placed in their ground pattern (apomorphy).

Maxillopoda

Autapomorphies (Figs. 60 → 11; 80 → 1)

- Trunk of ten segments and a telson with furca, with the following subdivisions.
  Thorax of seven segments. Thoracomeres 1–6 with six pairs of locomotory appendages (in the plesiomorphic manifestation as swimmerets). Thoracomere 7 with copulatory organ – a transformation product of the seventh pair of thoracopods.
  Abdomen of three appendageless segments before the telson.
- Nauplius eye of three pigment-cup ocelli.
  Apomorphous in contrast to the four ocelli that are postulated for the nauplius eye in the ground pattern of the Crustacea.

Since their establishment by DAHL (1956) there has been unending controversy over the validity of a high-ranking taxon Maxillopoda (Copepodoidea BEKLEMISCHEW 1958). In the above construction and division of the trunk, I find a convincing autapomorphy for judging the Maxillopoda

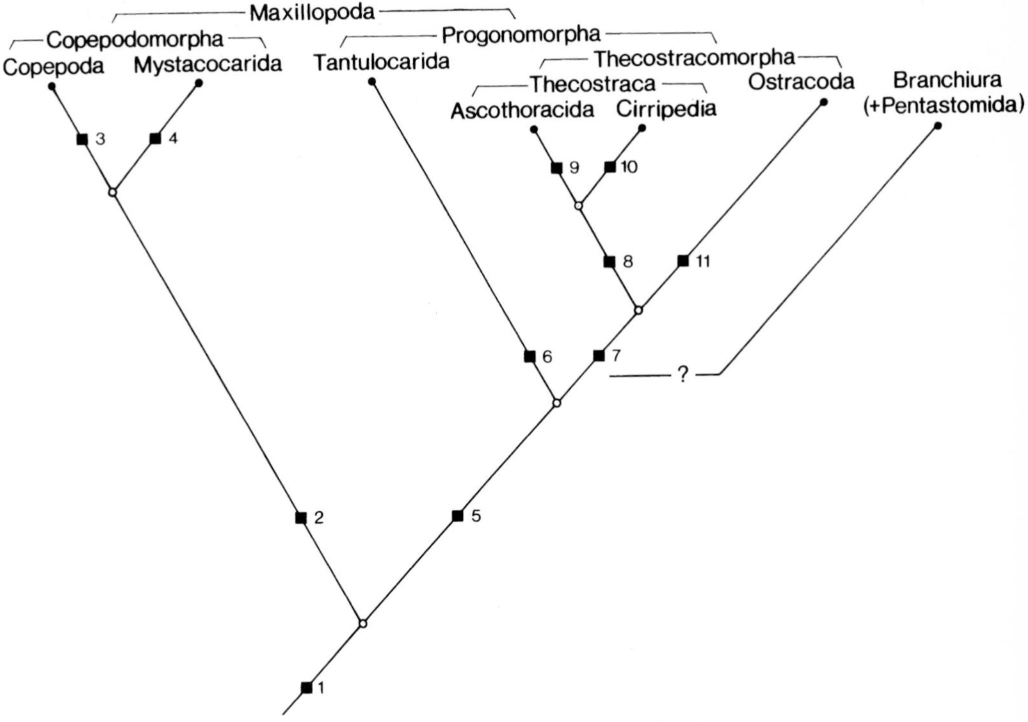

Fig. 80. Diagram of the phylogenetic relationships within the Maxillopoda

to be a monophylum. In contrast to the much larger number of thoracic and abdominal segments in the sister group Thoracopoda (p. 141), the **reduction to ten trunk segments** must be interpreted as an apomorphous condition.

The Copepoda, Mystacocarida, Tantulocarida, Ascothoracida, Cirripedia and Ostracoda can each be traced back to a **common stem species with seven thoracic and three abdominal segments**. Only with the Branchiura are there problems.

The outline of **a phylogenetic system for the Maxillopoda** is emerging (BOXSHALL & HUYS 1989; NEWMAN 1992; HUYS, BOXSHALL & LINCOLN 1993). I orient myself with the concepts of these authors using my own lines of argument.

Two large unities – the Copepodomorpha and the Progonomorpha – can be hypothesized to be the highest-ranking sister groups of the Maxillopoda.

For the **Copepodomorpha** the absence of a compound eye is judged to be an autapomorphy. The most parsimonious interpretation must postulate a unique reduction of the compound eye in a common stem lineage of the free-living Copepoda and Mystacocarida.

The essential autapomorphy of the **Progonomorpha** is the position of the female genital pores on the first thoracic segment – realized in the

Tantulocarida, Ascothoracida and Cirripedia. This is the most forward displacement of the female genital openings found in the Crustacea. Within the Progonomorpha a bivalved carapace evolved in a common stem lineage of the Ascothoracida, Cirripedia and Ostracoda.

The Branchiura remain problematical candidates. They possess four thoracic segments and an undivided abdomen; male and female gonopores are located on the fourth thoracomere and they have a flat, disc-shaped carapace.

Copepodomorpha – Progonomorpha

Copepodomorpha [16]

Autapomorphy (Fig. 80 → 2)

- Absence of compound eyes in all stages of the life cycle.

In the free-living Maxillopoda compound eyes are completely absent only in the Copepoda and Mystacocarida. This state is drawn on to establish a monophylum Copepodomorpha encompassing the two unities.

The absence of compound eyes in the parasitic Tantulocarida, which belong in the monophylum Progonomorpha, must be seen as a case of convergence.

Additionally, the Copepoda and Mystacocarida do not have a carapace. In accordance with the out-group comparison with the primary non-carapaced Thoracopoda this is a plesiomorphic, original state.

Since the two unities are alternately characterized by numerous autapomorphies, the Copepoda and Mystacocarida can be interpreted as sister groups.

Copepoda – Mystacocarida

Copepoda

With ca. 10,000 species and immense population densities the Copepoda form an important link in the food chain of marine and fresh-water ecosystems. From the mm-sized free-living species, parasitic Copepoda – in particular fish parasites – evolved in diverse lineages (KABATA 1979); this was connected with an enormous increase in body size. The maximum reached is 32 cm in the females of *Penella balaenoptera*, a parasite on Baleen whales.

---

[16] tax. nov.

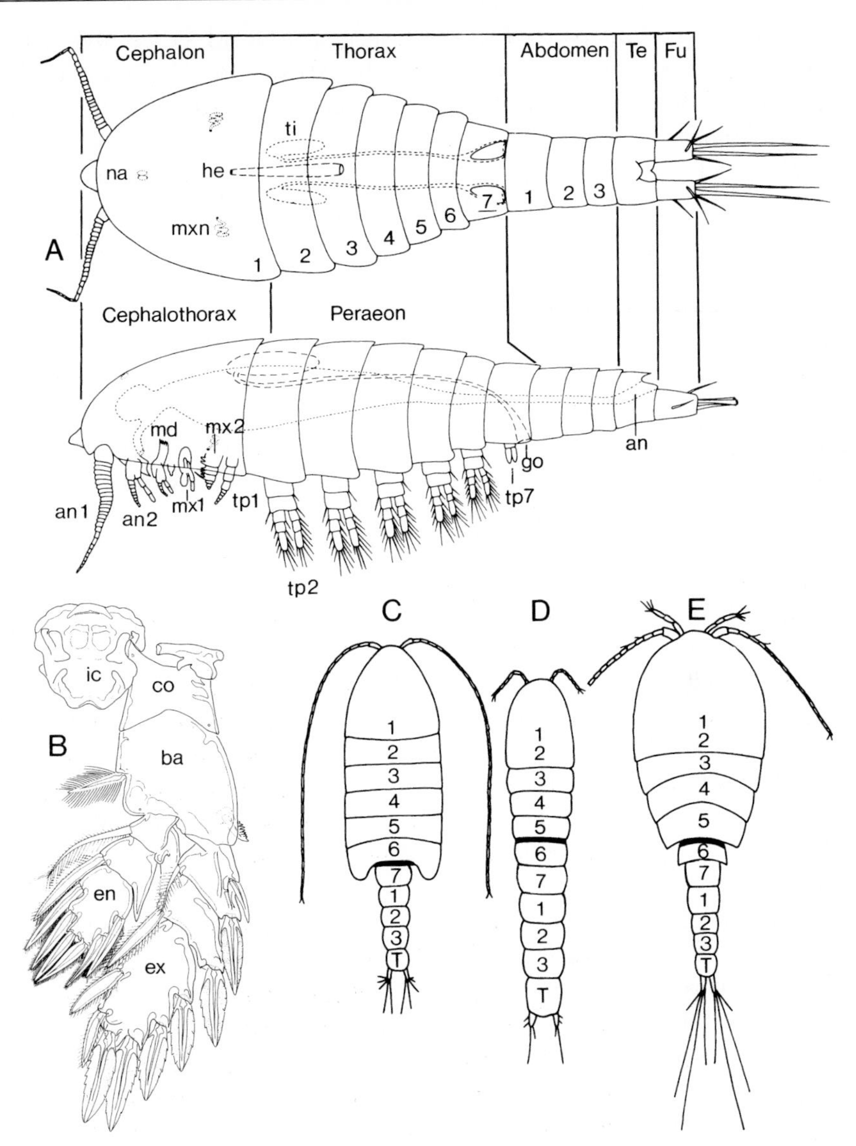

Fig. 81. Copepoda. **A** Ground pattern in dorsal and side view: Features of the free-living stem species of recent Copepoda. **B** *Platycopia inornata* (Platycopioida). Male. Thoracopod 6 with intercoxal plate on the coxa. **C-E** Position of the joint (*thick black line*) that subdivides the body into two mobile connected sections (morphological segmentation of the trunk into seven thoracic and three abdominal segments). **C** Calanoida with a joint between thoracomeres 6 and 7. **D** Harpacticoida. **E** Cyclopoida with a joint between thoracomeres 5 and 6. In the Calanoida the Cephalothorax consists of head+one thoracic segment, in the Harpacticoida and Cyclopoida of head+two thoracic segments. *an* Antenna; *as* anus; *ba* basis of the protopodite; *co* coxa; *en* endopodite; *ex* exopodite; *fu* furca; *go* gonopore; *he* heart; *ic* intercoxal plate; *md* mandible; *mx* maxilla; *mxn* maxillary nephridium; *na* nauplius eye; *te* (*t*) telson; *ti* testis; *tp* thoracopod. (**A** Boxshall 1983; **B** Huys & Boxshall 1991; **C–E** Gruner 1993)

From the **ground pattern** of the Copepoda (Fig. 81 A) we highlight the following features (BOXSHALL 1983; BOXSHALL et al. 1984). The body consists of the head, ten trunk segments and the telson with furca. There is no carapace.

The head is fused with the first thoracic segment to form a **cephalothorax** (cephalosoma). The associated thoracopods 1 follow the mouthparts as small maxillipeds. A comparable cephalothorax formation does not exist within the Maxillopoda.

The **subsequent trunk section consists of nine segments**, if we – as in the other Crustacea – do not count the telson, but rather see this together with the furca as a separate and terminal body part. With this we come to a discussion of the subdivision of the trunk into a thorax with appendages and an appendageless abdomen. Five thoracic segments follow the cephalothorax with **five pairs** of roughly equally sized **thoracopods**; these have three-part endo- and exopodites. An **intercoxal plate** connects the two appendages of a segment; as swimming legs they are thereby moved at the same time. Morphologically we are dealing with thoracopods 2–6. A **seventh pair of small thoracopods** obviously also belong in the ground pattern of the Copepoda; HUYS & BOXSHALL (1991) judge parts of the genital apparatus on the sixth free trunk segment (genital segment; morphologically thoracic segment 7) to be appendage derivatives. The appendageless **abdomen** then consists of three metameres – segments 8, 9 and 10.

If one wishes to refer to the free thorax following the cephalothorax as a pereon and its appendages as pereopods (p. 139), the Copepoda have a pereon of six segments with six pairs of pereopods (thoracomeres 2–7) in their ground pattern.

A special **joint** through which the body is divided into two mutually highly mobile sections can be highlighted as a prominent feature. We outline its states in three well-known taxa of free-living Copepoda (Fig. 81 C-E). In the Calanoida the joint lies between the sixth and seventh thoracic segments, which HUYS & BOXSHALL (1991) place as the original state in the ground pattern of the Copepoda. In contrast, in the Harpacticoida and Cyclopoida, the joint is found between thoracomeres 5 and 6; this is seen as a secondary displacement to the front and consequently assessed as the apomorphic state.

Compound eyes are absent in the ground pattern of the Copepoda. A nauplius eye with three pigment-cups is carried over into the adult from developmental stages. Antennal and maxillary nephridia stemming from the ground pattern of the Crustacea appear, however, at different stages in the development. The antennal nephridia developed in the nauplius are usually replaced by maxillary nephridia – the excretory organ of the adult.

Further ground pattern features result from the complex: **sexual organs – reproduction – development.** The stem species of the Copepoda was a

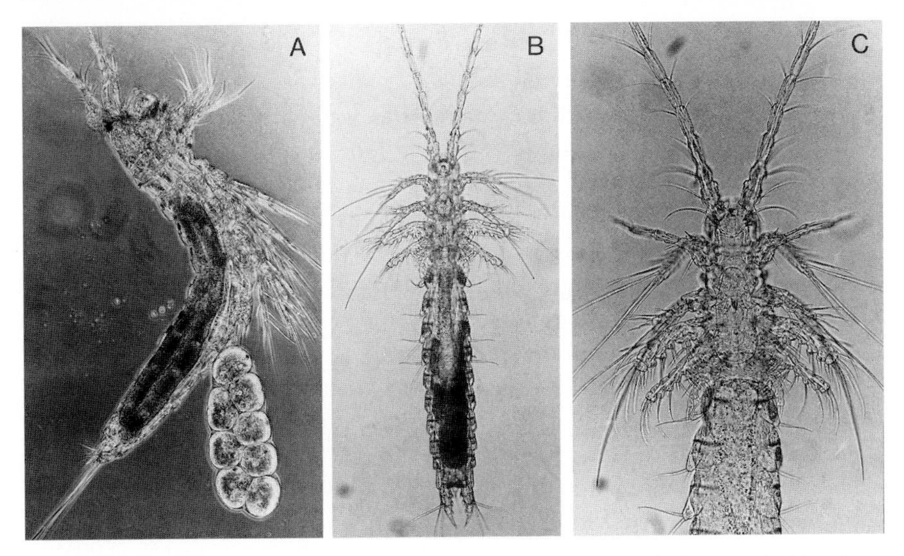

Fig. 82. Copepoda and Mystacocarida. **A** *Paraleptastacus spinicauda* (Harpacticoida, Copepoda). Female with egg-sac on the seventh thoracic segment (possible autapomorphy of the Copepoda). **B–C** *Derocheilocaris remanei* (Mystacocarida). **B** Habitus. **C** Anterior half of the body. Following the two antennae, the mandibles and maxillae 1 and 2 are impressive large laterally projecting appendages (autapomorphy of the Mystacocarida). (Originals: **A** Sandy Beach, Lappvik, Gulf of Finland; **B, C** Sandy Beach, Canet Plage, Mediterranean)

gonochoristic organism with testes or ovaries in the anterior thoracic section. In both sexes the gonoducts open in the last pereon segment (thoracic segment 7).

In the male the first antennae form a grasping organ for clasping females in the transfer of spermatophores. This takes place through a tube-shaped sperm container which is formed at the end section of the sperm ducts and which is applied to the vicinity of the female genital opening. The females possess paired seminal receptacles next to the genital pores into which the sperm from the spermatophores is transferred. The eggs are grouped together and deposited in egg-sacs which are carried on the body (Fig. 82 A). There are probably two sacs on the openings of the oviducts in the ground pattern.

The development with 11 stages demonstrates original traits. After six nauplii (resp. metanauplii) follow five copepodite stages with an organization and mode of life similar to the adult; the twelfth stage is the adult without further molting.

Autapomorphies (Fig. 80 → 3)

- Cephalothorax (Cephalosoma).
  Fusion of cephalon and thoracomere 1, as well as the transformation of thoracopod 1 into a uniramous grasping foot (maxilliped).
- Intercoxal plate.
  Median connection between the coxae of the thoracopods of a segment (Fig. 81 B).
- Antenna 1 in the male a grasping organ.
- Formation of spermatophores.
- Seminal receptacle for sperm storage.
- Formation of egg-sacs through gland secretions of the oviducts (Fig. 82 A).
  According to HUYS & BOXSHALL (1991), it is questionable as to whether the phenomenon belongs in the ground pattern of the Copepoda or first evolved within the Copepoda.

A **phylogenetic system of the Copepoda** has recently been presented by HUYS & BOXSHALL (1991) with the slight drawback that ten taxa are arranged as equally ranked "orders" which is not possible. In this system the species-poor **Platycopioida** form the adelphotaxon of all others which are grouped under the name **Neocopepoda**. Within the Neocopepoda the Calanoida and Podoplea are hypothesized to be the highest-ranking sister groups. The latter are characterized by the above-mentioned tagmosis with a body joint between the fifth and sixth thoracic segments.

In the Podoplea the Harpacticoida form a kinship group with the Mormonilloida, Poecilostomatoida, Siphonostomatoida and Monstrilloida. The opposing sister group includes the Cyclopoida, Gellyelloida and Misophrioida. We cannot follow this further here, but want to emphatically emphasize one point. The three taxa of free-living Copepoda familiar to every zoologist, that is the Calanoida (*Calanus, Diaptomus*), the Harpacticoida (*Harpacticus, Canthocamptus*) and the Cyclopoida (*Cyclops, Notodelphis*), do not form a monophyletic unity within the Copepoda; rather, they are each more closely related to other taxa than to one another.

Mystacocarida

The Mystacocarida are characteristic interstitial inhabitants of marine sands. The few known species are divided into two supraspecific taxa *Derocheilocaris* and *Ctenocheilocaris*.

Derocheilocaris species colonize eulitoral and supralitoral areas, i.e., high on the beach. Examples have been closely analyzed from both sides of the Atlantic, but may be distributed worldwide in warm water areas. In contrast the two *Ctenocheilocaris* species live in sublitoral zones; they have been found in South America (Chili, Brazil) at ca. 25 m water depth (DAHL 1952; RENAUD-MORNANT 1976).

Derocheilocaris typicus and *Derocheilocaris remanei* show corresponding distribution patterns in tidal zones on the east coast of the USA and the French Atlantic coast, respectively (RENAUD-DEBYSER 1963; HALL 1972). Their specific biotope is the mid to upper intertidal region. Both species colonize at ca. 10–40 cm depth in sandy substrates, and they remain here at low tide in the damp sand zone above the ground water.

Derocheilocaris remanei live in high population densities in the tideless Mediterranean in the supralitoral zone landwards of the retreating surf (DELAMARE DEBOUTTEVILLE 1954; JANSSON 1966). In the subterranean damp sand zone above the ground water level a maximum of ca. 1300 individuals/100 cm^3 sand have been counted (AX 1969).

The organization and mode of life of the Mystacocarida are constrained by the **need to fit into the minute spaces between grains of sand.**

The vermiform body measures about 0.5 mm; it can reach a maximum of 1 mm in length. The division into head, ten free trunk segments and telson with furca is original. There is no cephalothorax; a carapace is absent.

A third of the way along the long head a conspicuous transverse furrow is found. This separates off a frontal part with the first antennae. Four separated ocelli lie dorsally on this section; they are interpreted as secondarily separated elements of a nauplius eye with three eye-cups (DAHL 1952); two ocelli correspond to the lateral cups; the two other ocelli may have originated from a separation of the median eye-cup.

A large tongue-shaped labrum which covers the bases of both maxillae and the first thoracopods inserts ventrally in the second section of the head.

We now come to the appendages. The first antennae as uniramous multiarticulate elements do not show any peculiarities. In contrast, the second antennae and the mandibles are of great interest. These are unusual large branched appendages with endo- and exopodites that are carried over from the nauplial organization into the adult. This circumstance has a convincing functional explanation. The **second antennae** and the **mandibles** are the typical **locomotory organs** of the Mystacocarida (LOMBARDI & RUPPERT 1982). The two appendages carry out coordinated cyclic movements. The exopodites are pressed dorsolaterally against the sediment and produce a downward-directed force with which the endopodites are pressed ventrally against a sand grain; their cyclic forward and backward strokes push the creature jerkily through the interstices between the sand grains.

Both maxillae and the **first thoracopod** form together a second functional unit **used in feeding.** The maxillae are again very large, but unira-

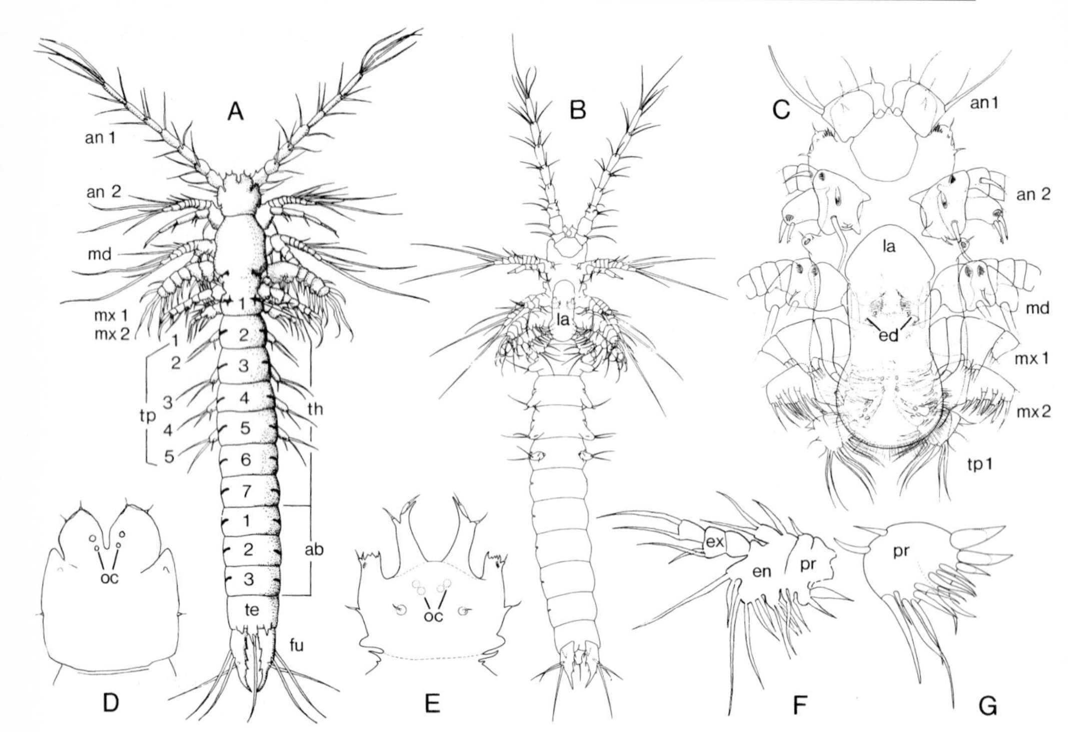

Fig. 83. Mystacocarida. *Derocheilocaris* and *Ctenocheilocaris*. **A** *Derocheilocaris remanei*. Dorsal view. **B** *Derocheilocaris typicus*. Ventral view. **C** *Derocheilocaris remanei*. Anterior body in ventral view. Insertion of the mouthparts and of the first thoracopods. A huge labrum covers the bases of the mandibles, the two maxillae and the first thoracopods. Mandibles with endites, maxillae and thoracopods 1 with medially directed setae under the labrum. **D** *Derocheilocaris remanei*. Anterior head section with four separate ocelli. **E** *Ctenocheilocaris claudiae*. Head with prominent rostral outgrowths; likewise four ocelli. **F** *Derocheilocaris remanei*. First thoracopod with endo- and exopodite. **G** *Ctenocheilocaris claudiae*. First thoracopod consists only of the protopodites. *ab* Abdomen; *an* antenna; *ed* endite; *en* endopodite; *ex* exopodite; *fu* furca; *la* labrum; *md* mandible; *mx* maxilla; *oc* ocellus; *pr* protopodite; *te* telson; *th* thorax; *tp* thoracopod. (**A** Delamare Deboutteville 1954; **B** Hessler 1969a; **C** Cals, Delamare Deboutteville & Renaud-Mornant 1968; **D** Noodt 1954; **E, G** Renaud-Mornant 1976; **F** Delamare Deboutteville & Chappuis 1954)

mous, appendages. In *Derocheilocaris* the first thoracopod is a branched appendage with endo- and exopodites (ground pattern); in *Ctenocheilocaris* it is reduced except for the protopodite. Food uptake is closely connected with locomotion. The medial setae of the maxillae and the first thoracopod scrape single-celled algae and bacteria from the sand grains while creeping through the interstices and convey them to underneath the large labrum; from here they are transported by the endites of the mandibles to the mouth.

From the trunk follow segments 2–5 with four pairs of tiny, undivided plates; these appendages have little more than a bracing function in movement. However, appendage muscles are found not only in segments 2–5 but also in the following appendageless segments 6 and 7 (HESSLER 1964; HUYS 1991).

From these facts the following statement can be made. The Mystacocarida have **a thorax of seven thoracomeres** that originated from seven appendage-bearing segments and an **abdomen of three** primary appendageless **segments.** With their immigration into the interstitial sand system thoracic legs 2–5 were transformed into small plates and thoracic legs 6 and 7 were fully reduced in the stem lineage of the Mystacocarida.

In this interpretation the copulatory organ on thoracomere 7 is secondarily absent. Eggs are deposited between the sand grains. Insemination takes place in the interstices. Post-embryonic development begins with a metanauplius; in *Derocheilocaris remanei* there are 11 stages before the adult.

From the organization outlined I highlight the following essential apomorphic characteristics and add two further features.

Autapomorphies (Fig. 80 → 4)

- Four separate ocelli on the head. Interpreted as being due to disintegration of the nauplius eye (Fig. 83 D).
- Long, tongue-shaped labrum reaching to the posterior end of the head (Fig. 83 C).
- Second antennae and mandibles branched appendages with endo- and exopodites (Figs. 82, 83). Locomotory appendages. The continuation of the nauplius state in the adult can be judged an autapomorphy.
- First maxilla, second maxilla and first thoracopod scraping and scratching structures for food uptake.
- Thoracic appendages 2–5 only developed as small undivided plates (Fig. 83A, B).
- Thoracic appendages 6 and 7 fully reduced.
- Location of the genital pores in the ♂♂ and ♀♀ on the fourth thoracomere.

  This state must be interpreted as the result of a secondary rostral displacement in comparison with the position of the pores on segment 7 in the Copepoda. The formally identical position in the Branchiura will be discussed later (p. 202).
- Lateral furrows in the posterior head section and on all trunk segments (Fig. 83 A).

  These are attachment sites for ventral longitudinal muscles (HESSLER 1964).

Progonomorpha[17]

Gonopores on thoracic segment 7 – as realized in the Copepoda – belong in the ground pattern of the Maxillopoda. The **position** of the **female gonopore on the first thoracic segment** must be seen as a derived state. This displacement to the anterior has long been known from the Ascothoracida and the Cirripedia and has recently been proven for the Tantulocarida (HUYS, BOXSHALL & LINCOLN 1993). For the Ostracoda, due to the shortening of the trunk, no assertion is possible; having a carapace, however, they belong in the subtaxon Thecostracomorpha.

Autapomorphy (Fig. 80 → 5)

- Location of the female genital pores on the first thoracic segment.

The Tantulocarida are, like the Copepoda and the Mystacocarida, primary non-carapaced Maxillopoda. This can be assumed despite large modifications as a consequence of the change to parasitism. In this interpretation the Tantulocarida are placed as the sister group opposite the Thecostracomorpha with carapace.

Tantulocarida – Thecostracomorpha

Tantulocarida

The Tantulocarida (BOXSHALL & LINCOLN 1983) are tiny ectoparasites on various Crustacea (Copepoda, Ostracoda, Malacostraca). The tantulus larva only reaches ca. 85–150 μm in length. Males and females of bisexual reproduction measure around 0.3–0.5 mm. Females with presumed parthenogenetic propagation can reach 1 mm.

Basipodella, Stygotantulus, Deoterthron, Microdajus.

We describe the above-mentioned stages in the life cycle of the Tantulocarida (BOXSHALL & LINCOLN 1989; BOXSHALL 1991; HUYS, BOXSHALL & LINCOLN 1993).

The **tantulus larva** (Fig. 84 A) represents the infection stage. The head has no appendages. The trunk consists of six segments with appendages (1–5 biramous; 6 uniramous) and an appendageless abdomen, whose segment number (including telson) within the unity can vary from between two to seven.

The tantulus larvae have a special attachment organ – a ventral oral disc at the anterior. This serves to permanently attach the animal to the host. With a stylet in the interior of the head a tiny hole (diameter 1–2 μm) is

[17] tax. nov.

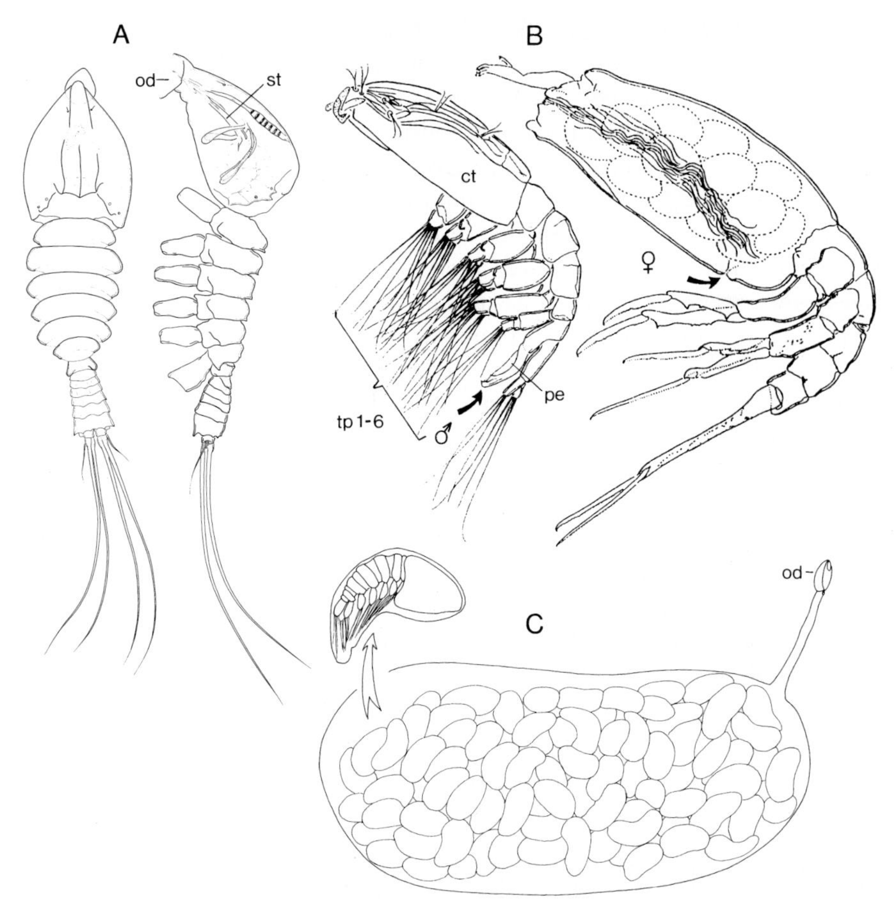

Fig. 84. Tantulocarida. **A** Tantulus larva from *Stygotantulus stocki* in dorsal and side view. Six thoracomeres with legs (rami of the branched appendages not shown) and seven abdominal segments (? incl. telson) without appendages. **B** Diagram of male and female from the bisexual reproductive cycle. **C** Female of *Microdajus langi* with parthenogenetic reproduction. Trunk-sac filled with tantulus larvae in egg envelopes. *ct* Cephalothorax; *od* oral disc; *pe* penis; *st* stylet; *tp* thoracopods. (**A** Boxshall & Huys 1989; **B** Huys, Boxshall & Lincoln 1993; **C** Boxshall 1991)

bored in the integument of the host – this is the only access to the host during the whole life cycle. The bodily fluids of the host organism are probably absorbed through root-like tissue processes of the larva; a gut is absent here as well as in the adult.

The **male and female of the bisexual cycle** (Fig. 84B) are astomous free-swimming organisms that do not feed. They develop through a far-reaching metamorphosis de novo in the interior of the tantulus larva. The process has been studied in the male. The posterior part of the larval thorax swells to a large trunk-sac in which a complete reorganization of the tissues takes place. From a dedifferentiated cell mass the male develops without molting into a sexually mature individual and breaks out of the sac; all external larval elements are left behind.

The adult male has the following divisions: cephalothorax with two integrated appendage-carrying thoracomeres – four free thoracic segments (3–6) with a further four pairs of swimming legs – thoracic segment 7 with a penis as a fusion product of the seventh thoracopods – apparently only one abdominal segment without appendages – telson with furca. Appendages are absent on the head as in the tantulus larva. Two rostral clusters of esthetascs are interpreted as remnants of the first antennae.

The adult female has similarly a cephalothorax of head and two thoracomeres; these, however, are without appendages. There then follows only two trunk segments with appendages and two trunk segments without appendages before the telson with furca. In contrast to the male the female has a pair of rod-shaped nonarticulated first antennae. Finally, the location of the genital pore ventrally on the cephalothorax in the region of the first thoracomere must be emphasized.

We now come to the second form of the **female with** presumed **parthenogenetic reproduction** (Fig. 84 C). These females possess long heads with an adhesive disc and a sac-shaped unsegmented trunk; using the adhesive disc they remain anchored onto the host during their entire life. The head originates from the head of the tantulus larva, the trunk develops from the larval trunk after molting and shedding of the larval appendages. Eggs develop in the trunk, from which the new tantulus larvae originate. The question concerning the ratio of the different females in the life cycle of the Tantulocarida is unanswered.

Autapomorphies (Fig. 80 → 6)

- Larval parasitism.
  The correlation of most features with this mode of life is reasonable – not, however, the number of more than three abdominal segments in the tantulus larva and less than three in the adult.
- Reduction of all head appendages except for antenna 1 in the female with sexual reproduction.
- Reduction of mouth and gut.
- Larva with adhesive disc to adhere to the host and with stylet to bore into the hosts skin. Uptake of the host's bodily fluids as nutrients (Fig. 84 A).
- In the male and female with bisexual reproduction the Cephalothorax consists of head and two thoracomeres.
- Male with one abdominal segment between the thorax and telson.
- Female with a further shortened trunk (only two pairs of appendages) (Fig. 84 B).

- Evolution of a second form of the female with parthenogenetic reproduction.

Thecostracomorpha[18]

Autapomorphy (Fig. 80 → 7)

- Bivalved carapace with adductor muscle within the framework of the Maxillopoda.

The placing of the Tantulocarida in the Maxillopoda as primary non-carapaced crustaceans forces the hypothesis of at least a double convergent evolution of a body enveloper known as a carapace – once in the stem lineage of the Phyllopoda + Malacostraca and once in the common stem lineage of the Ascothoracida, Cirripedia and Ostracoda. Furthermore we can unite the Ascothoracida and Cirripedia as Thecostraca and place them opposite the Ostracoda as the sister group.

Thecostraca – Ostracoda

Thecostraca

Autapomorphies (Fig. 80 → 8)

- Three-part crystalline cone in the ommatidia of the compound eye of the larva. Formation from three crystal cone cells (semper cells). Absence of compound eyes in the adult.
- Reduction of the second antenna (in the ground pattern of the Ascothoracida rudiments are present).
- Compared with the Ostracoda the carapace does not have a hinge margin.

Within the Crustacea the **formation of the three-part crystalline cone** is only known from the Ascothoracida, Cirripedia and Facetotecta (HALLBERG & ELOFSSON 1983; HALLBERG, ELOFSSON & GRYGIER 1985; GRYGIER 1987). This apomorphous state can, without opposition, be interpreted as an autapomorphy of a monophylum Thecostraca. This interpretation is consistent with the hypothesis of a single reduction of the second antenna in the stem lineage of the Thecostraca – obviously, convergent with the loss of the second antenna in the stem lineage of the Tracheata (p. 203).

[18] tax. nov.

GRYGIER (1987) took the name Thecostraca from the beginning of this century and extended it to include the three taxa Ascothoracida, Cirripedia and Facetotecta. Since the Facetotecta (GRYGIER 1985) have only been found as larvae (nauplius, cypris), they cannot be discussed here; the remaining Ascothoracida and Cirripedia are treated as adelphotaxa.

Ascothoracida – Cirripedia

Ascothoracida

Species-poor group of marine parasites on or in Anthozoa and Echinodermata.

Ectoparasites: *Waginella, Synagoga.*

Endoparasites: *Gorgonolaureus, Baccalaureus, Petrarca, Ascothorax, Dendrogaster.*

For the ordering of the essential ground pattern features only the ectoparasites with comparatively more original organizational traits are of use. We divide the relevant features directly into autapomorphies and plesiomorphies, as they can be postulated for the stem species of the Ascothoracida.

Autapomorphies (Fig. 80 → 9)

- First antenna terminal with pincer-like grasping organ for attaching to the host (Fig. 85D).
- Mouthparts (mandibles, first and second maxillae) transformed into piercing setae. The setae or stylets are enclosed by a tapered cone-shaped labrum (Fig. 85C, E).

Plesiomorphies compared with the Cirripedia.

- Adult with bivalved carapace.
- Division of the trunk into a thorax of seven segments, abdomen of three segments, telson with furca (Fig. 85C).
- Thoracomeres 1–6 with biramous swimming legs.
- Nauplius larva with round outline and without frontal side-horns.
- During development an ascothoracid stage with carapace that can be homologized with the cypris larva of the Cirripedia follows the nauplius larva. In contrast to the cypris larva, however, the ascothoracid stage is capable of nutrient uptake.

Cirripedia

"**Sessility is the crucial point**" in the mode of life and organization of the Cirripedia. Barnacles primarily colonize abiotic hard substrates. From here, various species have come to use other marine creatures as obligate,

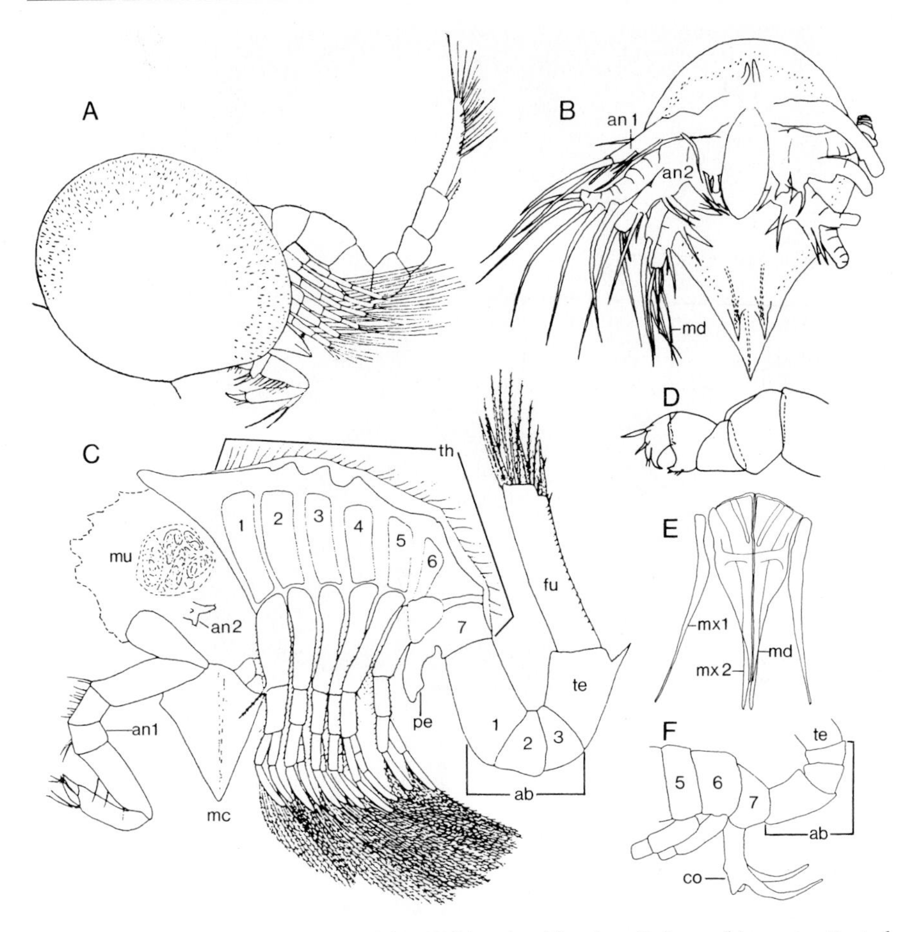

Fig. 85. Ascothoracida. **A** *Synagoga mira.* Habitus in side view. **B** *Laura bicornuta.* Ventral view of the nauplius. Setae on the appendages of the left-hand side of the body removed. **C** *Synagoga sandersi.* Female. Body in side view after removal of the carapace. With rudimentary second antennae and short equivalent of penis. **D** *Ascothorax ophioctenis.* Last articles of the first antenna forming a subchela. **E** *Synagoga sandersi.* Piercing stylets (mandibles, first and second maxillae) of the oral cone. **F** *Synagoga sandersi.* Male. Posterior end with copulatory organ on the seventh thoracomere. *ab* Abdomen; *an* antenna; *co* copulatory organ; *fu* furca; *mc* oral cone (labrum with piercing stylets inside); *md* mandible; *mu* closure muscle; *mx* maxilla; *pe* equivalent of penis; *te* telson; *th* thorax; *tp* thoracopod. (**A** Krüger 1940 a; **B** Grygier 1987; **C, E, F** Newman 1974; **D** Schram 1986)

mobile biotopes; *Coronula diadema* sits on whales and *Chelonibia testudinaria* on turtles. From such epizoa a change to parasitism took place which reaches its culmination in the familiar *Sacculina carcini.* The system of root-like processes of the adult female in *Carcinus maenas* show no trace of crustacean features; the developmental cycle through a nauplius larva with frontal horns and the cypris larva has been known since the

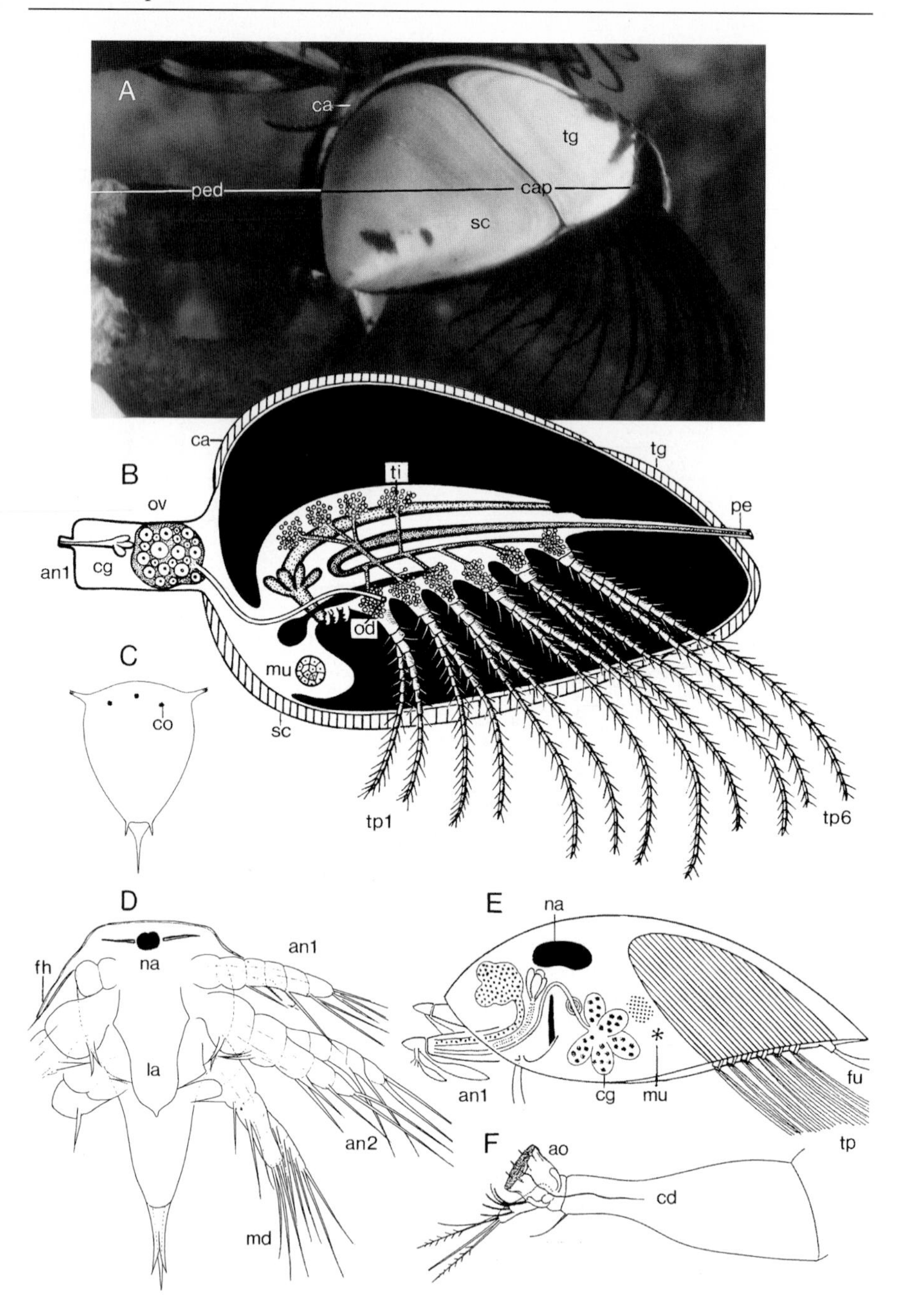

last century, the morphological differentiation of the male and female larva was only fully explained 100 years later (HOEG 1991).

We cannot follow the fantastic evolutionary paths within the monophylum Cirripedia further here, as they are not relevant for the elucidation of the

◀ **Fig. 86.** Cirripedia. **A** *Lepas anatifera* in morphological orientation. Anterior body with pedunculus to the left, posterior body with capitulum to the right. Carina dorsal, scuta and terga lateral. **B** *Lepas*. Diagram of organization. The head consists of the pedunculus (stalk) and the first quarter of the capitulum. The trunk with six pairs of cirri is the rest of the capitulum. **C** *Balanus improvisus*. Metanauplius with frontal side-horns, nauplius eye, compound eyes and caudal spine. **D** *Verruca stroemia*. Nauplius. **E** *Sacculina carcini*. Male cypris larva with carapace; first antenna with adhesive organ. **F** *Semibalanus balanoides*. First antenna of the cypris larva. Differentiation of the last but one article to an adhesive organ. *an* Antenna; *ao* adhesive organ; *ca* carina; *cap* capitulum; *cd* cement gland duct; *cg* cement gland; *co* compound eye; *fh* frontal horn; *fu* furca; *la* labrum; *md* mandible; *mu* closure muscle; *na* nauplius eye; *od* oviduct; *ov* ovary; *pe* penis; *ped* pedunculus; *sc* scutum; *tg* tergum; *ti* testis; *tp* thoracopod. (**A** Original; **B** Remane 1957; **C** Buchholz 1951; **D** Newman, Zullo & Withers 1969; **E** Hoeg 1987; **F** Nott & Foster 1969)

ground pattern. This must be derived from the comparatively original "Lepadomorpha" (*Lepas*) – a paraphylum within the Thoracica (GLENNER et al. 1995) – which have a stalk-shaped anterior body and sac-shaped posterior body as well as five calcareous plates developed in the carapace. Even the Balanomorpha (*Chelonibia, Balanus, Coronula*) are strongly derived through strong compression in the longitudinal axis of the body and the evolution of a crown-shaped wall of calcareous plates (eight plates in the ground pattern).

We highlight the **cirri** and **feeding methods** as prominent ground pattern features (Fig. 86 A, B). During the metamorphosis to adult the six pairs of thoracic swimming legs of the cypris larva are transformed into multiarticulate, thickly setose cirri; with these the Cirripedia fish and filter fine organic particles and plankton from the water. To assess this characteristic feature differentiations have to be made. The cirri and associated microphagous feeding are without doubt autapomorphies of the Cirripedia. However, the phenomenon of free nutrient uptake is itself a plesiomorphy in comparison with the pure parasitism of the sister group Ascothoracida. An abundance of other derived characteristic features are realized in all three stages of the life cycle – the nauplius, the cypris and the adult.

Autapomorphies (Fig. 80 → 10)

Nauplius larva

- Frontal horns with opening of glands of unknown function. With a caudal spine (Fig. 86 C, D).

  The horns of the nauplius larva are unique within the Crustacea. Already in the first larval stage, species of Cirripedia are identifiable as members of the unity – even parasites such as *Sacculina carcini* (Rhizocephala).

Cypris larva

- Carapace.

  Compared with the bivalved shell of the Ascothoracida the carapace of the cypris larva is dorsally undivided.

- Antenna 1 differentiated into an attachment organ with cement glands (Fig. 86 E, F).
- Boundary thorax – abdomen.
  An undivided penis structure sits caudally on the sixth thoracic segment. Since morphologically the penis belongs to the seventh thoracic segment, thoracomeres 6 and 7 may be fused. Two abdominal segments and the telson with furca follow (KRÜGER 1940 b).
- No nutrient uptake.

Adult
- Peduncle – capitulum.
  The division into stalk and capitulum is superimposed on the morphological body division (Fig. 86 A, B). The head consists of the peduncle and roughly the first quarter of the capitulum. The adjoining thorax extends over the rest of the capitulum.
- Carapace.
  Ventrally fused to form a sac-shape. Only in the posterior body does a slit remain for the extrusion of cirri and penis.
- Calcareous plates.
  Storage of calcite in the chitin cuticle in the form of separate plates. Five plates belong in the ground pattern of the Cirripedia – an unpaired dorsal carina and the paired scuta and terga on the sides of the body (Fig. 86 A).
- Thoracic appendages.
  Thoracopods 1–6 transformed into multiarticulate cirri for microphagous feeding.
  Thoracopods 7 form a long, mobile penis.
- Abdomen.
  In the course of the metamorphosis of the cypris larva to adult fully reduced.

Ostracoda

With several 1000 living species the Ostracoda are, after the Copepoda, the second largest taxon of the Maxillopoda. Like the Copepoda they are small organisms with an average length of 0.5 to 2 mm. Ostracoda have colonized all possible biotopes in marine and fresh waters.

We deal with certain **ground pattern** features significant for the assessment of kinship.

Of prime importance is the **carapace** with two large mussel-like valves, which fully enclose the body. The valves are each formed of two chitin lamellae. The outer calcareous lamellae are connected dorsally through an elastic chitinous membrane. The inner non-calcareous lamella (which is

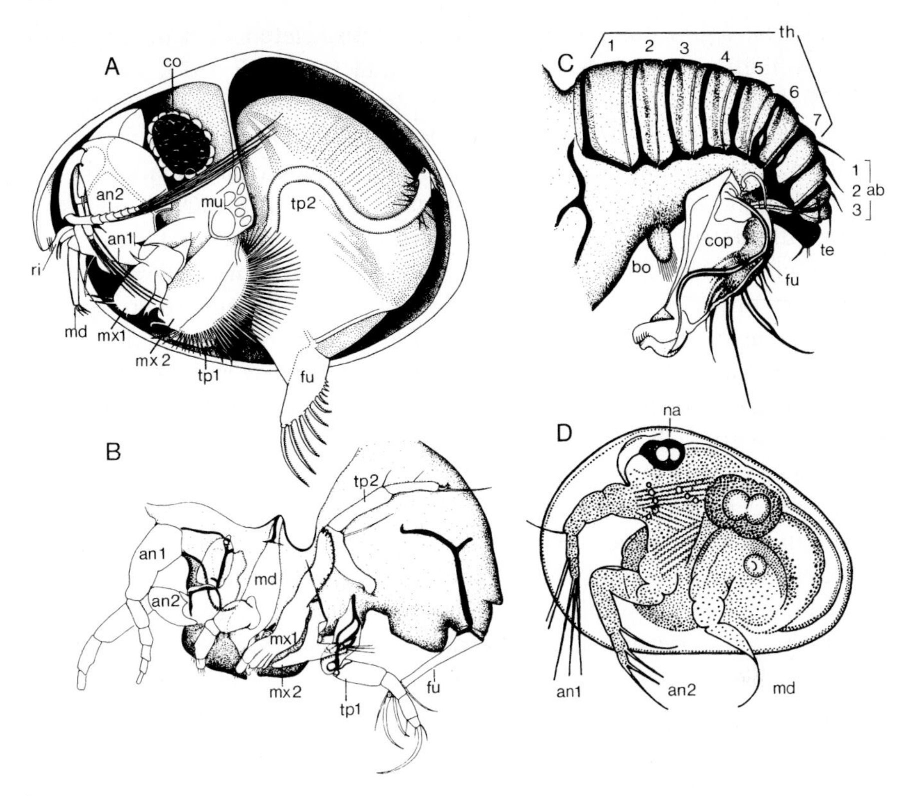

Fig. 87. Ostracoda. **A** *Deloris levis* (Myodocopida). Side view; left half of the carapace removed. Shell at the front with rostral incisor. Lamellae-like furca. Thoracopod 2 transformed into a cleaning leg. **B** *Heterocypris salina* (Podocopida). Macerated soft body in side view, female. Maxille 1 with respiratory plate (epipodite). Rod-shaped thoracopods (ground pattern of the Ostracoda). Rod-shaped furca. **C** *Cytherella* sp. (Podocopida). Lateral view of the trunk, male. Thoracopods removed. Chitinous skeleton in the wall of the trunk of regular longitudinal rods (black), which suggest a division into ten segments and a telson with furca. Since the copulatory organ can be assigned to segment 7, I count seven thoracic and three abdominal segments. **D** Nauplius (Cyprididae, Podocopida). With carapace. Second antennae and mandibles uniramous. *an* Antenna; *bo* brush-like organ; *co* compound eye; *cop* copulatory organ; *fu* furca; *md* mandible; *mu* closure muscle; *mx* maxilla; *ri* rostral incisor; *te* telson; *tp* thoracopod (**A, D** Gruner 1993; **B** Schulz 1975; **C** Schulz 1976, Tiemann 1984)

soft) continues into the body wall. A shell adductor muscle lies roughly in the middle of the body; it spans the body in the dorsal half.

Inside the shell, head and trunk are not sharply divided. Together they bear **seven pairs of appendages** and, in the male, **a paired copulatory organ.** Since on the head the full number of two pairs of antennae, the mandibles and two pairs of maxillae are developed, there are only two pairs of locomotory appendages on the trunk (Fig. 87 A, B). The paired copulatory organ lies ventrally a small distance from the end of the body. The furca is terminal.

Of special significance is the **chitinous ectoskeleton** of rods and apodemes which are embedded in the body wall (SCHULZ 1975, 1976; TIEMANN 1984). In *Cytherella* (Podocopida) ten longitudinal rods run at regular intervals over the trunk (Fig. 87 C); they are interpreted as the manifestation of former segment boundaries. The ectoskeleton provides then a foundation for tracing back the postcephalic body of the Ostracoda to a trunk with ten segments and telson – such as we hypothesize for the stem species of the Maxillopoda (p. 178). Since for this stem species a paired copulatory organ as an appendage derivative of the seventh trunk segment was postulated, the copulatory organ of the Ostracoda can without any problem be attributed to this metamere. In contrast it is not clear to which of the six previous thoracic segments the two thoracopods belong; they are purely conventionally counted as thoracic appendages 1 and 2. Their innervation also does not provide any clues. In *Cyprideis torosa* three nerve pairs extend from the ventral nerve cord into the trunk and swell here to form ganglia from which the two thoracopods and the copulatory organ are supplied (WEYGOLDT 1960). The innervation in the trunk is understandably limited to the three appendages which remain after the shortening of the body. Nevertheless clarification is found through the supply of the brush-like organs from the ganglion of the copulatory organs; these sensory organs realized in male Podocopida evidently also belong to the seventh thoracomere and are not the transformation product of the appendages of a further segment.

The male genital pores open in the copulatory organ and can therefore be placed in the seventh thoracic segment. In contrast, for the oviducts, which open in the region of the furca, no segment determination is possible.

The female Ostracoda have primary paired seminal receptacles next to the opening of the oviducts. A morphological comparison with the receptacles of the Copepoda has still to be made. Plesiomorphies in the ground pattern of the Ostracoda are paired compound eyes and a three-part nauplius eye.

Autapomorphies (Fig. 80 → 11)

- Extreme shortening of the trunk due to loss of original segmentation into ten metameres and a telson. Only two pairs of locomotory appendages (Fig. 87 A, B).
- Nauplius larva with carapace as well as uniramous second antennae and mandibles (Fig. 87 D).

The transformation of the trunk must have occurred in the stem lineage of the Ostracoda. The unsegmented state with two thoracopods for movement and the copulatory organ is postulated for the last common stem species of recent Ostracoda.

The Myodocopida and the Podocopida form the highest-ranking recent subtaxa in the traditional classification of the Ostracoda (HARTMANN 1966–1989). They can be justified as monophyla and proven to be adelphotaxa. I list for each unity some autapomorphies and plesiomorphies.

Myodocopida

Autapomorphies: Anterior edge of the carapace with rostral incisor for the protrusion of the antennae. Furca in the form of broad lamellae with strong spines (Fig. 87 A).

Plesiomorphies: Compound eyes and heart present. Antenna 2 with exopodite of 7–9 articles (swimming organ).

Cypridina, Philomedes, Cylindroleberis.

Podocopida

Autapomorphies: Exopodite of antenna 2 has a maximum of two articles. Maxilla 1 with large branchial plates. Compound eyes reduced. Heart and circulatory system absent (Fig. 87 B).

Plesiomorphies: Thoracopods 1 and 2 are stenopodia. Shell without a rostral incisor.

Cytherella, Cyprideis, Paradoxostoma (with stylet-like mandibles for piercing algal cells), *Cypris, Candona.*

Using the principle of parsimony we can hypothesize a homology of the Ostracoda shell with the Thecostraca carapace. Within the framework of this hypothesis the following fact is under discussion. In the Ostracoda the carapace is already present in the nauplius larva; it is formed here from the second antennal segment. This is unique among the Crustacea, but need not be an obstacle for the assumption of homology. The situation seen in the Ostracoda may be the result of a secondary forward displacement of the carapace construction in the earliest larval stage.

Branchiura (incl. Pentastomida)

Let us deal first with the Branchiura (fish lice) as they are traditionally presented, that is, without the Pentastomida. Fish lice are **temporary bloodsuckers** on marine and fresh-water fish, occasionally also on tadpoles. They unite in themselves two opposing life styles, which have consequently had a distinct influence on the construction of the dorsoventrally compressed body. (1) Fish lice are only periodic ectoparasites; in this capacity the head appendages function as grasping and attachment organs. (2) The rest of the time fish lice are elegant free swimmers, their locomotion is effected by the four appendage pairs of the thorax.

Fig. 88. Branchiura and Pentastomida. **A** *Argulus laticauda* (Branchiura). Male in ventral view. ▶ The mandibles lie in an oral pipe (proboscis). Before this is a preoral spine. Thoracopods 1 and 2 each bear a small flagellum for cleaning the inner wall of the carapace. **B** *Argulus foliaceus* (Branchiura). Compound eye in the carapace. The eye lies in a hemolymph lacuna that is surrounded by epidermis with cuticle; it can be moved by muscles in the region of the axon of the retinula cells. **C** *Cephalobaena tetrapoda* (Pentastomida). Total dorsal view (*left*) and anterior end with four short appendages (ventral view). **D** *Linguatula serrata* (Pentastomida) with noticeable secondary annulation. **E–F** Comparison of the sperm of **E** *Argulus foliaceus* (Branchiura) and **F** *Raillietiella* (Pentastomida). In the middle of the sperm from top to bottom: Longitudinal section through the transition region (centriol region) of pseudoacrosome and sperm body – cross section – longitudinal section through the sperm. *a* Axons of the retinula cells; *ab* abdomen; *an* antenna; *at* anterior; *ax* axoneme; *ca* carapace; *cc* crystalline cone; *ce* centriol; *cr* centriol region (transition region); *cu* cuticle; *do* dorsal; *dp* dorsal rod (hardening) of the pseudoacrosome; *dr* dorsal band in the sperm body; *ep* epidermis; *fl* flagellum; *fu* furca; *hl* hemolymph sinus; *mi* mitochondrium; *mu* muscle; *mx* maxilla; *nu* nucleus; *pa* pseudoacrosome; *pb* proboscis (oral tube); *ps* preoral spine; *rh* rhabdome; *tp* thoracopod; *ts* thoracic segment; *vp* ventral rod of the pseudoacrosome; *vr* ventral band in the sperm body. (**A** Hessler 1969b; **B** Hallberg 1982; **C, D** Gruner 1969; **E, F** Wingstrand 1972)

Dolops, Argulus, Chonopeltis.

Body length is usually between 5 and 20 mm. From the head with a full complement of appendages follows the thorax of four segments with the above-mentioned swimming legs (endo- and exopodites present) and an unsegmented abdomen with a tiny furca (Fig. 88 A). The number of segments fused in the abdomen is not known.

We can now turn to the specific derived features of the Branchiura.

Autapomorphies

- Flat carapace.
  The carapace itself can be seen as forming a plesiomorphic feature of the Branchiura. Definitely apomorphic, however, is its disc shape, which covers the dorsoventrally compressed body to the thorax – abdomen boundary – this being an adaptation for optimal attachment to the host.
- Antennae 1 and 2 transformed into short, uniramous hooks. Mandibles in *Dolops* tiny hooks close behind the mouth (ground pattern), in *Argulus* displaced into a long oral tube (proboscis) (apomorphy).
- Uniramous first maxillae. In *Dolops* with large grasping hooks (ground pattern), in *Argulus* with suckers (apomorphy).
- Uniramous second maxillae likewise with hooks.
- Sessile compound eyes.
  The existence of compound eyes is a plesiomorphy; their displacement under the body surface is apomorphic. The eyes, however, do not lie in an eye chamber (Phyllopoda) but rather in a hemolymph sinus (Fig. 88 B).
- Genital opening in the fourth thoracic segment.
- Attachment of eggs with secretion of the oviducts onto hard objects.
- Nauplius larva suppressed. Hatching at copepodite stage.

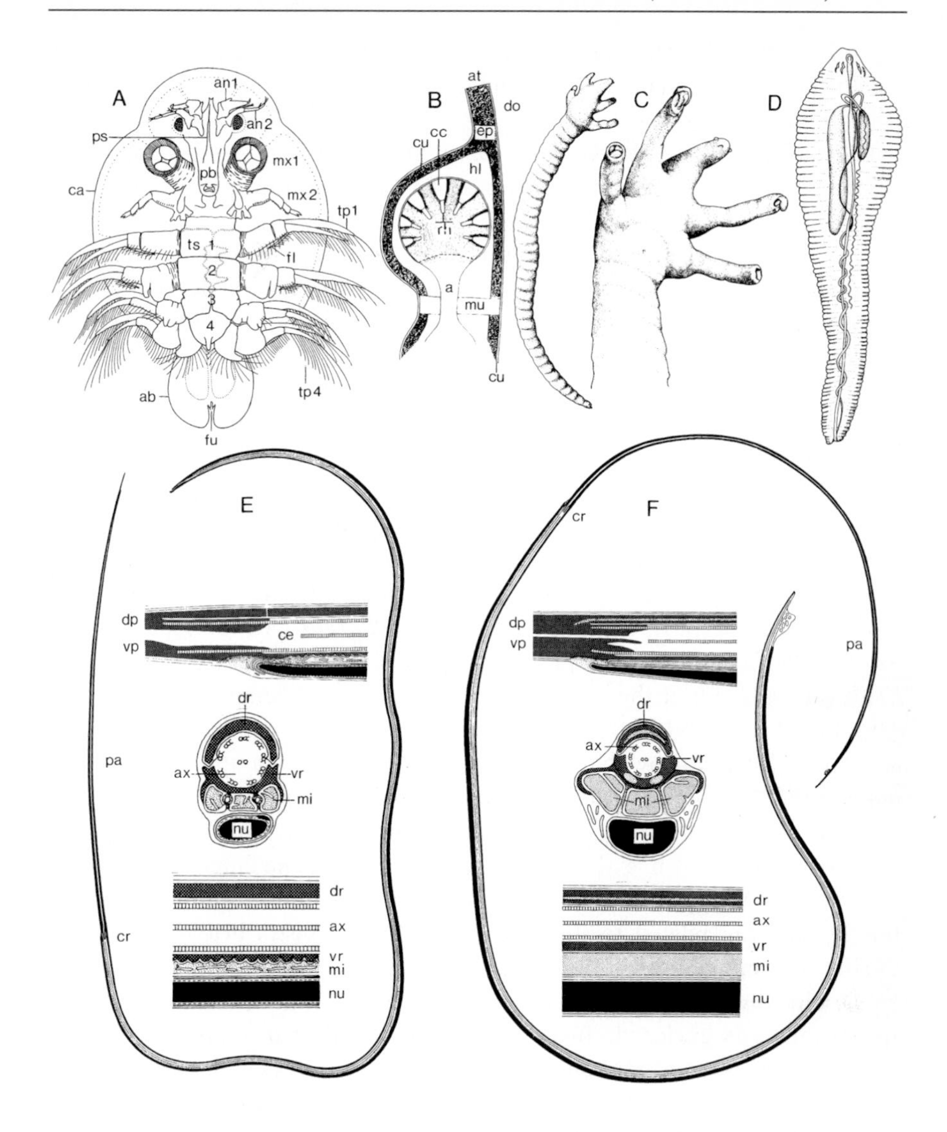

The placing of the Branchiura in the Maxillopoda creates certain problems. At first glance it appears very simple to view the trunk of four thoracomeres and an undivided abdomen as the result of reductions from the ground pattern of the Maxillopoda with ten trunk segments. However, in contrast to other Maxillopoda taxa, there are no convincing arguments for this to be found from the organization and development of the Branchiura. If the thorax of the Branchiura has only four segments secondarily, then the position of the genital pores in the fourth (last) segment would be understandable. The gonopores must have been shifted this far forwards from the seventh thoracomere as the thorax was reduced in the stem lineage of the Branchiura. From this perspective the identical position on the fourth thoracic segment in the Branchiura and the Mystacocarida becomes a meaningless congruence.

The flat disc on the back of the Branchiura could be a secondary state for the carapace of the Thecostracomorpha in connection with ectoparasitism. With a possible homology the Branchiura can be interpreted as members of this monophylum. However, there are also problems here. The Branchiura do not have the forward displacement of the female genital opening to the first thoracic segment, as was postulated for the stem species of the Progonomorpha.

Pentastomida

Regardless of the problematic kinship relations of the Branchiura, a challenge for phylogenetic systematics is given through comparative studies of the sperm ultrastructure in the Branchiura and the Pentastomida (Linguatulida) – endoparasites in the respiratory tracts of vertebrates (Fig. 88 C, D). The sperm of the fish lice *Argulus foliaceus* and the tongue-worm *Raillietiella hemidactyli* are practically identical in structure and development (Fig. 88 E, F); at the same time they represent a unique apomorphous state (WINGSTRAND 1972). This discovery has been confirmed (STORCH & JAMIESON 1992) by the study of other Pentastomida species (*Cephalobaena tetrapoda*, *Raillietiella* sp., *Porocephalus crotali*).

Some fundamental congruencies should be stressed. An extremely thin pseudoacrosome with a dorsal and ventral rod in the interior forms the first third of the sperm (total length 110–130 μm). This pseudoacrosome is not in any way related to the acrosome which appears briefly in *Argulus* during spermiogenesis. The centriolar region connects the pseudoacrosome with the rest of the sperm body that bears the nucleus and in which the axoneme is enclosed. Two ribbons join together in a ring around the axoneme; the ventral ribbon also encircles the two lateral mitochondria. The mitochondria are three adjacent filiform rods which are interpolated between the axoneme and the likewise filiform nucleus.

We would be abandoning the principle of parsimony central to phylogenetic systematics if, from these facts, we did not emphatically advocate the hypothesis of a single evolution of the sperm pattern presented above in a common stem lineage of the Branchiura and the Pentastomida. This results in the hypothesis of the monophyly of a unity of the Branchiura and the Pentastomida that moreover receives support on a molecular level (18S rRNA nucleotide sequences) (ABELE et al. 1989).

The validity of this hypothesis has not been changed by the finding of possible stem lineage representatives of the Pentastomida from the Late Cambrian (WALOSSEK & MÜLLER 1994). A corresponding provable hypothesis concerning the phylogenetic kinship of the Pentastomida to any other unity of the Arthropoda does not exist.

Tracheata

Autapomorphies (Figs. 36 → 7; 91 → 1)

- Mode of life.
  Terrestrial air-breathing stem species.
  Characteristic features connected with conquering the terrestrial realm.
- Tracheae with paired, lateral spiracles per segment.
  (Convergent to the evolution of tracheae within the Arachnida.)
- Malpighian tubules. A pair of outgrowths of the proctodaeum at the midgut–hindgut boundary.
  (Convergent to the evolution of homonymous excretory organs in the stem lineage of the Arachnida.)
- Reduction of metameric nephridia to two pairs in segments 4 and 5 (p. 80).
- Temporal organ.
  (Tömösvary's organ, pseudoculus, postantennal organ.)
  Paired organ on the side of the head behind the antennae (Fig. 102 A). Fine pores in the cuticle point to an olfactory function. In all the high-ranking subtaxa of the Tracheata. Probably homologous (HAUPT 1973).
- Indirect spermatophore transfer.
  The males deposit spermatophores on the substrate from where the females pick them up. No genital contact between ♂ and ♀.
  (Convergent to the evolution of indirect spermatophore transfer in the stem lineage of the Arachnida.)
  Characteristic features without justifiable connection to the terrestrial mode of life.
- Only one pair of antennae.
  From the antennal head segment follows an appendageless intercalary segment (Fig. 104 A, B). Compared with the two pairs of antennae in the sister group Crustacea the second antennae are absent; they were reduced in the stem lineage of the Tracheata.
  In the Tracheata the antennae homologous with the first antennae of the Crustacea are known simply as the antennae.
- Mandibles without palps.
  Complete reduction of the endopodites of the appendage.

From the apomorphic peculiarities of the Tracheata (Antennata, Atelocerata) we must examine the **air-breathing organs** more closely. Even in recent literature the existence of tracheae in the feature pattern of the terrestrial stem species of the Tracheata has been denied and their multiple convergent evolution in various stem lineages hypothesized (DOHLE 1985, 1988, 1996, 1997; KRAUS & KRAUS 1994; HILKEN 1997, 1998; KRAUS 1997). This opinion is based on the differing position of the spiracles and the structure of the tracheae in different subgroups of the Tracheata. We are, however, confronted with the following facts. In the three high-ranking taxa Chilopoda, Progoneata and Insecta, whose kinship will be analyzed in the next section, tracheal systems exist with paired segmental spiracles over practically the whole trunk. This is the case in the Geophilomorpha amongst the Chilopoda and in the Diplopoda within the Progoneata; furthermore, ten successive spiracle pairs on the trunk belong in the ground pattern of the Insecta.

In phylogenetic systematics the most parsimonious assessment of these facts can lead to only one conclusion. In the stem lineage of the Tracheata, during the change to land, a tracheal system evolved on the trunk with strongly segmentally arranged pairs of spiracles once only. This tracheal system was carried over from the last common stem species of recent Tracheata into their different stem lineages; it was realized in the stem species of the Chilopoda, Progoneata and Insecta. All patterns that differ from a tracheal system with segmental spiracle pairs must be interpreted as evolutionary modifications within the Tracheata. Whosoever wishes to postulate a multiple independent origin of tracheae in the Notostigmophora, Pleurostigmophora (Chilopoda), Symphyla, Dignatha (Progoneata) and Insecta, cannot call on the structural differences in the air-breathing systems. They must give reasons for setting aside the principle of parsimony, and plausibly explain how the – without exception – terrestrial stem species of the unities mentioned and their countless antecedents up to the stem species of the Tracheata could have lived without an air-breathing organ.

A tracheal system with a segmental manifestation of spiracles was integrated into the feature pattern of a **terrestrial stem species with continuous uniform, homonomous trunk segments**. This homonomy is seen in the Geophilomorpha within Chilopoda and the Diplopoda within the Progoneata – if one refrains from uniting homonomous segments to diplosomites. Paleozoic insects with multiarticulate appendages on the abdomen, such as *Dasyleptus* from the stem lineage of the Dicondylia, also show a strong homonomous segmentation of the trunk. Heteronomous states such as heterotergy within the Chilopoda and in the Progoneata (Symphyla, Pauropoda), or the subdivision of the trunk into thorax and abdomen in the Insecta, form apomorphies in the taxon Tracheata.

The question regarding the construction of the palpless **mandibles** in the ground pattern of the Tracheata is not of essential importance for the systematization that follows; it will therefore be touched on only briefly. In the Chilopoda and the Progoneata the mandibles are commonly divided into several sclerites. In contrast insect mandibles are usually uniform, undivided structures; the indication of traces of joints in the primary wingless Archaeognatha is disputed (KRAUS & KRAUS 1994, 1996; N.P. KRISTENSEN 1997a).

I conclude that the aquatic stem species of the Mandibulata, like the primary aquatic Crustacea, possessed a gnathobasic mandible. This mandible with an endite was then carried over not only into the stem lineage of the Crustacea, but also into the stem lineage of the Tracheata. From this perspective, the subdivision of the mandible into different sclerites in the Tracheata must be seen as an apomorphy. Did this phenomenon originate once only – and if so, was it already present in the stem lineage of all Tracheata (reduction in the Insecta) or did it first evolve in one stem lineage common only to the Chilopoda and Progoneata? If in the latter, a subdivided mandible could be counted as a further autapomorphy of the unity Myriapoda (p. 212).

Competing Kinship Hypotheses

The Chilopoda, Progoneata and Insecta form three high-ranking subtaxa of the Tracheata – all are well established as monophyla. Significant derived peculiarities of the Chilopoda are the maxillipeds on the first trunk segment, of the Progoneata the position of the genital openings in the anterior body and of the Insecta the division of the trunk into thorax (three segments) and abdomen (11 segments + telson).

There are in principle three hypotheses concerning the kinship relations of the three taxa to one another (Fig. 89).

A. **Myriapoda hypothesis**
   Chilopoda and Progoneata are adelphotaxa; they are united under the name Myriapoda. The Myriapoda and the Insecta then form sister groups one hierarchic level up.
B. **Labiophora hypothesis**
   Progoneata and Insecta are adelphotaxa. The Labiophora, comprising these two unities, is then the sister group of the Chilopoda.
C. **Opisthogoneata hypothesis**
   Chilopoda and Insecta are adelphotaxa; they are united as the Opisthogoneata. One system level up the Progoneata and Opisthogoneata are sister groups.

The Opisthogoneata hypothesis (POCOCK 1893; VERHOEFF 1925) is based on plesiomorphic congruencies such as the position of the genital apertures at the posterior – from which the group derives its name. Synapomorphic congruencies do not exist between the Chilopoda and the Insecta. Consequently, we will not follow this hypothesis further.

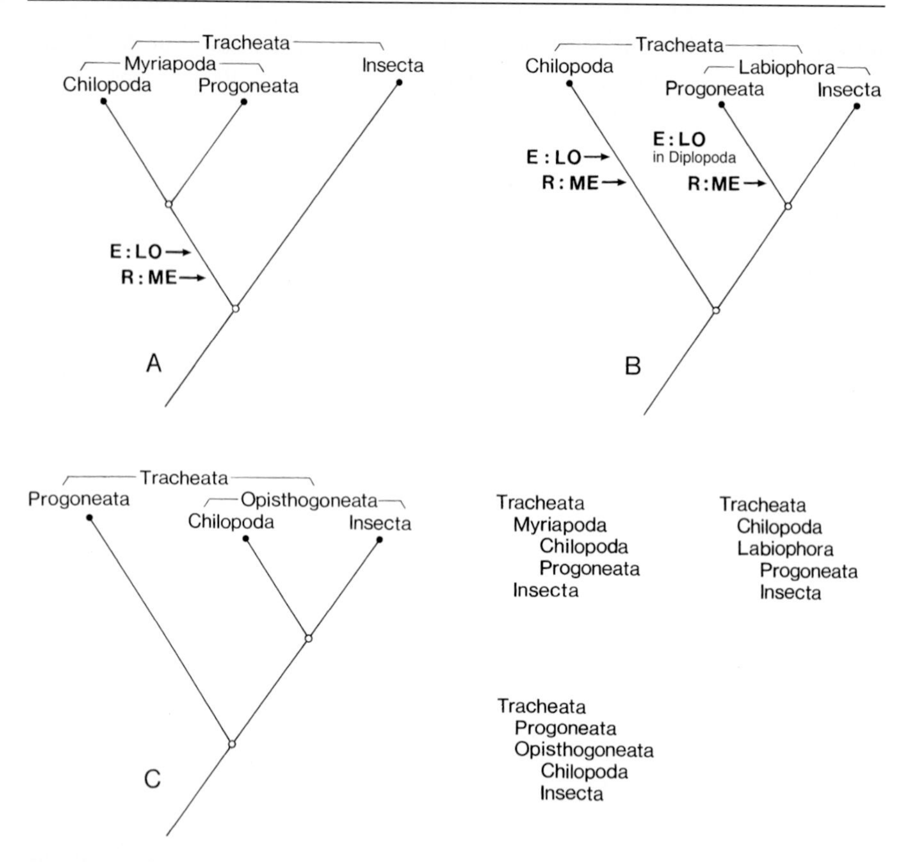

Fig. 89. The three possibilities of the phylogenetic relationships between the Chilopoda, Progoneata and Insecta and the corresponding systematization of the Tracheata. Presentation of the three hypotheses in kinship diagrams and hierarchical tabulations (*bottom right*). **A** Myriapoda hypothesis. **B** Labiophora hypothesis. **C** Opisthogoneata hypothesis. In the Myriapoda hypothesis a single reduction of the median eyes (R:ME) and a single evolution of lateral ocelli (E:LO) is postulated. This corresponds to the simplest explanation using the principle of parsimony. The Labiophora hypothesis forces the more expensive assumption of at least a double independent reduction of the median eyes in the stem lineages of the Chilopoda and the Progoneata and a double evolution of lateral ocelli (Chilopoda, Diplopoda)

The remaining kinship hypotheses are both advocated within the framework of phylogenetic systematics – the Myriapoda hypothesis by HENNIG (1969, 1981) and BOUDREAUX (1987) and the Labiophora hypothesis by KRAUS & KRAUS (1993a, 1994, 1996). In the following discussion names and features of certain subgroups are mentioned which will be dealt with later in context.

Myriapoda Hypothesis

For the Myriapoda hypothesis two facts in particular are advanced:
- The complete absence of median eyes in all species of the Chilopoda and Progoneata.
- Lateral eyes of comparatively few, loosely arranged ocelli without crystalline cone in the Chilopoda and Diplopoda (Fig. 90)[19].
  They replace the compound eyes of the Crustacea and Insecta with hundreds of tightly packed ommatidia.

Four median eyes belong as an autapomorphy in the ground pattern of the Euarthropoda; they are called nauplius eyes in the Crustacea and median or frontal ocelli in the Insecta. Furthermore, compound eyes with a primary construction of the ommatidia of a cornea, two corneagenous cells, a crystalline cone with four crystal cells and a rhabdome with eight retinula cells are identical in the Crustacea and the Insecta.

Crustacea and Tracheata are adelphotaxa of the monophylum Mandibulata. This well-established kinship hypothesis forces us to the following conclusions.
- The absence of median eyes in the Chilopoda and the Progoneata is a derived, secondary state in comparison with their existence in the Crustacea and Insecta.
- The lateral ocelli without a crystalline cone in the Chilopoda and the Diplopoda are apomorphic in contrast to the compound eyes of the Crustacea and Insecta.

If one looks for an evolutionary explanation for the changes in the light-sensory organs, then a connection may exist with the cryptic habitat of the Chilopoda and the Progoneata.

More significant to me, however, is the fact that the following decisive questions of phylogenetic systematics are logically independent of the interpretation of the underlying evolutionary processes. These are:
- Were the median eyes reduced once in the stem lineage of a unity Myriapoda, lost twice convergently in the stem lineages of the Chilopoda and Progoneata, or perhaps even lost several times independently?
- Did lateral ocelli originate through the loss of the crystalline cone once in the stem lineage of a unity Myriapoda or did they evolve several times independently of one another – for example in the Chilopoda and in the Diplopoda (Progoneata)?

---

[19] The "pseudocompound eye" in *Scutigera* (Scutigeromorpha) will be discussed later (p. 225).

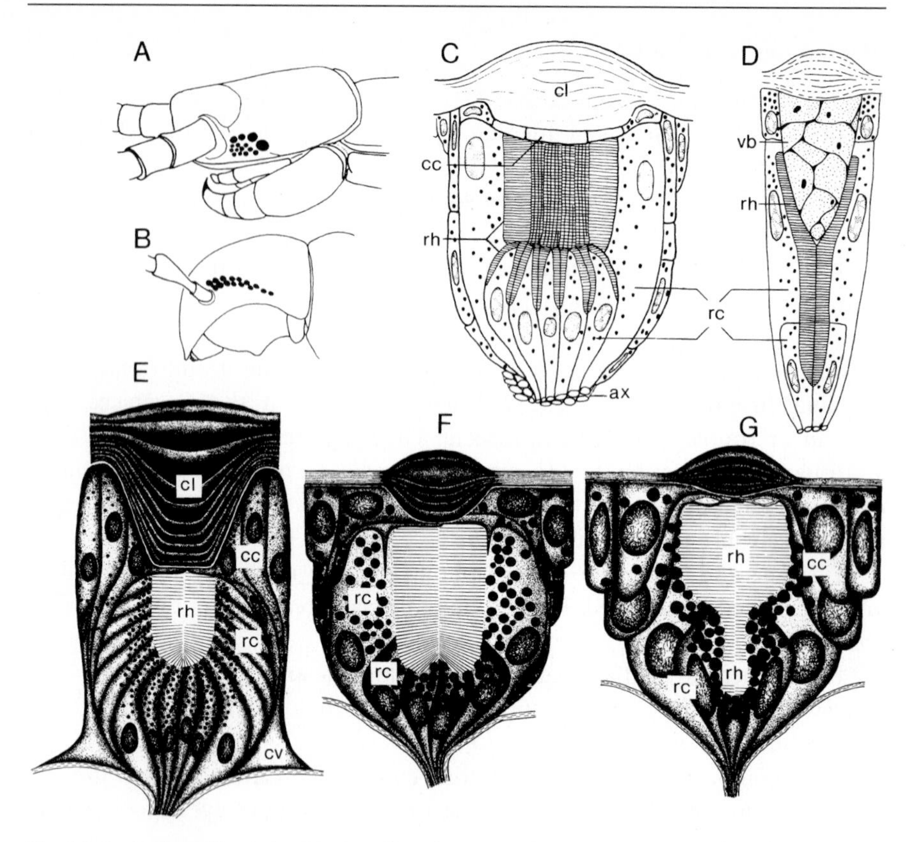

Fig. 90. Lateral ocelli of the Myriapoda – interpreted as an autapomorphy of the descent community. **A** *Lithobius* (Chilopoda). Head with 15 ocelli of different sizes. **B** *Proteroiulus* (Diplopoda). Head of a juvenile individual with 15 ocelli. **C** *Lithobius forficatus* (Chilopoda). Ocellus with biconcave corneal lens. Corneagenous cells form a single layer of modified epidermis cells beneath the cornea. Between 35 and 110 retinula cells arranged in two layers. Flat epithelium of protective cells on the optic cells. **D** *Scutigera coleoptrata* (Chilopoda). Ommatidium of the pseudocompound eye. Beneath the cornea a vitreous body, the secretion product of numerous peripheral pigment cells. Two-stage rhabdome with many retinula cells. **E** *Ommatoiulus sabulosus* (Diplopoda). On each side of the body 30–40 strongly pigmented ocelli. Corneal lenses of the ocellus sunk conically. Rhabdome formed from 300–400 retinula cells arranged in several layers. Covered by protective cells in the periphery. **F** *Polyzonium germanicum* (Diplopoda). Lateral eyes each of 6–8 ocelli. Biconvex corneal lens. 25–30 retinula cells in three layers. **G** *Polyxenus lagurus* (Diplopoda). Only five small ocelli on each side of the head. Biconvex corneal lens. Rhabdome of seven retinula cells in two layers. The low number of optic cells and their arrangement is correlated with the evolutionary diminuation of the ocelli. The assumption of an independent origin of the ocelli of *P. lagurus* from single ommatidia of a compound eye (SPIES 1981) is unproven. *ax* Axon; *cc* corneagenous cell; *cl* corneal lens; *cv* protective cell; *rc* retinula cell; *rh* rhabdome; *vb* vitreous body. (**A–D** Paulus 1986; **E–G** Spies 1981)

Let us begin with an answer to the first question. The persistent repeated hints of multiple convergent reduction in other Arthropoda unities, which should speak in favor of the second alternative, are irrelevant or at the very least of little help in the actual case in question. Here we are dealing with the interpretation of a circumstance in which **median eyes** are absent in all species of two of the three high-ranking subtaxa of the Tracheata. For this, the principle of parsimony dictates a clear answer. We must postulate a **single reduction of median eyes** in a common stem lineage of the Chilopoda and Progoneata. We must then test this hypothesis as to its compatibility or conflict with the assessment of other features.

The hypothesis is compatible with the answer to the second question. As a result of the evolution of **lateral ocelli** besides (1) the loss of the crystalline cone, other **significant positive congruencies** appear between the Chilopoda and the Diplopoda; these are (2) the secretion of the ocellus lens from a multicellular layer of epidermis cells, and (3) the formation of a multi-layered retinula of numerous cells (Fig. 90).

There is an astounding variety in the number of retinula cells per ocellus: 1000 in *Scolopendra*, 35–100 in *Lithobius forficatus* (Chilopoda), 450–500 in *Craspedosoma simile*, 300–400 in *Ommatoiulus sabulosus*, 25–50 in *Polyzonium germanicum* and only 7 in the few ocelli of *Polyxenus lagurus* (Diplopoda) (BÄHR 1971; PAULUS 1979; SPIES 1981).

Of the explanations for the evolutionary processes leading to the lateral ocelli, fusion of ommatidia through formation of larger more efficient lenses and reduction of the then “superfluous” crystalline cone is just as valid as an increase and rearrangement of elements of the former ommatidia. However, independent of the validity of such considerations, for the interpretation of the above-mentioned congruencies, phylogenetic systematics must again turn to the principle of parsimony and postulate a single evolution of the lateral ocelli in the Chilopoda and Progoneata – that is, it must hypothesize their evolution in the stem lineage of a unity Myriapoda. This hypothesis then forces us to the assumption of a convergent reduction of the lateral ocelli in some subtaxa of the Chilopoda (Geophilomorpha, *Cryptops* in the Scolopendromorpha) and the Progoneata (Symphyla, Pauropoda, Polydesmida in the Diplopoda).

The two clear congruencies in the light-sensory organs interpreted as synapomorphies of the Chilopoda and Progoneata conflict with the Labiophora hypothesis.

Labiophora Hypothesis

The following congruencies are referred to in comparing the Progoneata and the Insecta (KRAUS & KRAUS 1993b, 1994, 1996; KRAUS 1997):

- Common possession of a maxillary plate from basal parts of the second maxillae. This forms the ventral (posterior) closure of the buccal cavity.
- Existence of coxal organs and styli.

Within the Progoneata only in the Symphyla do maxillae 1 and 2 insert as two pairs of distinct separated appendage derivatives on the head capsule. This corresponds to the situation seen in the Insecta. In their place in the Dignatha (Pauropoda + Diplopoda) one structural and functional unit is found – without question an apomorphy within the Progoneata, regardless of whether we are dealing here with only the first maxillae or a union of maxillae 1 and 2 (p. 230).

Given this, a comparison of even just the **second maxillae of the Symphyla and Insecta** is of relevance to the problem at hand. In both taxa the second maxillae amalgamate and partially grow together to form an unpaired structure that is named the labium. Further congruencies do not exist. The labium of the Symphyla has two sclerotized plates between which a median longitudinal suture is found (Fig. 100 G); each plate bears distally three small cones with sensory structures. In contrast, in the Insecta, maxillae 2 are proximally fused seamlessly. This stem part of the labium is then transversely divided into the mentum and the prementum. The latter bears centrally as endites the glossae and the paraglossae and laterally the multiarticulate labial palp (Fig. 109). In each case, the labia of the Symphyla and Insecta are apomorphies in contrast to the primary state of separate second maxillae. Justifiable identifications between certain elements of the two labia are, however, not possible (DOHLE 1965, 1980, 1988). There are no special indications which can be used to homologize the labium of the Progoneata and the labium of the Insecta and regard them as a synapomorphy of the two taxa.

Segmental paired **coxal organs** – membranous, projecting bubbles for the uptake of moisture from thin moisture-films – can be placed with good justification in the ground patterns of the Progoneata and the Insecta. In addition there is good agreement in the number of ventral coxal organs between the Symphyla (trunk segments 3–10) and the Archaeognatha (trunk segments 4–10) among the primary wingless insects.

Coxal organs are completely lacking, however, in some of the Diplopoda and in most Insecta (Protura, Pterygota). Due to their systematic position as subtaxa of the Progoneata and Insecta, this lack can only be the result of convergent reduction. The absence of coxal organs in the Chilopoda remains to be explained.

HENNIG (1969, 1981) interpreted the coxal organs as an autapomorphy of the Tracheata. Their evolution could be connected with the change to a terrestrial mode of life – in the sense of a first utilization of the minimal water sources in the soil. With further advances into the terrestrial realm through strengthening of the cuticle, coxal organs were reduced several times convergently – in the above-mentioned subtaxa of the Progoneata and Insecta and also in the stem lineage of the Chilopoda. Certainly, this is nothing more than a possible scenario – but perhaps also a timely reminder – that the coxal organs in their present distribution in the Tracheata should not blindly be assumed to be a synapomorphy of the Progoneata and Insecta.

Styli in the form of short pegs are found, like the coxal organs in the Progoneata and Insecta, in a serial arrangement on the ventral surface of the trunk. Today, however, they are assessed very differently to the coxal organs. According to new results from fossils, styli cannot be placed in the ground pattern of the Insecta (p. 255); they must have originated independently in different stem lineages of the Insecta. Therefore, there can be no homologization with the styli of the Progoneata.

To date, molecular data is contradictory. According to WHEELER (1997), the Myriapoda form a paraphyletic species group. However, in the analyses of ZRAVY et al. (1997), they are always shown to be monophyletic. Sequence information from nuclear ribosomal DNA by FRIEDRICH & TAUTZ (1995) gives support for the monophyly of the Myriapoda.

From my own appraisal, the **Myriapoda hypothesis** has the **better arguments**. It can call on two congruencies in the light-sensory organs that compatibly can be interpreted as synapomorphies of the Chilopoda and Progoneata. For the three congruencies between the Progoneata and Insecta – labium, coxal organs and styli – established objections exist for each against their interpretation as synapomorphies. Therefore as a basis for the following systematization I draw on the Myriapoda hypothesis.

Systematization

Tracheata
- **Myriapoda**
  - **Chilopoda**
  - **Progoneata**
    - **Symphyla**
    - **Dignatha**
      - **Pauropoda**
      - **Diplopoda**
- **Insecta**

Myriapoda – Insecta

Myriapoda

Autapomorphies (Fig. 91 → 2)

- Loss of median eyes.
  Absent in all species of the Chilopoda and Progoneata.
- Ocelli.
  Modification of the compound eyes into groups of loosely clustered lateral ocelli with the following features: absence of a crystalline cone; secretion of the lens from several epidermis cells; formation of a multilayered retinula of numerous cells (Fig. 90).

Chilopoda – Progoneata

Compared to their adelphotaxon, the Chilopoda have as a plesiomorphy the terminal position of the genital pores at the posterior, and, as an autapomorphy, the differentiation of the first trunk leg pair into maxillipeds with poison glands.

Conversely, the anterior position of the genital apertures is an autapomorphy in the ground pattern of the Progoneata, and the normal manifestation of the appendages of trunk segment 1 as walking legs a plesiomorphy.

Chilopoda

Autapomorphies (Figs. 91 → 4; 93 → 1)

- Maxillipeds (fangs).
  The first trunk segment appendages are transformed into large claws or fangs (Fig. 92 A, B). Each claw contains a poison gland, which opens just before the tip of the last article.
- Raptorial feeding. The prey is pierced by the fangs. Paralysis or death of the prey occurs due to the poison.
- Hatching spine on the second maxillae of the embryo to rip open the egg membrane (Fig. 94 B).
- Sperm with striated cylinder and mantle.
  The axoneme of the thread-like sperm is closely surrounded by a striated cylinder of fibrous material. Following a gap there is an extensive mantle; this consists of membranous elements that are separated by spiral septa or helices (Fig. 92 D, E).

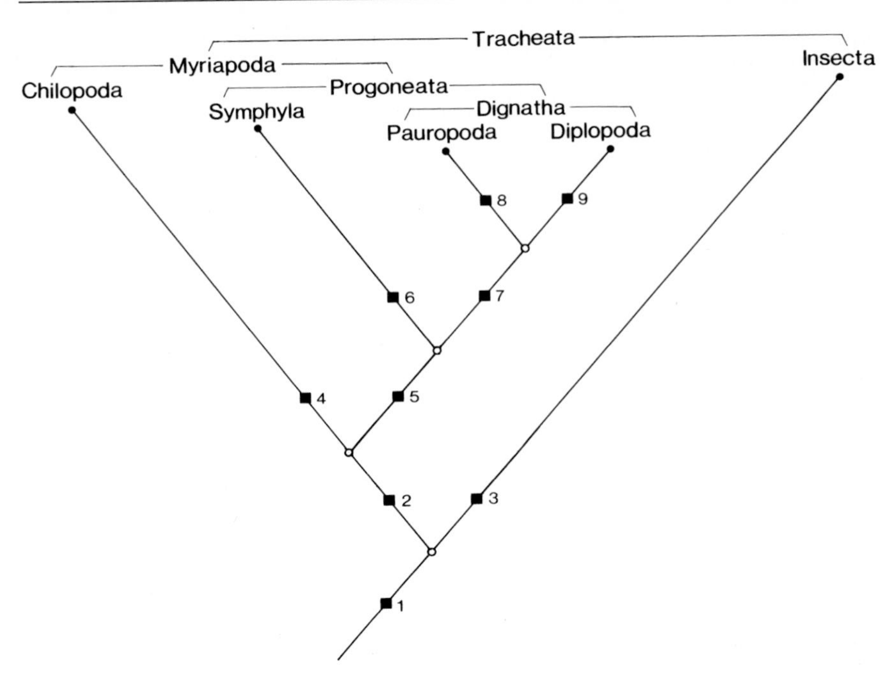

Fig. 91. Diagram of the phylogenetic relationships between high-ranking subgroups of the Tracheata

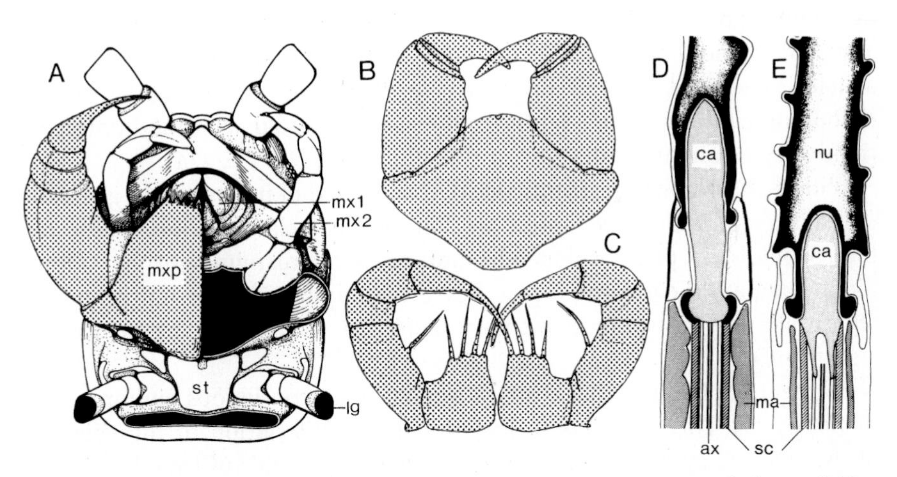

Fig. 92. Autapomorphies of the Chilopoda. **A–C** Maxillipeds. **A** *Lithobius forficatus* (Lithobiomorpha). Anterior end in ventral view. Maxilliped from the left side removed. **B** *Cryptops hortensis* (Scolopendromorpha). Coxae fused. **C** *Scutigera coleoptrata* (Scutigeromorpha). Maxillipeds with separate coxae. **D–E** Sperm ultrastructure **D** *Lithobius forficatus*. **E** *Clinopodes linearis* (Geophilomorpha). Longitudinal section of the sperm with proximal section of the nucleus, a connecting section with capitulum and main part with axoneme, striated cylinder and mantle. *ax* Axoneme; *ca* capitulum; *lg* first locomotory leg pair of the trunk; *ma* mantle; *mx* maxilla; *mxp* maxilliped; *nu* nucleus; *sc* striated cylinder; *st* sternum. (**A** Rilling 1968; **B, C** Borucki 1996; **D, E** Jamieson 1987)

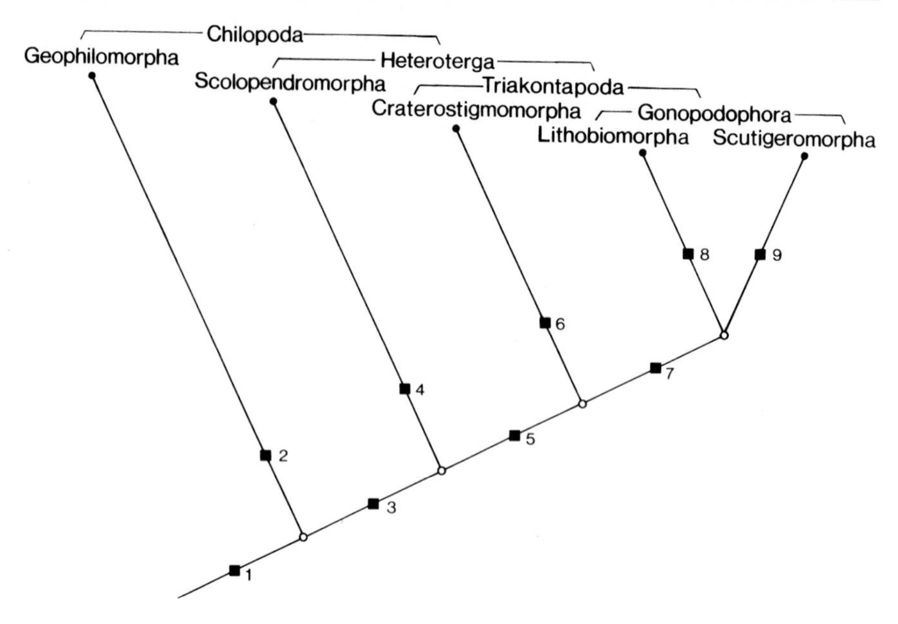

Fig. 93. Diagram of the phylogenetic relationships between high-ranking subgroups of the Chilopoda

The Chilopoda are without question established as a monophylum if only through the **evolution of the maxillipeds** that functionally are assigned to the mouthparts of the head. I consider the fusion of the coxal articles to a rigid plate as original (Fig. 92 B), and the formation of free mobile maxillipeds in the Scutigeromorpha connected with the hunting of rapid prey as an autapomorphy of this unity (Fig. 92 C; p. 224).

Within the Chilopoda (centipedes) five high-ranking taxa are generally distinguished, which are orders in conventional classification. These are the Geophilomorpha, Scolopendromorpha, Craterostigmomorpha, Lithobiomorpha and Scutigeromorpha. For the grouping of these unities to sister groups the following elementary facts appear to me to be of decisive significance.

In the Geophilomorpha a strict sequence of continuous uniform segments exists. This **homonomy** includes the manifestation of the spiracles of the tracheal system. Almost every segment has a pair of lateral spiracles; only the maxilliped segment, the metameres of the first and last leg pair, and the genital region do not possess spiracles.

In the Scolopendromorpha, Craterostigmomorpha, Lithobiomorpha and Scutigeromorpha, homonomy is only realized on the ventral side of the trunk segments. This is opposed by a dorsal heteronomy in the form of the **heterotergites.** These consist of a sequence of trunk segments with tergites of differing lengths; normally, alternating long and short tergites. A constant excep-

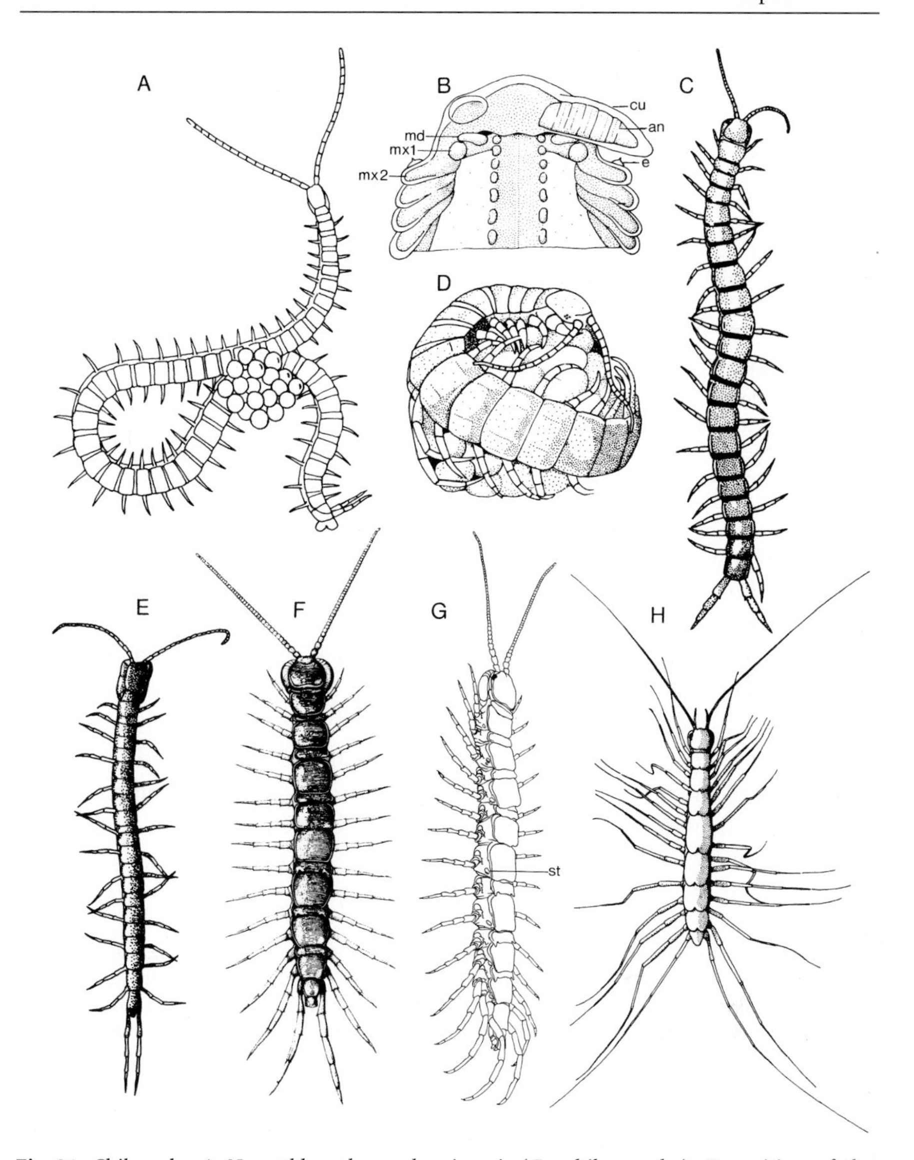

Fig. 94. Chilopoda. **A** *Necrophloeophagus longicornis* (Geophilomorpha). Deposition of the eggs in a clump on the back, for which the female forms a sling. **B** *Scolopendra cingulata* (Scolopendromorpha). First embryonic stadium. The egg tooth on the second maxilla splits the egg shell (autapomorphy of the Chilopoda). **C** *Scolopendra morsitans*. **D** *Scolopendra cingulata*. Brood care by a female. The ventral side is turned towards the clutch of eggs. **E** *Craterostigmus tasmanianus* (Craterostigmomorpha). **F** *Lithobius forficatus* (Lithobiomorpha). **G** *Lithobius forficatus*. Side view with spiracles. **H** *Scutigera coleoptrata* (Scutigeromorpha). *an* Antenna; *cu* cuticle; *e* egg tooth; *md* mandible; *mx1* first maxilla; *mx2* second maxilla; *st* spiracle. (**A–D, H** Lewis 1981; **E** Dunger 1993; **F** Rilling 1968; **G** Rilling 1960)

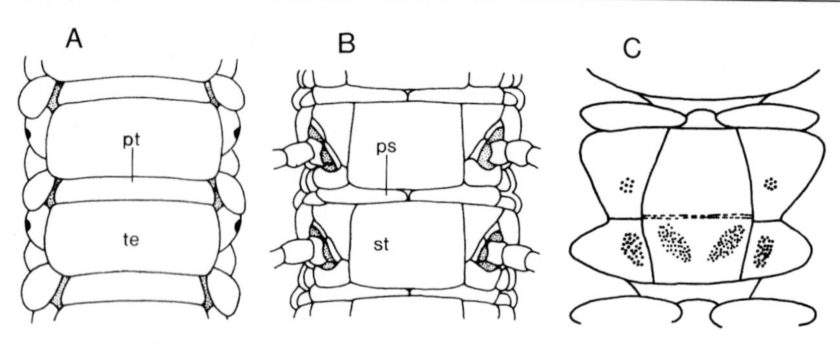

Fig. 95. Geophilomorpha. **A–B** *Necrophloeophagus longicornis.* Sixth and seventh trunk segments in **A** dorsal view and **B** ventral view. Demonstration of the separation of pretergites and presternites (probable ground pattern feature of the Chilopoda). **C** *Strigamia acuminata.* Pore field of ventral defense glands in the sternites and pleurites (autapomorphy of the Geophilomorpha). *ps* Presternite; *pt* pretergite; *st* sternite; *te* tergite. (**A, B** Lewis 1981; **C** Dunger 1993)

tion to this is seen in trunk segments 8 and 9 (walking leg segments 7 and 8). Here two long tergites are found together; this leads to a change in the sequence of short and long tergites in the second half of the body.

A specific segment-jumping **spiracle pattern** is correlated with this heterotergy (Fig. 96). Normally segments with long tergites and spiracles alternate with segments with short tergites and no spiracles. The exception is found precisely in the region of the interruption of the regular heterotergy by the two long tergites. Here the spiracles make a double spring; two segments without spiracles are found between the sixth and ninth trunk segments.

Fig. 96. Trunk segments of Heteroterga (Chilopoda). Demonstration of the increasing manifestation of heterotergy from the Scolopendromorpha to the Scutigeromorpha and of the segment-jumping arrangement of the spiracles. The numbering refers to the trunk segments following the head (*regular dots*). The maxilliped segment is thereby trunk segment 1, the first walking leg pair segment=trunk segment 2,...the 15th walking leg pair segment= trunk segment 16. In the numbering, the figures of the segments with long tergites are in the middle of the segments, the figures of the segments with short tergites are to the right-hand side. The fusion of short and long tergites (Scutigeromorpha) is discussed below. The spiracles are symbolized by *black dots.* The paired lateral spiracles of the Scolopendromorpha, Craterostigmomorpha and Lithobiomorpha are set at the edges of the tergites. The seven unpaired spiracles of the Scutigeromorpha are in the appropriate positions on the back. **A** Scolopendromorpha: *Alipes grandidieri.* Autapomorphy: The tergites of trunk segments 1 and 2 (maxillipeds, first walking leg pair) are fused. **B** Craterostigmomorpha. *Craterostigmus tasmanianus.* Autapomorphy: Secondary transverse division of long tergites 4, 6, 8, 9, 11, 13 (stippled areas). **C** Lithobiomorpha. *Lithobius forficatus.* In the ground pattern of the Lithobiomorpha there is also a spiracle pair in the second trunk segment. Thereby, there is complete agreement with the Scutigeromorpha in the pattern of the segmental manifestation of spiracles. **D** Scutigeromorpha. *Scutigera coleoptrata.* Autapomorphy: Following the tergite of trunk segment 2, six new large tergites develop through fusion of successive short and long tergites. Additionally long tergites 8+9 also fuse. *tg* tergite of the genital region. (Borucki 1996, modified) ▶

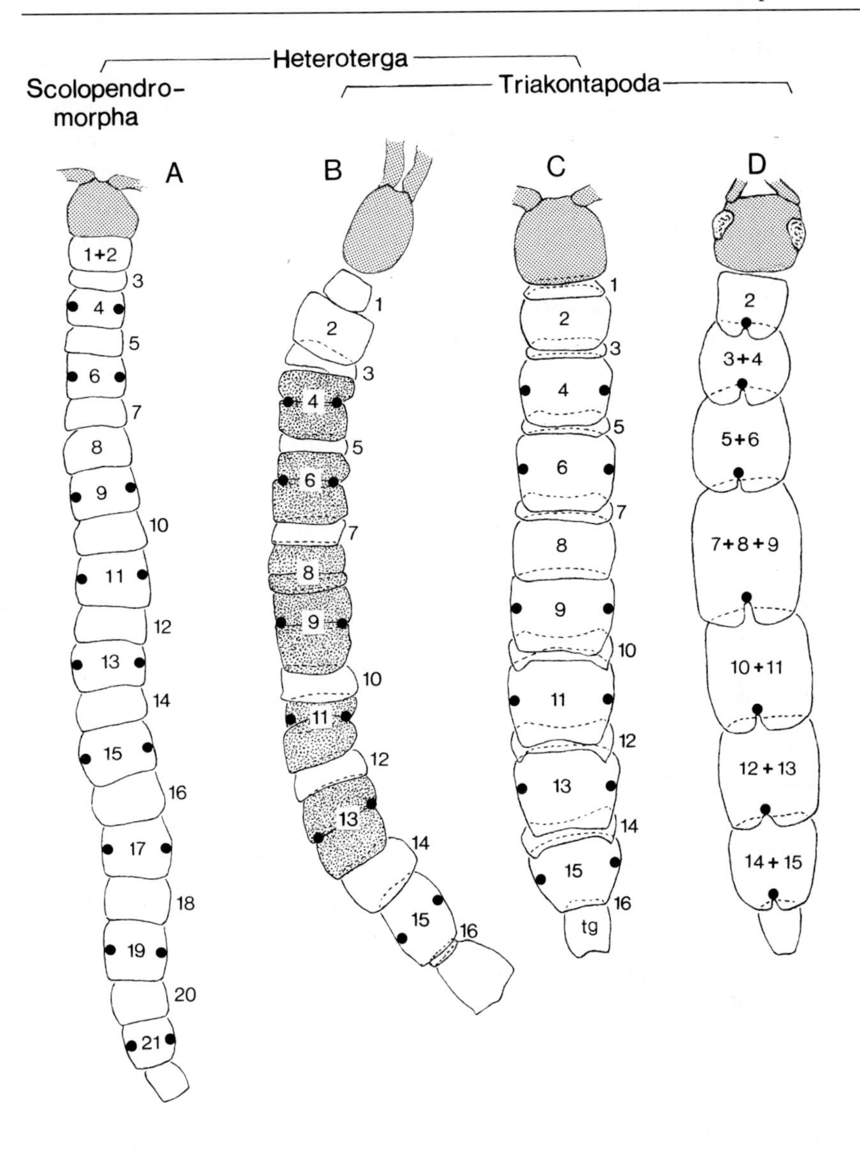

This pattern holds true for the paired spiracles of the "pleurostigmophoric" Scolopendromorpha, Craterostigmomorpha and Lithobiomorpha, as well as for the unpaired spiracles of the "notostigmophoric" Scutigeromorpha. Two examples demonstrate this. *Lithobius forficatus* (Lithobiomorpha) has six pairs of pleural spiracles in trunk segments 4, 6, 9, 11, 13 and 15. In *Scutigera coleoptrata* (Scutigeromorpha) there are seven unpaired spiracles dorsally in segments 2, 4, 6, 9, 11, 13 and 15 – to be precise they are found in indentations on the posterior edge of the seven large tergites.

Systematization

As a consequence of a systematization in which the Notostigmophora (=Scutigeromorpha) and Pleurostigmophora (Lithobiomorpha + Craterostigmomorpha + Scolopendromorpha + Geophilomorpha) are hypothesized to be sister groups (DOHLE 1985, 1996; HENNIG 1994; BORUCKI 1996; KRAUS 1997), heterotergy and the segment-jumping manifestation of spiracles must belong in the ground pattern of the Chilopoda. I cannot accept this concept. Through a comparison of the Geophilomorpha with the Diplopoda and the Insecta I have reached the conclusion that the stem species of the Tracheata was a homonomous organism with similar segments on the trunk and paired segmental spiracles (p. 204). I find no argument for tracing back the Geophilomorpha to a heteronomous organism. Heteronomy in the sequence of dissimilar tergites and in the arrangement of spiracles which miss out certain segments must be assessed as apomorphies of the Chilopoda. The simplest interpretation that they are the product of a single evolution results in a systematization with the following adelphotaxa: Geophilomorpha and Heteroterga – a unity comprising all other Chilopoda.

The repercussions connected with this for the assessment of other feature alternatives will be dealt with later.

Chilopoda
- **Geophilomorpha**
- **Heteroterga**
  - **Scolopendromorpha**
  - **Triakontapoda**
    - **Craterostigmomorpha**
    - **Gonopodophora**
      - **Lithobiomorpha**
      - **Scutigeromorpha**

Geophilomorpha – Heteroterga

Geophilomorpha

Autapomorphies (Fig. 93 → 2)

- Absence of lateral ocelli.
- Repugnatorial glands.

  Existence of glands with a segmental sequence on the ventral side of the trunk (Fig. 95 C).

  Opening in pore areas on the sternite. Production of sticky secretion and cyanide derivatives (HOPKIN & READ 1992; TURCATO et al. 1995).

The number of leg-bearing segments varies from 29 in *Dinogeophilus oligopodus* to 191 in *Gonibregmatus plurimipes*. In addition, segment number commonly varies between individuals of a single species.

In the homonomous segments a pretergite is separated from the main tergite; correspondingly, a presternite lies ventrally before the main sternite (Fig. 95 A, B). Since comparable proportions are realized in the Scolopendromorpha and ventrally in *Craterostigmus* the phenomenon may form a ground pattern feature of the Chilopoda. At any rate, it is not connected with the evolution of segmentally successive long and short tergites in the stem lineage of the Heteroterga.

The Geophilomorpha represent the Chilopoda in deep, **unlit soil layers.** The complete reduction of the lateral light-sensory organs is correlated with this apomorphic mode of life, probably also the evolution of an extremely flat head for burrowing into the soil.

Strigamia, Haplophilus, Geophilus, Schendyla.

Heteroterga[20]

Autapomorphies (Fig. 93 → 3)

- Heterotergy.
  Dorsal heteronomous trunk. Regular alternation of short and long tergites. Disruption of this pattern in trunk segments 8 and 9; here two long tergites are found together. Because of this a change in the alternation of short and long tergites occurs (Fig. 96).
- Anisostigmophory.
  Regular succession of segments with and without spiracles. Disruption as with the tergites. Trunk segments 7 and 8 form metameres without spiracles (Fig. 96).

Scolopendromorpha – Triakontapoda

Scolopendromorpha

Autapomorphies (Fig. 93 → 4)

- Absence of an independent tergite on the first trunk segment.
  The tergite of the maxilliped segment fuses with the tergite of the first walking leg segment (Figs. 94 C, 96 A).
- Musculature on the spiracle pouches.
  “Respiratory muscles” on the spiracle pouches that are interposed between the spiracles and the inner atria. No influence on the closure of

[20] tax. nov.

the spiracles; they probably serve to expand the atria (gas exchange) (HILKEN 1998).
Other Chilopoda do not possess comparable muscles.

Compared with the sister group Triakontapoda **heterotergy in the Scolopendromorpha** is decidedly **weakly developed**; this must be assessed as a plesiomorphic state. In any case, however, a succession of short and long tergites is realized and trunk segments 8 and 9 with two long tergites can be recognized.

Plutonium zwierleinii (Cryptopidae) apparently show no heterotergy and possess segmental successive spiracles. This observation and the systematic position need to be examined.

The **strongly manifested heterotergy of the Triakontapoda** is correlated with a constant number of **15 walking leg segments**, which may be based on a functional interdependence. Therefore, the number of 21 (or 23) segments with walking legs in the Scolopendromorpha within the Heteroterga can be seen as the plesiomorphy.

Scolopendromorpha of the tropics and subtropics are the giants of the Chilopoda. Even *Scolopendra cingulata* from the Mediterranean region can attain a length of 17 cm. The Scolopendromorpha are also represented in northern Europe by the small *Cryptops hortensis* (2.5 cm).

Triakontapoda[21]

Autapomorphies (Fig. 93 → 5)

- Extreme heterotergy.
  In the ground pattern of the Triakontapoda the short tergites form only narrow strips between the long tergites (Craterostigmomorpha, Lithobiomorpha). Eventually they disappear altogether through fusion with the following long tergite (Scutigeromorpha).
- Limiting of the locomotory appendages to a constant number of 15 pairs (Fig. 94E–H).

Craterostigmomorpha – Gonopodophora

Craterostigmomorpha

Craterostigmus tasmanianus (Fig. 94E). Tasmania, New Zealand. Length 4.6 cm. Two species (?). Found as fossils from the Devonian. The taxon is distinguished within the Triakontapoda through numerous characteristic features (DOHLE 1990).

[21] tax. nov.

Autapomorphies (Fig. 93 → 6)

- The six long tergites 4, 6, 8, 9, 11, 13 are each divided into two plates by transverse joint membranes (Fig. 96 B).
- On the corresponding segments 4, 6, 8, 9, 11, 13 lateral pleural glands of unknown function open.
- Trunk segment 16 (bearing the 15th leg pair) a strongly sclerotized ring without any demarcation between tergite, pleurite and sternite.
- Posterior with a bivalvular piece (anogenital capsule); internally with a meshwork of chitin ridges.
- Only one ocellus on each side of the headshield.

Gonopodophora[22]

Autapomorphies (Fig. 93 → 7)

- Gonopod pincers used in egg deposition.
  Females with a pair of gonopods behind the walking legs; in the ground pattern probably of three articles (Fig. 97 A).
  The absence of female gonopods in the other Chilopoda can be interpreted as an original feature. The expression of the genetic information for these appendages only in the Lithobiomorpha and Scutigeromorpha can be assessed as a synapomorphy of the two taxa.

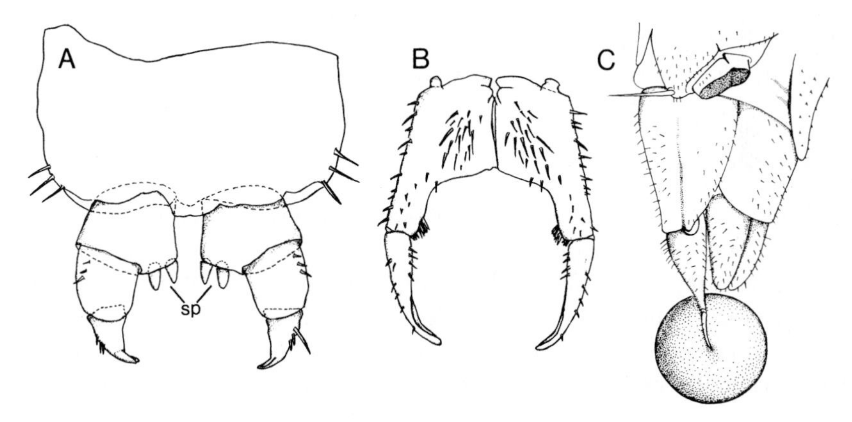

Fig. 97. Gonopodophora. **A** *Lithobius forficatus* (Lithobiomorpha). **B** *Scutigera coleoptrata* (Scutigeromorpha). Female gonopods with pincers. Spines on the basal segment (*sp*) may be an autapomorphy of the Lithobiomorpha. **C** *Scutigera coleoptrata*. Female, posterior end. Laid egg is held between the gonopods. (**A, B** Borucki 1996; **C** Dohle 1970)

[22] tax. nov.

- Deposition of single eggs.
  The eggs are at first carried between the gonopods (Fig. 97 C); here they are repeatedly moistened by the secretion of accessory genital glands and pressed into the ground. The egg, covered in and camouflaged by particles of earth, is then pushed into spaces in the soil. A comparison with the deposition of clusters of eggs will be discussed later (p. 229).
- Pinnate setae on the first maxillae.
  According to BORUCKI (1996), it is questionable as to whether these are homologous structures in the Lithobiomorpha and Scutigeromorpha.

Lithobiomorpha – Scutigeromorpha

Lithobiomorpha

Lithobius forficatus (length ca. 3 cm) is one of the most common Chilopoda to be found in Europe. The species is recognized at a glance by every zoologist and is pointed out on field trips as one of the most representative members of the Chilopoda when individuals are produced from beneath stones or tree bark (Fig. 94 F, G).

However, it is difficult to justify a taxon Lithobiomorpha, which turns up in all classifications, by clear autapomorphies.

Autapomorphies (Fig. 93 → 8)

- Basal articles of the female gonopods have prominent, terminal rounded spines (Fig. 97 A). Discussed as a possible characteristic feature (DOHLE 1985; BORUCKI 1996).
- ? Unpaired testes. Disputed as an autapomorphy.

Scutigeromorpha

As *Lithobius forficatus*, a secretive crevice inhabitant, is a good example of the Lithobiomorpha, we can view *Scutigera coleoptrata* (length 2.5 cm) as a Mediterranean representative of the Scutigeromorpha (Fig. 94 H) living on the soil surface; it lies in wait motionless for its prey, hunts other arthropods with great rapidity, and encircles them lasso-like with its long appendages. This mode of life is connected with a whole series of autapomorphies.

Autapomorphies (Fig. 93 → 9)

- Long, annulated antennae.
  From the biarticulated scapus follows a flagellum of two articles, which are subdivided into numerous annuli.

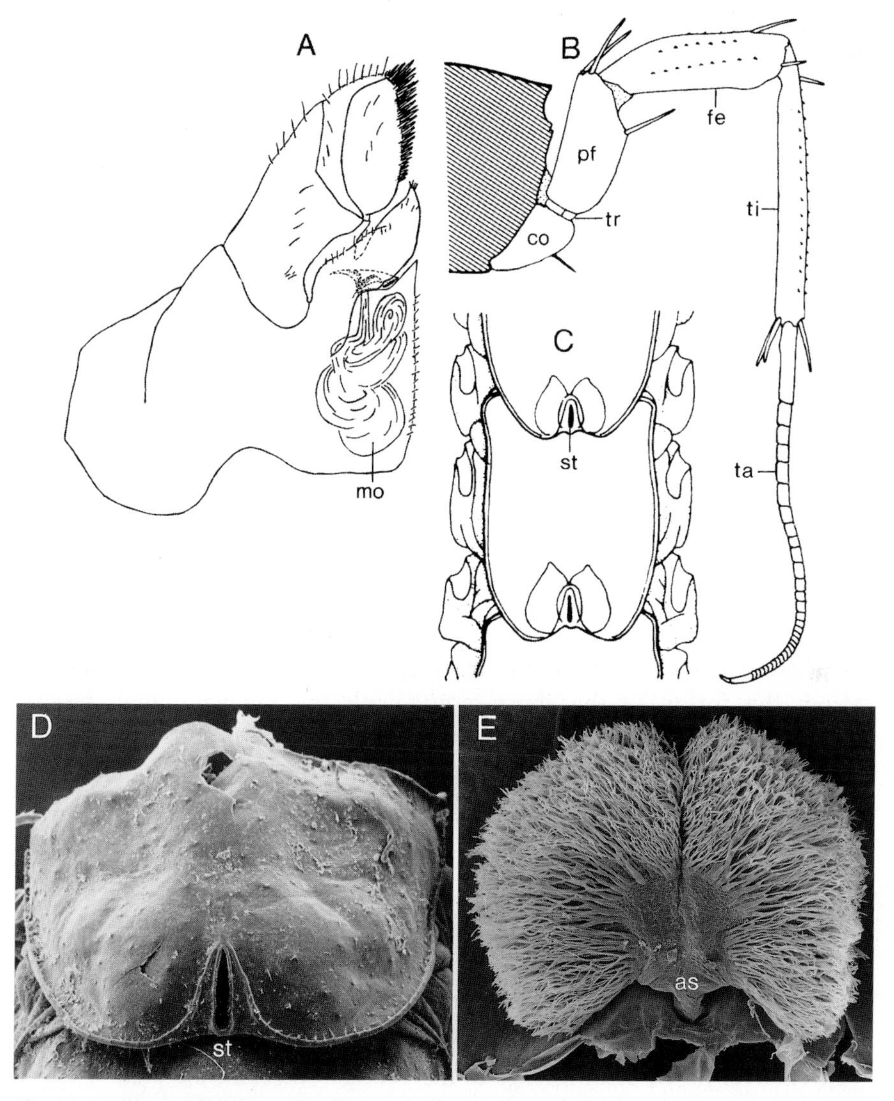

Fig. 98. *Scutigera coleoptrata* (Scutigeromorpha). Autapomorphies. **A** Maxillae 1 with maxillary organ. **B** Cross section through the trunk with walking leg. Secondary annulation of the tarsus. **C** Unpaired, dorsal spiracles on the posterior margin of the tergites. **D** Tergite with indentation at the posterior in which the spiracle lies. **E** Preparation of a segmental tracheal system from which two bundles with hundreds of tracheal stems originate; the stems branch into short, thin tracheae. *as* Atrium sac; *co* coxa; *fe* femur; *mo* maxillary organ; *pf* prefemur; *st* spiracle; *ta* tarsus; *ti* tibia; *tr* trochanter. (**A** Borucki 1996; **B,C** Lewis 1981; **D,E** Originals G. Hilken, Essen)

(Convergent to the annulate antennae of the Ectognatha within the Insecta.)

- Maxillary organ.
  Paired complex organ in the coxae of the first maxillae (Fig. 98 A). Protruding clusters of spindle and thread-shaped appendages. Function unknown (BORUCKI 1996).
- Maxillipeds with separate coxae (Fig. 92 C).
  Separate mobile maxillae very probably belong in the functional sphere of hunting agile prey. Contrary to common opinion I interpret this state therefore as an autapomorphy in comparison with the united coxae of the other Chilopoda.
- Long appendages.
  Subdivision of both tarsi as in the two distal antennal articles (Fig. 98 B). The tarsi consist of around 40 annuli.
- Two-part gonopods in the female (Fig. 97 B).
  Compared with the three-part gonopods in the Lithobiomorpha this may be a derived state.
- Seven large tergites (Fig. 96 D).
  From the small tergite of the maxillary segment (trunk segment 1) follow seven large tergites. With the exception of tergite 1 they are formed from the fusion of the short tergite with the following long tergite. Moreover the two long tergites on trunk segments 8 and 9 also unite. The following morphological arrangement of the seven large tergites to the trunk segments results from this process: T1 = S2, T2 = S3+4, T3 = S5+6, T4 = S7+8+9, T5 = S10+11, T6 = S12+13, T7 = S14+15.
- Seven unpaired spiracles – tracheal lungs.
  Seven dorsal slit-like spiracles[23] lie at the posterior end of the seven large tergites. They are developed exactly in those segments on which in the Lithobiomorpha laterally paired spiracles are found. This identical segment-jumping sequence (p. 216) clearly points to homology. Therefore, the state in the Scutigeromorpha must be seen as the apomorphous alternative. In the stem lineage of the Scutigeromorpha the pleural spiracles were moved dorsalwards and there united into unpaired pores. (Fig. 98 C, D).
  An original paired development is also suggested by the construction of the tracheal system (Fig. 98 E). A short tube leads from the spiracle into an atrium sac. From this pairs of several hundred tracheal stems originate; these then branch two to four times into short tracheae (HILKEN 1994, 1998). As opposed to other Chilopoda the tracheae of the Scutiger-

[23] The name Notostigmophora derived from these and the name Scutigeromorpha are synonyms; they are concerned with one and the same taxon. Consequently the name Notostigmorpha should be dropped.

omorpha do not lead to effector organs; on the contrary they are bathed by hemolymph, which takes up oxygen through the respiratory pigment hemocyanin and transports it throughout the body. One can therefore refer to a tracheal lung organ.

- Pseudocompound eyes (Fig. 90 D).
  The Scutigeromorpha are the only taxon of the Myriapoda with compound eyes of 200–250 closely packed ommatidia. Beneath the cornea of the ommatidia there is a vitreous body – exactly where the crystalline cone is found in the Insecta and Crustacea. Clear differences exist, however, in the ultrastructure of the dioptric apparatus and in the number of retinula cells. The vitreous body of *Scutigera coleoptrata* is a secretion of 8–13 peripheral pigment cells; it consists of numerous hyaline parts – irregularly divided and without nuclei. Two layers of retinula cells form the rhabdome; distally there are 9–23, proximally a constant number of four cells. The compound eye of the Scutigeromorpha is not homologous with the compound eye of other Mandibulata. On the contrary, the pseudocompound eye is interpreted as an extremely evolved lateral eye of the Chilopoda through formation of a new vitreous body (PAULUS 1979, 1986).

Possible Conflicts Within the Argument

The aforementioned systematization regarding the Chilopoda is confronted with alternatives in the mode of development and reproductive behavior, the prevailing assessment of which leads to conflict.

Anamorphosis – Epimorphosis

In anamorphosis the young hatch with an incomplete number of segments; the missing segments are added post-embryonically. In contrast, in epimorphosis, the full adult complement of segments are formed during embryogenesis.

In comparing the Tracheata taxa Chilopoda, Progoneata and Insecta the following facts need to be discussed. Within the Chilopoda anamorphosis is realized in the Triakontapoda (Craterostigmomorpha, Lithobiomorpha, Scutigeromorpha) as well as in all Progoneata. Epimorphosis is found in the Geophilomorpha and Scolopendromorpha within the Chilopoda; this mode also belongs in the ground pattern of the Insecta.

If one postulates anamorphosis for the stem species of the Tracheata, then epimorphosis must have originated convergently in parts of the Chilopoda and in the Insecta. On the other hand, if one postulates epimorphosis for the ground pattern of the Tracheata, then an independent evolution of anamorphosis in the Triakontapoda and the Progoneata is the consequence. With regard to convergence a numerical parity exists between the two hypotheses.

According to DOHLE (1985) and BORUCKI (1996), anamorphosis is a plesiomorphy of the Chilopoda. MINELLI & BORTOLETTO (1988, 1990) argue that the timing of the developmental processes in the case of anamorphosis requires greater control; thus they view epimorphosis as the plesiomorphic developmental mode of the Chilopoda. The evolution of anamorphosis appears to be documented in stages in the Triakontapoda. *Craterostigmus tasmanianus* hatches with 12, *Lithobius forficatus* with seven and *Scutigera coleoptrata* with four leg pairs.

Reproductive Behavior

Geophilomorpha, Scolopendromorpha and Craterostigmomorpha deposit all their eggs at one time on their bodies, but conduct themselves very differently during the following brood care period. Geophilomorpha females carry the eggs on their backs (Fig. 94 A); *Scolopendra* and *Craterostigmus* wrap themselves around the ventrally deposited eggs (Fig. 94 D).

In contrast, *Lithobius* and *Scutigera* deposit their eggs singly. Eggs are at first carried between the gonopods of the female, moistened several times with secretions from the genital opening, and encrusted with particles from the ground before being pushed into cracks and crevices in the soil (p. 222).

Contrary to prevailing opinion (DOHLE 1985; BORUCKI 1996), I interpret the behavior of the Lithobiomorpha and Scutigeromorpha as the apomorphous alternative. I assess the camouflaging of eggs and their insertion in the ground as an optimal adaptation for guarding against predators. The production of unconcealed clusters of eggs leads to brood care which represents a dangerous immobile phase in the life cycle. An avoidance through a change to the embedding of eggs in the earth with the aid of the gonopods appears plausible to me, not however vice versa.

Progoneata

Autapomorphies (Fig. 91 → 5)

- Position of the genital opening in the anterior body.
- Trichobothria with a bulbous extension of the sensory hair near the root (Fig. 99).
- Maxilla 1 without a palp.

In the Symphyla the unpaired genital pores lie in the sternum of the fourth trunk segment; in the sister group Dignatha (Pauropoda + Diplopoda) there are paired gonopores in the second leg pair segment. Sensilla known as trichobothria exist in all three high-ranking subtaxa of the Progoneata. One pair of trichobothria at the posterior of the Symphyla, five pairs of corre-

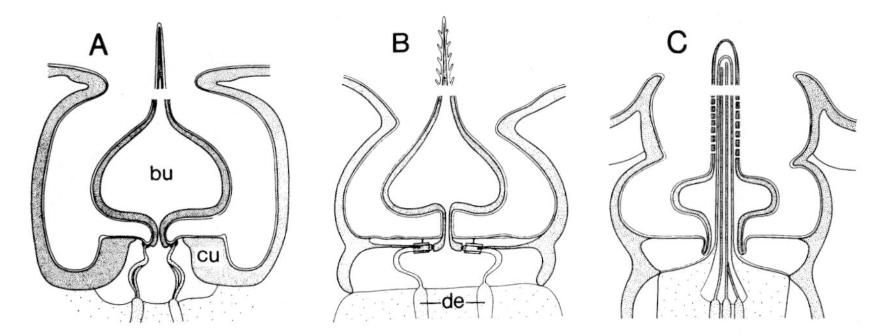

Fig. 99. Trichobothria of the Progoneata. **A** *Scutigerella immaculata* (Symphyla). **B** *Allopauropus* (Pauropoda). **C** *Polyxenus lagurus* (Penicillata, Diplopoda). Each is a longitudinal section through the proximal part of the sensory hair with jointed insertion in a cup-shaped pit of the cuticle. In the three taxa there is congruence in the expansion of the hair to form a hollow bulb between the stalk-like base and the distal hair shaft. *bu* Bulb; *cu* cuticle; *de* dendrite. (Haupt 1979)

sponding sensilla on certain tergites of the Pauropoda and a few sensory hairs on the head of the Penicillata (Diplopoda).

Trichobothria are lacking in the Chilopoda. The question of a possible homology with the like-named sensory hairs in the Insecta is unanswered. In any case, trichobothria with a conspicuous basal bulb of the hair can be seen as a significant derived feature of the Progoneata.

Symphyla – Dignatha

Symphyla

Autapomorphies (Fig. 91 → 6)

- Labium.
  Fusion product of the second maxillae. Two sclerotized plates with a median longitudinal suture. Each plate has three distal sensory cones (Fig. 100 G; p. 210).
- Unpaired genital opening.
  The position in the fourth leg pair segment may be original in comparison with the sister group Dignatha. The existence of a single pore between the sternal sclerites must, however, be an apomorphy in contrast to the paired gonopores in the ground pattern of the Dignatha.
- Paired head spiracles.
  Only one pair of tracheae in the head region with spiracles above the mandibles.
  Paired segmental tracheae on the trunk are hypothesized for the ground pattern of the Tracheata (p. 204). Consequently, the state in the Symphy-

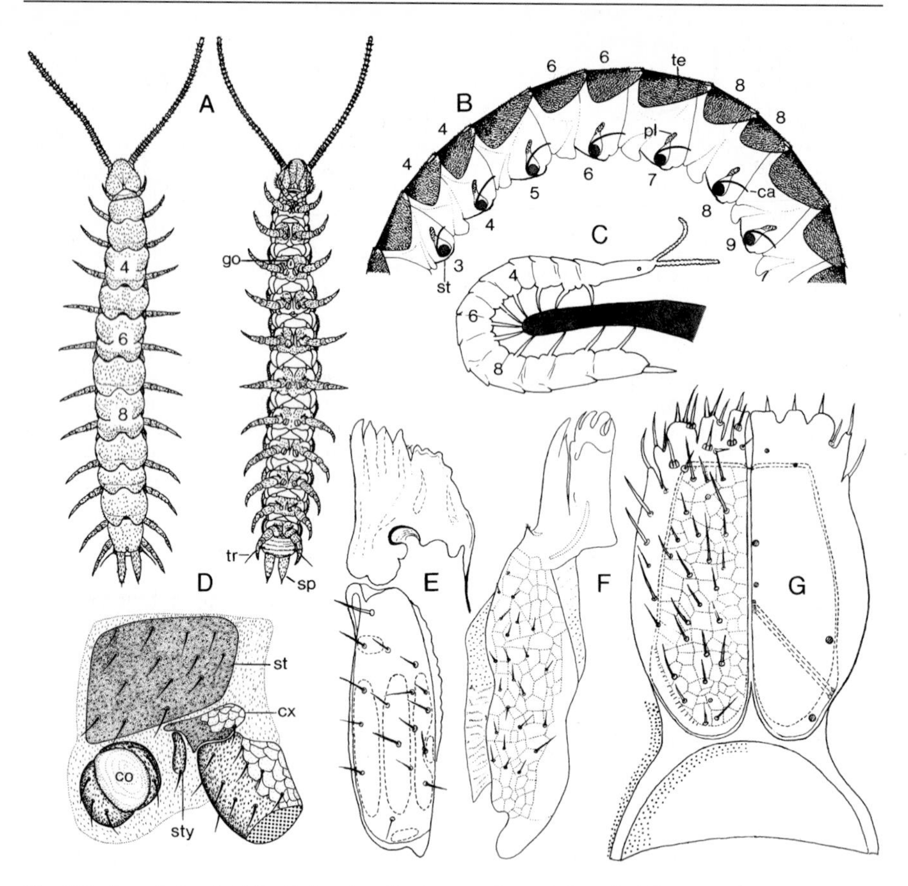

Fig. 100. Symphyla. **A–B, D–G** *Scutigerella immaculata*. **C** *Symphylella*. **A** Habitus, organization from dorsal and ventral. **B** Mid-trunk in side view. Tergites stippled, pleurites white, sternites black, coxal apodemes black. **C** Juvenile animal creeping around the edge of a leaf showing the intersegmental flexibility in segments 4, 6 and 8 with two tergites. **D** Right side of a trunk segment from ventral with sternite, coxal organ with everted bubble, stylus and coxa. **E** Divided mandible with base and cutting part (ventral view). **F** Maxilla 1 (ventral view). **G** Labium (maxilla 2) ventral. Setae only shown on the right-hand side. Following the sclerotized plates the membranous center is flanked by firm strips. *an* Anus; *ca* coxal apodeme; *co* coxal organ; *cx* coxa; *go* gonopore; *pl* pleurite; *sp* spinneret; *st* sternite; *sty* stylus; *te* tergite; *tr* trichobothrium. (**A** Dunger 1993; **B, C** Manton 1977; **D** Eisenbeis & Wichard 1985; **E–G** Ravoux 1975)

la can be regarded as the result of a far-reaching reduction and shifting of a spiracle pair to the head.

The hypothesis of a displacement of one spiracle pair surely offers a simpler explanation than the common assumption of a complete reduction of the tracheal system and a consequent new formation of tracheae with spiracles in the head.

The interpretation of segmental coxal apodemes (Fig. 100B) as remnants of a tracheal system is discussed in the Dignatha.

- Increase in tergites.
  During development two tergites form on each of the trunk segments 4, 6 and 8 (Fig. 100 A–C). This means there are 15 tergites in the ground pattern of the Symphyla. This division obviously leads to an increase in flexibility of the body for moving on and within the soil (MANTON 1977).
- Absence of lateral ocelli.
- Spinnerets.
  Paired spin glands (lateral glands) open in processes of the preanal segment. A homology with the cerci of the Insecta is disputed.
- Uptake of sperm into buccal pouches in the female's oral cavity.
  The whole mechanism of sperm transfer and fertilization forms a unique apomorphic peculiarity of the Symphyla.
  Deposition of stalked spermatophores from the male → uptake of the drops of sperm by the female with the mouth parts and storage of the sperm in seminal receptacles of the oral cavity → deposition of the eggs in moss → fertilization with the sperm from the oral cavity.

Due to their weak sclerotization the Symphyla are restricted to moist ground habitats. They can be up to 8 mm in length.

The trunk consists of 14 segments and a tiny telson around the anal opening. The first 12 segments bear appendages. In the Symphyla ground pattern these segments are covered by 15 tergites (*Scutigerella*, *Hanseniella*); an increase to up to 24 tergites is possible (*Ribautiella*).

Trunk segment 13 bears the spinnerets; segment 14 the paired trichobothria with long sensory hairs. Styli on trunk segments 3–12 and coxal organs on segments 3–10 also belong in the ground pattern (Fig. 100 D).

The post-embryonic development is named hemianamorphosis. Young hatch with 6 or 7 leg pairs. At each molt a further appendage pair is developed.

Dignatha

Autapomorphies (Fig. 91 → 7)

- Dignathy.
  Apart from the mandibles only one further structural and functional unit exists in the mouthparts. This forms the posterior termination of the preoral chamber – as a lower lip in the Pauropoda and as a gnathochilarium in the Diplopoda (Fig. 101 A, B).
- Penes.
  The paired genital openings of the male lie ventrally on the second trunk segment near the base of the appendages. They open on the tip of cone-shaped, mobile protuberances (Fig. 101 C, D).

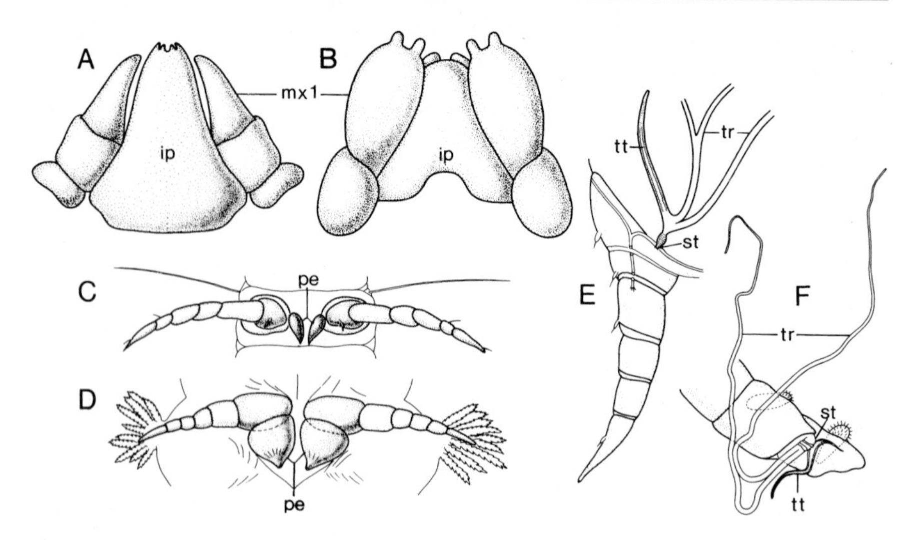

Fig. 101. Autapomorphies of the Dignatha. **A–B** Caudal termination of the preoral cavity. **A** Lower lip of *Pauropus silvaticus* (Pauropoda). **B** Gnathochilarium of *Glomeris marginata*. **C–D** Ventral formation of the penes on the second trunk segment. **C** *Pauropus silvaticus* (Pauropoda). **D** *Polyxenus lagurus* (Diplopoda). **E–F** Position of the spiracles in the vicinity of appendage bases and formation of sealed tracheal pouches. **E** *Millotauropus latiramosus* (Pauropoda). **F** *Polyxenus lagurus* (Diplopoda). Each a section from the region of the insertion of the first leg pair. *ip* Intermaxillary plate; *mx1* first maxilla; *pe* penis; *st* spiracle; *tr* trachea; *tt* tracheal pouch (apodeme). (Dohle 1980)

- Position of the spiracles.
  The spiracles of the tracheae lie close to the coxae of the appendages (Fig. 101 E, F).
- Tracheal pouches.
  Every spiracle continues internally into a sealed tracheal pouch; tracheal branches arise from these. The pouches themselves form apodemes for the attachment of trunk musculature (Fig. 101 E, F).
  In the Symphyla coxal apodemes (Fig. 100) exist that can be interpreted as the remnants of tracheal reduction. These may therefore be homologous to the tracheal pouches of the Dignatha (DOHLE 1980). In this case, the tracheal system just outlined with a ventral position of the spiracles and tracheal pouches would be an apomorphous ground pattern feature of the Progoneata.

The outstanding apomorphic peculiarity of the Dignatha lies in the construction of the mouthparts. While in the adelphotaxa Symphyla two pairs of maxillae are present, the Dignatha possess in their place only one structural and functional unit. There are two interpretations regarding the morphological assessment of this caudal boundary of the preoral chamber.

According to the first interpretation, the **lower lip of the Pauropoda** and the **gnathochilarium of the Diplopoda** develop from the first maxillae segment only. In both unities an unpaired intermaxillary plate is interposed between the appendage buds of the maxillae. This plate originates from the sternum of the segment. The first maxillae themselves lie laterally against it (TIEGS 1947; DOHLE 1964, 1980, 1996, 1997).

In the Dignatha the second maxillary segment does not form appendage buds (Fig. 104 A, B). As in the Tracheata, where the second antennae are absent because the intercalary segment does not develop appendages, this finding speaks for a reduction of the second maxillae in the stem lineage of the Dignatha. The type of innervation supports this view. The gnathochilarium and its total musculature are only supplied by one pair of maxillary nerves (Fig. 104 D) that, following the mandibular nerves, emerge from the subesophageal ganglion (FECHTER 1961).

The second hypothesis regards the **lower lip of the Pauropoda** and the **gnathochilarium of the Diplopoda** as a combined product of elements from **maxillae 1 and 2** (HILKEN & KRAUS 1994; KRAUS & KRAUS 1994). According to this interpretation, during evolution the second maxillae pushed between the first maxillae. The intermaxillary plate in the ontogeny of the Pauropoda and Diplopoda and the lamellae lingualis of the gnathochilarium differentiated from this are hypothesized to be the second maxillae.

For the systematization of the Progoneata this question is irrelevant. Either way the phenomenon of dignathy with only one structural and functional unit following the mandibles can be established as a synapomorphy of the Pauropoda and Diplopoda.

Pauropoda – Diplopoda

Pauropoda

Autapomorphies (Fig. 91 → 8)

- Trunk segmentation.
  12 segments + telson with 12 tergites and 11 pairs of appendages.
- Structure of the antennae.
  Six articles. Distal with forked branches from which “flagella” (antennal filaments) develop (Fig. 102).
- Tracheal system.
  Only one pair of tracheae with one pair of spiracles on the bases of the first leg pair.
- Absence of lateral ocelli.

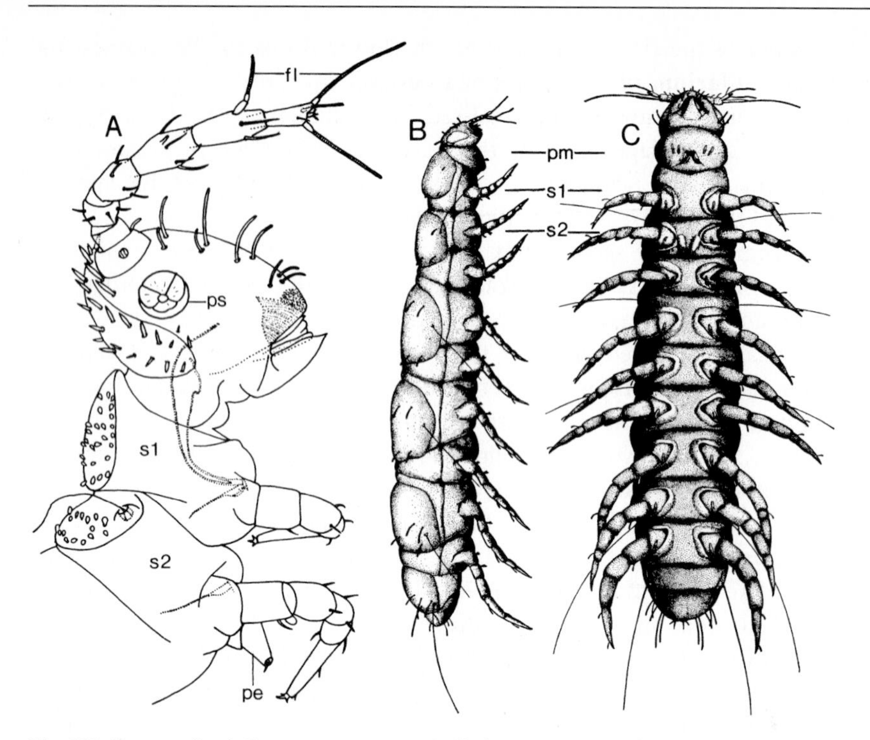

Fig. 102. Pauropoda. **A** *Rosettauropus temporalis* (Hexamerocerata). Anterior body of a male. Trunk segment 1 follows the head after a neck-like tapering. On segment 2 penes are developed. Length ♀ 1.15–1.3 mm; ♂ 1.25–1.4 mm. **B–C** *Pauropus silvaticus* (Tetramerocerata). **B** Side view. **C** Ventral view of male. From the small head a "post-maxillary segment" with eversible bubbles of unknown nature is separated. Length 1–1.2 mm. *fl* Flagellum (antennal filament); *pe* penis; *pm* post maxillary segment; *ps* pseudoculus (temporal organ); *s* trunk segment. (**A** Hüther 1968; **B, C** Tiegs 1947)

- Unpaired genital pore in the female.
  Compared with the paired openings in the sister group Diplopoda and in the male Pauropoda this can be assessed as an apomorphy.

The Pauropoda contain the smallest Progoneata species. By means of the manifestation of gas exchange organs, this is clearly interpretable as the result of regressive development. From the homologous, segmental tracheae of the Diplopoda only in the tropical Hexamerocerata with a length of between 1.3 and 1.8 mm is a single pair retained. With a further decrease in size to 1 mm or less tracheae completely disappeared in the Tetramerocerata.

From the few species of **Hexamerocerata** (*Millotauropus*: Africa, Madagascar; *Rosettauropus*: South America) (REMY 1950, 1953; HÜTHER 1968) the ground pattern of the Pauropoda can be deduced. The trunk consists of 12 segments + telson with 12 tergites and 10–11 leg pairs. From the normally developed head follows directly the first trunk segment bearing the first pair of appendages and the tracheae (Fig. 102 A). The antennae have six articles, whereby "flagella" are developed on the fifth and sixth articles. I have found no details concerning the question of the monophyly of a unity Hexamerocerata.

The cosmopolitan **Tetramerocerata** (*Pauropus, Allopauropus*) have many derived characteristics. From the tiny head the last segment with a part of the brain is shifted rearwards (DOHLE 1964). Only after this "post-maxillary segment" follows the first trunk segment with the first appendage pair (Fig. 102 B,C). These are restricted to nine (occasionally ten) pairs. Usually only six tergite plates are found dorsally. The antennae have only four articles, "flagella" are found on the last article.

Diplopoda

Autapomorphies (Figs. 91 → 9; 103 → 1)

- Gnathochilarium.
  Interpretation as either the exclusive equivalent of the first maxillae or as the fusion product of maxillae 1 and 2 disputed (p. 231).
- Diplosegments (diplosomites).
  With the exception of the four simple segments behind the head diplosomites distinguish the construction of the trunk (Fig. 106 K).
  In the ground pattern five sclerite plates belong to a diplosegment – one dorsal diplotergite, two lateral diplopleurites (one on each side) and two ventral sternites one behind the other. The sclerites are separated primarily by membranes; within the Diplopoda they fuse to form rigid, uniform annuli (Juliformia, Polydesmida) (Fig. 106 C,H,J).
  Both sternites of a diplosegment each bear a pair of appendages and possess a pair of spiracles. The segmental structures of the internal organization are similarly present in a doubled condition per diplosomite – the nervous system with two pairs of ganglia as well as the heart with two pairs of ostia, two pairs of (lateral) arteries and two pairs of wing muscles.
- Antennae.
  Four sensory cones are found on the last of the eight antennal articles (Fig. 104 F).
- Aciliary sperm.
  The sperm of the Chilopoda, Symphyla and Pauropoda have axonemata with the original 9×2+2 microtubuli pattern (JAMIESON 1987). In the stem lineage of the Diplopoda an immobile sperm evolved through loss of the axoneme.

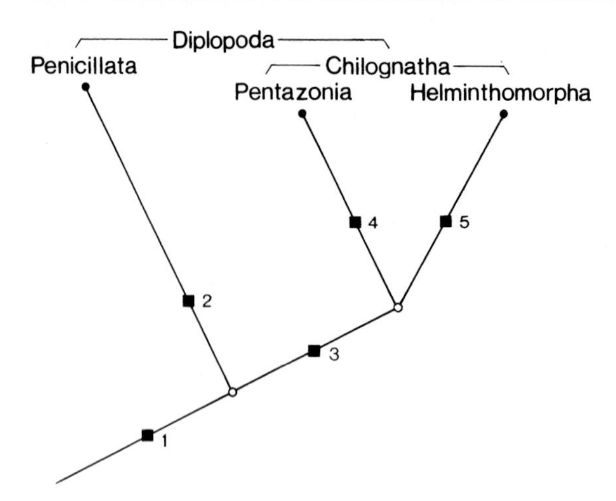

Fig. 103. Diagram of the phylogenetic relationships between the highest-ranking subgroups of the Diplopoda.

Different opinions exist regarding the **nature of the first four trunk segments.** Are we dealing morphologically with primary simple segments or with modified diplosegments?

In the skin shed after molting two separate tergites and pleurites are found in each of the anterior segments (KRAUS 1990, 1997). Embryology shows them, however, as single segments with no trace of doubling

Fig. 104. Diplopoda. **A–B** *Glomeris marginata.* Segmentation of the germ band with head and first two trunk segments. **A** Early developmental stage with segment construction. **B** Late stage with differentiation of appendage buds. No buds on the intercalary segment (= segment of the second antennae) and of the post-maxillary segment (= segment of the second maxillae). **C** *Cylindroiulus teutonicus.* Ventral view of the head. The gnathochilarium is flanked laterally by the triarticulate mandibles; only the two basal articles are visible. The occipital region is below. **D** *Cylindroiulus teutonicus.* Gnathochilarium. View of the inner surface with nervous system. All lobes and the total musculature are innervated by one nerve pair, which emerges from the subesophagael ganglion following the mandibular nerve. The nerve pair is assigned to the first maxillary segment. **E** *Glomeris distichella.* Gnathochilarium – In hypothesis 1 the whole gnathochilarium of the Diplopoda is a product of the maxillary segment (segment of the first maxillae). In hypothesis 2 cardines and stipites are interpreted as derivatives of maxillae 1, and the lamellae lingualis as derivatives of maxillae 2. Partially corresponding names for parts of the gnathochilarium with parts of the mouthparts of other Arthropoda do not make statements about homology. **F** *Glomeridesmus.* Tip of the antenna with four sensory cones on the last article. *an* Antenna; *ans* antennal segment; *ca* cardo; *gu* gula (derivative from the sternite of the post-maxillary segment); *ics* intercalary segment; *l* leg pair; *la* labrum; *le* lobus exterior; *li* lobus interior; *ll* lamella lingualis; *lm* lobus medius; *ls* leg pair segment; *md* mandible; *mdd* distal mandible base article; *mdp* proximal mandible base article; *mds* mandible segment; *me* mentum; *mx* maxilla; *mxs* maxillary segment; *ncv* ventral collum nerve (first trunk segment); *nmd* mandibular nerve; *nmx* maxillary nerve (first maxilla); *pms* post-maxillary segment; *sc* pharyngeal connective; *st* stipes. (**A, B** Dohle 1974; **C, D** Fechter 1961; **E, F** Enghoff 1990, labels from Hilken & Kraus 1994) ▶

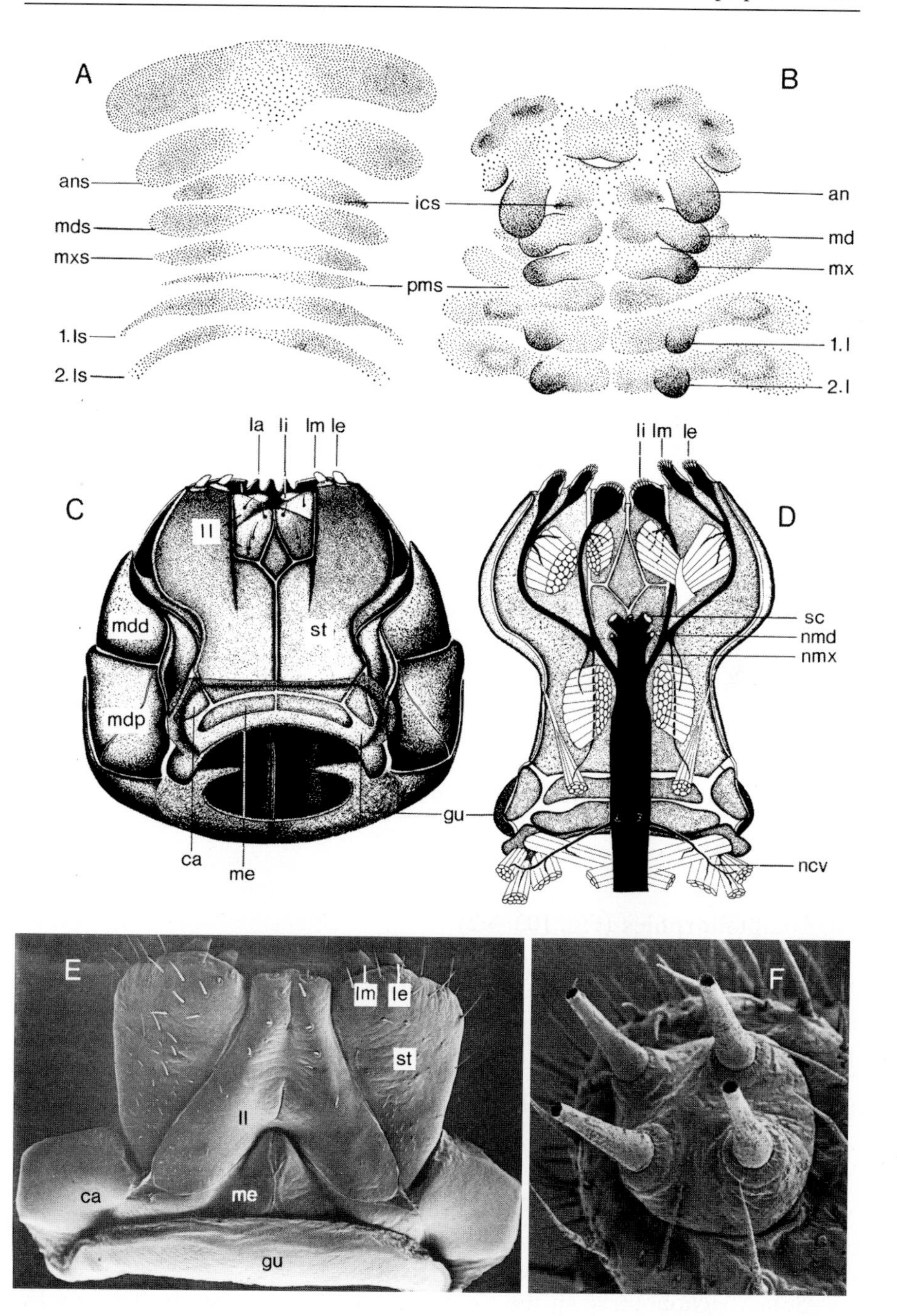
A
ans
mds
mxs
1. ls
2. ls
ics
pms
B
an
md
mx
1. l
2. l
C
la
li
lm
le
ll
mdd
st
mdp
ca
me
gu
D
li
lm
le
sc
nmd
nmx
ncv
E
lm
le
st
ll
ca
me
gu
F

(DOHLE 1964, 1974, 1997) and with the following arrangement of the appendages. The first leg pair belongs to the first trunk segment with the collum (cervical sclerite). The three following leg pairs each belong to a segment with one tergite. Appendage pairs 5 and 6 become part of the first doubled segment with a diplotergite.

Complications arise in species with uniform annuli (see above). Here the sternites with their appendage pairs are shifted one segment backwards during ontogeny. Because of this the first leg pair is taken from the collum segment and assigned to the second segment, the second leg pair to segment 3 and the third leg pair to segment 4. Finally the fourth leg pair of the last of the single segments is integrated as the foremost element into the first diplosegment.

Systematization

Diplopoda
- **Penicillata**
- **Chilognatha**
  - **Pentazonia**
  - **Helminthomorpha**

Penicillata – Chilognatha

Penicillata

Autapomorphies (Fig. 103 → 2)

- Lateral eyes reduced to a few isolated ocelli with a limited number of retinula cells.
  (Complete reduction within the unity.)
- Segmental tufts and rows of serrated setae (trichomes). Additionally a tuft of long setae on the telson (Fig. 105).

Compared with the sister group Chilognatha the Penicillata (Pselaphognatha) possess a whole series of original features.

Plesiomorphies
- Non-calcareous flexible cuticle.
- Some trichobothria on the head. The Pselaphognatha are the only Diplopoda with these sensilla from the ground pattern of the Progoneata.
- Absence of repugnatorial glands.
- Indirect sperm transfer through spermatophores that are deposited for the ♀♀ as drops on a web with signal threads.

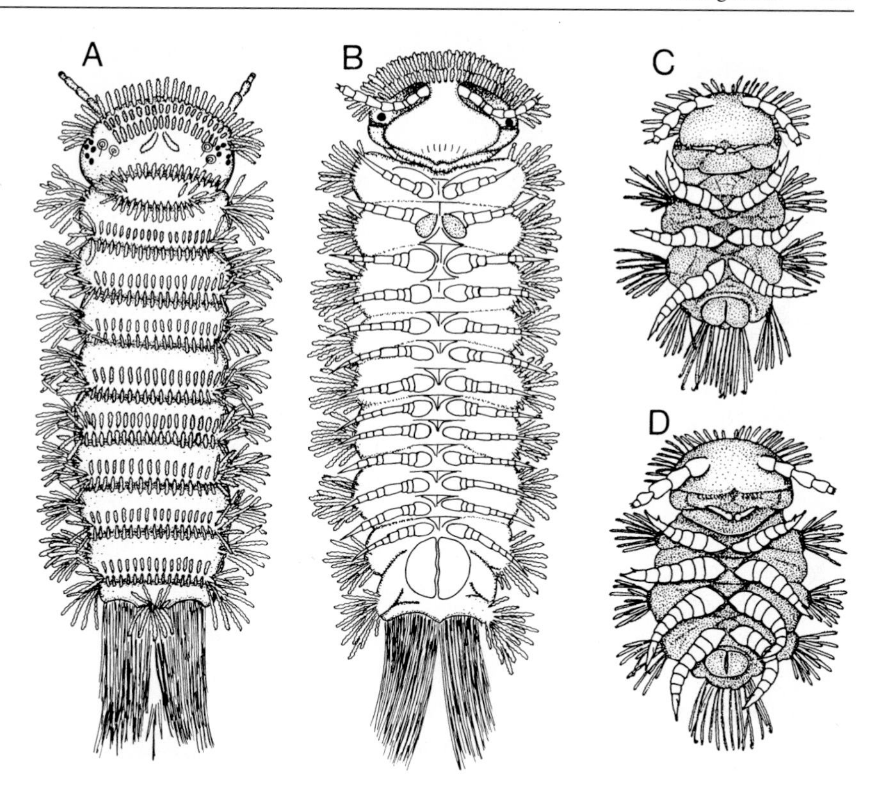

Fig. 105. Penicillata (Diplopoda). *Polyxenus lagurus*. **A** Adult, dorsal view. Few lateral ocelli and trichobothria on the head. **B** Adult, ventral view. With 13 leg pairs. Genital papillae in the second trunk segment. **C** Stage I of the development with three leg pairs (anamorphosis) **D** Stage II with the four leg pairs of the first four trunk segments. (**A,B** Eisenbeis & Wichard 1985; **C,D** Enghoff, Dohle & Blower 1993)

The Penicillata can reach a length of nearly 4 mm. They have a maximum of 17 leg pairs (*Phryssonotus*).

Polyxenus lagurus found in Germany possesses 13 leg pairs and ten tergites.

Chilognatha

Autapomorphies (Fig. 103 → 3)

- Storage of calcium salts in the cuticle.
- Complete absence of trichobothria.
- One pair of repugnatorial glands with poisonous secretions per diplosomite. Dorsolateral opening of the gland pores (Fig. 106 F).
- ? Direct sperm transfer.

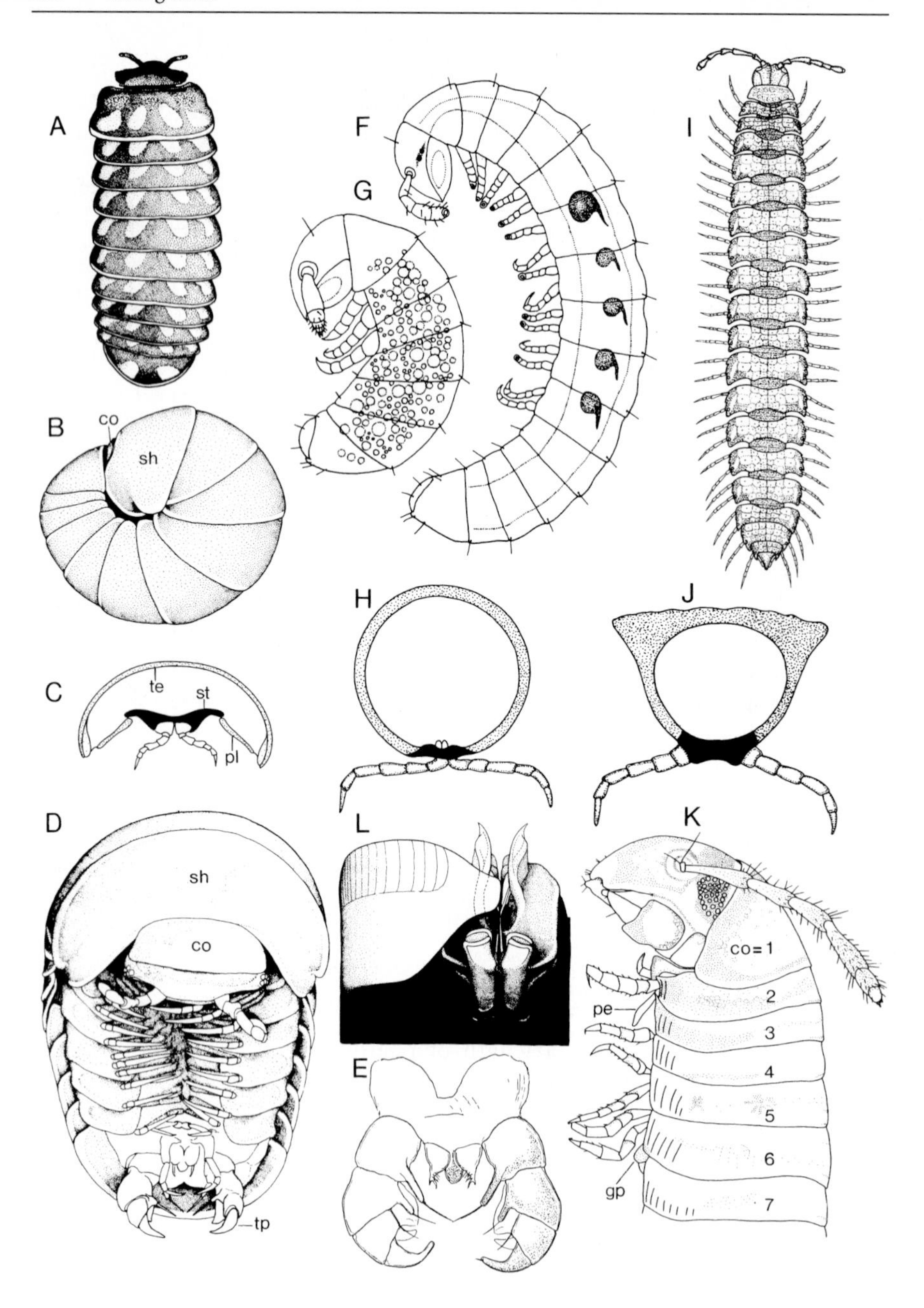

In all Chilognatha **sperm** is transferred directly from the male to the female using **modified appendages**. The position of the appendages on the body and their significance in copulation are very different in the Pentazonia (Opisthandria) and the Helminthomorpha (Proterandria) and are therefore also rightly named differently. The **telopods** on the last segment

◀ **Fig. 106.** Chilognatha (Diplopoda). **A–E** Pentazonia. **A** *Glomeris marginata.* Juvenile animal. **B** *Glomeris marginata.* Partially rolled-up animal; a small part of the collum is still visible. **C** *Glomeris marginata.* Body cross section. Mobile connection of the tergite and the stiff sternal region (insertion of the legs) through the ventrolateral pleurites. **D** *Glomeris marginata.* Ventral view of male. Large telopods at the posterior (leg pair 19). **E** *Glomeris helvetica.* Telopods. **F–L** Helminthomorpha. **F** *Proteroiulus fuscus* (Juliformia). Developmental stage I with three leg pairs. **G** *Proteroiulus fuscus.* Developmental stage III. Development of the defense glands starts in the sixth segment. **H** *Cylindroiulus caeruleocinctus.* Body cross section with stiff uniform ring. **I** *Polydesmus angustus* (Polydesmida). Band habitus through winglike extensions of the tergites. **J** *Polydesmus complanatus.* Body cross section. Dorsolateral wings on the stiff ring. **K** *Cylindroiulus waldeni* (Juliformia). Male. Head and anterior body with segments 1–7. **L** *Cylindroiulus latestriatus.* Gonopods in ventral view. The left part of the pleurotergal ring removed. *co* Collum (tergite 1); *gp* gonopod; *pe* penis; *pl* pleurite; *sh* shield (tergite 2); *te* tergum; *tp* telopod. (**A, B, D, L** Blower 1985; **C, H, I, J** Eisenbeis & Wichard 1985; **E** Schubart 1934; **F, G** Enghoff, Dohle & Blower 1993; **K** Hopkin & Read 1992)

of the male **Pentazonia** are primary claspers for grasping the female. The **gonopods** in the anterior of the **Helminthomorpha** are in contrast copulatory feet, which take sperm from the genital opening and insert it into the vulva of the female.

There is no legitimate position from which to trace back the telopods and the gonopods to a common apomorphic state in the ground pattern of the Chilognatha (ENGHOFF 1984, 1990). The resulting hypothesis postulates their independent evolution in the stem lineages of the Pentazonia and the Helminthomorpha and interprets them as autapomorphic alternatives of these taxa.

For the stem species of the Chilognatha two possibilities of sperm transfer thus remain.

1. In the stem species there was a direct sperm transfer, whose mechanism is not known. This is without doubt an unsatisfactory interpretation.
2. The stem species possessed, as in the Penicillata, the original mode of indirect sperm transfer. This mode was taken over into the stem lineages of the Pentazonia and the Helminthomorpha and was replaced by the evolution of telopods on the one hand and of gonopods on the other.

Pentazonia – Helminthomorpha

Pentazonia

Autapomorphies (Fig. 103 → 4)

- Telopods.
  Last pair of appendages in the male greatly enlarged and differentiated into grasping organs (Fig. 106 D, E). To transfer sperm the male grasps the vulva of the female with the telopods.

- Repugnatorial glands with unpaired median dorsal pores.
  Only found in some Pentazonia. Since they appear in pairs during the development of *Glomeris* (DOHLE 1964), the formation of an unpaired pore can be interpreted as an autapomorphy.

The Pentazonia (Opisthandria) have a short body with a maximum of 22 tergites and 37 leg pairs in the tropical Glomeridesmida. The widespread Glomerida in Europe (*Glomeris*) are even smaller; they possess only 10–11 tergites together with 17 (♀) or 19 (♂) leg pairs. The Pentazonia can roll into a ball concealing head and legs (Fig. 106 A–E).

Helminthomorpha

Autapomorphy (Fig. 103 → 5)

- Gonopods.
  In the male, at least one leg pair of the seventh trunk segment is transformed into copulatory feet (Fig. 106 L). The male takes sperm from his penes with the gonopods and transfers it into the vulvae of the female.

In the Helminthomorpha (Proterandria) the Diplopoda reach their maximum size with a length of 30 cm. *Illacme plenipes* from California has up to 350 leg pairs.

The Juliformia (*Julus*, *Tachypodoiulus*, *Unciger*) represent the cylindrical habitus (Fig. 106 F–H). The Polydesmida (*Polydesmus*) acquire a flat shape through the development of dorsolateral keels on the tergites (Fig. 106 I, J).

Insecta

Autapomorphies (Figs. 91 → 3; 113 → 1)

- Labium.
  Fusion product of the second maxillae. Grown together seamlessly proximally and primarily divided transversely into mentum and prementum. Prementum medially with glossae and paraglossae, laterally with multiarticulate labial palps (Fig. 109).
- Division of the trunk into a thorax of three segments and an abdomen of 11 segments + telson.
- Abdominal segment 10 without appendages.
- Ommatidia of the compound eyes with two primary pigment cells.
  The two pigment cells (main pigment cells) evolved from the two corneagenous cells of the original compound eyes of the Mandibulata (Fig. 39 C).

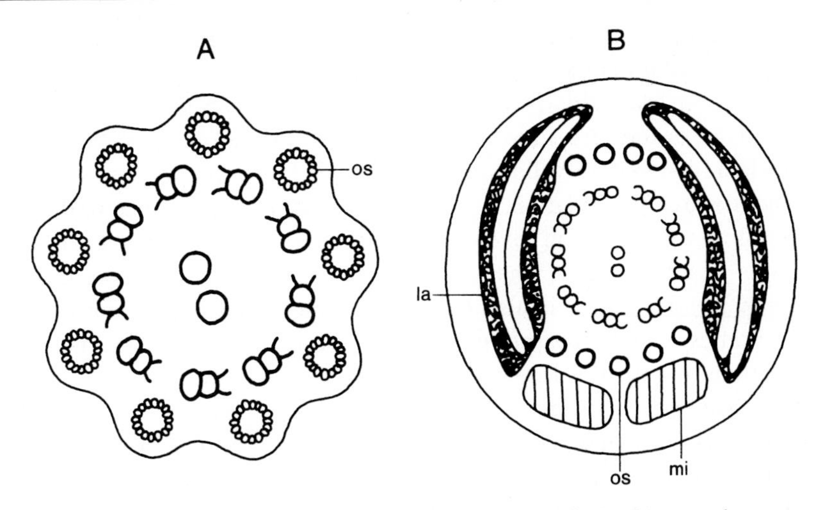

Fig. 107. Sperm axoneme with "9+9+2" structure as an autapomorphy in the ground pattern of the Insecta (cross sections). **A** *Neuronice* (Trichoptera). A regular external crown of nine single tubuli surrounds the nine double tubuli with dynein arms. In the middle there are two central tubuli. **B** *Petrobius brevistylis* (Archaeognatha). The outer crown is divided into two groups of four and five single tubuli. *la* Lateral body; *mi* mitochondrium; *os* outer single tubuli. (**A** Jamieson 1987; **B** N.P. Kristensen 1991)

- Axoneme of the sperm with "9+9+2" pattern.
  In contrast to the original 9+2 structure of the sperm axoneme in the ground pattern of the Metazoa (nine peripheral double tubuli and two central tubuli), there is a further outer crown of nine single tubuli. Within the Tracheata this 9+9+2 pattern (Fig. 107) is only known from the Insecta. Due to its widespread appearance in primary wingless taxa, and in the Pterygota, it can be postulated for the ground pattern of the Insecta (JAMIESON 1987; N.P. KRISTENSEN 1995).
- Differentiation of the ovary into a germarium and a vitellarium.
  Germocytes (oocytes) capable of development only form in the end section of the ovarioles (p. 257).

Overview

To begin with we present a hierarchic tabulation of the phylogenetic system of the Insecta up to the level of those unities which in traditional classifications are ranked as orders (KAESTNER 1973); for emphasis, these are marked in boldface (p. 243).

Advocates of the traditional Linnean hierarchy with categories concede that such labels denote different things in different system unities. An order of the Insecta cannot be compared with an order of the Amphibia or the Mammalia. The categories have validity however, they argue for denoting hierarchic levels in a certain taxon – e.g. within the taxa Insecta, Amphibia or Mammalia.

However, even this position is untenable. The 33 taxa of the Insecta labeled as orders are indeed, with two exceptions, well-established monophyla and therefore valid unities of the phylogenetic system of the Insecta. The hierarchic tabulation, however, shows at a single glance that these taxa stand at very different levels – even when some of them appear at certain levels as adelphotaxa. Their labeling as "orders" of Insecta is totally arbitrary. In addition, it is also harmful, as they feign equal rank between the 33 unities, which does not exist and which clouds the phylogenetic kinship relations within the Insecta. Categories as labels of taxa are of no use in the phylogenetic system of organisms (Vol. I, p. 18).

In this overview the conventional division of the Blattodea into the "Blattaria" (cockroaches) and Isoptera (termites) is absent for the following reason. The "Blattaria" are identified as a paraphylum and thus eliminated. The monophyletic Isoptera merely form a subordinate subtaxon within the Blattodea (p. 291). Furthermore, the "Psocoptera" are probably also a paraphylum.

The Highest Ranking Adelphotaxa Entognatha – Ectognatha and the Kinship of the Primary Apterygotic Insecta

The Insecta (=Hexapoda)[24] comprises five monophyla of primary wingless organisms – the Diplura, Protura, Collembola, Archaeognatha and Zygentoma – as well as the winged Pterygota.

We will deal first with the problem of the basic systematization and then turn to a more detailed analysis of the ground pattern of the Insecta.

The traditional typological classification places the Diplura, Protura, Collembola and "Thysanura"[25] with identical rank side-by-side in a taxon "Apterygota". This, however, is an artificial grouping established due to a

[24] We follow the common equalization of the names Insecta and Hexapoda for a taxon comprising all insects. Limiting the name Insecta to some of the insects, such as the Ectognatha (N.P. KRISTENSEN 1991; BITSCH 1994) or Diplura+Ectognatha (KUKALOVÁ-PECK 1987, 1991), can lead to confusion.

[25] The name "Thysanura" comprises in typological divisions the Archaeognatha and the Zygentoma. A taxon with these limitations is a further paraphylum in the paraphylum "Apterygota" and has no place in the phylogenetic system of the Insecta. Additionally the practice of using the name Thysanura sensu stricto for the Zygentoma should be discontinued.

- Insecta
  - Entognatha
    - **Diplura**
    - Ellipura
      - **Protura**
      - Collembola
  - Ectognatha
    - **Archaeognatha**
    - Dicondylia
      - **Zygentoma**
      - Pterygota
        - Palaeoptera
          - **Ephemeroptera**
          - **Odonata**
        - Neoptera
          - **Plecoptera**
          - N. N.
            - Paurometabola
              - **Embioptera**
              - Orthopteromorpha
                - Blattopteriformia
                  - **Notoptera**
                  - N. N.
                    - **Dermaptera**
                    - Dictyoptera
                      - **Mantodea**
                      - Blattodea
                - Orthopteroidea
                  - Saltatoria
                    - **Ensifera**
                    - **Caelifera**
                  - **Phasmatodea**
            - Eumetabola
              - Paraneoptera
                - **Zoraptera**
                - Acercaria
                  - Psocodea
                    - **"Psocoptera"**
                    - **Phthiraptera**
                  - Condylognatha
                    - **Thysanoptera**
                    - **Hemiptera**
                      - Sternorrhyncha
                      - Euhemiptera
                        - Cicadomorpha
                        - N. N.
                          - Fulgoromorpha
                          - Heteropteroidea
                            - Coleorrhyncha
                            - Heteroptera
              - Holometabola
                - Neuropteriformia
                  - Neuropteroidea
                    - **Planipennia**
                    - N. N.
                      - **Megaloptera**
                      - **Raphidioptera**
                  - Coleopteroidea
                    - **Coleoptera**
                    - **Strepsiptera**
                - Mecopteriformia
                  - **Hymenoptera**
                  - Mecopteroidea
                    - Amphiesmenoptera
                      - **Trichoptera**
                      - **Lepidoptera**
                    - Antliophora
                    - N. N.
                      - **Mecoptera**
                      - **Siphonaptera**
                    - **Diptera**

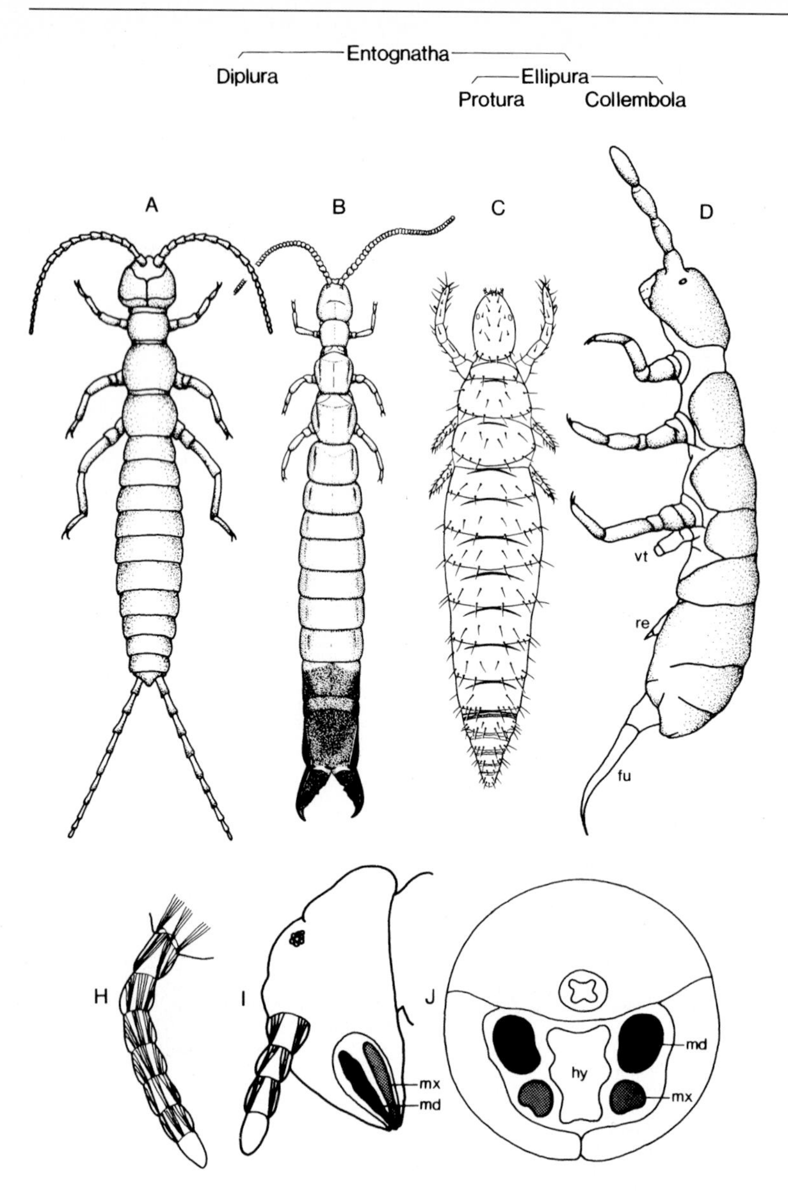

Fig. 108. *Top row:* Representatives of the taxa of primary wingless Insecta and one species of the winged Ephemeroptera. **A–B** Diplura. **A** *Campodea.* **B** *Heterojapyx evansi.* **C** Protura. *Australentulus tillyardi.* **D** Collembola. *Folsomotoma.* **E** Archaeognatha. *Allomachilis froggatti.* **F** Zygentoma. *Acrotelsella devriesiana.* **G** Ephemeroptera (Pterygota). *Ataloblephia. Bottom row*: Plesiomorphic and apomorphic feature states in the construction of the antennae and mouthparts in the Entognatha and Ectognatha. **H** Articulated antenna with musculature up to the last but one article (plesiomorphy). **I** *Podura aquatica* (Collembola). Head with articulated antenna (plesiomorphy) as well as mandibles and maxillae in pouches of a head capsule (autapomorphy of the Entognatha). **J** Collembola. Head cross section to further illustrate entognathy. Hypopharynx between the mandibles and maxillae. **K** *Nesomachilis australica* (Archaeognatha). Monocondylic mandible (plesiomorphy). **L** *Ctenolepisma longicaudata* (Zygentoma). Di-

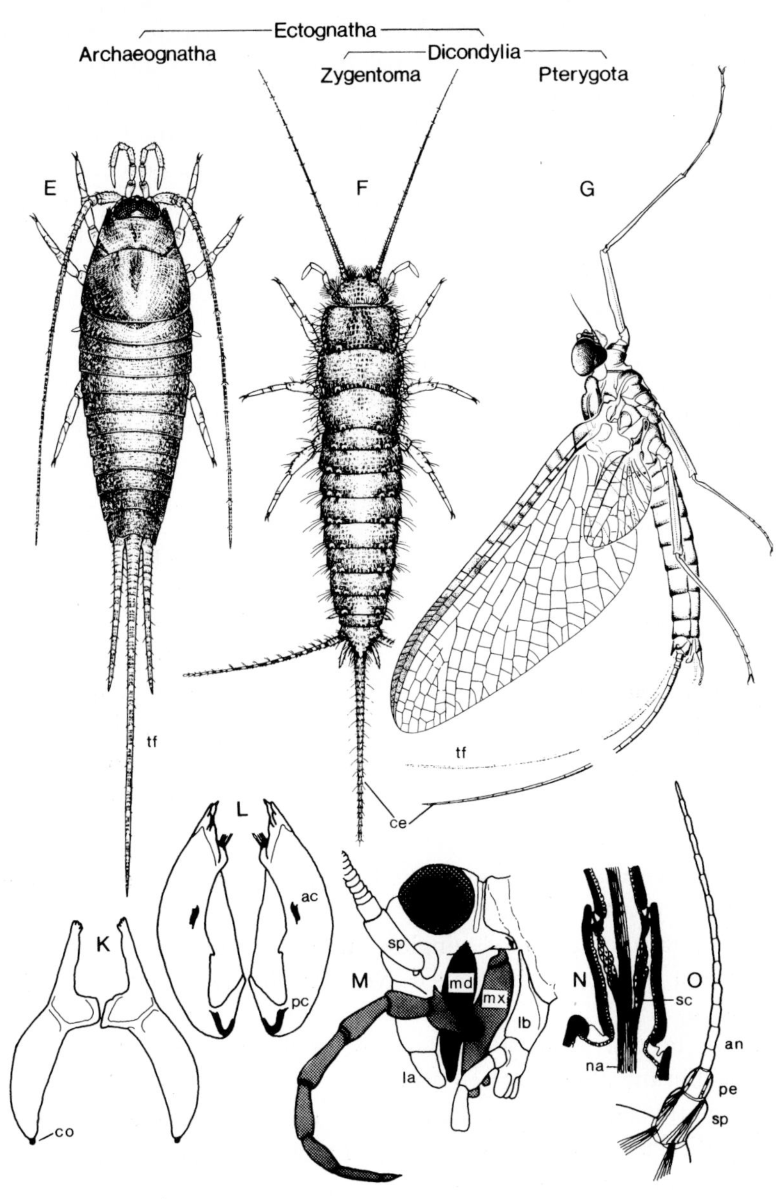

condylic mandible (autapomorphy of the Dicondylia). **M** *Nesomachilis maoricus* (Archaeognatha). Head in side view. Free articulation of mandibles and maxillae (plesiomorphy). **N** *Melolontha melolontha* (Coleoptera). Johnston's organ in the pedicellus (autapomorphy of the Ectognatha). **O** Annulated antennae with scapus, pedicellus and muscle-free anellus (autapomorphy of the Ectognatha). *ac* Anterior condyle; *an* anellus; *ce* cercus; *co* condyle (monocondylic mandible); *fu* furca; *hy* hypopharynx; *la* labrum; *lb* labium; *md* mandible; *mx* maxilla; *na* antennal nerve; *pc* posterior condyle; *pe* pedicellus; *re* retinaculum; *sc* scolopidia; *sp* scapus; *t* terminal filament; *vt* ventral tube. (**A** Eidmann and Kühlhorn 1970; **B** Condé & Pagés 1991; **C** Imadaté 1991; **D** Greenslade 1991; **E, F, K, L** Watson & Smith 1991 a, b; **G** Peters & Campbell 1991; **H, I, O** Weber & Weidner 1974; **J** Wolter 1963; **M** Boudreaux 1979; **N** Seifert 1995) plesiomorphy – the primary lack of wings. Phylogenetic systematics must consequently reject the traditional division into "Apterygota" and Pterygota. It has developed two competing hypotheses concerning the highest-ranking adelphotaxa relationships of the Insecta. In the first hypothesis the Entognatha (Diplura + Ellipura) and the Ectognatha (Archaeognatha + Dicondylia) are presented as sister groups, in the second hypothesis the Ellipura (Protura + Collembola) and the Insecta (Diplura + Ectognatha). The hypotheses diverge on one point – the assessment of the kinship of the Diplura.

Hypothesis 1
(HENNIG 1953, 1969)
Insecta (Hexapoda)
- **Entognatha**
  - **Diplura**
  - **Ellipura**
    - **Protura**
    - **Collembola**
- **Ectognatha**
  - **Archaeognatha**
  - **Dicondylia**
    - **Zygentoma**
    - **Pterygota**

Hypothesis 2
(KUKALOVÁ-PECK 1987, 1991)
Hexapoda
- **Ellipura**
  - **Protura**
  - **Collembola**
- **Insecta**
  - **Diplura**
  - **Ectognatha**
    - **Archaeognatha**
    - **Dicondylia**
      - **Zygentoma**
      - **Pterygota**

Basic congruencies between the two kinship hypotheses are found in the following points.

a. The taxa of primary wingless insects do not form a system unity; rather they are distributed between the two highest-ranking sister groups.
b. The taxa of primary wingless insects occupy very different rank levels within the phylogenetic system.
c. The transformation of the antennae to annulated antennae with scapus, pedicellus and anellus is interpreted as an autapomorphy of a monophylum Ectognatha.
d. The Zygentoma as a unity of primary wingless insects form the adelphotaxon of the Pterygota.

For the determination of the **position of the Diplura** I look to the following considerations.

1. Entognathy.
   In the Diplura, Protura and Collembola the mandibles and maxillae are surrounded by lateral mouth folds; they lie in deep pouches of the

head capsule (Fig. 108 I, J). Related to this, the maxillary and labial palps are greatly shortened. Compared with the free articulation of the mouthparts on the head of the Ectognatha (Fig. 108 M), this phenomenon of entognathy is without doubt an apomorphy. Using the principle of parsimony, we postulate a single evolution of entognathy in one stem lineage common to the Diplura and Ellipura (Protura + Collembola) (hypothesis 1). The emphasizing of certain differences between the entognathous states of the Ellipura and the Diplura (KOCH 1997; KRAUS 1997) requires detailed documentation to track the more complex, a priori less probable, assumption of a double evolution of entognathy on the basis of the Insecta (hypothesis 2).

2. Integration of leg parts into the lateral surfaces of abdominal segments. According to KUKALOVÁ-PECK (1987, 1991), during the evolution of the Insecta, different numbers of appendages were incorporated into the formation of the pleura of the abdominal segments – two (subcoxa and coxa) in the Ellipura, and three (subcoxa, coxa, trochanter) in all other insects. The latter is considered to be an autapomorphy of a taxon consisting of the Diplura + Ectognatha (hypothesis 2). This interpretation is strongly criticized (p. 252).
3. Synapomorphies between recent Diplura and Ectognatha?
   Only the presentation of congruencies which are convincingly interpretable as synapomorphies between the Diplura and Ectognatha can, a posteriori, compel the hypothesis of a convergent origin of entognathy in the stem lineages of the Ellipura and the Diplura. Such congruencies, however, are not known. The common epimorphosis in the Diplura and Ectognatha is, contrary to KOCH (1997), a plesiomorphy – carried over from the ground pattern of the Insecta. Within the Insecta only the Protura show a post-embryonic increase in abdominal segments from 8 to 11. This has long been interpreted as the delay of a procedure completed in embryogenesis in the other insects (WEBER & WEIDNER 1974). Within the Insecta therefore anamorphosis in the Protura is the apomorphous alternative.

At the present state of debate, it appears to me that the continuation of hypothesis 1 with a systematization of the Insecta into the adelphotaxa Entognatha and Ectognatha is more valid (Fig. 108).

Fig. 109. Head of the Insecta. Unpaired formations and paired appendage derivatives. **A** Diagram of the head in sagittal section. The unpaired hypopharynx divides the oral cavity into the dorsal (anterior) cibarium and the ventral (posterior) salivarium. **B** *Petrobius brevistylis* (Archaeognatha). Biting-chewing mouthparts with monocondylic mandible and paired superlinguae on the hypopharynx. **C** *Periplaneta americana* (Blattodea). Biting-chewing orthopteroid mouthparts. Mandible with two condyles; division into an anterior cutting part and behind this a chewing part. **D** *Micropterix callunae* (Micropterigoidea, Lepidoptera). Original biting-chewing mouthparts in the butterflies. Labrum with lateral wings (pilifer). The glossae on the labium are united to form a uniform plate. *ac* Anterior condyle; *ca* cardo; *ci* cibarium; *cl* clypeus; *co* condyle of the monocondylic mandible; *fr* frons; *ga* galea; *gl* glossa; *hy* hypopharynx; *la* labrum; *lb* labium; *lc* lacinia; *m* mouth; *md* mandible; *mt* mentum; *mx* maxilla; *pc* posterior condyle; *pg* paraglossa; *pi* pilifer; *pl* palpus labialis; *pm* palpus maxillaris; *pmt* prementum; *sa* salivarium; *sg* salivary gland duct; *sl* superlingua; *smt* submentum; *st* stipes. (**A** Bitsch 1973; **B** Imms 1957; **C,D** Seifert 1995)

Ground Pattern

Up to this point only an outline of a few assured autapomorphies of the Insecta has been presented. We now introduce further elements from the feature pattern of the last common stem species of all recent insects that are of importance for the systematization of the taxon (BITSCH 1994; N.P. KRISTENSEN 1997a).

Connected with this we must mention new results from Paleozoic insects that call into question ideas which have up to now been widespread. The singular observations regarding the primary wingless taxa *Testajapyx thomasi* (Diplura), *Ramsdelepidion* (Zygentoma) and *Dasyleptus* (Dicondylia), as well as basic members of the Pterygota (KUKALOVÁ-PECK 1987, 1991, 1992) (Fig. 110), are, however, strongly disputed and urgently require verification on more substantial material.

Number of Trunk Segments

Fourteen trunk segments + telson agree exactly with the segment number of the Symphyla within the Progoneata. However, no argument exists for hypothesizing 14 trunk segments for the ground pattern of the Tracheata. It remains an open question as to whether a trunk of 14 segments is a plesiomorphy or an autapomorphy of the Insecta.

Thorax – Abdomen

In the stem lineage of the Insecta a division of the trunk into a **thorax of three segments** and an **abdomen of 11 segments + telson** took place. The latter is well developed as an independent body section only in the Protura; otherwise the telson forms a membrane ring with tiny sclerites around the anus.

The differentiation of the trunk into thorax and abdomen is a fundamental autapomorphy of the Insecta.

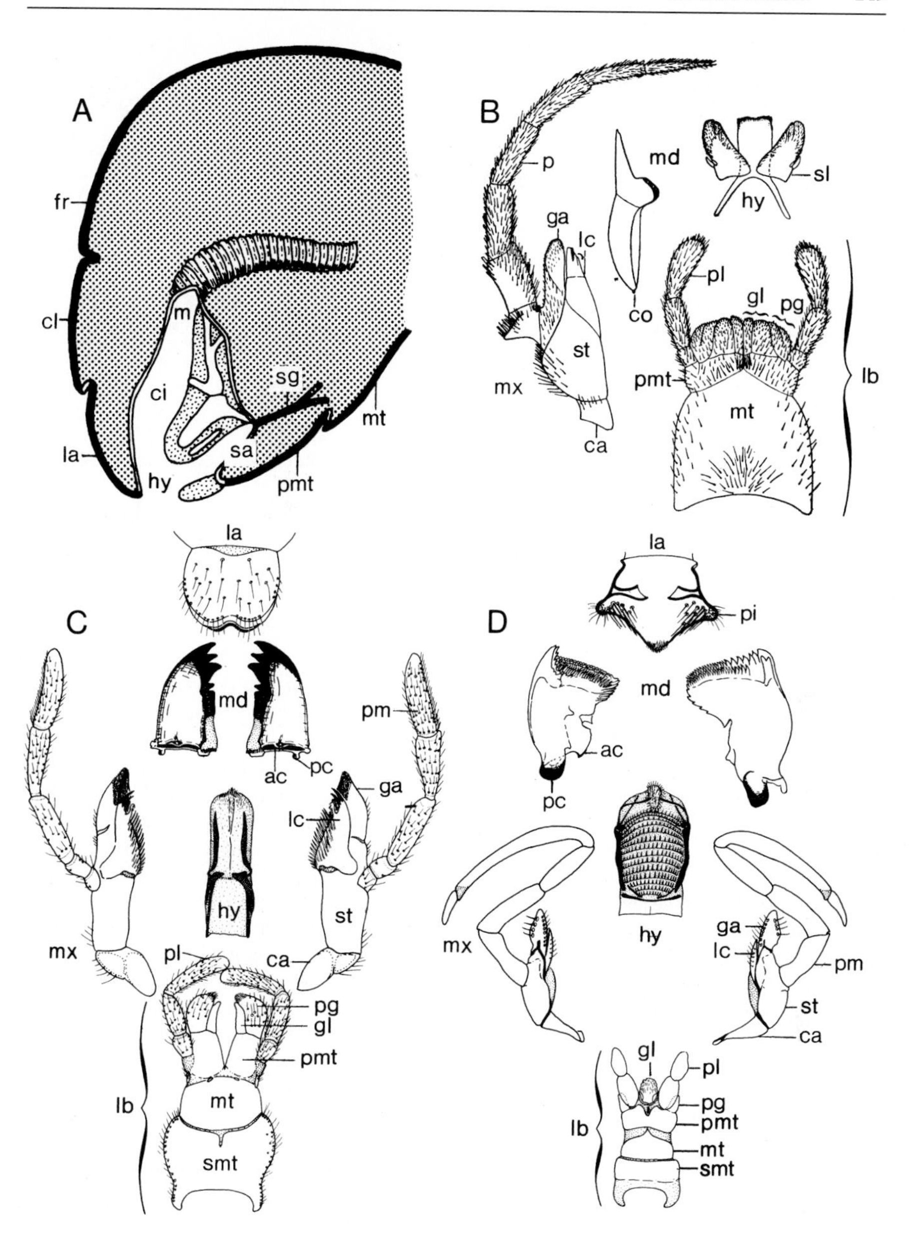
A
fr
cl
m
ci
sg
mt
la
sa
hy
pmt
B
p
md
sl
hy
ga
lc
co
pl
gl
pg
st
pmt
lb
mx
mt
ca
C
la
md
pm
ac
pc
ga
lc
hy
st
mx
pl
ca
pg
gl
pmt
lb
mt
smt
D
la
pi
md
ac
pc
hy
mx
ga
lc
pm
st
ca
gl
pl
pg
pmt
lb
mt
smt

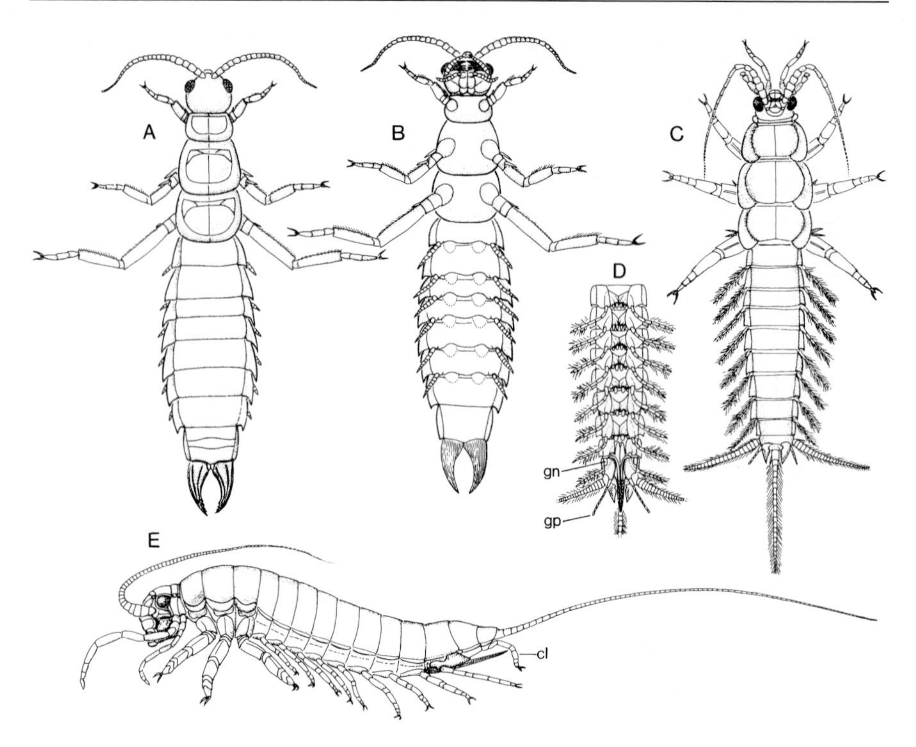

Fig. 110. Primary wingless insects from the Paleozoic. **A–B** †*Testajapyx thomasi*. Dorsal and ventral view. Reconstructions from the holotype. Length without antennae 47.5 mm. Assignment to the Diplura disputed. Upper Carboniferous, Illinois. **C–D** †*Ramsdelepidion schusteri*. **C** Dorsal reconstruction, chiefly from the holotype. **D** Ventral view of the abdomen; articulated appendages and two pairs of coxal organs on several segments. Length without appendages 60 mm. Stem lineage representative of the Zygentoma. Upper Carboniferous, Illinois. **E** †"*Dasyleptus*" with dicondylic mandible and gonangulum which prove the animal to be a stem lineage representative of the Dicondylia. Reconstruction. Length without antennae and terminal filament 35 mm. Abdomen with ten pairs of multiarticulate appendages that bear paired claws. Such an appendage takes the place of the cerci on segment 11. Upper Carboniferous, Illinois. *cl* Cercal leg; *gn* gonangulum; *gp* gonopodium. (Kukalová-Peck 1987)

Head Appendages

One pair of **antennae** was carried over from the ground pattern of the Tracheata into the stem species of the Insecta. The **articulated (segmented) antennae** represent the plesiomorphic condition. They consist of numerous, similar articles which are equipped with muscles, an exception being the last article (Fig. 108 H). In the stem lineage of the Ectognatha a transformation into the apomorphous state of **annulated antennae** with scapus, pedicellus and anellus took place (Fig. 108 N, O).

Biting-chewing mouthparts are widespread among the Insecta. They are formed from three pairs of appendages – the **mandibles**, the **maxillae** and the **labium** being a fusion product of the second maxillae. In most textbooks biting-chewing mouthparts (Fig. 109) are usually dealt with as characterizing the Orthopteromorpha and are therefore also known as orthopteroid mouthparts. They are realized in the Pterygota up to subtaxa of Holometabola and even belong in the ground pattern of the Lepidoptera. Within the framework of our discussion, however, far more important is that primary wingless insects with monocondylic mandibles are also provided with biting-chewing mouthparts. They characterize the Archaeognatha (*Machilis*, *Petrobius*) in the Ectognatha and can even be postulated for the ground pattern of the Entognatha, in which the labium is incorporated into the formation of the jaw pouches of the head capsule. Thus, most Collembola have biting-chewing mouthparts. From the distribution outlined above we can draw the conclusion that the stem species of the Insecta possessed biting-chewing orthopteroid mouthparts. In the sections that follow we deal with the essential elements from the primary state. Transformations into apomorphous manifestations will be discussed where they are applicable as features for the characterization of high-ranking subtaxa of the Insecta.

The **mandible** is an appendage consisting of a single unit; in the ground pattern of the Insecta it has a incisor region (pars incisiva) and a molar region (pars molaris). The mandible is primarily connected to the cranium by a single condyle (Entognatha, Archaeognatha) (Fig. 108 K). In the stem lineage of the Dicondylia a second more anterior articulation point evolved on this monocondylic mandible; the dicondylic mandible originated (Fig. 108 L).

The **maxilla** is connected to the head capsule by the basal cardo. The following stipes bear laterally a multiarticulated maxillary palp; distally the galea as an outer and the lacinia as an inner process are found.

The unpaired **labium** results from fusion of the last head segment appendages. The two primary regions are the proximal postmentum and the distal prementum. On the postmentum two sclerite plates can be secondarily subdivided; these are then known as the submentum and mentum. The prementum corresponds to the two stipites of the maxillae and possesses correspondingly similar appendages. These are here first the lateral labial palps consisting of only a few articles. Inwards on both sides are then found the paraglossa and the glossa; they appear in the place of the galea and lacinia of the maxilla.

Unpaired Structures on the Head

Clypeolabrum and **hypopharynx** represent unpaired formations on the head that cannot be traced back to appendages.

The **labrum** is the tip of the acron in the Euarthropoda. The plate-like organ inserts onto the clypeus and covers the bases of the mandibles from the anterior. The inner wall of the labrum is known as the epipharynx; it forms the anterior side of the buccal or pre-buccal cavity.

The **hypopharynx** is formed from the medioventral walls of a few postoral segments. The hypopharynx divides the buccal cavity into a dorsal cibarium through which nutrients are led into the pharynx and a ventral salivarium as the opening place of the salivary glands (Fig. 109 A). The superlinguae as paired lateral protuberances on the hypopharynx – somewhat as they are realized in the Archaeognatha – probably belong in the ground pattern of the Insecta (Fig. 109 B). They are absent in the majority of insects.

Appendages with a Locomotory Function

Locomotory appendages exist in recent insects only on the three thoracic segments. This is, however, not an autapomorphy of the Insecta. On the contrary the three pairs of thoracic appendages were taken over from the stem species of the Tracheata into the stem lineage of the Insecta; they are a plesiomorphy in their ground pattern.

As possible ground pattern autapomorphies, however, two structural features of the trunk appendages from recent Insecta can be considered: (a) a low number of six articles in the thorax appendages – coxa, trochanter, femur, tibia, tarsus (in the Ectognatha subdivided into several annuli) and pretarsus; and (b) the absence of comparable legs with a locomotory function on the abdominal segments.

However, the fossil finds mentioned earlier appear to refute the existence of these two features in the feature pattern of the last common stem species of recent insects. Let us consider the problem of the appendages on the thorax and abdomen separately.

Thorax. For the first stem species of the Insecta at the beginning of their stem lineage KUKALOVÁ-PECK reconstructed appendages with 11 articles – epicoxa, subcoxa, coxa, trochanter, prefemur, femur, patella, tibia, basitarsus, tarsus (subdivided into two annuli) and pretarsus.

In Paleozoic Pterygota epicoxa and coxa should be assimilated into the wall of the thoracic segments, leaving appendages with nine free articles. For the wingless *Testajapyx*, *Dasyleptus* and *Ramsdelepidion* from the Carboniferous thoracic appendages with a larger number of articles are also depicted (Fig. 110).

The details and concepts of KUKALOVÁ-PECK have been, however, vigorously disputed (BITSCH 1994; N.P. KRISTENSEN 1997 a; WILLMANN 1997 b); they force the hypothesis of a multiple convergent evolution of thoracic appendages with six articles in the primary wingless Insecta (Ellipura, Diplura, Archaeognatha, Zygentoma) and within the Pterygota

(WILLMANN 1997b). This is extremely unlikely. Much more likely is a secondary increase in the number of articles from the ground pattern number of six – as they are still occasionally to be seen in recent Insecta (Odonata, Hymenoptera: WILLMANN op. cit.).

Abdomen. *Dasyleptus* sp. from the stem lineage of the Dicondylia had according to KUKALOVÁ-PECK ten pairs of appendages with eight articles on abdominal segments 1–9 and 11 (Fig. 110E). Furthermore, multiarticulate legs on the abdomen are postulated for *Testajapyx*, *Ramsdelepidion* and Paleozoic Pterygota. According to these observations the last stem species of recent insects must have possessed appendages of several articles with a locomotory function on practically all abdominal segments. Their reduction must have occurred several times independently in diverse stem lineages of the Insecta (Diplura, Ellipura, Archaeognatha, Zygentoma, Pterygota). With regard to the uninspiring conclusions which are also forced here, a re-examination of the Paleozoic insects is sorely needed.

Absence of appendages on segment 10.

Derivatives of **appendages are absent on segment 10 of the abdomen** in all recent insects. In addition, in fossil finds no appendages are observed on this segment. The hypothesis of a single, early reduction of the appendages of abdominal segment 10 in the stem lineage of the Insecta is without doubt valid. Their lack is placed as an autapomorphy in the ground pattern of the Insecta.

Cerci

Paired thread-like feeler structures of numerous articles exist in the Diplura, Archaeognatha, Zygentoma (Fig. 111) and some basal taxa of the Pterygota on the eleventh abdominal segment. It is tempting to homologize these and place the possession of cerci as an autapomorphy in the ground pattern of the Insecta. Against this argument are the observations of KUKALOVÁ-PECK on *Dasyleptus* sp. from the Carboniferous. In the depiction of the fossil, locomotory appendages with claws are shown on the eleventh abdominal segment (Fig. 110E). Accordingly, cerci as non-clawed, thread-like feeler organs must have originated convergently several times in the Insecta – in the Diplura as well as in the stem lineages of the Archaeognatha and Dicondylia.

However, just as with the thoracic and abdominal appendages, the hypothesis of a multiple, independent evolution to an identical result within the Insecta is again very questionable; it is therefore not accepted (N.P. KRISTENSEN 1997a).

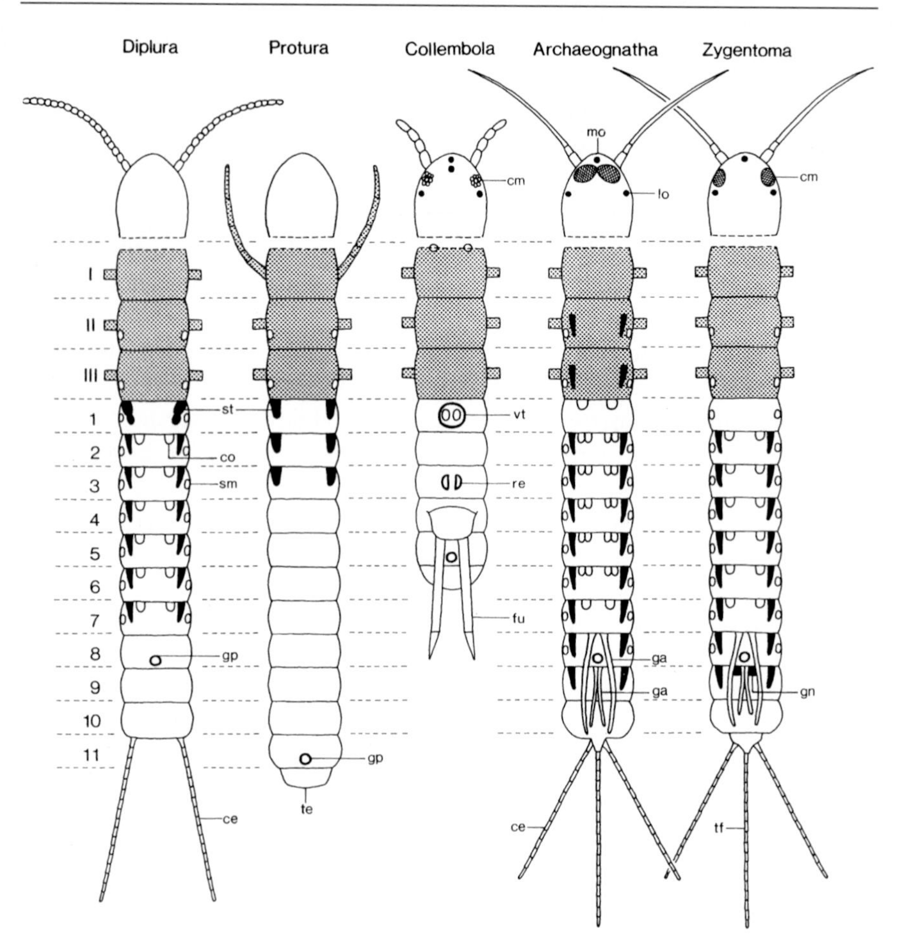

Fig. 111. Development of certain systems such as antennae, light-sensory organs, tracheal spiracles, styli, coxal organs, genital openings (♀), gonapophyses, cerci and terminal filament amongst others in the ground pattern of the individual taxa of primary wingless insects. Diagrammatic presentation. Dorsal view of the head; ventral view of the trunk. *I–III* Thoracic segments; *1–11* abdominal segments. *ce* cercus; *cm* compound eye; *co* coxal organ; *fu* furca; *ga* gonapophyse; *gn* gonangulum; *gp* genital pore; *lo* lateral ocellus; *mo* median ocellus; *re* retinaculum; *sm* spiracle; *st* stylus; *te* telson; *tf* terminal filament; *vt* ventral tube. (Original)

Terminal Filament

Along with the cerci an unpaired terminal filament as a median attachment on the eleventh abdominal segment has been discussed for the ground pattern of the Insecta. The terminal filament (paracercus), however, is found in both recent and fossil Insecta only in the Ectognatha (Figs. 108, 111). There are no convincing arguments to postulate its existence in the stem species of all recent Insecta. We interpret the terminal thread as an autapomorphy of the Ectognatha (p. 266).

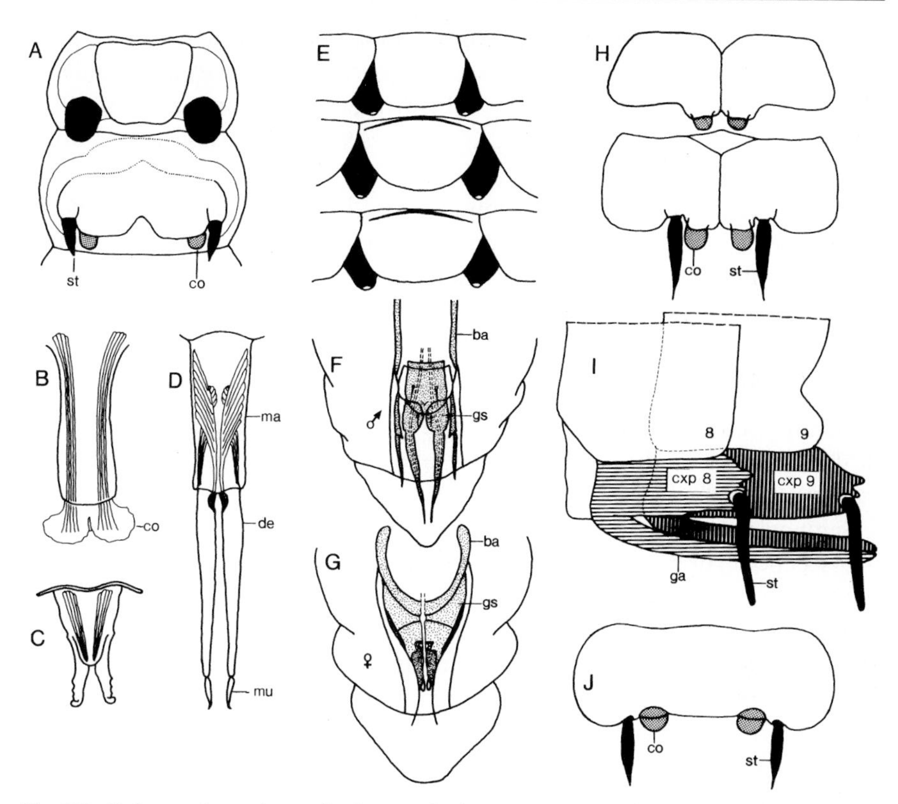

Fig. 112. Abdomen formations of primary wingless insects. **A** Diplura. *Metriocampa.* Ventral view of the first and second abdominal segments. **B–D** Collembola. *Tomocerus vulgaris.* **B** Ventral tube. **C** Retinaculum. **D** Furca. **E** Protura. *Australentulus tillyardi.* Ventral view of abdominal segments 1–3. **F–G** Protura. *Eosentomon.* Genital apparatus of male and female with openings in the 11th abdominal segment. **H** Archaeognatha. *Nesomachilis australica.* Ventral view of abdominal segments 1 and 2. **I** Archaeognatha. *Machilis.* Lateral view of female abdominal segments 8 and 9 with ovipositor. **J** Zygentoma. *Atopatelura hartmeyeri.* Ventral view of abdominal segment 6. *ba* Basal apodeme; *co* coxal organ; *cxp* coxopodite; *de* dens; *ga* gonapophyse; *gs* gonostylus; *ma* manubrium; *mu* mucro; *st* stylus. (**A** Condé & Pagés 1991; **B–D, I** Seifert 1995; **E–G** Imadaté 1991; **H, J** Watson & Smith 1991 a, b)

Styli

Styli are appendage derivatives. In the form of short pegs, they appear ventrally on the trunk in the Symphyla and Insecta. Their distribution among the primary wingless insects is shown in Figs. 111 and 112.

The styli of the Symphyla are subdivisions of the coxae of trunk segment appendages 3–10. A similar arrangement of styli is found in the Archaeognatha on the meso- and metathorax. Styli also appear, however, in recent insects on abdominal segments 1–9, which bear no appendages. What interpretation do fossil specimens offer? Neither in *Testajapyx* nor *Dasyleptus* and *Ramsdelepidion* are styli observed in the coxal region of the abdominal legs. Abdominal styli are thus obviously not a division of abdominal legs but rather the remains of articles of reduced appendages (BITSCH 1994). For the two-part styli on the first abdominal segment of recent Diplura, KUKALOVÁ-PECK (1983, 1991) proposes a derivation from a multiarticulate appendage.

Peg-like abdominal styli – as found today in recent Insecta – probably do not belong in the feature pattern of their stem species. A homologization with the like-named styli of corresponding trunk segments in the Symphyla is not possible.

Coxal Organs

Protruding coxal organs which serve in the uptake of moisture are found in the Symphyla and Insecta with the following wide-ranging congruencies (p. 210).
- Symphyla = trunk segments 3–10
- Insecta = trunk segments 4–10 (abdomen 1–7)

The evolution of coxal organs can without doubt be postulated for the stem lineage of the Tracheata, and this is independent of attempts to derive them from the coxa and trochanter of appendages. In any case, coxal organs form original features (plesiomorphies) in the ground pattern of the Insecta.

Within the Ectognatha there are in some cases two pairs of coxal organs per segment. In *Ramsdelepidion* from the stem lineage of the Zygentoma this appears to be the case on abdominal segments 1–7; in recent Archaeognatha they are found on segments 2–6 (Fig. 111). This is probably a secondary state (BITSCH 1994).

Light-Sensory Organs

Four frontal ocelli belong as a plesiomorphy in the ground pattern of the Insecta. With two median ocelli and two lateral ocelli they are realized in their original number in the Collembola.

In the stem lineage of the Ectognatha the two median ocelli fused; three frontal ocelli are thus an autapomorphy in the ground pattern of the Ectognatha. In the stem lineages of the Protura and the Diplura ocelli were completely reduced (Fig. 111).

Compound eyes were taken over from the ground pattern of the Mandibulata; they form therefore a plesiomorphy of the Insecta. A derived feature in the formation of the ommatidia is, however, the transformation of both corneagenous cells into the main pigment cells (Fig. 39C).

Tracheal System with Spiracles

Spiracles in a lateral position lead into the tracheal system. Two pairs of spiracles on the mesothorax and metathorax and eight spiracle pairs on abdominal segments 1–8 are generally interpreted as the original state in the Pterygota. To achieve the ground pattern of the Insecta in their totality, we must include the primary wingless insects (Fig. 111). The Zygentoma have, like the Pterygota, two thoracic and eight abdominal spiracle pairs. In the Archaeognatha only the spiracle pair on the first abdominal segment is missing from this pattern. In the Diplura two pairs of spiracles on the thorax also belong in the ground pattern; on the abdomen the eighth spiracle pair is missing. The Protura have an abdomen free of spiracles, but even here there remain two spiracle pairs on the meso- and metathorax. Finally, within the Collembola, there is only partially one pair of ventral "throat spiracles" between head and thorax. With the exception therefore of the Collembola all the above-mentioned taxa have two thoracic spiracle pairs. Furthermore, with the exception of the Collembola and the Protura, spiracle pairs are realized on abdominal segments 1–8 (Pterygota, Zygentoma), 2–8 (Archaeognatha) and 1–7 (Diplura).

From this distribution we can draw the following conclusions. The stem species of the Insecta possessed a tracheal system with two spiracle pairs on the mesothorax and metathorax as well as eight spiracle pairs on the adjoining abdominal segments 1–8. Thereby no longitudinal or transverse connections existed between the tracheae leading from the single spiracles into the body interior.

Excretory Organs

Midgut with six Malpighian tubules (N.P. KRISTENSEN 1997 a).

Genital Organs and Reproduction

Gonads

Paired testes and paired ovaries belong in the ground pattern of the Insecta – more precisely testes divided into several follicles and the ovaries into ovarioles (egg tubes).

The differentiation of the single ovarioles into two sections, known as the germarium and vitellarium, is an autapomorphy of the Insecta. There are two elementary forms of this differentiation. Panoistic ovaries are interpreted as original. Oocytes (germocytes) originate in the germarium on the tip of the ovariole (end chamber); these migrate distalwards and are surrounded by a single layered follicle epithelium in the adjoining vitellarium. In contrast meroistic ovaries are in two essential features derived. (1) In the end chamber of the ovariole, egg cells (germocytes) and nutrient cells (vitellocytes) develop. (2) The nutrient cells are transported to the developing egg cells in the vitellarium via special pathways.

BÜNING (1994, 1998) reported extensively on the diversity in the manifestation of the ovaries in species from different high-ranking taxa of the Insecta. Their application in explaining kinship relations, however, appears to be limited.

Genital Openings

Gonopores on the tenth abdominal segment in the male and on the eighth segment in the female probably belong in the ground pattern of the Insecta (Fig. 111; BITSCH 1979; N.P. KRISTENSEN 1997 a).

Gonopods

The Collembola and the Diplura do not possess gonopods – neither copulatory organs nor egg-laying apparatus.

In the Protura complicated genital structures in both sexes are realized. They are retracted into deep genital pouches in the region of abdominal segments 9–11 (Fig. 112 F, G). In the Ectognatha abdominal segments 8 and 9 are the bearers of two pairs of exterior gonapophyses as appendage derivatives (Fig. 112 I). There is no agreement with the genital apparatus of the Protura.

These facts lead to the following most parsimonious explanation. Gonopods do not belong in the ground pattern of the Insecta. Their absence in the Collembola and Diplura is a plesiomorphy. The genital apparatus of the Protura forms an autapomorphy of this taxon. Exterior gonapophyses (♀♀) in the region of abdominal segments 8 and 9 evolved first in the stem lineage of the Ectognatha. In the stem lineage of the Dicondylia the gonangulum then developed as a transformation of a sclerite joint on the ninth abdominal segment (Figs. 111, 140). Difficulties arise over the interpretation of the styli on genital segments 8 and 9 in the Ectognatha; they are known as gonostyli. The Archaeognatha and the Zygentoma each possess two pairs of gonostyli. On the ovipositor of the Pterygota styli are absent on the eighth abdominal segment. It is tempting to homologize the gonostyli of the Archaeognatha, Zygentoma, and Pterygota and, for the latter, postulate a reduction on genital segment 8. Due to the disputed assessment of the abdominal styli in recent Insecta, however (p. 255), caution is advised (WILLMANN 1997 b).

Sperm Transfer

Indirect sperm transfer with the deposition of spermatophores on the substrate is – as far as is known – widespread among the primary wingless insects. An exception may be the Protura. For the rest, the change to copulation took place in the stem lineage of the Pterygota. Here, as well as a direct transfer of free sperm, a direct transfer of spermatophores is also possible – up to the Holometabola (Coleoptera, Lepidoptera).

The indirect transfer of sperm through spermatophores is a plesiomorphy in the ground pattern of the Insecta. This mode was taken over from the ground pattern of the Tracheata into the stem species of the Insecta.

Cleavage

A short phase of total cleavage can be placed in the ground pattern of the Insecta (N.P. KRISTENSEN 1997 a).

Systematization

The Insecta are the Metazoan taxon with which HENNIG first put the theory of phylogenetic systematics (1950, 1966) into practice (1953, 1969, 1981). Today, numerous subtaxa have been successfully treated – however, today, as before, unsolved problems still exist in the systematization of the Insecta (N.P. KRISTENSEN 1991, 1995, 1997a; WILLMANN 1997 a, b). Following the controversial division into the Entognatha and Ectognatha discussed above, let us investigate other questions case by case.

To establish the monophyly of high-ranking subtaxa of the Insecta and their kinship relations, I have divided the system depicted (see p. 243) into five sections. In each section, the hierarchic tabulation and the complementary relationship diagram will be shown together.

Systematization: Section 1

Insecta
- **Entognatha**
  - **Diplura**
  - **Ellipura**
    - **Protura**
    - **Collembola**
- **Ectognatha**
  - **Archaeognatha**
  - **Dicondylia**
    - **Zygentoma**
    - **Pterygota**

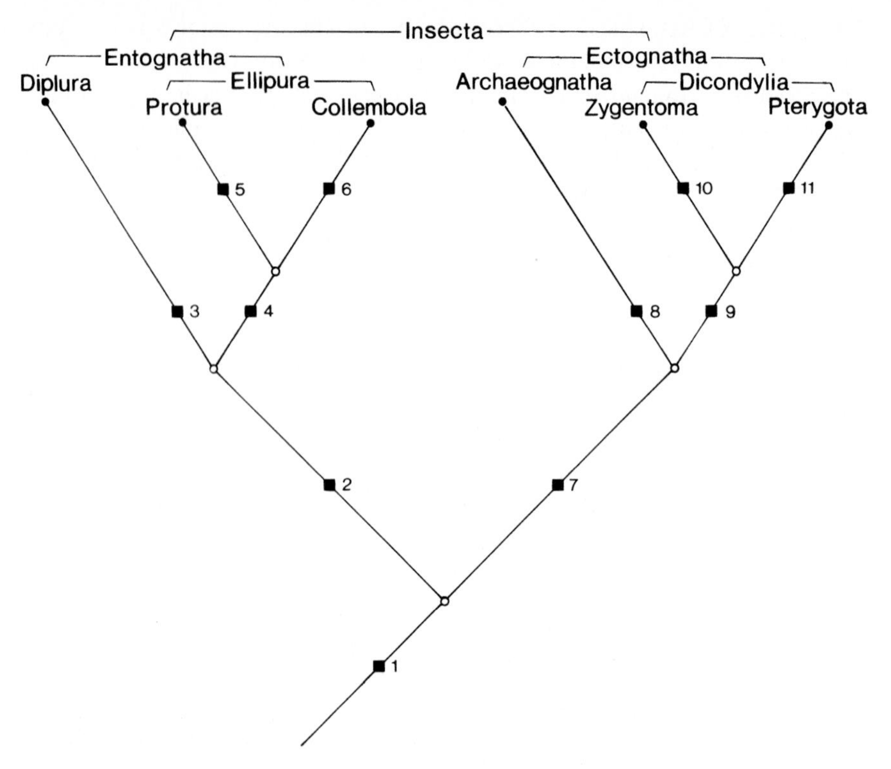

Fig. 113. Diagram of phylogenetic relationships of the Insecta with the Entognatha and Ectognatha as sister groups. Section 1 of the systematization of the Insecta

Entognatha - Ectognatha

Entognatha

Autapomorphies (Fig. 113 → 2)

- Entognathy.
  Covering of the mandibles and maxillae by lateral oral folds of the head capsule fused behind with the labium. Through this process mandibles and maxillae are incorporated into special jaw pouches (Fig. 108 I, J). Palps of the maxillae and labium strongly reduced.
- Reduction of the Malpighian tubules.

The articulated antennae are emphasized as a plesiomorphy of the entognathous unities Diplura and Collembola. These antennae consist primitively of numerous sequential uniform articles that, with the exception of the last article, are equipped with musculature (Fig. 108 H).

Diplura – Ellipura

Diplura

A limited number of derived apomorphic peculiarities are opposed by a large number of plesiomorphic features in the Diplura.

Autapomorphies (Fig. 113 → 3)

- Fusion of abdominal segments 10 and 11 during embryogenesis (BITSCH 1994).
- Absence of spiracles on abdominal segment 8 (Fig. 111).
- In both sexes gonopores between abdominal segments 8 and 9 (Fig. 111).
- Complete absence of compound eyes and ocelli.
  Reduction in the stem lineage of the Diplura, convergent with the loss of eyes in the Protura and within the Collembola.

Plesiomorphies

- The manifestation of the cerci in the Campodeoidea (*Campodea*) in the form of long multi-articulated structures (Fig. 108 A) is generally thought to be the plesiomorphic state (BITSCH 1994) – taken over from the ground pattern of the Insecta (p. 253). The likewise multi-articulated but short cerci of the Projapygoidea (*Projapyx*) appear in contrast to be relatively apomorphic and the single articled forceps-like cerci of the Japygoidea (*Japyx*, *Metajapyx*) may represent the most strongly derived state (Fig. 108 B). Because of its posterior grasping claws (Fig. 110 A, B) *Testajapyx thomasi* from the Carboniferous is assigned to the Japygoidea (KUKALOVÁ-PECK 1987) – i.e., placed in a subgroup of the Diplura. Since *Testajapyx* has well-developed compound eyes, the assumption that eyes were reduced twice in different stem lineages within the Diplura is a consequence of this measure. In other words, the absence of light-sensory organs could no longer be used as an autapomorphy of the Diplura as a whole. The position postulated by KUKALOVÁ-PECK has, however, been emphatically questioned. Hooks on the abdomen (? 8th segment) do not alone suffice to characterize *Testajapyx thomasi* as a representative of the Diplura (BITSCH 1994).
- Articulated antennae with numerous segments.
- All thoracic appendages with a joint between the tibia and tarsus.
- Abdomen with the original number of 11 segments (10 and 11 fuse during development).
- Coxal organs on abdominal segments 2–7 (Fig. 111).
- Well-developed tracheal system with two pairs of spiracles in the thorax (*Parajapyx*, *Anajapyx*: PACLT 1956) and seven pairs of spiracles on the

following abdominal segments. Only the loss of spiracles on abdominal segment 8 is an autapomorphy in the ground pattern of the Diplura. There are, however, considerable evolutionary modifications within the Diplura. The number of thoracic spiracles is increased to three pairs (*Campodea, Projapyx*) or four pairs (*Japyx*). In *Campodea* the abdominal spiracles are completely reduced.

The Diplura are generally small, secretive soil inhabitants. *Campodea staphylinus* (central Europe) has a length of 4.5 mm. *Heterojapyx* species (Australia) with a length of 5 cm are real giants.

Ellipura

The establishment of the Ellipura as a monophylum is primarily supported by a number of reductional features which usually carry less weight as new evolutionary acquisitions. Nevertheless, there are here several common features of the Protura and Collembola that are without doubt interpretable as products of single evolutionary processes. Of these, N.P. KRISTENSEN (1991) refers to special evolutionary novelties in the construction of the head.

Autapomorphies (Fig. 113 → 4)

- Reduction of the antennae.
  Maximum of four articles in the Collembola; completely reduced in the Protura (Fig. 108 C, D).
  The simplest explanation for this is that antennal reduction began in a common stem lineage of the Collembola + Protura.
- Reduction of the compound eye.
  In the Collembola there are species with a maximum of eight ommatidia; the Protura are eyeless. The simplest interpretation here is also to postulate the beginning of eye reduction in a common stem lineage.
- Wide-ranging reduction of the tracheal system.
  There are no spiracles in the abdomen.
- Pretarsus with unpaired claw.
  Paired claws (Diplura, Ectognatha) are probably a plesiomorphy in the ground pattern of the Insecta since they are also found in the Symphyla and some of the Chilopoda (Scutigeromorpha). A homology is, however, questionable (N.P. KRISTENSEN 1997 a).
- Fusion of tibia and tarsus in the meso- and metathorax appendages.
- Absence of cerci – insofar as these form a ground pattern feature of the Insecta (p. 253).

Protura - Collembola

An outstanding evolutionary novelty of the Protura is the change of function of the prothorax appendages. To replace the lost antennae they are raised from the ground and directed forward as new antennae. Through this process the Protura evolved into functional "Tetrapoda" within the Hexapoda.

A unique autapomorphy of the Collembola is the jumping apparatus with retinaculum and furcula on abdominal segments 3 and 4.

Protura

Autapomorphies (Fig. 113 → 5)

- Absence of antennae (Fig. 108 C).
- Absence of compound eyes and ocelli.
  The complete reduction of antennae and light-sensory organs can only have occurred after the separation of the Collembola and Protura in the stem lineage of the Protura.
- Evolution of the prothorax appendages into new sensory organs.
- On hatching eight abdominal segments. Increase to 11 segments + telson during post-embryonic development (p. 247).
- Genital apparatus sunk in a deep body pocket. Opening of the pocket at the posterior of the 11th abdominal segment. In both sexes the apparatus consists of a pair of basal apodemes and distal gonostyli (Fig. 112 F, G).

A complete loss of spiracles on the abdomen had probably already occurred in the stem lineage of the Ellipura. The spiracle pairs found on the meso- and metathorax are plesiomorphies of the Protura - carried over from the ground pattern of the Insecta. Within the Protura there is a complete reduction of the tracheal system.

Plesiomorphic congruencies with the Diplura are found in the high number of abdominal segments and in the primitive tibia-tarsal joint of the prothorax appendages.

Small animals with a length of between 0.5 and 2 mm. Hygrophilous inhabitants of the upper soil layers. *Eosentomon transitorium* is a common species in central Europe.

Collembola

Autapomorphies (Fig. 113 → 6)

- Shortening of the abdomen to six segments (Fig. 108 D).
- Concentration of the trunk ganglia to three ganglion pairs in the thorax. The abdominal ganglia become part of the metathorax ganglia.

- Tracheal system.
  Only one pair of spiracles in the throat region with a few tracheae originating from these in some of the Actaletidae and Sminthuridae.
  Otherwise complete absence within the Collembola.[26]
- Ventral tube (Fig. 112 B).
  Derivative of the first abdominal segment appendages. Pipe-shaped structure. Distally with paired vesicles or two tubes (coxal organs), eversible through an increase in hemolymph pressure. Different possible functions, such as anchoring onto the substrate, moisture uptake, osmoregulation and gas exchange.
- Jumping apparatus of retinaculum and furcula (Fig. 112 C, D).
  Derivative of the appendages of abdominal segments 3 and 4. The small retinaculum at the posterior of the third segment consists of an unpaired basal article (corpus) and two toothed branches (rami). The furcula of the fourth segment is formed from an unpaired basal manubrium and two rods, which themselves are divided into dens and mucro. The furcula is anchored on the teeth of the retinaculum with the rods and held under tension; when suddenly released through muscle pressure it strikes the ground and propels the animal into the air.
- Loss of the tibio-tarsal joint on the prothorax appendages.

As well as the large number of autapomorphies the ground pattern of the Collembola includes several characteristic plesiomorphies. In possessing articulated antennae the Collembola are more original than the Protura; in possessing compound eyes more primitive than the Protura and Diplura. Thereby the loose arrangement of a maximum of eight ommatidia per eye with a round corneal lens is apomorphic in contrast to the honeycomb pattern of numerous tightly packed ommatidia (PAULUS 1972 a, 1985).

Of special interest is the **possession of frontal ocelli with the plesiomorphic number of two median ocelli and two lateral ocelli.** Fully reduced in the Diplura and Protura, three ocelli are found the ground pattern of the Ectognatha. The unpaired median ocellus of the Ectognatha is a fusion product of primary paired median eyes; the lateral ocelli of the Ectognatha correspond to those of the Collembola (PAULUS 1972 b, 1979).

An elongated body with clear segmentation of thorax and abdomen also belongs in the ground pattern of the Collembola. The grouping of species with this original habitus into a unity "**Arthropleona**" leads, however, to a

[26] There is no proof for a widely held view according to which tracheae were totally reduced in the ground pattern of the Collembola and developed as evolutionary novelties in the taxa named above. The assumption that the tracheae realized within the Collembola are remnants remaining after a reduction process, whereby one spiracle pair is shifted from the thorax to the head-thorax boundary, is much simpler.

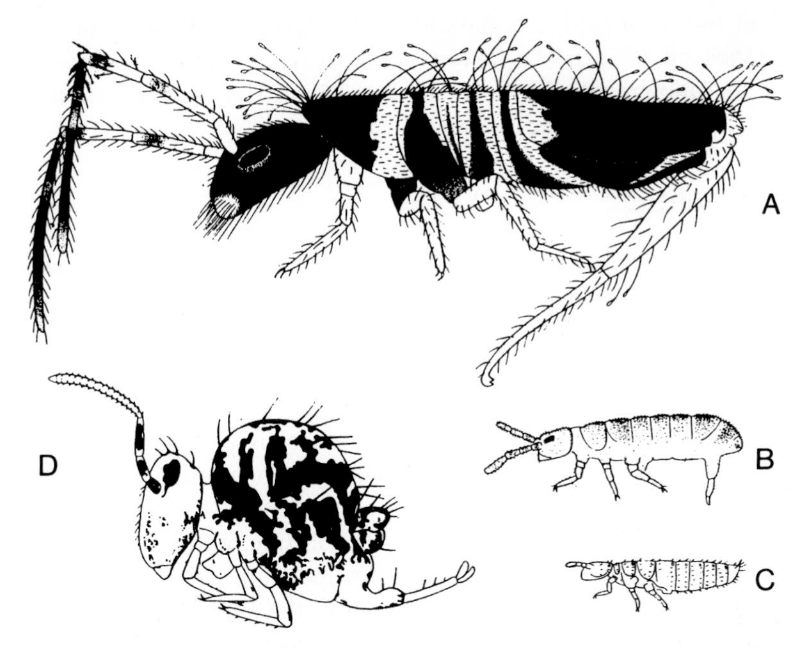

Fig. 114. Collembola. Evolutionary adaptations to modes of colonization in different soil horizons. **A** *Entomobrya*. Inhabitant of the soil surface with a body length of 3.5 mm. **B** *Proisotoma* from the upper soil horizons. Unpigmented organisms 1.5 mm in length. Short antennae and short appendages. **C** *Tullbergia* from a depth of 30 cm. Body length around 1 mm; compound eyes and furca reduced. **D** *Sminthurus* (Symphypleona). Representative of the springtails from the soil surface. Body length 1.5 mm. Pigmented, with compound eyes, long antennae and appendages. (Schaller 1962)

paraphylum. In contrast the species group having as an autapomorphy a globular swelling of the trunk form a monophylum **Symphypleona**.

Collembola colonize soils from the surface to 30 cm depth with high species diversities and population densities. Evolutionary modifications manifest themselves in strong differences between species from different soil horizons (Fig. 114).

Inhabitants of the soil surface with elongated bodies (*Orchesella, Entomobrya, Tomocerus*) reach a length of 5 mm. Surface dwellers have long antennae, long appendages, a long furcula and long setae on the body. They are strongly pigmented and have compound eyes of eight ommatidia.

Already in the upper subsurface layers there is approximately a halving of body length to 1.5–2 mm (*Proisotoma, Folsomia*), a loss of pigmentation, a shortening of the antennae and appendages, as well as a reduction in the number of ommatidia.

Finally, at depths of around 30 cm, tiny species 0.6–0.8 mm in length (*Tullbergia*) are found. These are blind, colorless "arthropleonic" springtails, which pull themselves through the substrate with short appendages.

In the Symphypleona evolution leads from the 1.5-mm-long pigmented species on the surface (*Sminthurus*) to the smallest Collembola of all – blind, unpigmented animals of 0.3–0.5 mm body length found at unlit depths (*Megalothorax*, *Neelus*).

Ectognatha

With the adelphotaxa Archaeognatha and Dicondylia (Zygentoma + Pterygota) the Ectognatha are well established as a monophylum – independent of whether a unity Entognatha or only the Diplura form the sister group.

Autapomorphies (Fig. 113 → 7)

- Annulated antennae with scapus, pedicellus and anellus (Fig. 108 N, O). The scapus (first article) is the bearer of the antenna. Only this article is equipped with musculature.
  The pedicellus (second article) is the location of Johnston's organ. Numerous scolopidia form a peripheral cylinder in the pedicellus (Fig. 108 N). The chordotonale organ is anchored in the joint integument between the pedicellus and anellus; it registers movements of the flagellum.
  The anellus (third article) is differentiated through secondary subdivisions (annuli) into a long flagellum.
- Three frontal ocelli.
  An unpaired median ocellus formed from the fusion of primary paired median eyes and paired lateral ocelli (pp. 256, 264).
- Gonopods (♀♀).
  Appendage derivatives of abdominal segments 8 and 9 used in reproduction. Two pairs of gonapophyses (valvulae) as rod-shaped processes on the coxopodites form the female ovipositor (Fig. 112 I).
- Terminal filament (paracercus).
  Unpaired annulated tail thread. Formed from the tergum of abdominal segment 11 (BOUDREAUX 1979; N.P. KRISTENSEN 1991: ? plus telson) or from the telson (KUKALOVÁ-PECK 1991). The caudal filament is only developed in the Archaeognatha, Zygentoma and Ephemeroptera (Pterygota) (Fig. 108 E–G); in all other winged insects the tail thread is secondarily missing.
- Subdivided tarsus. Pretarsus ring reduced; the claws insert on the distal tarsomere.
- Formation of an amniotic cavity in embryogenesis. Originally with an open pore.

Compared with the Entognatha the free articulation of the mandibles and maxillae on the head capsule is a plesiomorphy from the ground pattern of the Insecta (Fig. 108 M).

Archaeognatha – Dicondylia

Archaeognatha

Autapomorphies (Fig. 113 → 8)

- Enlarged compound eyes, medially contiguous (Fig. 108 E).
  The displacement of the eyes from the original lateral position to the upper surface of the head is judged to be an apomorphy.
- Hypertrophic maxillary palps.
  With a high number of articles, larger than the locomotory appendages of the thorax (N.P. KRISTENSEN 1997 a).
- Absence of spiracles on the first abdominal segment (Fig. 111).
  Interpreted as a reduction compared with their existence in the Dicondylia.

Plesiomorphies

- Mandibles with one condyle (Fig. 108 K).
  As in the Ellipura and Diplura mandibles monocondylic.
- Styli.
  Found on the appendages of the meso- and metathorax (Fig. 111).

The jumping bristletails (*Machilis*, *Petrobius*) live above all on stony substrates and can even be found on seacoasts. *Petrobius brevistylis* colonizes Lower Triassic gravels on Heligoland (Germany) just above the high water mark.

Dicondylia

Autapomorphies (Fig. 113 → 9)

- Mandibles with two condyles.
  Evolution of a second joint between the mandible and the head capsule in the stem lineage of the Dicondylia (Fig. 108 L). Correlated with this is a restructuring of the mandible musculature.
- Gonangulum.
  Basis of the ovipositor with a new sclerite. Articulation of the gonangulum with the tergum of the ninth abdominal segment and attachment onto the first gonapophyse (Fig. 140). Leads to an increase in efficiency of the ovipositor when pushing the gonapophyses into the substrate.
- Tarsus probably of five articles.
- Abdominal tracheae with longitudinal and transverse connections.
- Superficial cleavage.
- Pore of the amniotic cavity at least partly closed.

Zygentoma – Pterygota

Zygentoma

Autapomorphies (Fig. 113 → 10)

- Labial palp with a clearly broadened final article.
- Absence of superlinguae on the hypopharynx.
- Paired sperm.
  A firm lateral connection is established between two ripe spermatozoa at the end of spermiogenesis; the cell membranes remain separate. Sperm pairing appears to be of importance for movement, single sperm are immobile (DALLAI & AFZELIUS 1984; JAMIESON 1987).

Compared with the sister group Pterygota the ground pattern of the Dicondylia is obviously less modified in the taxon Zygentoma.

Compound eyes are usually weakly developed in the Zygentoma and can be totally absent, but belong assuredly in their ground pattern. *Tricholepidion gertschi* has 40–50 ommatidia per eye. The compound eyes of *Lepisma saccharina* and *Thermobia domestica* each have 12 ommatidia. The firebrat *Atelura formicaria* is eyeless.

Only in *Tricholepidion gertschi* are the three frontal ocelli from the ground pattern of the Ectognatha retained; the species is probably the adelphotaxon of all other Zygentoma (N.P. KRISTENSEN 1997 a).

Styli are found in *Tricholepidion* and in the Nicoletiidae on abdominal segments 2–9 (Fig. 111). Reductions within the Zygentoma go down to one pair in segment 9. Only *Tricholepidion* and *Nicoletia* possess high numbers of coxal organs on abdominal segments 2–7 (Fig. 111); they are lacking in most Zygentoma.

The reduction of styli and coxal organs occurred within the taxon Zygentoma convergent with a similar reduction in the sister group Pterygota (BOUDREAUX 1979).

Zygentoma are above all known as synanthropic animals. The silverfish *Lepisma saccharina* with shiny, silvery scales is cosmopolitan; the thermophilic species is found in northern Europe in houses and lives there in damp rooms. The even more strongly thermophilic *Thermobia domestica* can appear in bakeries.

Pterygota

Autapomorphies (Figs. 113 → 11; 119 → 1)

- Two pairs of wings on the pterothorax (meso- and metathorax).
  The wings originate from evaginations of the epidermis and are found dorsolaterally between the terga and pleura of the two thoracic segments. During development tracheal stems – derivatives of the leg tracheae of the meso- and metathorax – migrate into the wings; here, they form the longitudinal veins and their branches. The longitudinal veins are primitively linked together by means of numerous transverse veins without tracheae (archaedictyon) (Fig. 115 A).
  The median part of the wing joint lies under the edge of the tergum; more laterally the joint rests on a wing process of the pleura. Lowering and raising of the tergum through indirect wing muscles leads to the upward and downward strokes of the wing (Fig. 116).
- Pleura of the wing-bearing segment elevated and strongly sclerotized.
- In the wing-bearing segments there is also a thoracic endoskeleton for strengthening purposes and muscle attachment.
- Absence of coxal organs.
  Coxal organs belong in the ground pattern of the Insecta and are realized in the sister group Zygentoma. Their lack in the Pterygota must therefore be assessed as a secondary state.

Styli are lacking in the Pterygota on the pregenital abdominal segments. Due to the state of the controversial interpretation of insect styli it must remain open as to whether this is a further autapomorphy or a plesiomorphy.

The **Ephemeroptera** (may flies) and the **Odonata** (dragonflies and damselflies) are insects with uniform wing surfaces (Figs. 115 A, 118 A, C); they cannot fold their wings and lay them over the abdomen. At rest in the Ephemeroptera the wings are held upright with the dorsal surfaces together rather like a sail. Similar behavior is found in the Zygoptera (damselflies) within the Odonata, whereas the Anisoptera (dragonflies) spread their wings laterally.

An "archaedictyon" is connected with this primitive paleoptery in the wings of the Ephemeroptera and Odonata. During the later evolution of the winged insects, this primitive network of fine veins has been reduced to a few transverse veins.

In the **Neoptera** – the overwhelming majority of the Pterygota – the free wing surface is divided into three fields – remigium, vannus and neala (Fig. 115 B, C). Between the fields there are flexible folds. The wings can be folded together and laid flat over the abdomen by means of a muscle in-

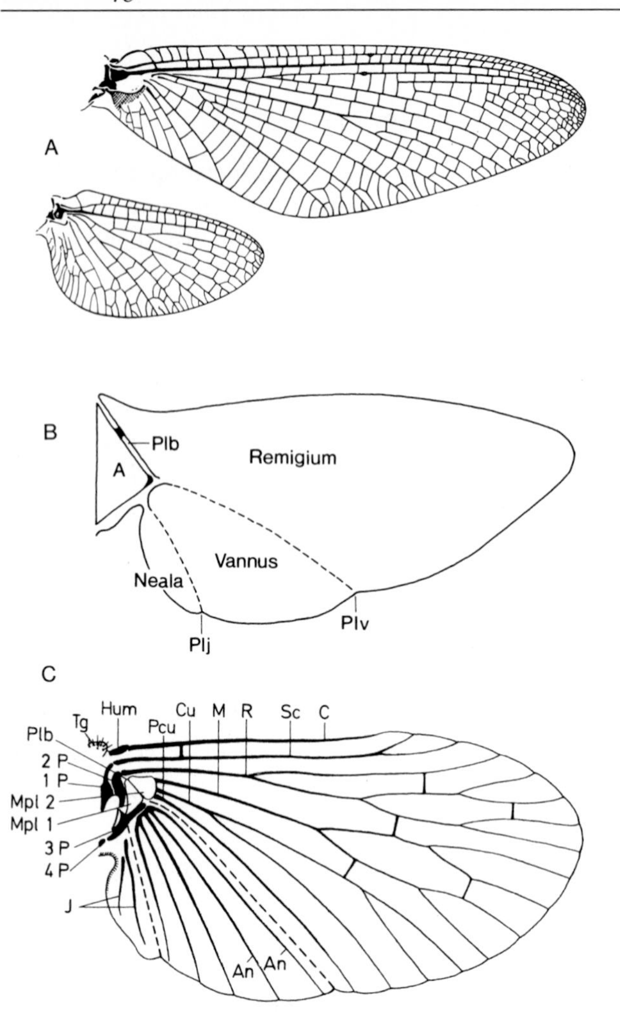

Fig. 115. Wings. **A** Ephemeroptera. Fore- and hindwing of *Mirawara*. Uniform wing surface with a network of fine veins. **B** Neoptera. Division of the wing into the joint or axillary region (*A*) and the free fields remigium (costal field), vannus (anal field) and neala (jugal field). The fields are separated by the plica vannalis (*Plv*) and plica jugalis (*Plj*). The plica basalis (*Plb*) is between the joint and remigium. **C** Neoptera. Diagram of the wing showing the veins in the three fields and sclerites of the axillary region. The costa (*C*) forms the anterior margin. In the costal field there are the subcosta (*Sc*), radius (*R*), media (*M*), cubitus (*Cu*) and postcubitus (*Pcu*) which branch at the periphery. The anal field with anales (*An*) and the jugal field with jugales (*J*) follow. In the axillary region there are four pteralia (*1P–4P*) (=axillaria) and mid-plate 1 (*Mp 1*). On the other side of the plica basalis lies mid-plate 2 (*Mp 2*) which is connected to the roots of the media, cubitus and postcubitus. Connected to the costa are the humerus sclerite (*Hum*) and tegula (*Tg*) with hair sensilla. (**A** Peters & Campbell 1991; **B, C** Seifert 1995)

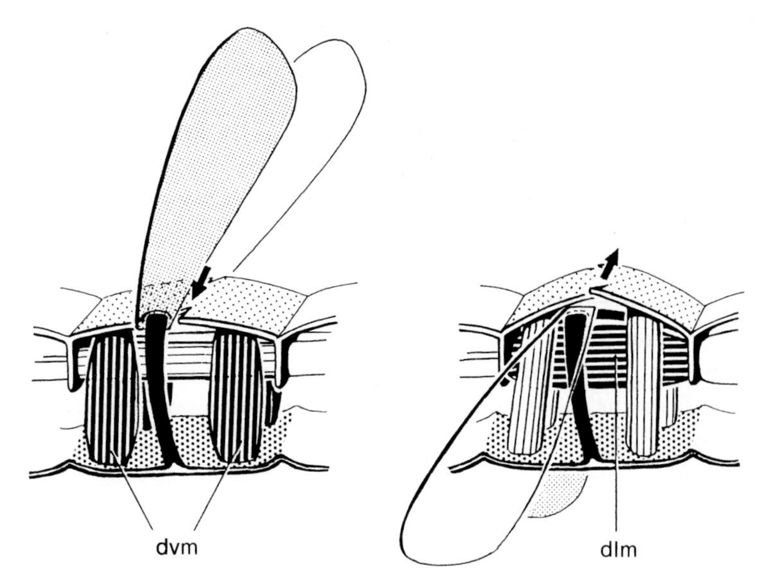

Fig. 116. Principle of the tergal-arching mechanism (Ephemeroptera, Neoptera). Contraction of the dorsoventral muscles (*dvm*) leads to a flattening of the tergum (upbeat of the wing); contraction of the dorsal longitudinal muscles (*dlm*) leads to its arching in the longitudinal direction of the animal (downbeat of wing). Membranous lateral notches in the tergum (*arrows*) improve the arching of the tergum and with this the lever effect for beating the wings. *Light stippling* tergum, *heavy stippling* sternum, *black* pleural ledge. Only indirect muscles shown. (Pfau 1986)

serted on the third pterale. This phenomenon is known as **neoptery** and interpreted as an autapomorphy of the Neoptera.

We now come to the **wing propulsion system**. PFAU (1986) differentiated between the tergal-arching mechanism of the Ephemeroptera and Neoptera and the tergal-plate mechanism of the Odonata.

The tergal-arching mechanism operates by raising and depressing the tergum which causes the up and down stroke of the wings (Fig. 116). The raising and lowering of the tergum is effected mainly by dorsoventral muscles attached to the tergum and sternum (indirect elevators) as well as dorsal, purely tergal longitudinal muscles (indirect lowerers). To this fundamental "indirect-indirect" mechanism direct lowerers as synergists of the dorsal longitudinal muscles can also be added; however, these usually have additional functions.

In the tergal-plate mechanism the tergum is raised and lowered as one piece during the beating of the wings. Dorsoventral muscles between the tergum and sternum again function as propulsion muscles (indirect raisers); however, basal and subalar muscles attached to the wing cause downward movement of the wing (direct lowerers). The Odonata thus fly with an "indirect-direct" beat mechanism.

According to PFAU (1986), a flight propulsion apparatus with a combination of elements from the tergal-arching and tergal-plate mechanisms belongs in the ground pattern of the Pterygota. From this apparatus the systems of the Ephemeroptera, Odonata and Neoptera apparently developed separately from one another. From analyses of the mechanisms of wing movement, PFAU (1986) sees no possibility of making a contribution to solving the problem of the basal phylogenetic branching of the Pterygota.

With this we come to the **question** of the **phylogenetic kinship between the Ephemeroptera, Odonata and Neoptera.**

The Ephemeroptera and Odonata with non-flexing wings are each established as monophyla through convincing derived peculiarities (pp. 275, 278). For the Neoptera neoptery as discussed above forms the basic autapomorphy (p. 280).

Using these facts, in principle, three possibilities for the phylogenetic kinship between the three monophyla exist: (A) Ephemeroptera and Odonata, (B) Odonata and Neoptera or (C) Ephemeroptera and Neoptera are adelphotaxa.

All three possibilities have appeared in the phylogenetic literature (overview in KLAUSNITZER & RICHTER 1981; PFAU 1991; KLAUSNITZER 1996). The resulting three systematizations are depicted together with their corresponding hierarchic tabulations (Fig. 117).

In considering the competing kinship hypotheses an **evolutionary assessment** of the **aquatic larvae** of the Ephemeroptera, Odonata and Plecoptera as a "basal" unit of the Neoptera is necessary (Fig. 118).

According to KUKALOVÁ-PECK (1983), aquatic juveniles belong in the ground pattern of the Insecta and, furthermore, in the ground pattern of the Tracheata. Moreover, KUKALOVÁ-PECK (1983) considers the closed tracheal system of the aquatic larvae as representing the original state, and hence the entry of air into the tracheae through spiracles as a secondary evolutionary phenomenon. These hypotheses are improbable. The tracheal system of the Tracheata can hardly have originated to disseminate oxygen that diffuses out of the water into the body, but rather for the uptake of oxygen from the air through spiracles and its distribution in a land-living organism. A clear vote for this argument results from the following facts. Neither the Chilopoda and Progoneata nor the primary wingless insects and the majority of Pterygota have aquatic larvae. They must therefore have been reduced several times convergently within the Tracheata; however, there is no evidence for this. In contrast, clear examples exist for a secondary origin of aquatic larvae with tracheal gills within the winged insects. Within the Holometabola with a terrestrial stem species these are the larvae of the Megaloptera and the majority of the caddis fly larvae (Trichoptera).

Therefore the stem species of recent Tracheata (p. 203) and Insecta can be seen with good justification as organisms with terrestrial developmental

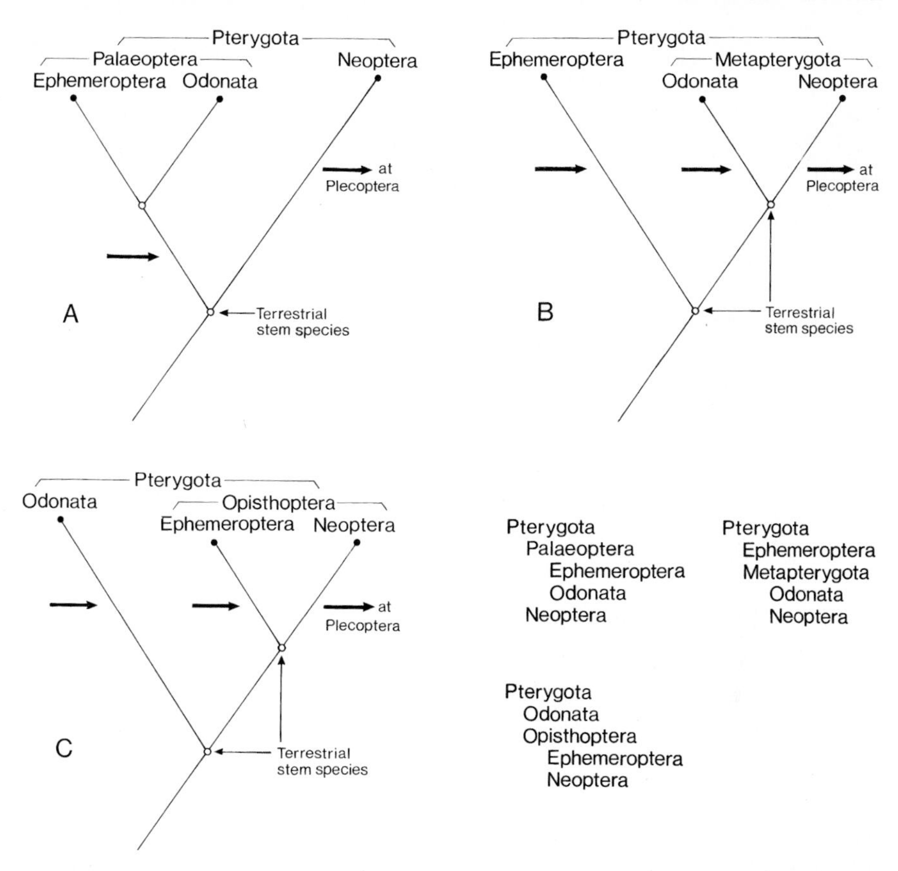

Fig. 117. The three possibilities of phylogenetic relationships between the Ephemeroptera, Odonata and Neoptera as well as the corresponding systematizations of the Pterygota. Given in the form of kinship diagrams and hierarchic tabulations (*bottom right*). If terrestrial stem species for the Insecta, the Pterygota and the Neoptera are assumed, aquatic larvae must have evolved convergently within the "basal" Pterygota. Hypothesis A offers the simplest explanation: Evolution of aquatic larvae in the stem lineage of the Palaeoptera (synapomorphy between the Ephemeroptera and Odonata) and at the Plecoptera within the Neoptera. According to kinship hypotheses B and C a triple independent evolution of aquatic larvae is necessary – in the stem lineages of the Ephemeroptera, the Odonata and the Plecoptera

stages. This assumption also holds true for the stem species of the Pterygota; as realized in the sister group Zygentoma, this must have been a species with a terrestrial life cycle.

If we postulate from this basis that the **aquatic larvae** of the **Ephemeroptera, Odonata and Plecoptera are secondary acquisitions**, then the principle of parsimony must be considered. Kinship hypotheses B and C force the assumption of a triple independent evolution of aquatic juveniles in the stem lineages of the Ephemeroptera, the Odonata and the Plecoptera within the Neoptera.

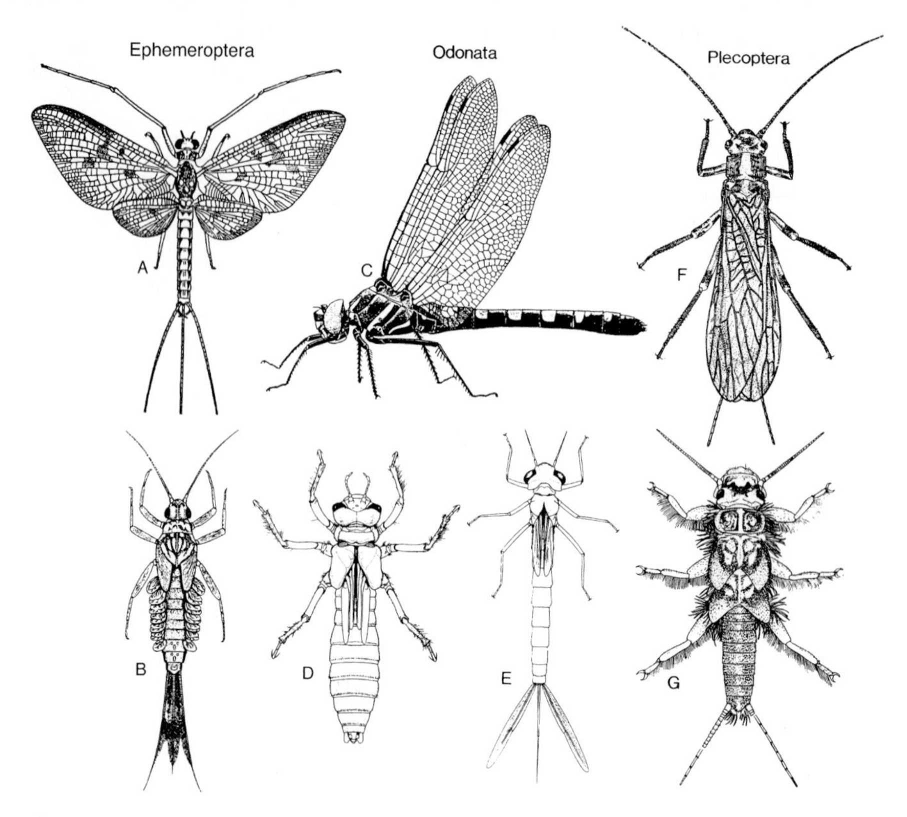

Fig. 118. Ephemeroptera, Odonata and Plecoptera as the three "basic" taxa of the Pterygota with aquatic larvae. **A** *Ephemera vulgata* (Ephemeroptera). Imago. Cerci and terminal filament only shown to a third of their length. **B** *Cloeon dipterum* (Ephemeroptera). Larva. **C** *Macromia magnifica* (Odonata, Anisoptera). **D** *Petalura hesperia* (Odonata, Anisoptera). Larva. **E** *Caliagrion billinghursti* (Odonata, Zygoptera). Larva. **F** *Isoperla confusa* (Plecoptera). Imago. **G** *Perla placida* (Plecoptera). Larva. (**A** Eidmann & Kühlhorn 1970; **B, C** Wesenberg-Lund 1943; **D, E** Watson & O'Farrel 1991; **F, G** Despax 1949)

In kinship hypothesis A, however, only a double evolution of a corresponding larvae is necessary – once in the stem lineage of a unity Palaeoptera (Ephemeroptera + Odonata) and once in the phylogeny of the Plecoptera.

Accordingly, the Palaeoptera hypothesis should be favored. It is also supported by paleontological studies (KUKALOVÁ-PECK 1983, 1991, 1994), although in comparing recent Ephemeroptera and Odonata only a moderate confirmation is found through weak apomorphic congruencies (p. 275). We follow the systematization of the Pterygota into the Palaeoptera and Neoptera as sister groups therefore with corresponding reserve. In the view of N.P. KRISTENSEN (1995), the question regarding the basal branching of the Pterygota should be left open until a thorough study of all pertinent features has been carried out.

Systematization: Section 2

Pterygota
  Palaeoptera
    Ephemeroptera
    Odonata
  Neoptera
    Plecoptera
    N. N.
      Paurometabola
      Eumetabola

Palaeoptera – Neoptera

Palaeoptera

Autapomorphies (Fig. 119 → 2)

- Aquatic larva with closed tracheal system.
- Fusion of galea and lacinia in the maxillae of the larva (Fig. 120 E).
- Short bristle-like flagellum of the annulated antennae in the imago (cf. WILLMANN 1997 a, b).

Ephemeroptera – Odonata

Ephemeroptera

Autapomorphies (Fig. 119 → 3)

- Small hindwings.
  The hindwings of the Ephemeroptera are always more weakly developed than the forewings (Figs. 108 G, 118 A). Within the unity total reduction takes place, e.g. in *Cloeon dipterum.*
- No food uptake in the imago.
  Correlated to the short life span of the may flies – from hours to days – the mouthparts are vestigial; the adults do not feed.
- Aerostatic organ.
  Function change of the intestine. The midgut which is not needed in its primary function is filled with air and closed against the proctodaeum and stomodaeum. Together with the firm connection of thorax and abdomen an elastic axis as a support against the static load during flight is formed (ILLIES 1968).

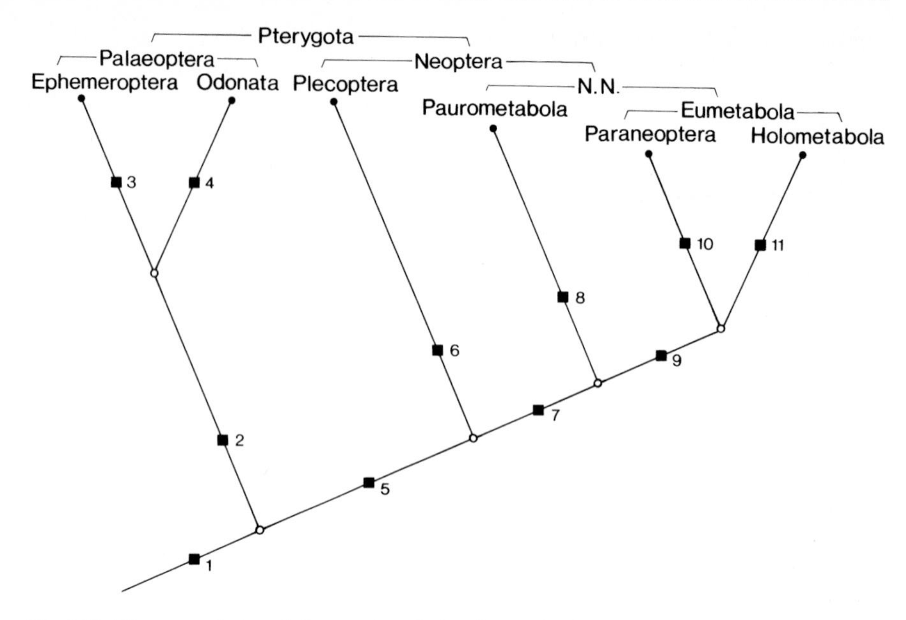

Fig. 119. Diagram of the phylogenetic relationships of the Pterygota with the Palaeoptera and Neoptera as sister groups. Section 2 of the systematization of the Insecta

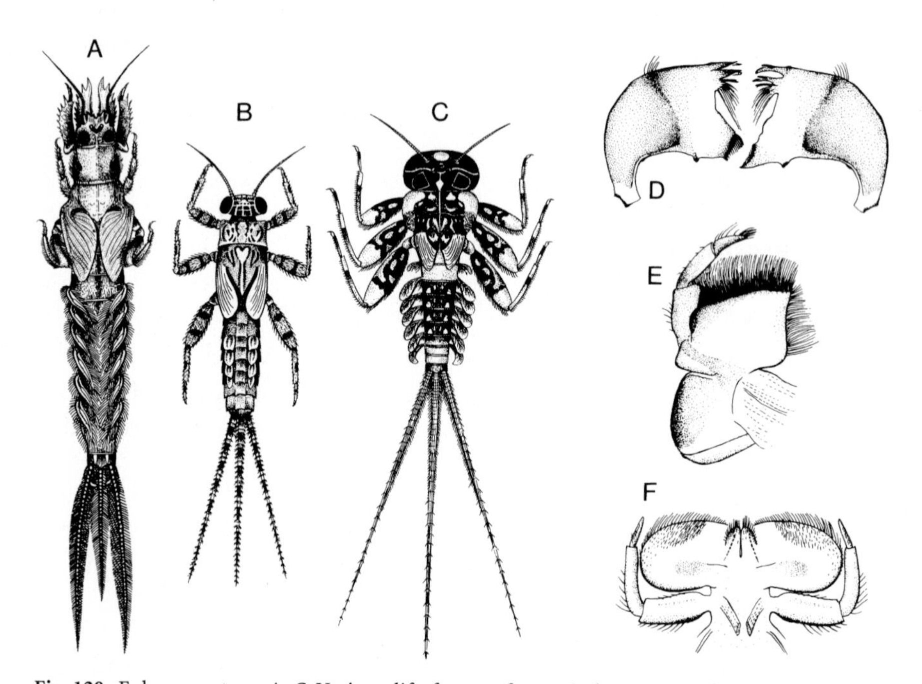

Fig. 120. Ephemeroptera. **A–C** Various life forms of aquatic larvae. **A** *Palingenia longicauda* (burrowing larva). **B** *Ephemerella ignita* (creeping larva). **C** *Ecdyonurus forcipula* (rheophile larva). **D–F** Mouthparts of the larva of *Atalophlebioides*. **D** Mandibles. **E** Maxilla (galea and lacinia fused to form a lobe). **F** Labium. (**A–C** Günther 1969a; **D–F** Peters & Campbell 1991)

- Tracheal gills.
  The aquatic larvae have breathing organs in the form of tracheal gills on abdominal segments 1–7 (Fig. 118B). They are interpreted to be derivatives of abdominal appendages.
- No ovipositor.
  Since an egg-laying apparatus was postulated for the ground pattern of the Ectognatha, and is also developed in the probable sister group Odonata, its absence in the female Ephemeroptera must be seen as an autapomorphy.
- Paired genital openings.
- Male with paired penes.

Plesiomorphies

The Ephemeroptera are distinguished by a number of original features. Within the Pterygota the two following plesiomorphies are found only in this taxon.

- Subimago.
  A stage in the life cycle which is capable of flight but which is not sexually mature. The subimago hatches from the last larva and molts within a short time to the imago.
- Terminal filament (paracercus).
  Taken over from the ground pattern of the Ectognatha together with the cerci. Three long, filiform projections are found at the posterior of the larva and imago (Fig. 118A, B).

In the **larval stage** the Ephemeroptera demonstrate marked **evolutionary adaptations** to different fresh-water biotopes (Fig. 120). In burrowing species (*Palingenia*, *Ephemera*) the first leg pair can be differentiated into veritable excavating tools; they excavate burrows in the substrate. To achieve a narrow body the other thorax legs are held close to the body and the gills borne on the back. Conversely, the tracheal gills of open-water species project outwards (*Cloeon*); here, together with the long setose posterior appendages, they serve in the swimming, leaping locomotion. Creeping larvae (*Ephemerella*) tend towards extremely torrenticolous life forms (*Ecdyonurus*) with a dorsoventrally flattened body with which they press themselves firmly onto stones in foaming mountain springs, whereby the gills can be modified to anchoring structures (WESENBERG-LUND 1943).

Odonata

Both larvae and adults of the Odonata are predacious. Impressive characteristic features with which the Odonata can without doubt be established as a monophylum have resulted from the spheres of prey acquisition and reproduction.

Autapomorphies (Fig. 119 → 4)

- Wing propulsion apparatus with an "indirect-direct" flight mechanism (p. 271).
- Capture of flying prey with a "capturing basket".
  Odonata hunt their prey in flight, capture it with their legs and often consume it on the wing.
  The pterothorax (meso-and metathorax) is held at an oblique angle to the small mobile prothorax; consequently, the leg bases are shifted forward. The thorax appendages form a "capturing basket"; because of this construction the imago is not well suited to walking.
- Secondary copulatory apparatus on the sternites of abdominal segments 2 and 3.
  The new copulatory organ of the ♂ consists of a penis, seminal vesicle and hooks. To transfer sperm the ♂ grasps a ♀ in flight with his legs and anchors himself with abdominal claspers of cerci and paraprocta onto the females thorax (ground pattern realized in the Zygoptera). After formation of this tandem the ♂ fills the secondary copulatory organ with sperm from the caudal genital opening by curving in the abdomen. Finally, the ♀ bends her abdomen forwards and brings her own genital opening in contact with the secondary copulatory organ of the ♂. The familiar copulation wheel is formed (Fig. 121 A, B).
- Larva with labial grasping mask (Fig. 121 C–E).
  The larval labium lies under the head and in the resting position "masks" the other mouthparts. To capture prey the extensible grasping organ, which is furnished with a hinge joint, is shot rapidly forwards and the living prey is captured with terminal forceps-like arms.
- Larvae with weakly developed rectal gills.
  Three bulges protrude into the intestinal lumen: they function as tracheal or blood gills; this state is manifested in the Zygoptera.

LOHMANN (1996) developed a strict phylogenetic system of the Odonata with the Zygoptera and Epiprocta at the highest level of subordination.

Odonata
- **Zygoptera**
- **Epiprocta**
  - **Epiophlebioptera**
  - **Anisoptera**

From the autapomorphies compiled by LOHMANN (1996) I describe a few textbook-like clear features for the four monophyla of the Odonata.

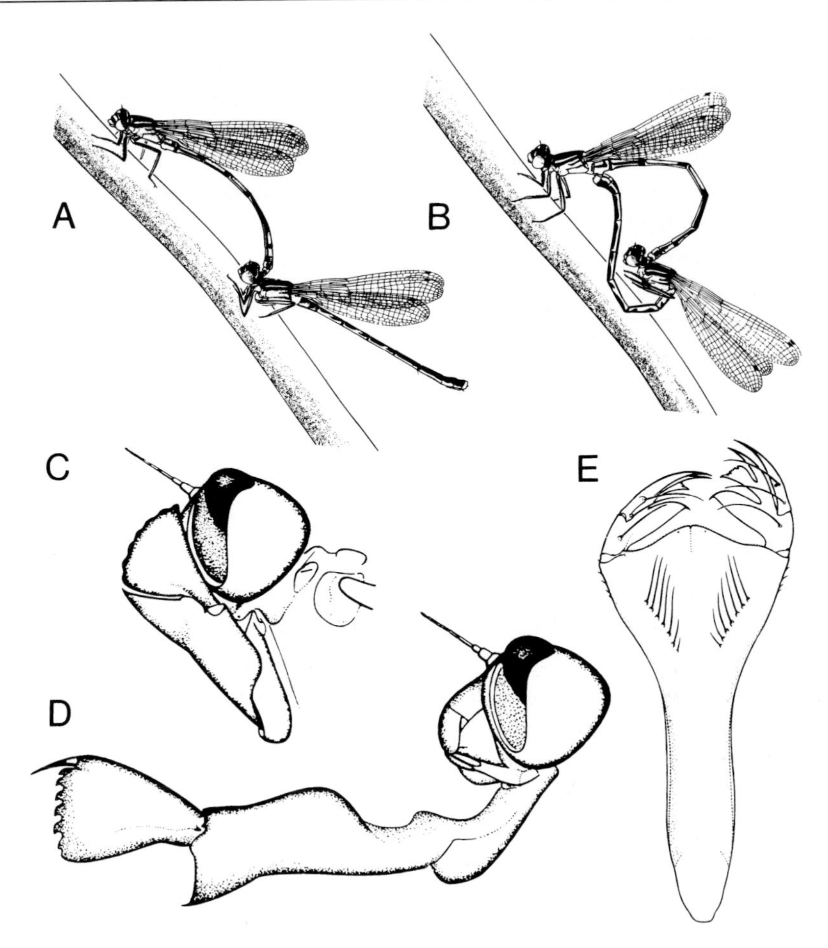

Fig. 121. Odonata. **A–B** Copulation in *Coenagrion lyelli* (Zygoptera). **A** Formation of the tandem. The ♂ anchors himself with abdominal claspers onto the thorax of the ♀. **B** Formation of the copulatory wheel. The ♀ bends her abdomen forwards and brings her genital opening into contact with the copulatory organ of the ♂ at the beginning of the posterior body. **C–E** Labium (capturing mask) of the larva. **C–D** *Synthemis eustalacta* (Anisoptera). Head in side view with retracted and extended labium. **E** *Austrolestes analis* (Zygoptera). Distal section from above the labium with terminal forceps. (Watson & O'Farrel 1991)

Zygoptera (damselflies)

Autapomorphies: Broad barrel-shaped head with compound eyes set far apart. Pterothorax with an extreme oblique position. Larva with three caudal lamellae (tracheal gills) as projections of the anal valves of the 11th abdominal segment (epiproct, paired paraprocta) (Fig. 118E).

Plesiomorphy: similar fore- and hindwings.

Calopteryx, *Lestes*, *Pyrrhosoma*, *Coenagrion*.

Epiprocta

Autapomorphies: Eyes enlarged and close to one another. Abdominal claspers in the ♂ of three elements (cerci + epiproct) for anchoring onto the head of the ♀ (ground pattern of the Odonata with four elements: cerci + paraprocta as well as anchoring onto the thorax). Further development of the rectal gills (ground pattern Odonata) to complex hindgut gills → tracheal gill organ composed of hindgut-folds in a rectal chamber (hindgut breathing).

Epiophlebioptera

Autapomorphies: Pedicellus of the antennae flat, egg-shaped. Larvae with stridulating organs on abdominal segments 3–6.

Epiophlebia (Japan, Himalaya).

Anisoptera (Dragonflies)

Autapomorphies: Dissimilar wings (Fig. 118 C). Hindwing especially at the base clearly larger than the forewing (expansion of the cubito anal field). Dorsal longitudinal muscles of the flight apparatus reduced. Four-part penis on the secondary copulatory apparatus. Larvae with transverse muscles in the anterior abdominal segments.

Aeshna, Gomphus, Cordulia, Libellula, Sympetrum.

Neoptera

Autapomorphies (Fig. 119 → 5)

- Division of the free wing surface (especially the hindwing) into the fields remigium, vannus and neala by two flexible folds – jugal and anal folds (Fig. 115).
- Roof-like arrangement of the folded wings over the abdomen through a pleural muscle inserted on the third pterale (axillare) of the wing joint (Fig. 118 F).

The evolution of neoptery must be seen as an essential step in the phylogeny of winged insects. The ability to lay their wings flat enabled the Neoptera to acquire food by entering narrow soil spaces and to escape from predators in minute hiding places beneath bark and between leaves or stones.

Plecoptera – N. N. (All Other Neoptera)

Plecoptera

Autapomorphies (Fig. 119 → 6) (cf. WILLMANN 1997b)

- Aquatic larvae with tracheal gills (Fig. 118G).
  The tracheal gills of the stone flies are not comparable to the homonymous formations in the Ephemeroptera and Odonata. In the Plecoptera there are short, flat or cylindrical tracheal gills, which are developed singly or in bundles on very different body regions – on the head (mentum, submentum), ventrally on the thorax, on the abdomen, in the anal region and also on the appendages (THEISCHINGER 1991).
- The larva has in its thorax and abdomen "on each side of the central nervous system a band of segment-jumping muscle cords" (HENNIG 1994). If the larva as such form an autapomorphy of the Plecoptera, this feature cannot be a separate autapomorphy.
- Gonads in both sexes joined at the anterior.
- Male without gonopods and without phallic organ on abdominal segment 9.
- Female without ovipositor.
- Tarsi with three articles.

For several of these features their interpretation as autapomorphies of the Plecoptera seems somewhat uncertain (N.P. KRISTENSEN 1991). Phallic organs are also absent in the Embioptera. The lack of male gonostyli and ovipositor and the reduction of tarsal articles is also seen in the Embioptera, Dermaptera and Zoraptera.

The long multiarticulated cerci found in both larva and imago form a conspicuous plesiomorphy of the Plecoptera (Fig. 118F, G).

Perla, Nemoura, Leuctra.

N. N.: All Other Neoptera

In a series of special feature alternatives the Plecoptera are possibly more primitive than the other Neoptera. This leads to the hypothesis of an adelphotaxa relationship between the Plecoptera and a unity comprising all other Neoptera (ZWICK 1981; N.P. KRISTENSEN 1991). I list here the possible autapomorphies of such a group followed by the corresponding plesiomorphies of the Plecoptera (P).

Autapomorphies (Fig. 119 → 7)

- Lateral wall of the prothorax a uniform plate.
  (P = division into anapleurite and coxopleurite.)
- Uniform testes.
  (P = metameric structure of the testes.)
- One male genital pore.
  (P = two separate openings.)
- Stipes without transverse muscle.
  (P = stipes with transverse muscle.)
- Terminal filament reduced.
  (P = larvae of some Plecoptera species with a caudal filament.)

The above-mentioned apomorphies are the result of relatively simple evolutionary changes. They have only a limited value for establishing the monophyly of a unity of the Neoptera after exclusion of the Plecoptera. Consequently the unity has remained unnamed.

Paurometabola - Eumetabola

Two large unities known as the Paurometabola and Eumetabola may form the adelphotaxa of the Neoptera after exclusion of the Plecoptera. However, the monophyly of these unities also appears to be only weakly justified.

Paurometabola

Autapomorphy (Figs. 119 → 8; 122 → 1)

- Grape-like accessory genital glands in the male.

Compared with the norm in the Insecta of one pair of accessory genital glands the above-mentioned feature is without doubt an apomorphy. However, whether it forms an autapomorphy for a monophylum Paurometabola or not is problematic, as in the Notoptera only two pairs of accessory genital glands are developed. Since the Notoptera are a subtaxon of the Blattopteriformia within the Paurometabola, a secondary modification of a grape-like gland must, a posteriori, be considered.

The paurometabolous development in which the larva resembles more closely the imago with each molt and the wings develop step-by-step is obviously a plesiomorphy.

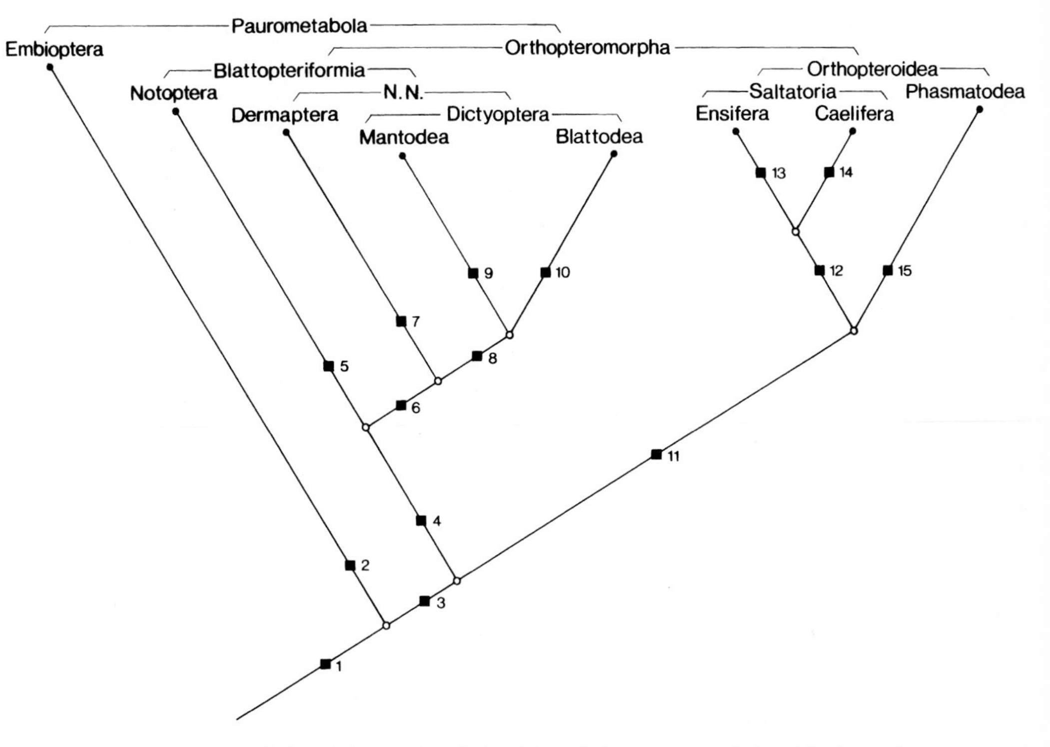

Fig. 122. Diagram of the phylogenetic relationships of the Paurometabola with the Embioptera and Orthopteromorpha as sister groups. Section 3 of the systematization of the Insecta

Systematization: Section 3

Paurometabola
- **Embioptera**
- **Orthopteromorpha**
  - **Blattopteriformia**
    - **Notoptera**
    - **N. N.**
      - **Dermaptera**
      - **Dictyoptera**
        - **Mantodea**
        - **Blattodea**
  - **Orthopteroidea**
    - **Saltatoria**
      - **Ensifera**
      - **Caelifera**
    - **Phasmatodea**

Eumetabola

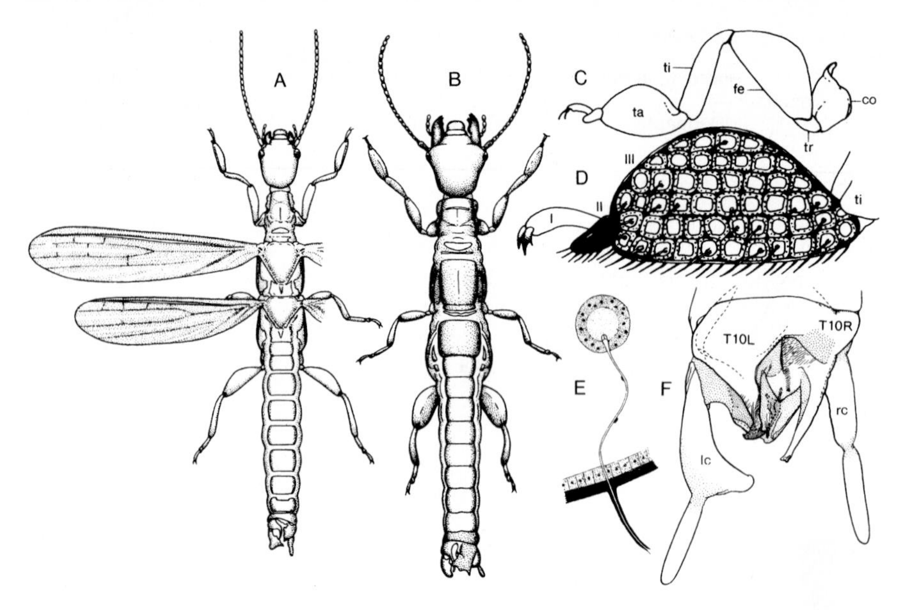

Fig. 123. Embioptera. **A** *Notoligotoma nitens.* Dorsal view of the male. **B** *Metoligotoma* sp. Wingless male. **C** *Oligotoma saundersii.* Foreleg with tarsus of three articles, female. **D** Longitudinal section through the tarsus of the Embioptera with silk glands in article III and gland ducts on the ventral side. **E** *Embia ramburi.* Single gland with canal and hollow gland hair of the cuticle. **F** *Aposthonia glauerti.* Posterior end of the ♂ with asymmetrically developed cerci as a clasping apparatus. *co* Coxa; *fe* femur; *lc* left cercus; *rc* right cercus; *ta* tarsus; *ti* tibia; *tr* trochanter; *T10L*, *T10R* left and right half tergites from abdominal segment 10. (**A, B, D, F** Ross 1991; **C** Weber & Weidner 1974; **E** Kaltenbach 1968)

Embioptera – Orthopteromorpha

Embioptera

Autapomorphies (Fig. 122 → 2)

- Production of webs.
- Production of silk in the third tarsal article of the prothorax appendages. The article is greatly swollen (Fig. 123 C–E).
- Prognathous head without ocelli (Fig. 123 A, B).
- Vannus (anal field) of the hindwing reduced in the male (Fig. 123 A).
- Female totally wingless.
- Femur of the metathorax appendages enlarged (Fig. 123 A, B).
- Cerci with two articles.

A small group of slender Neoptera from 1.5–2 mm in length. Tropical and subtropical inhabitants. The Embioptera live in silk-like webs which they construct as tubular burrows or shallow webs. Correlated with this apomorphic life style are basic derived peculiarities in structure.

On the third tarsal article of the anterior legs there are up to 200 tightly packed circular silk glands. The silk is secreted through hollow setae on the ventral side of the tarsal articles and brushed onto the surrounding substrate (Fig. 123 E).

Adaptations to this self-created biotope are the slender body with prognathous head and short appendages, as well as the reduction of the wings in the female. The strongly developed tibial depressor muscle in the posterior appendages effects a rapid return into the tunnel system as a reaction against predators. Sensitive cerci function as tactile sensory organs at the posterior. The cerci also serve as grasping hooks during copulation. For this the left cercus can be bent inwards in a hook-like manner (Fig. 123 F); the two articles can also fuse.

Species of the taxa *Haploembia* and *Monotylota* are found in the Mediterranean region. In these species the male is also wingless – a phenomenon that is likewise realized in other diverse taxa of the Embioptera (Fig. 123 B).

Orthopteromorpha

Autapomorphy (Fig. 122 → 3)

- Forewings braced to form more or less rigid tegmina.
  This change in construction is evidently connected with the strongly soilbound life style (HENNIG 1969, 1981).

The validity of the feature as an autapomorphy of a unity of all Paurometabola after branching off of the Embioptera is questionable because non-sclerotized anterior wings are realized in several subtaxa – such as the termites (N.P. KRISTENSEN 1975); otherwise this must be shown to be a secondary state within the Orthopteromorpha.

The next lowest hierarchic level with the roach-like Blattopteriformia and grasshopper-like Orthopteroidea as possible adelphotaxa is according to N.P. KRISTENSEN (1975) also only weakly justified. Due to lack of a reliable alternative, however, we follow with some reserve the systematization of HENNIG.

Blattopteriformia – Orthopteroidea

Blattopteriformia

Autapomorphy (Fig. 122 → 4)

- Overlapping of the pro-, meso-, and metasternum.

The significance of this feature is disputed (N.P. KRISTENSEN 1975) – and therefore the grouping of all roach-like Paurometabola into a taxon Blattopteriformia is not sufficiently justified.

Notoptera – N. N. (All Other Roach-Like Paurometabola)

Notoptera

Autapomorphies (Fig. 122 → 5)

- Prognathous head.
- Compound eyes modified. Low numbers of ommatidia (40–70 per eye) separated from one another.
- No ocelli.
- No wings (Fig. 124 A).
- Sternite of the first abdominal segment with an extrusible sac.
- Asymmetric gonopods in the male.

Plesiomorphic features are a free ovipositor, tarsi with five articles and long cerci of 5–9 articles (Fig. 124 A).

The Notoptera (Grylloblattodea) with 25 species (2–3.5 cm long) are only found in cold climate zones of the northern hemisphere. They are soilbound mountain inhabitants colonizing above the tree line.

Grylloblatta, *Galloisiana*.

N. N.: All Other Roach-Like Paurometabola

Autapomorphy (Fig. 122 → 6)

- Female genital opening and the strongly reduced ovipositor lie in a genital chamber, which is limited ventrally by the subgenital plate (sternite of abdominal segment 7).

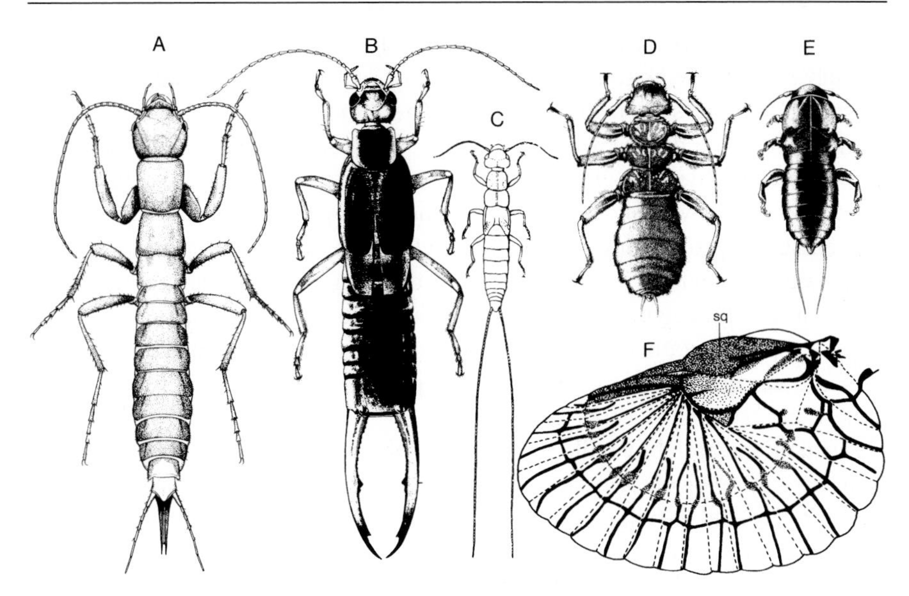

Fig. 124. **A** Notoptera. **B–F** Dermaptera. **A** *Grylloblatta campodeiformis* (Notoptera). Female. **B** *Labidura truncata* (Dermaptera). Male. **C** *Diplatis longisetosa* (Dermaptera). Larva with long multiarticulated cerci. **D** *Arixenia esau* (Dermaptera). Female. Epizoon on bats. **E** *Hemimerus vosseleri* (Dermaptera). Female. Ectoparasite on rodents. **F** *Forficula auricularia* (Dermaptera). Fan-shaped left hindwing with large expansion of the anal field. Sclerotization dark lines. *sq* Squama. (**A** Rentz 1991a; **B, D, E** Rentz & Kevan 1991; **C** Günther & Herter 1974; **F** Kleinow 1966)

Dermaptera – Dictyoptera

Dermaptera

Autapomorphies (Fig. 122 → 7)

- Prognathous head.
- Forewings modified into short, rigid wing covers (elytra) (Fig. 124B).
- Large hindwings due to strong development of the anal field (vannus) which is broadened into a fan shape (Fig. 124F).
- Long single-articled cerci differentiated into forceps-like structures (Fig. 124B).
- Ocelli reduced.
- Tarsus with three articles.
- Glossae and paraglossae fused in the labium.

The dark-brown to black shiny earwigs are markedly distinguished by the first four features.

On the dorsoventrally flattened body the forewings are modified into extremely short, strongly sclerotized wing covers; the distal ends are trimmed transversely over the back. The hind pair alone functions as wings. These are differentiated into large, semicircular leaves. The posterior wings are folded fan-like together along a rich pattern of fold lines and are then folded over inwards twice. In the resting position only small sclerotized scales project out beneath the elytra. Especially conspicuous are the large, single-articled cerci; in the ground pattern of the Dermaptera they form forceps-like structures. The cerci can be used for capturing prey, defense against predators and for copulation. During development, the larvae of some earwigs (*Diplatis*, *Bormansia*) pass through the original state of long, multiarticulated cerci (Fig. 124C).

Dermaptera are above all tropical and subtropical inhabitants. The European earwig *Forficula auricularia* is cosmopolitan in cooler regions.

The nocturnal earwigs hide from the light under stones and in spaces in the soil, beneath the bark of dead trees or in rotting substrates. *Arixenia* species live as epizoa on bats in southeast Asia; *Hemimerus* species are ectoparasites on African rodents (*Cricetomys*, *Beamys*). In these, eyes and wings are convergently reduced; the cerci are secondarily rod-shaped (Fig. 124D,E).

Dictyoptera

KLASS (1995, 1997) set forth eight synapomorphies between the Mantodea and Blattodea, whereby some of them were based on subtle morphological details. I describe two of the features.

Autapomorphies (Fig. 122 → 8)

- Ootheca formation.
  Cementation of several eggs into a packet (ootheca) with the secretion of accessory genital glands that open in the vestibulum (Fig. 125C).
- Genital pouch.
  Female with two-part genital pouch of genital chamber and vestibulum. Ventrally, the seventh abdominal sternite forms a shell-like subgenital plate. Lateral sternites of the eighth abdominal segment are tilted into the ventral wall of the genital chamber on both sides of the oviduct opening.
  The genital pouch of the Dictyoptera is possibly a further development of states realized in the Dermaptera.

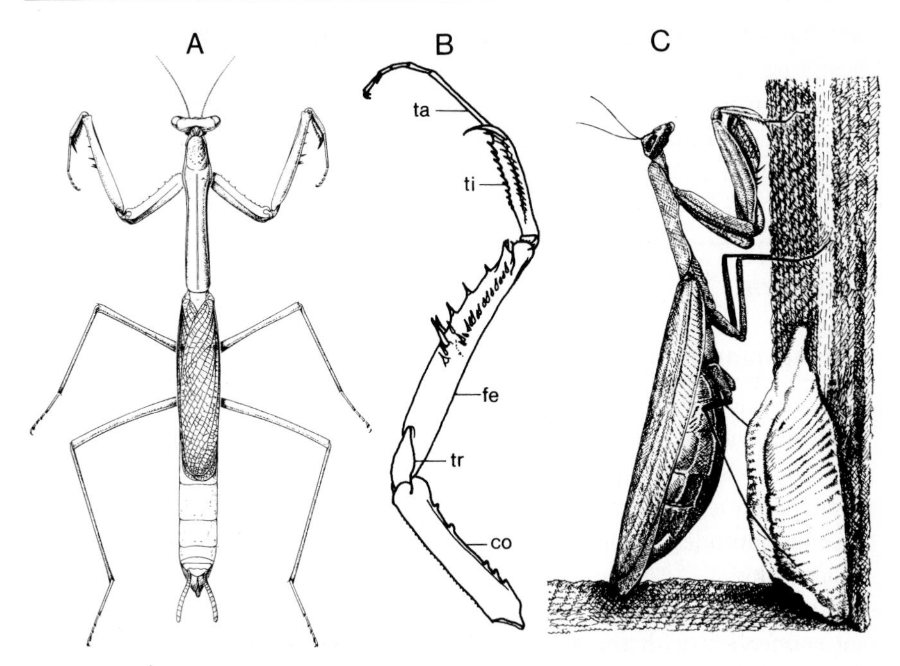

Fig. 125. Mantodea. **A** *Archimantis latistylus*. Female. **B** Right capturing leg of *Archimantis latistylus*. View oblique from underneath: Long mobile coxa. Femur and tibia each have two rows of spines. Inner row of the tibia with a large terminal claw; by folding the articles this fits into a pit of the femur. **C** Preying mantis *Mantis religiosa*. Female. *Right*, packet of eggs (ootheca) with numerous eggs. *co* Coxa; *fe* femur; *ta* tarsus; *ti* tibia; *tr* trochanter. (**A, B** Balderson 1991; **C** Jacobs & Renner 1974)

Mantodea – Blattodea

Mantodea

Autapomorphies (Fig. 122 → 9)

- Head freely movable against the prothorax.
- Strongly elongated narrow prothorax with a mobile connection to the mesothorax (Fig. 125 A).
- Anterior appendages differentiated into predacious legs. Long coxa, femur and tibia ventrally with rows of spines. Tibia closes against the femur like a jack-knife (Fig. 125 B).

The **three prominent characteristic features** of the preying mantids form a related feature complex which serve in **predacious feeding.** In the waiting position the head and prothorax are stretched upwards. With the capturing legs oriented forwards and placed together the Mantodea look rather like they are praying – hence their name. After capture, prey (insects, small vertebrates) is clamped between the spines of the femur and tibia and passed to the mouth.

Plesiomorphic ground pattern features are the tarsi with five articles, three ocelli and multiarticulated cerci.

Species-rich taxon with a high diversity in the tropics. *Mantis religiosa* is common in the Mediterranean region.

Blattodea

Autapomorphies (Fig. 122 → 10)

- Loss of the frontal ocellus in the imago.
  Only the two lateral ocelli are present. The existence of three ocelli in the Mantodea is the plesiomorphic alternative.
- Gonapophyses end in the vestibulum.
  In the Mantodea the long ovipositor projects out of the vestibulum (plesiomorphy).
- Ootheca formation in vestibulum.
  In connection with the short ovipositor in the Blattodea the egg packet (ootheca) is formed in the vestibulum; in the Mantodea with a long ovipositor it is formed outside the genital pouch (plesiomorphy).

In conventional classifications "Blattaria" (cockroaches) and Isoptera (termites) are placed opposite one another as unities with the same rank. However, through phylogenetic systematics the cockroaches (Fig. 126 A, C),

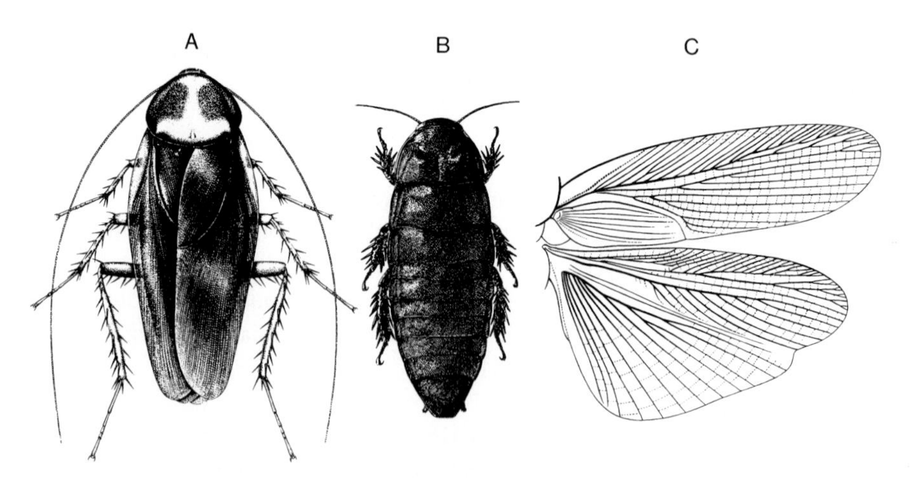

Fig. 126. Blattodea. **A** *Periplaneta brunnea*. Male. **B** *Cryptocercus punctulatus*. Representative of the taxon *Cryptocercus* as sister group of the Isoptera. **C** *Neotemnopteryx australis*. Right fore- and hindwing. Well-developed anal field on the hindwing (plesiomorphy). (**A, C** Roth 1991; **B** Chopard 1949)

with such well-known representatives as *Ectobius lapponica* (forest inhabitant) or *Blatta orientalis* and *Periplaneta americana* (thermophilic synanthropic animals), are known to be a paraphylum. The "**Blattaria**" possess no derived features; they were therefore phased out by KLASS (1995). On the other hand, the Isoptera are only a subordinate subtaxon of the Blattodea. In the place of the "Blattaria" and Isoptera there are now the highest-ranking adelphotaxa **Blattomorpha** (*Archiblatta*, *Blatta*, *Periplaneta* amongst others) and **Blattellomorpha** - the grouping of all other Blattodea.

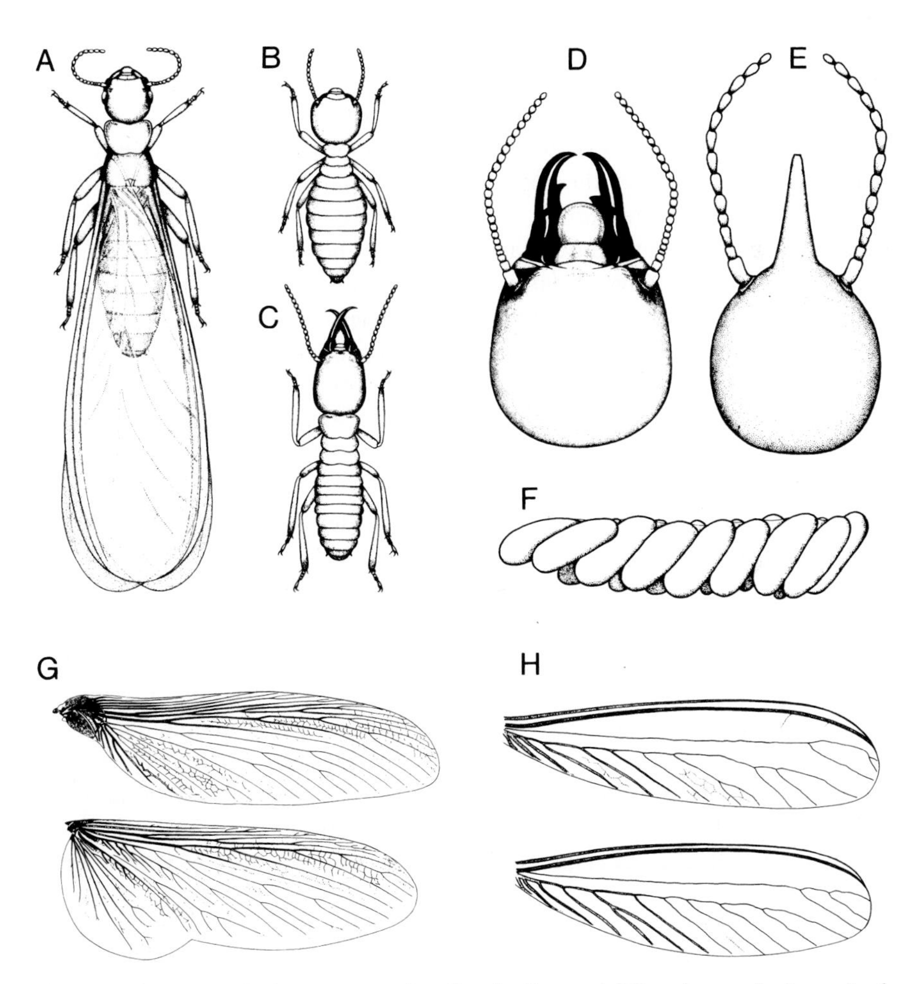

Fig. 127. Isoptera. **A–C** *Coptotermes acinaciformis*. Castes. **A** Winged reproductive animal. **B** Worker. **C** Soldier. **D** *Mastotermes darwiniensis*. Head of a soldier with large mandibles. **E** *Tumulitermes recalvus*. Head of a soldier with frontal cone. **F** *Mastotermes darwiniensis*. Egg packet. **G** *Mastotermes darwiniensis*. Fore- and hindwing; the latter with a broad anal field. **H** *Nasutitermes dixoni*. Similar fore- and hindwing. (Watson & Gay 1991)

Within the Blattellomorpha the taxon **Cryptocercus** (three species from the northern hemisphere) is identified as the **sister group of the Isoptera** (Fig. 126 B). Apart from similarities in the proventriculus morphology (KLASS 1995), I point to impressive congruencies in the biology. *Cryptocercus punctulatus* (North America) live in families of one female, a few males and up to 35 larvae; this is similar to a newly formed termite colony (SEELINGER & SEELINGER 1983). Living in social groups similar behavior patterns exist, such as recognition of group members and fighting against non-members; the latter trigger as an alarm a rhythmic twitching against the body of family members. *Cryptocercus* and the Isoptera consume wood. Species of both taxa have single-celled Hypermastigida and Oxymonadina in the intestine, which take up particles of wood and digest cellulose intracellularly. In the social group the symbionts are probably passed on through anal trophallaxis (see below).

The **Isoptera** are well established as a monophylum through several **autapomorphies:** wings of the reproductive animal with preformed sutures. They are shed after the nuptial flight. Sclerotization of the thorax strongly reduced; extended membranous areas between the sclerite plates. Caste formation within a colony – Alatae=primary winged reproductives; soldiers and workers=wingless, sterile males and females.

We now come to the highest-ranking division of the Isoptera. *Mastotermes darwiniensis*, a single recent species from Australia, forms the adelphotaxon of all other termites – this grouping is as yet unnamed. We show here some of the feature alternatives.

Mastotermes darwiniensis **Plesiomorphies**	**N. N. (other Isoptera)** **Apomorphies**
Hindwings with well-developed vannus (anal field) taken over from the ground pattern of the Blattodea (Fig. 127 G).	Reduction of the anal field of the hindwing. Similar fore- and hindwings (Fig. 127 H) (name of the taxon).
12–15 Malpighian tubules.	Maximum of eight Malpighian tubules.
Five tarsal articles (ground pattern of the Blattodea).	First and second tarsal articles fused with one another.
Formation of clutches of eggs (Fig. 127 F).	Change to deposition of single eggs.

Termite colonies are founded by one pair of reproductive individuals. At first, in the young colony, only soldiers and workers are produced; winged reproductives only appear after some years. Both Queen and King can live many years, whereby the female is repeatedly fertilized by the male. The soldiers have a strongly sclerotized head. They are equipped either with large mandibles or possess frontal cones. Through these a secretion from numerous frontal glands is sprayed onto enemies.

The workers take up cellulose-rich nutrients of plant origin. Their own cellulase is obviously of primary importance for the utilization of cellulose. A supplement is provided by the metabolite of single-celled symbionts that colonize the hindgut of most termite taxa (WATSON & GAY 1991). Certain Termitidae, that are secondarily free from symbionts, eat fungi which they cultivate in fungi gardens.

Soldiers, larvae and reproductives are fed by the workers through transmission of regurgitated food or through feces (trophallaxis).

Termites are found mainly in the tropics and subtropics. *Reticulitermes* sp. are occasionally carried to port towns of central and northern Europe.

Orthopteroidea

Autapomorphies (Fig. 122 → 11)

- Cerci of only one to two articles.
- Fusion of the labial stipites to a plate.

There may be a sister group kinship between the jumping grasshoppers (Saltatoria) and the stick insects (Phasmatodea) (HENNIG 1969, 1981, 1994). As a possible autapomorphy of a unity Orthopteroidea the wide-ranging reduction of the cerci is set forth. The hypothesis is only weakly justified; there are, however, no convincing alternatives.

Saltatoria – Phasmatodea

Saltatoria

Autapomorphies (Fig. 122 → 12)

- Cryptopleury.
  Lateral lobes of the pronotum cover most of the pleura (Fig. 128 A, C, D). Correspondingly, the sides of the segment are strongly desclerotized.
- Jumping legs.
  The differentiation of the posterior appendages to jumping legs is characterized by the following features: (1) the femur is greatly enlarged to take up the extensor muscle of the tibia. (2) Straightening of the joint between the femur and tibia (Fig. 128).
- Tibia of the posterior appendage with two dorsal rows of teeth.
- Spiracles of the tracheal system horizontally subdivided in the prothorax. Consequently, two inner separated spiracle spaces develop.

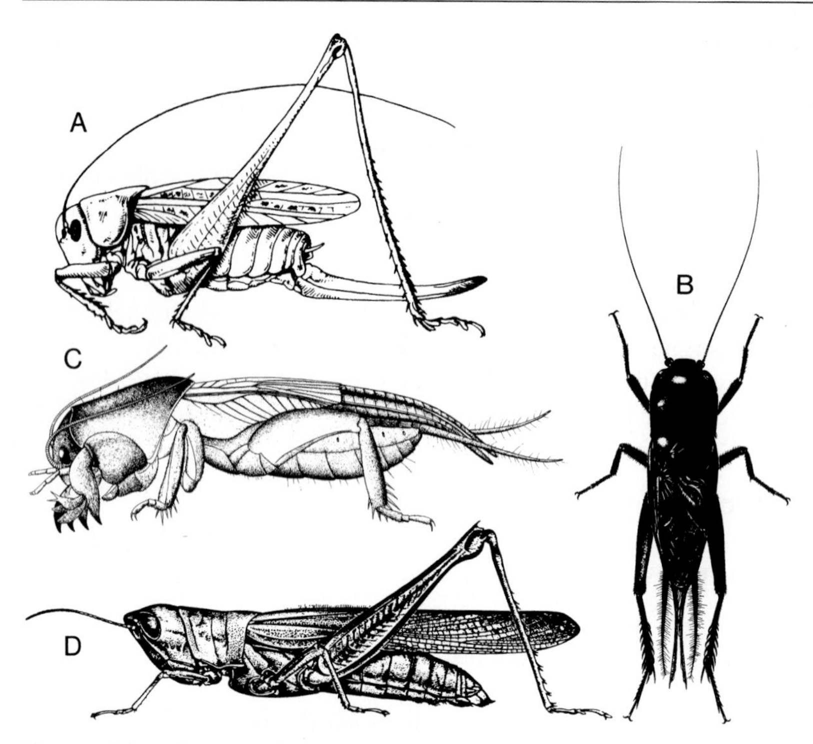

Fig. 128. Saltatoria. **A** *Decticus verrucivorus* (Ensifera, Tettigonioidea) with long antennae and long ovipositor. **B** *Teleogryllus commodus* (Ensifera, Grylloidea). Hindwings project over the abdomen in the form of long points; with long cerci. **C** *Gryllotalpa gryllotalpa* (Ensifera, Grylloidea). Appendages of the prothorax differentiated to digging shovels. **D** *Bermiella acuta* (Caelifera, Acrididae) with short antennae and short ovipositor. (**A** Jacobs & Seidel 1975; **B,D** Rentz 1991b; **C** Eisenbeis & Wichard 1985)

- Tarsi with a maximum of four segments, originated through fusion of segments 1 and 2.

With the above characteristics the Saltatoria (Orthoptera) are a well-established monophylum in which the Ensifera (long-horned grasshoppers) and the Caelifera (short-horned grasshoppers) probably form the adelphotaxa.

The well-known sound produced by the grasshoppers and crickets with stridulating organs and an associated auditory organ belongs, however, neither in the ground pattern of the Saltatoria nor in the ground patterns of the Ensifera and the Caelifera; these capabilities were first developed within the Ensifera and Caelifera by very different routes.

Ensifera - Caelifera

Ensifera

Autapomorphies (Fig. 122 → 13)

- Proventricle in the foregut with armature of six longitudinal rows of complex attachments, separated by sclerotized sections.
- During rest the anal field (vannus) and the cubital region of the hindwings are folded underneath the remigium.
- Male accessory glands open into paired or unpaired gland sacs.

Compared with the Caelifera the Ensifera remain plesiomorphic in the following features:
- Long antennae. Anellus with usually more than 30 articles (Fig. 128A).
- In the ground pattern tarsi with four articles.
- The gonapophyses of abdominal segments 8 and 9 form a long ovipositor that usually projects freely over the end of the body (Fig. 128A). To deposit eggs the egg-laying apparatus is thrust deep into the substrate.

Stridulating organs and auditory organs only exist in the subtaxa Tettigonioidea (katydids: *Tettigonia*, *Decticus*) and Grylloidea (crickets: *Gryllus*, *Acheta*, *Gryllotalpa*).

Their singing is achieved by rubbing together the forewings. In the Tettigonioidea a scraper of the left wing (cubitus posterior) rubs over a stridulatory file of the right wing. In the Grylloidea scrapers and stridulatory files are equally developed on the two elytra.

The auditory organ of the katydids and crickets lies in the tibia of the anterior legs near to the joint with the femur. The tympanic membranes are thin cuticular areas supported by tracheal vesicles (amplifiers). Primarily the tympanic membrane lies free on the surface of the tibia. Mostly, however, it is largely concealed by an integumental fold; a slit then forms the only entrance to the tympanicum.

Caelifera

Autapomorphies (Fig. 122 → 14)

- Short antenna. Less than 30 articles (Fig. 128D).
- Tarsus with a maximum of three articles.
- Male without gonopods.
- Intestine without proventriculus.
- In the female the gonapophyses form a short, stocky ovipositor (Fig. 128D). After loosening the soil with the gonapophyses the female bores a deep hole with her abdomen into which a clutch of eggs is laid.

In short-horned grasshoppers sound production and detection are only realized in the species-rich taxon Acrididae (*Locusta*, *Stenobothrus*, *Psophus*).

Several different routes were taken in the evolution of stridulating organs. The manifestation of a stridulating file on the inner edge of the femur of the saltatory legs is widespread; stridulation takes place by rubbing the hind femora against the outer surface of the forewings.

Within the Acrididae tympanic membranes are uniformly found left and right on the first abdominal segment.

Phasmatodea

Autapomorphies (Fig. 122 → 15)

- Paired repugnatorial glands in the prothorax. Opening in the anterior corners of the pronotum.
- Posterior section of the midgut with pear- or skittle-shaped glands that are furnished with a tube-like appendage.
- The single-articled cerci in the ♂♂ become very long and can be differentiated into grasping organs.
- Eggs thickly shelled, with operculum (Fig. 129 C).
- Sperm secondarily without mitochondria. Axoneme surrounded by two dense sheets (JAMIESON 1987).

Compared with the presumed sister group Saltatoria the Phasmatodea are primitive in lacking jumping legs. Furthermore the manifestation of tarsi with five articles is plesiomorphic.

The main geographic range of the stick insects is in the tropics. Walking sticks and leaf insects form the two extreme habitus forms (Fig. 129 A, B). A stick-like extension to a length of over 30 cm, reduction of the wings, formation of spines and brown pigmentation present a picture of a thin twig. A well-known example, a favorite of biology labs, is *Carausius morosus*. At the other extreme, with lamellae-like outgrowths on the flattened abdomen and appendages, the taxon *Phyllium* is a perfect copy of tree leaves.

Eumetabola

Autapomorphies (Fig. 119 → 9)

- Sclerotized clasp (jugal bar) on the jugal field of the hindwing.
- Morphogenesis of the male genitalia from the ventral side of the tenth abdominal segment. According to BOUDREAUX (1979) an apomorphy in contrast to its usual origin on the ninth abdominal segment.

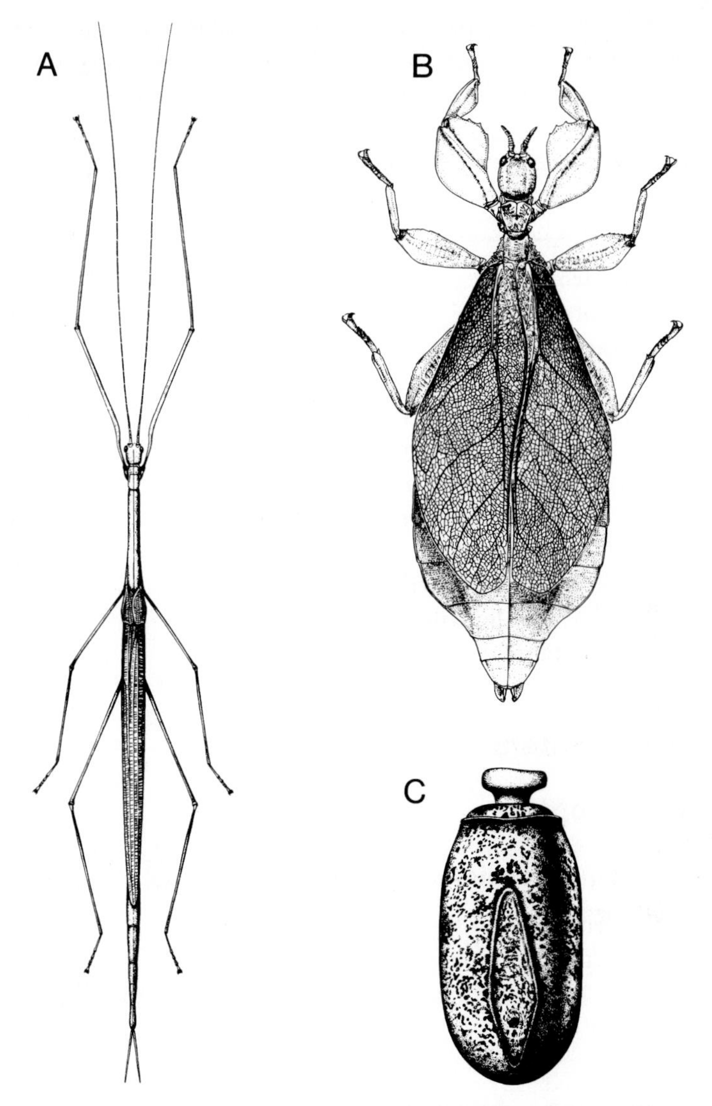

Fig. 129. Phasmatodea. **A** *Sipyloidea filiformis* (walking stick). Female. **B** *Phyllium siccifolium* (leaf insect). Female. **C** *Didymuria violescens*. Egg with operculum. (Key 1991)

- In the ovarioles a nutritive cord is formed from the cytoskeleton of the anterior oocytes (germocytes) (N.P. KRISTENSEN 1995; BÜNING 1998).

The absence of ocelli is common. This cannot, however, be an apomorphy of a unity Eumetabola as ocelli are found in the nymph of the Zoraptera (N.P. KRISTENSEN 1975).

The hypothesis that the unity Eumetabola is a monophylum containing the adelphotaxa Paraneoptera (including Zoraptera) and Holometabola is only weakly justified. However, no other competing alternative exists.

Systematization: Section 4

Eumetabola
- **Paraneoptera**
  - **Zoraptera**
  - **Acercaria**
    - **Psocodea**
      - **"Psocoptera"**
      - **Phthiraptera**
    - **Condylognatha**
      - **Thysanoptera**
      - **Hemiptera**
        - **Sternorrhyncha**
        - **Euhemiptera**
- **Holometabola**

Paraneoptera – Holometabola

Paraneoptera

Autapomorphies (Figs. 119 → 10; 130 → 1)

- Concentration of the abdominal ventral nerves to two ganglion masses.
- Reduction in the number of Malpighian tubules to six.
- Reduction in the number of tarsal articles to three per appendage.
- Sperm with two cilia (N.P. KRISTENSEN 1991).

Zoraptera – Acercaria

According to HENNIG, there is a sister group relationship between the species-poor tropical Zoraptera and the giant army of lice and bug-like Acercaria found worldwide. The kinship hypothesis is, however, strongly disputed.

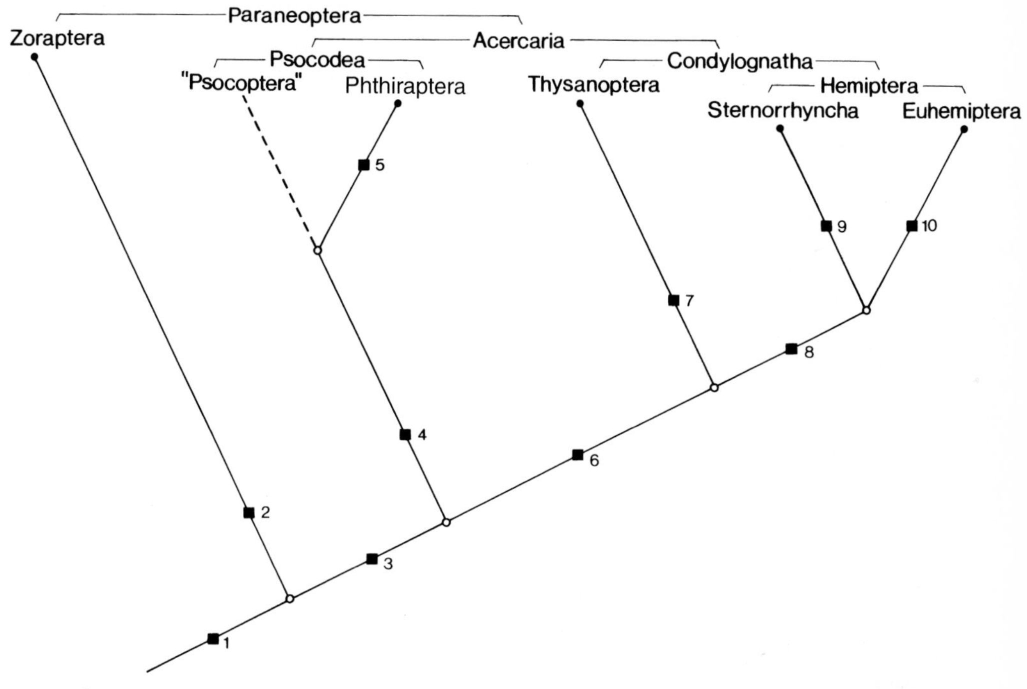

Fig. 130. Diagram of the phylogenetic relationships of the Paraneoptera with the Zoraptera and Acercaria as sister groups. Section 4 of the systematization of the Insecta

Zoraptera

Autapomorphies (Fig. 130 → 2)

- Tarsi with only two articles.
- Wings with greatly reduced simple venation (Fig. 131 B).
- Hindwing clearly smaller than the forewing.
- Ovipositor reduced.

Only ca. 30 species from the tropics known; all placed in one supraspecific taxon *Zorotypus*. Soil-dwelling animals less than 3 mm in length.

For most species two adult forms are described – a pigmented form with eyes, ocelli and wings, and a colorless blind form without wings (Fig. 131).

Compared with the Acercaria the Zoraptera are characterized by the following plesiomorphic features:

- Possession of cerci; these are, however, short and nonarticulated.
- Biting mouthparts with mandibles and maxillae as chewing structures.
- Two free ganglia in the abdomen. They form the primitive state of the "concentration of the abdominal nervous system" as an autapomorphy of the Paraneoptera.

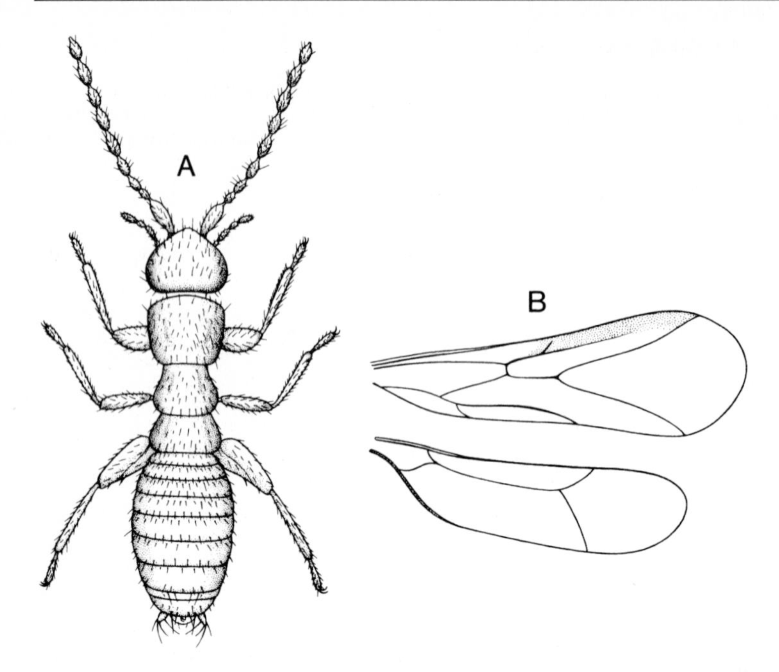

Fig. 131. Zoraptera. **A** *Zorotypus hubbardi.* Male. Blind, wingless form. **B** *Zorotypus hubbardi.* Male. Winged form; venation of the wings greatly reduced. (Smithers 1991 a)

Acercaria

Autapomorphies (Fig. 130 → 3)

- Loss of the cerci.
- Further concentration of the abdominal ventral nerve cords. Fusion to one abdominal ganglion located close to the metathoracic ganglion.
- Further reduction in the number of Malpighian tubules to four (MAHNER 1993).
- Transformation of the mouthparts towards a piercing-sucking apparatus → lacinia of the maxilla rod-shaped elongated (Fig. 132 C, D).
- Acrosome of the sperm without perforatorium (JAMIESON 1987).

Within the Acercaria there is probably an adelphotaxa kinship between the Psocodea ("Psocoptera" + Phthiraptera) and the Condylognatha (Thysanoptera + Hemiptera). Recently, a sister group relationship between the Psocodea and Thysanoptera has also been discussed.

Psocodea – Condylognatha

Psocodea

Autapomorphies (Fig. 130 → 4)

- Apparatus for the oral uptake of water vapor from the atmosphere (Fig. 132 F–H).
- Rupture-facilitating mechanism for the antenna through cuticular modifications on the basis of the flagellum.
- Anchoring of the fore- and hindwings through a binding device (p. 303; Fig. 132 E).

The outstanding characteristic of the Psocodea is an **apparatus** in the preoral cavity for the **uptake of water vapor from the atmosphere** (RUDOLPH 1982, 1983; RUDOLPH & KNÜLLE 1982). We outline its structure from studies on the free-living "Psocoptera" (Fig. 132 F–H). In the resting position the oral cavity is closed to the front and behind by the labrum (upper lip) and labium (lower lip), and bound laterally by the mandibles and maxillae. The preoral cavity itself is divided into an anterior cibarium and posterior salivarium by the hypopharynx which is pushed far forwards. At the entrance to the mouth the cibarium bears cuticular hardenings – the cibarial sclerite and the epipharyngeal sclerite – the latter is connected to a large clypeo-epipharyngeal muscle which radiates dorsalwards. In the salivarium pairs of dorsal and ventral labial glands open. Ventrally the hypopharynx possesses a pair of long, oval lingual sclerites; the sclerites project into the salivarium. In appropriate air humidity conditions the salivarium is opened. The lingual sclerites extend and push between the margins of the labium and labrum. The surface of the sclerites is coated with a fluid with hygroscopic properties that probably stems from the dorsal labial glands. Water vapor condenses on the secreted film.

The lingual sclerites and the cibarial sclerite are connected anteriorly in the hypopharynx through two thin cuticular tubes (tubular filaments). Regular contractions lead to a rhythmic movement of the cibarial and epipharyngeal sclerite complex. Like a suction pump they pull up the condensate through the tubes of the hypopharynx into the mouth. At the end of the absorption phase contraction of paired hypopharynx retractors brings the lingual sclerites back into the salivarium.

This mechanism of oral uptake of water vapor is not only found in numerous "Psocoptera" but also assuredly belongs in the ground pattern of the Phthiraptera (lice) – realized here in species of the Amblycera and Ischnocera.

Moisture absorption is absent, however, in the Rhynchophthirina and the Anoplura as obligate bloodsuckers. It is tempting to connect the new

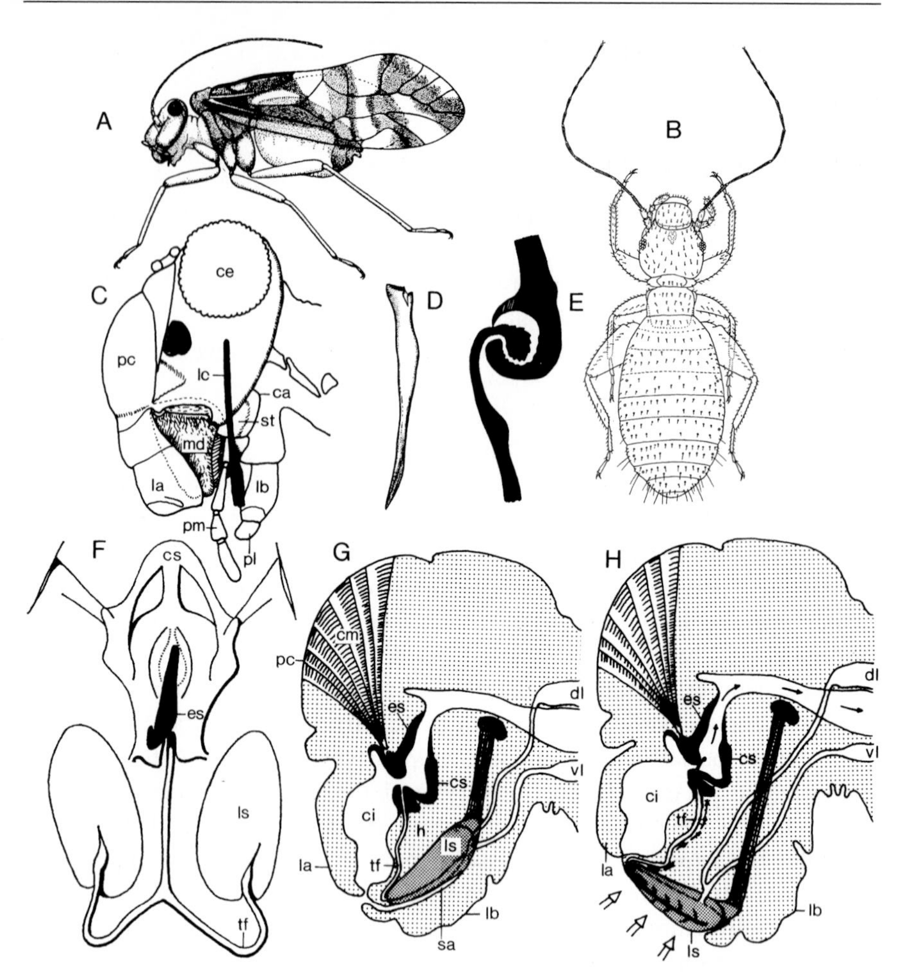

Fig. 132. "Psocoptera". **A** *Pentacladus eucalypti.* Male. **B** Book louse *Liposcelis divinatorius.* Wingless species, without ocelli. **C** *Psococerastis gibbosus.* Head from the side. Rod-shaped lacinia shown in black. **D** *Myopsocus australis.* Rod-shaped lacinia (autapomorphy of the Acercaria). **E** Coupling structure between the forewing (above) and hindwing (below). **F–H** Apparatus for oral uptake of water vapor from the atmosphere (autapomorphy of the Psocodea). **F** Sclerites in the preoral cavity of the book louse *Liposcelis.* Top view. **G–H** "Psocoptera" head. Sagittal section with those structures that are involved in the uptake of water vapor. **G** Position of the hypopharynx at rest, with the position of the tongue sclerites in the preoral cavity. **H** During uptake of water vapor the tongue sclerites are pushed out between the labrum and labium. The probable path of condensed water from the tongue sclerites through the tubular filaments to the cibarial sclerite and further to the mouth is marked by *arrows. ca* Cardo; *ce* compound eye; *ci* cibarium; *cm* clypeoepipharyngeal muscle; *cs* cibarial sclerite; *dl* dorsal labial gland; *es* epipharyngeal sclerite; *h* hypopharynx; *la* labrum; *lb* labium; *lc* lacinia; *ls* tongue sclerite; *md* mandible; *pc* postclypeus; *pl* palpus labialis; *pm* palpus maxillaris; *sa* salivarium; *st* stipes; *tf* tubular filament; *vl* ventral labial gland. (**A,D** Smithers 1991b; **B,C** Weidner 1972; **E** Jacobs & Renner 1974; **F** Haub 1972; **G,H** Rudolph & Knülle 1982)

uptake of blood fluid with the reduction of the original mode of water vapor uptake.

"Psocoptera" – Phthiraptera

"Psocoptera"

Compared with the parasitic Phthiraptera the "Psocoptera" (psocids, booklice) probably form a paraphylum of primary free-living species of Psocodea. There are a few possible derived peculiarities in their development (thin chorion with absence of micropyles, unusual dorsal position of the embryo in the egg). In a genuine conflict of evidence these are opposed by a number of apomorphic features which the subtaxon Liposcelidae within the "Psocoptera" (Fig. 132 B) have in common with the Phthiraptera (no wings, no ocelli, only two ommatidia, no spiracles on abdominal segment 1 or 2 and only three pairs of testis follicles). The Liposcelidae may be the adelphotaxon of the Phthiraptera (LYAL 1985).

The "Psocoptera" (Copeognatha) are soilbound insects a few mm in length, with synanthropic species such as the book louse *Liposcelis simulans* and the dust louse *Trogium pulsatorium.*

At rest winged species carry their wings sloping roof-like over their backs (Fig. 132 A). Also noteworthy is an anchoring of the fore- and hindwings during flight. A groove-like structure of the forewing fits over the facing curved part of the hindwing (Fig. 132 E). Although this is a very characteristic binding formation, it can in no way be an autapomorphy of the "Psocoptera". The following interpretation is much more likely. Just as wingless species originated in various subtaxa of the "Psocoptera", the Phthiraptera can be traced back to a stem species in which wings with such a binding structure were reduced. In other words, such wings very probably belong as a derived feature in the ground pattern of the Psocodea.

Phthiraptera

Lice are ectoparasites. Their biotope is the plumage of birds or the coats of mammals. Numerous apomorphic peculiarities evolved in the stem lineage of the Phthiraptera connected with parasitism. Through these the Phthiraptera are clearly established as a monophylum.

Autapomorphies (Fig. 130 → 5)

- Ectoparasite on a homoiothermic vertebrate.
- Unification of the metathoracic ganglion and abdominal ganglion to a single ventral ganglion. End point of a step-wise concentration of the ventral ganglion in the stem lineages of the Paraneoptera and Acercaria.

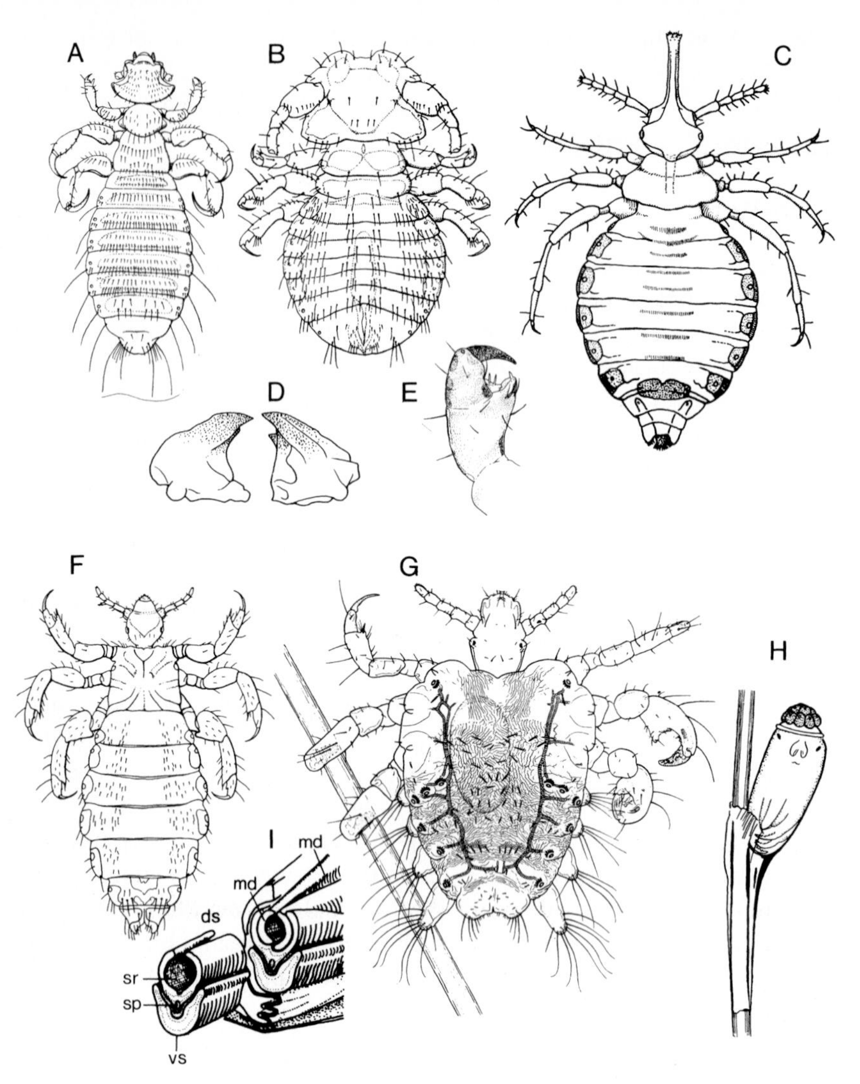

Fig. 133. Phthiraptera. **A** *Gyropus ovalis* (Amblycera). **B** *Trichodectes melis* ("Ischnocera"). **C** *Haematomyzus elephantis* (Rhynchophthirina). **D** *Laemobothrion* (Amblycera). Mandibles. **E** *Haematopinus asini* (Anoplura). Distal part of the anterior appendage. Clasping apparatus formed from the tarsus and a process of the tibia. **F** *Pediculus humanus*, body louse (Anoplura). **G** *Phthirus pubis*, pubic louse (Anoplura). (**F** and **G** show the fusion of the thoracic segments in the Anoplura.) **H** *Phthirus pubis* (Anoplura). Egg with the pointed pole of the egg cemented onto a hair: lid region free. **I** *Haematopinus suis*, hog louse (Anoplura). Distal region of the mouthparts. Sections through the piercing apparatus with a dorsal stylet (*ds*) [=sucking pipe (*sr*)+salivary pipe (*sp*)] and a ventral stylet (*vs*) (=piercing stylet). In the sucking pipe the tips of the mandibles (*md*) push in from behind and form a overflow furrow for the blood. (**A–C** Séguy 1951 a; **D, E** Calaby & Murray 1991; **F** Séguy 1951 b; **G, H** Martini 1952; **I** Ramcke 1965)

- Absence of wings.
- Absence of ocelli.
- Reduction of the compound eyes to two ommatidia (further reduction to one ommatidium and complete loss within the monophylum).
- Appendages with hook structures for anchoring between feathers and in fur or hair (Fig. 133 E).
- Reduction of the spiracles of the tracheal system in the metathorax and in abdominal segments 1 and 2. From the ground pattern of the Pterygota (two thoracic and eight abdominal spiracle pairs) one pair of spiracles in the mesothorax and six pairs on abdominal segments 3–8 remain.
- Deposition of the eggs on the host. Attachment with cement-like substance onto feathers or hairs (Fig. 133 H).
- Development of an egg cover (operculum).
- Ontogeny with reduction to three larval stages.

If the **Liposcelidae** are the sister group of the Phthiraptera then the common apomorphies mentioned earlier (p. 303) must be removed from the list of the Phthiraptera autapomorphies.

The Phthiraptera are traditionally divided into two equal ranking unities – the "Mallophaga" (biting-chewing lice) and the Anoplura (sucking lice). In insects biting is carried out by the mandibles – and so the common name of biting lice for the "Mallophaga" should give one pause for thought; it refers to a basic plesiomorphy, the manifestation of the mandibles as biting and chewing structures. Correspondingly, primitive in comparison with the Anoplura is the type of feeding; the biting-chewing lice feed mainly on keratin from feathers, hair and skin scales as well as the secretion from skin glands.

No autapomorphies exist to define a taxon Mallophaga. We have once again the common case of a paraphylum ("Mallophaga") placed opposite a monophylum (Anoplura). Moreover it has been possible to totally eliminate the "Mallophaga" – an instructive example of the strength and transparency of phylogenetic systematics. For this reason, I include this example here, even though the comprehensive arguments of LYAL (1985) can only be briefly presented. We describe first the four subtaxa of the Phthiraptera of which the first two (sometimes three) are traditionally grouped together as the "Mallophaga".

Amblycera (Fig. 133 A, D)

Autapomorphies: Pedicellus of the antenna swollen into a club shape. At rest the antenna is safely lodged in a pit of the head capsule.

Example: the poultry louse *Menopon gallinae*.

"Ischnocera" (Fig. 133 B)

No autapomorphies are known for certain.
Example: The dog louse *Trichodectes canis*.

Rhynchophthirina (Fig. 133 C)

Autapomorphies: Elongation of the head in a rostrum. Mandibles terminal, turned 180°. *Haematomyzus* with two species – on *Elephas* (both elephant species) and *Phacochoerus* (wart hog).

Anoplura (Fig. 133 F–I)

Autapomorphies: Correlated to the uptake of mammalian blood, mouthparts unique within the insects are realized. Two stylets lie in the head capsule on top of one another. The dorsal sucking and salivary tube is an extension of the hypopharynx. The ventral piercing seta originates from the prementum of the labium. Projecting into the sucking tube are apical processes of the mandibles which act as a transfer channel by means of which the blood is extracted. Fusion of the pronotum and meso-metanotum. Terga of mesothorax and metathorax very small, so that the pleura are elongated dorsalwards.

Examples: *Haematopinus suis* on wild and domestic pigs. *Pediculus capitis* (head louse). *Pediculus humanus* (body louse) and *Phthirus pubis* (pubic louse) on humans.

As a next stage, it is necessary to postulate step-by-step more comprehensive kinship groupings between the four taxa. LYAL (1985) favors two hypotheses that lead to the following systematization.

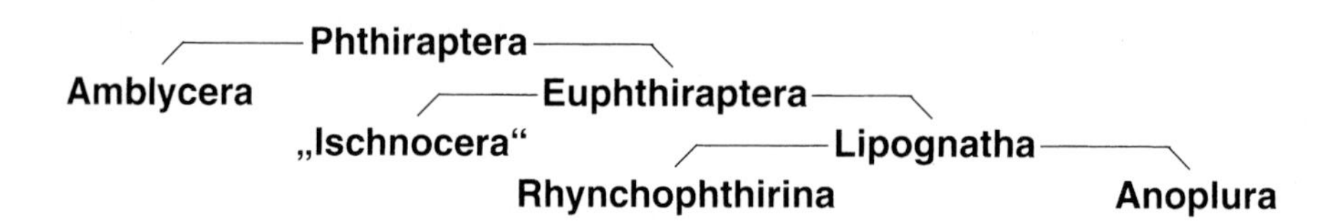

Lipognatha

Rhynchophthirina and Anoplura are adelphotaxa.

Choice of possible **autapomorphies:** Prognathy with a firm connection between the head and thorax. Loss of articulation between the prothorax and coxae of the prothorax appendages. Absence of lacinia glands. Tarsus with only one claw.

Euphthiraptera

"Ischnocera" and Lipognatha form together a monophylum.

Choice of possible **autapomorphies:** Antenna with disc-shaped sensorium on both the last articles. Only one ommatidium (complete loss occurs within the unity). Fusion of meso- and metanotum. Reduction of testes follicles to two pairs. Possession of a spiraculum gland.

Despite the advanced kinship analyses a gap remains. The "Ischnocera" appear at the present state of investigation neither justifiable as a monophylum nor are they removable. So at present it cannot be decided as to whether all or perhaps only some "Ischnocera" are the sister group of the Lipognatha.

With this last point we come back to the start. For the assessment of the Phthiraptera as a whole there are no problems. Their monophyly is assured through the large number of derived peculiarities mentioned at the beginning.

Condylognatha

Autapomorphies (Fig. 130 → 6)

- Continued development of the mouthparts into a piercing-sucking apparatus. As well as the lacinia from the ground pattern of the Acercaria the mandibles are now also differentiated into piercing setae (Fig. 134B, D).
- Mandibular stylet with only one condyle.

Thysanoptera – Hemiptera

Thysanoptera

Autapomorphies (Fig. 130 → 7)

- Asymmetric mouthparts. Right mandible reduced to a small sclerite.
- Wings transformed into narrow bands the edges of which have long setae (fringes). Wing venation greatly simplified (Fig. 134A).
- Reduction in the number of tarsal articles to two.
- Pretarsus with protrusible adhesive arolium.
- Claws in the adult atrophied.
- Only two pairs of spiracles present in the abdomen. One pair on the first and the other on the eighth abdominal segment.
- Last larval stage inactive and with extensive internal reorganization (convergence to the Holometabola).

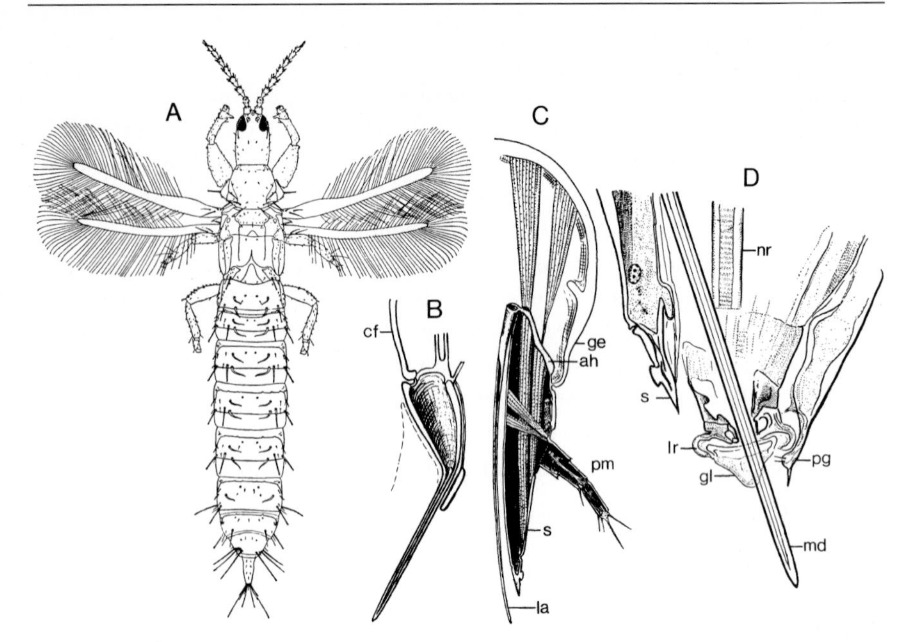

Fig. 134. Thysanoptera. **A** *Podothrips graminum.* Female. **B–D** Mouthparts of *Thrips physapus.* **B** Left mandible in longitudinal section. **C** Left maxilla. Reconstruction of a frontal section series with a dorsal view in the stipes. **D** Distal part of the oral cone showing the tip of the stipes and tip of the mandible. *ah* Articulation lever of the lacinia; *cf* clypeofrons; *ge* gena; *gl* glossa; *la* lacinia; *lr* labrum; *md* mandible; *nr* nutrient furrow of the hypopharynx; *pg* paraglossa; *pm* palpus maxillaris; *s* stipes. (**A** Priesner 1968; **B–D** Risler 1957)

Of the numerous characteristic features that firmly establish the monophyly of the Thysanoptera (thrips), we emphasize two phenomenon – wings and mouthparts.

The habitus is characterized by the narrow wings with their thick fringes of long setae. The setae of the forewings can unite with the curved setae of the hindwings to form one functional unit.

The mouthparts are joined to form a movable cone on the ventral side of the head (Fig. 134B–D). The front is formed by the labrum, the sides by the stipites with palps and the back wall by the labium. The left mandible consists of a pyramid-like basal part and a long hollow spine; this projects terminally out of the oral cone. The right mandible is reduced to a small basal sclerite. The paired maxillae with cardo, stipes and lacinia are also piercing setae. The left mandible and the maxillary stylets lie on the sides of the hypopharynx.

In Thysanoptera, which feed on plant sap and pollen, the mandibular stylet bores a hole through which the maxillary stylet enters the plant.

Together the stylets form a pipe through which a cibarial pump sucks up the nutrients (MOUND & HEMING 1991).

The majority of thrips only reach a length of a few mm. Common thrips in central Europe are species from the taxa *Aeolothrips*, *Thrips*, *Taeniothrips*, *Megathrips*.

Hemiptera

Judged by species numbers alone the Hemiptera (Rhynchota) is one of the great "successful" monophyla of the insects. Compared with the less than 1000 species of thrips, evolution in the sister group, the bugs, has resulted in over 75,000 species.

Autapomorphies (Fig. 130 → 8)

- The maxillae (laciniae) are folded into one another and form a double tube (Fig. 135 B).
- The labium is differentiated into a rostrum of four articles. The tubes enclose the stylet apparatus of the two maxillae and mandibles (Fig. 135 A, C).
- Maxillary and labial palps reduced.
- Anal field of the forewings separated from the rest of the wing as a "clavus".

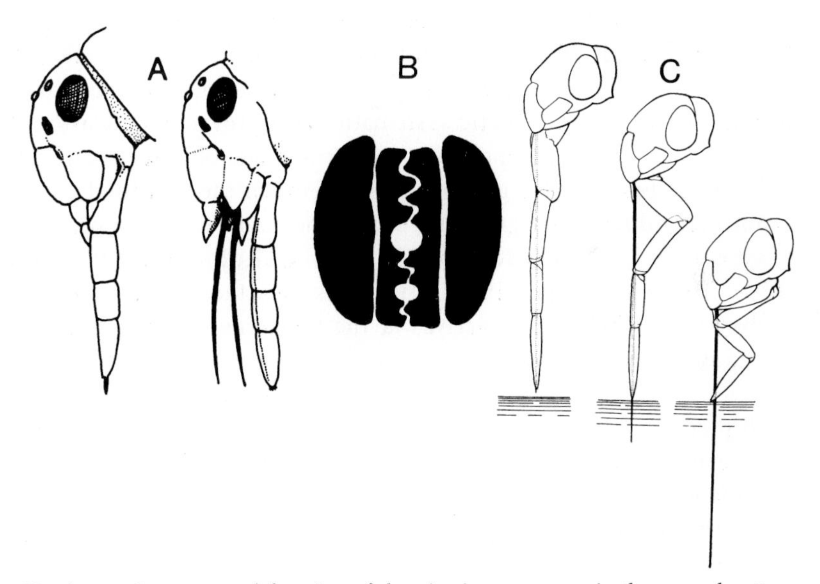

Fig. 135. Hemiptera. Structure and function of the piercing apparatus in the ground pattern of the bugs. **A** Head in lateral view. *Left* Piercing setae (stylets) surrounded by tube-shaped labium (rostrum). *Right* Piercing setae taken out of the labium through the anterior gap and separated; only two of the four stylets shown. **B** Cross section through the piercing setae bundle. The mandibles lie laterally. In the middle, the laciniae (maxillae) are folded together with one another; between them a nutrient canal (*above*) and a salivary passage (*below*) remain open. **C** Piercing mechanism. The labium with the enclosed setae is set on an object of prey. In piercing the labium bends at the joints and springs back (Miridae, Heteroptera). (**A** Weber 1933; **B** Pesson 1951; **C** Poisson 1951)

If laciniae as piercing setae belong in the ground pattern of the Acercaria, and furthermore mandibles as piercing setae in the ground pattern of the Condylognatha, then logically these stylets cannot be autapomorphies of the Hemiptera. On the contrary, the existence of two mandibles is plesiomorphic compared with the lack of the right mandible in the sister group Thysanoptera. Certainly, however, the **specific construction of the laciniae** is interpretable as an autapomorphy of the Hemiptera. The interiors of the laciniae have two semicircular indentations. Because of their close proximity and the folding of the laciniae, two separate pipes are created which serve to introduce saliva into the puncture wound (salivary canal) and to extract fluids from the prey (nutrient canal).

A further derived feature is the construction of the labium as a pipe through reduction of the palps. The labium together with the four piercing setae is placed on the object of prey but does not enter the wound made by them.

Through the influence of HENNIG (1969, 1981) the traditional typological division of the bugs into the "Homoptera" (Paraphylum) and Heteroptera (monophylum) has finally been discontinued. A division into three subtaxa has found widespread acceptance – the Sternorrhyncha (phytophagous lice), Auchenorrhyncha (cicadas) and the Heteropteroidea (true bugs). However, this systematization also has its weaknesses: (1) the monophyly of a unity Auchenorrhyncha was difficult to justify, and (2) the "equally ranked" arrangement of the three subtaxa next to one another can only be countenanced in phylogenetic systematics as a provisional measure. It must be replaced by precise hypotheses concerning adelphotaxa relationships.

Indeed, lately, a sister group kinship between the Sternorrhyncha and a taxon Euhemiptera as a grouping of all other bugs has been proposed from various sides (WOOTTON & BETTS 1986; CARVER et al. 1991; MAHNER 1993). The latter author in addition dissolves the Auchenorrhyncha as a probable paraphylum and arrives at the following phylogenetic **systematization of the Hemiptera.**

Hemiptera
- **Sternorrhyncha**
- **Euhemiptera**
  - **Cicadomorpha**
  - **N. N.**
    - **Fulgoromorpha**
    - **Heteropteroidea**
      - **Coleorrhyncha**
      - **Heteroptera**
        - **Gymnocerata**
        - **Cryptocerata**

Sternorrhyncha – Euhemiptera

Sternorrhyncha

Autapomorphies (Fig. 130 → 9)

- Base of the proboscis shifted back to between the front coxae
- Reduction in the number of tarsal articles on all appendages to two.
- Basal sections of the wing veins radius, media and cubitus fused (Fig. 136C).
- Embryo with frontal carina to break egg-shell.

The Sternorrhyncha comprises the sister groups Aphidomorpha and Psyllomorpha which themselves can be subdivided into adelphotaxa (SCHLEE 1969a, b; HENNIG 1969, 1981, 1994). Due to space restrictions, explanations are kept to a minimum in the following overview.

Aphidomorpha – Psyllomorpha

Aphidomorpha

Autapomorphies

- Stems of the radius, media and cubitus widely fused; their branches originate from the radius like the teeth of a comb.
- Egg-laying apparatus absent; the eggs are freely deposited.

Aphidina (aphids). **Autapomorphies:** Alternation of hosts and generations (Fig. 136A). No Malpighian tubules.

Coccina (scale insects). **Autapomorphies:** Sexual dimorphism. Female wingless (Fig. 136B). Single-segmented tarsus with one claw.

Psyllomorpha

Autapomorphies

- Sperm pump at the entrance of the ductus ejaculatorius.
- Abdomen stalk-like displaced.
- Coxae of the posterior legs are broadened and lie close together.

Psyllina (jumping plant lice) (Fig. 136C). **Autapomorphies:** Coupling of the fore- and hindwing in flight. Coxae of the posterior legs (jumping legs) fused with the thorax.

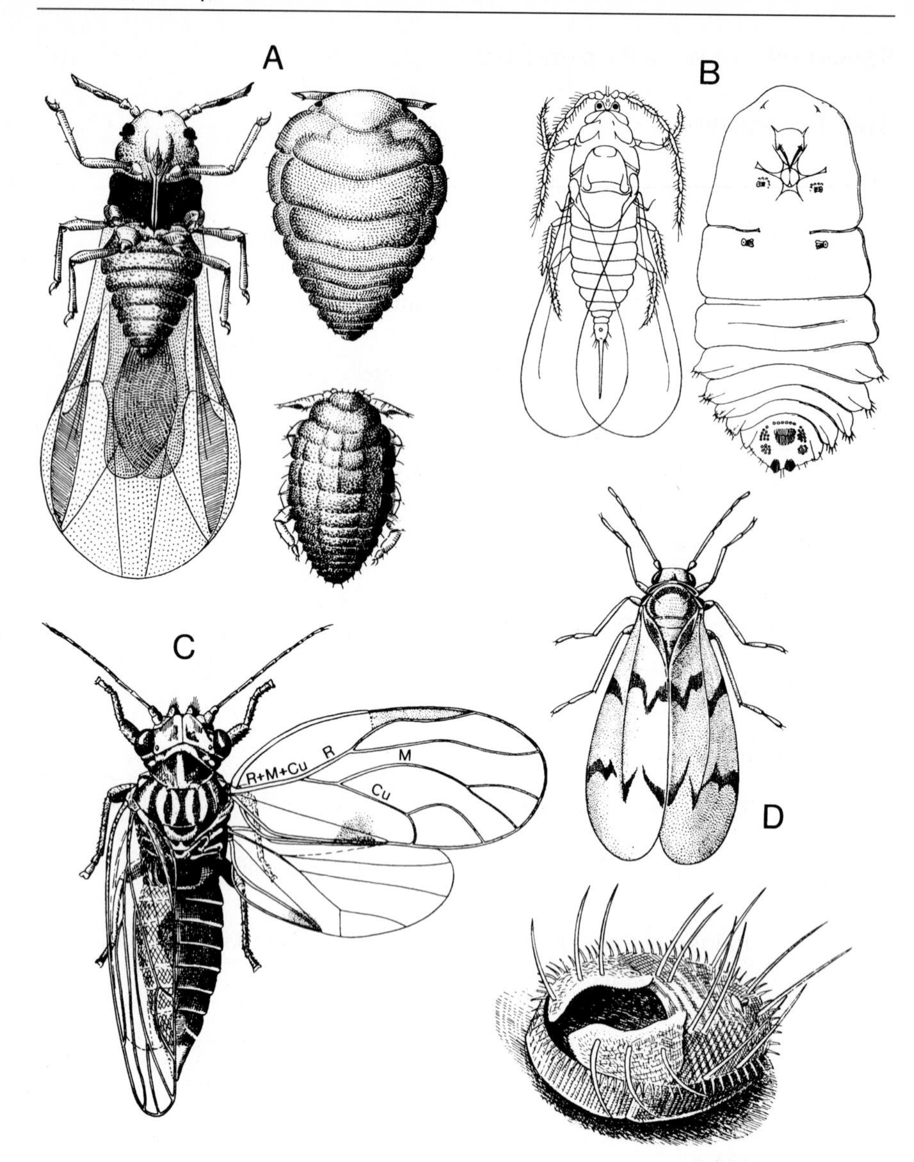

Fig. 136. Sternorrhyncha. **A** Aphidina. *Viteus vitifolii* with various individuals (morphs) from the "alternation of generations". *Left*: Winged sexupara, female. *Top right*: Wingless fundatrix, female. *Bottom right*: Root inhabiting individual (wintering root gall louse). **B** Coccina. *Mytilaspis fulva*. *Left*: Winged male. *Right*: Wingless female. Antennae and appendages reduced. **C** Psyllina. *Psylla pyricola*, female. In the forewing basal fusion of radius, media and cubitus visible (autapomorphy of the Sternorrhyncha). **D** Aleyrodina. *Aleyrodes abutilonea*. Female. *Bottom*: *Trialeurodes vaporariorum*. Puparium (cuticle) of the immobile fourth larval stage from which the imago hatches. *Cu* Cubitus; *M* media; *R* radius. (Pesson 1951)

Aleyrodina (white flies) (Fig. 136D). **Autapomorphies:** Wax glands ventrally on the basis of the abdomen. Anterior ocellus reduced. Abdominal spiracles only on segments 2–8. Only two Malpighian tubules (as opposed to four in the Psyllina).

Euhemiptera

Autapomorphies (Fig. 130 → 10)

- Posterior lobes of the pronotum partly cover the anterior margin of the mesonotum.
- Lateral mesoscutellum process connected to the mesosubulare (LARSEN 1945).

Within the confines of the Euhemiptera the hypothesis of the monophyly of a unity Auchenorrhyncha (Cicadomorpha + Fulgoromorpha) can still exist. In support of the hypothesis is the apomorphous congruence of a short bristle-like flagellum of the antennae. Given this it would be possible to regard the Auchenorrhyncha and the Heteropteroidea as sister groups (CARVER et al. 1991; WHEELER et al. 1993). Against this hypothesis, however, is the fact that the common differentiation of the penis into phallosoma and endosoma occurs only in the Fulgoromorpha and Heteropteroidea (Fig. 137E). MAHNER (1993) judged this to be a synapomorphy of these two taxa; we give this very marked and complex congruence special weight which leads to the assumption of the following sister group relationship.

Cicadomorpha – N. N. (Fulgoromorpha + Heteropteroidea)

Cicadomorpha

Autapomorphy

- Manifestation of a marginal vein in the fore- and hindwings which runs around the whole edge of the wing (CARVER et al. 1991).

After the Ensifera and the Caelifera the Cicadidae (cicadas) (Fig. 137A) is the third insect taxon with species in which sound-producing and auditory organs appear together in one individual.

Sound production is caused by a pair of tymbals (♂) on the sides of the tergum of the first abdominal segment. These are discs which through muscle action are dented inwards and then rebound to the starting position in rapid succession. Ventral to the tymbals in the male (and in the

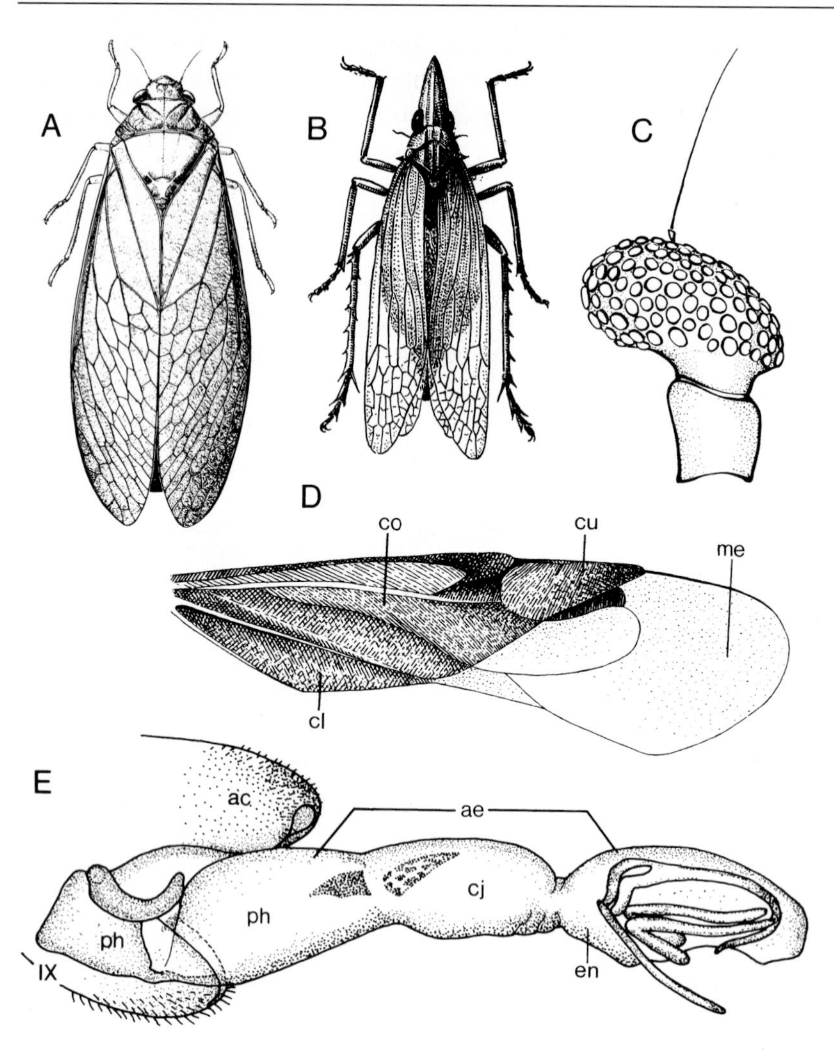

Fig. 137. Cicadomorpha, Fulgoromorpha, Heteroptera. **A** *Cystosoma saundersii* (Cicadomorpha, Cicadidae). **B** *Dictyophara europaea* (Fulgoromorpha). Frontal process between the eyes. **C** *Desudaba psittacus* (Fulgoromorpha). Antenna with large, club-shaped pedicellus and short anellus. **D** *Calocoris striatellus* (Heteroptera). Hemielytra. **E** *Trepobates pictus* (Heteroptera, Gerridae). Copulatory organ with division into phallobasis (phallosoma) and endosoma; the latter bears sclerites. *ac* Anal cone; *ae* aedeagus; *cj* conjunctiva; *cl* clavus; *co* corium; *cu* cuneus; *en* endosoma; *me* membrane; *ph* phallobasis; *IX* abdominal segment 9. (**A,C** Carver et al. 1991; **B** Pesson 1951; **D** Günther 1969b; **E** Poisson 1951)

corresponding position in the female) there is a pair of tympana with tympanic membranes. Paired sound-producing organs are widespread among the Cicadomorpha. Unpaired organs are found in the Fulgoromorpha dorsally on the first and second abdominal segments.

Within the Cicadomorpha tympanic organs are limited to the Cicadidae. They appear, however, again convergently in the Hemiptera, on the mesothorax, metathorax and abdominal segment 1 of the Cryptocerata (aquatic bugs).

A comparative analysis of the tympanic and sound-producing organs of the Hemiptera for the purposes of phylogenetic systematics is urgently needed.

N. N.: Fulgoromorpha + Heteropteroidea

Autapomorphies

- Differentiation of the penis into two sections – phallosoma and endosoma (Fig. 137E).
- Features of the forewing venation such as the apical fusion of the two anal veins to form a Y vein.

Fulgoromorpha

Autapomorphies

- Coxae of the metathorax appendages immobile.
- Antennae lie beneath the compound eyes.
- Lateral ocelli adjacent to the antennae and compound eyes.
- Pedicellus of the antennae greatly enlarged; equipped with closely packed sensory structures (Fig. 137C).

Well-known species of the Fulgoromorpha are the large tropical lantern flies (*Fulgora*, *Laternaria*) in which the head extends as a long process between the eyes. Also found in Europe (Fig. 137B).

Heteropteroidea

Since the fundamental research by SCHLEE (1969c) the union of the Coleorrhyncha and Heteroptera with a adelphotaxa kinship is generally acknowledged.

Autapomorphies

- Head erect (prognathous); rostrum and piercing stylets shifted forwards.
- From the circular cross section of the abdomen in the ground pattern of the Hemiptera the tergites have been reshaped into flat elements, which gives the characteristic plate-like habitus of the true bugs. Additional sclerites link the tergites and sternites together to a lateralwards overhanging edge. The spiracles point downwards.

- The forewings lie flat on the flattened body.
- "One piece" anal cone; formed from two fused segments.
- Horse-shoe-shaped sclerite at the basis of the aedeagus (copulatory organ).
- Reduction in the number of antennal articles to four.

Coleorrhyncha - Heteroptera

Coleorrhyncha

Autapomorphies

- Antenna inserted hidden between the eyes and oral opening (Fig. 138 A). Propleural antennal sheath.
- Structure of the frontal lobes (SCHLEE 1969 c).
- Only two tarsal articles.

Small taxon (25 species) in the southern hemisphere. Body length only 2–5 mm. Secretive species in moist moss of the cooler rain forests; found especially in *Nothofagus* forests.

Compared with the adelphotaxon Heteroptera the absence of a gula is a plesiomorphic feature.

Peloridium hammoniorum is the only species with flight capability. In the adult there are macroptere and submacroptere individuals; the former have hindwings and two ocelli. All other Coleorrhyncha species cannot fly; hindwings and ocelli are secondarily absent.

Heteroptera

Autapomorphies

- Formation of stink glands. In the imago they open over the coxae on the metathorax; in the larva on abdominal tergites 4–6.
- Prognathous head with a sclerotized gular region.
- Hemielytra. Forewings differentiated into a strongly sclerotized, leathery corium and a distal skin-like membrane (Fig. 137 D). Disputed as an autapomorphy of the Heteroptera because the feature is absent in the subtaxon Enicocephalomorpha (Fig. 138 B); this could, however, be a secondary state (MAHNER 1993).

The prolonged debate concerning the phylogenetic systematization of the bugs culminates surprisingly with the following kinship hypothesis of MAHNER (1993). The conventional taxa from the last century Gymnocera-

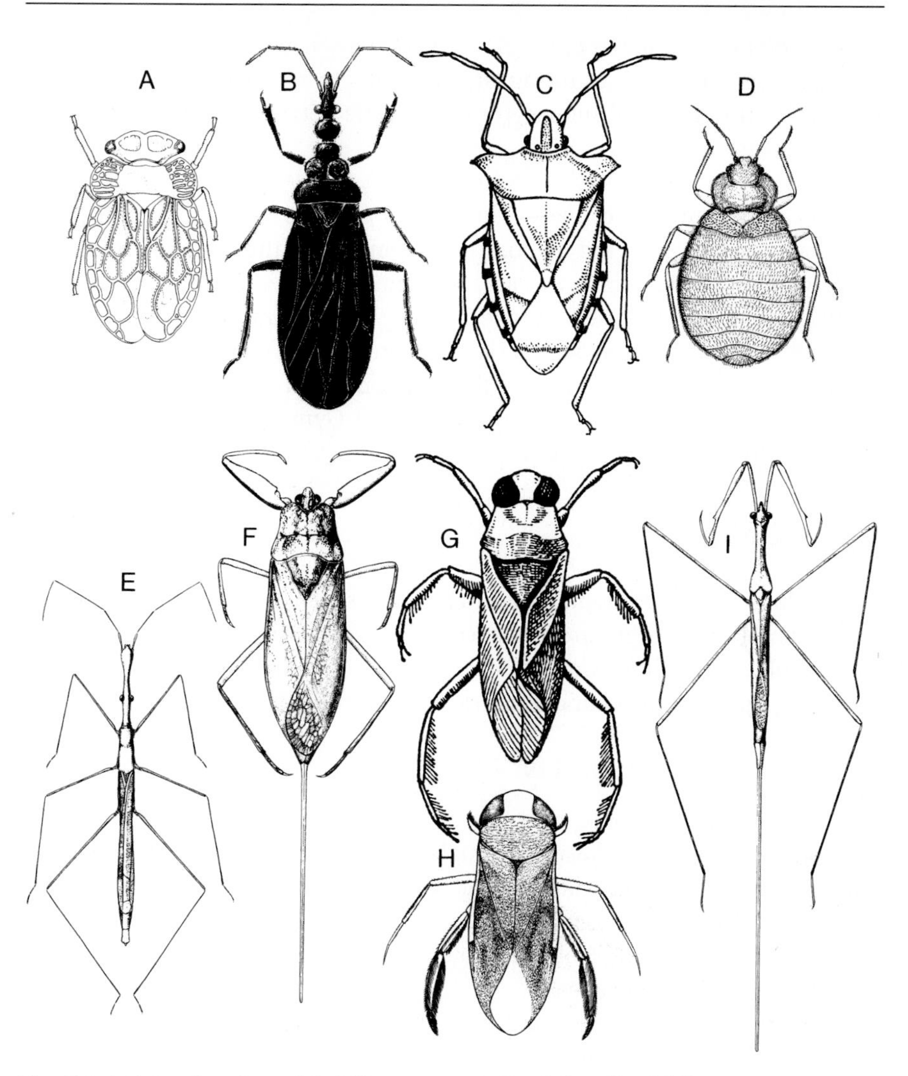

Fig. 138. **A** Coleorrhyncha and **B–I** Heteroptera. **A** *Hemiodoecellus fidelis.* **B–E** Gymnocerata (terrestrial bugs). **B** *Oncylocotis* (Enicocephalidae). Wings not developed as hemielytra. **C** *Pentatoma rufipes* (Pentatomidae). **D** Bed bug *Cimex lectularius* (Cimicidae). Wingless species. **E** *Hydrometra strigosa* (Hydrometridae). **F–I** Cryptocerata (aquatic bugs). **F** *Laccotrephes tristis* (Nepidae). **G** *Notonecta glauca* (Notonectidae). **H** *Agraptocorixa eurynome* (Corixidae). *Ranatra dispar* (Nepidae). (**A, B, D–F, H, I** Carver, Gross & Woodward 1991; **C, G** Eidmann & Kühlhorn 1970)

ta (terrestrial bugs) and Cryptocerata (aquatic bugs) are monophyla; they become adelphotaxa. The extensive justification for this position can only be briefly given here.

Gymnocerata. Autapomorphies: Shortened heart with few ostia only in abdominal segments 5–8. Limiting of the ileum to the pylorus region. Corpora allata fused. Paired intersegmental sclerite between the two last antennal segments.

Cryptocerata. Autapomorphies: Three pairs of tympanic organs in the mesothorax, metathorax and abdominal segment 1 (the only insect taxon with tympanic organs in serial arrangement). Subesophageal ganglion and prothoracic ganglion fused. Aquatic life style with air-bubble respiration. Spiracles 1–3 with sieve plates (spiracle membrane); no spiracle closure apparatus. Very small antennae with insertion under or behind the eyes.

Examples of some species from the subtaxa Gymnocerata and Cryptocerata are depicted in Fig. 138 B–I.

Holometabola

The Holometabola (Endopterygota) form the sister group of the Paraneoptera. They include over three-quarters of all recent insect species. The Holometabola are without doubt established as a monophylum through the following apomorphic characteristics.

Autapomorphies (Figs. 119 → 11; 139 → 1)

- Holometabolous development.
  "Complete metamorphosis" through formation of a pupa as a resting stage between the last larval stage and the imago. The pupa does not feed; furthermore, immobility of the pupa can be postulated for the ground pattern of the Holometabola.
- Endopterygote stem species.
  Development of the wings and genital appendages beneath the body wall in sunken cavities (imaginal discs). Evagination first occurs at the penultimate molt during pupation. The pupa is the first stage with prominent external wing structures.
- Degeneration of the larval stemmata and new formation of the imaginal compound eyes.
  The lateral eyes (stemmata) of the Holometabola larva are compound eyes modified to a greater or lesser degree (PAULUS 1979); they degenerate during pupation. New compound eyes develop in the pupa from independent imaginal discs.

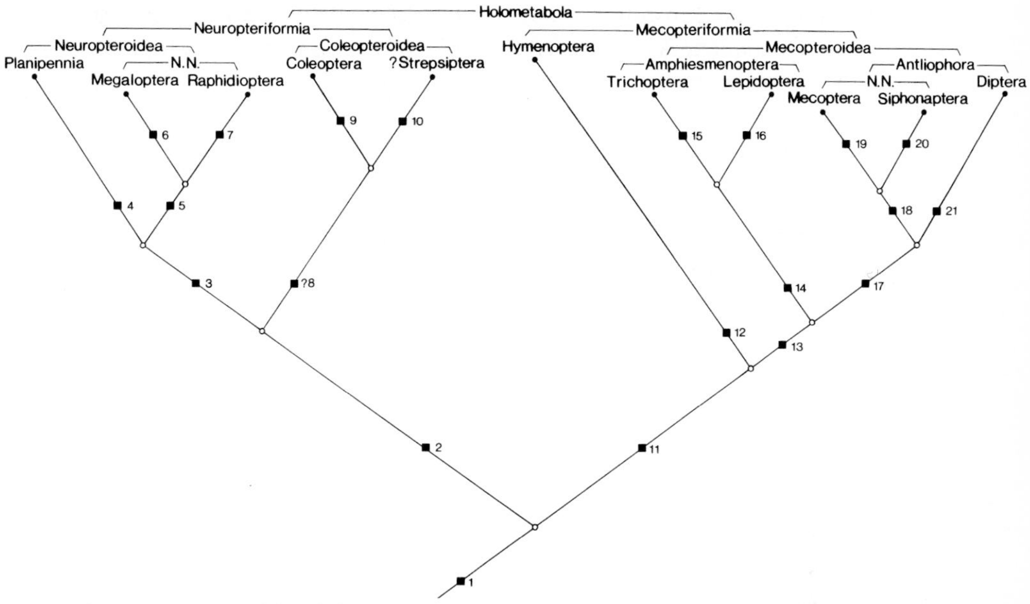

Fig. 139. Diagram of the phylogenetic relationships of the Holometabola with the Neuropteriformia and Mecopteriformia as sister groups. Section 5 of the systematization of the Insecta

Systematization: Section 5

Holometabola
- **Neuropteriformia**
  - **Neuropteroidea**
    - **Planipennia**
    - **N. N.**
      - **Megaloptera**
      - **Raphidioptera**
  - **Coleopteroidea**
    - **Coleoptera**
    - **? Strepsiptera**
- **Mecopteriformia**
  - **Hymenoptera**
  - **Mecopteroidea**
    - **Amphiesmenoptera**
      - **Trichoptera**
      - **Lepidoptera**
    - **Antliophora**
      - **N. N.**
        - **Mecoptera**
        - **Siphonaptera**
      - **Diptera**

An analysis of the kinship relations of the Holometabola by means of ribosomal DNA sequences (WHITING et al. 1997) results in a systematization with extensive similarities to the one above. An essential difference concerns the disputed Strepsiptera problem (N.P. KRISTENSEN 1981). The Strepsiptera are not seen as adelphotaxa of the Coleoptera, but interpreted as the sister group of the Diptera.

Neuropteriformia – Mecopteriformia

The hitherto unnamed unities **Neuropteriformia** (Neuropteroidea + Coleopteroidea) and **Mecopteriformia** (Hymenoptera + Mecopteroidea) are widely discussed as being adelphotaxa of the Holometabola. A convincing autapomorphy of the Neuropteriformia is the transformation of the gonostyli of the female genital organs into vaginal palps with a sensory function. On the other hand, the production of the pupal cocoon from the secretion of modified salivary glands (labial glands) is seen as a derived feature of the Mecopteriformia.

Neuropteriformia[27]

Autapomorphies (Fig. 139 → 2)

- Gula.
  Development of a median sclerite plate on the underside of the head. Interpolated between the occipital foramen and the base of the labium.
- Ovipositor with vaginal palps.
  In the ground pattern of the Holometabola there is a ovipositor formed from the gonapophyses of abdominal segments 8 and 9. The ovipositor is flanked by the long gonostyli of segment 9 (Fig. 140 A). In the Neuropteriformia gonapophyses 8 are shortened and fused, gonapophyses 9 reduced and gonostyli 9 transformed into vaginal palps with a sensory function (Fig. 140 B; MICKOLEIT 1973).
- Cerci.
  Wide-ranging reduction; no jointed insertion. The phenomenon is probably connected with the evolution of the vaginal palps.

[27] tax. nov.

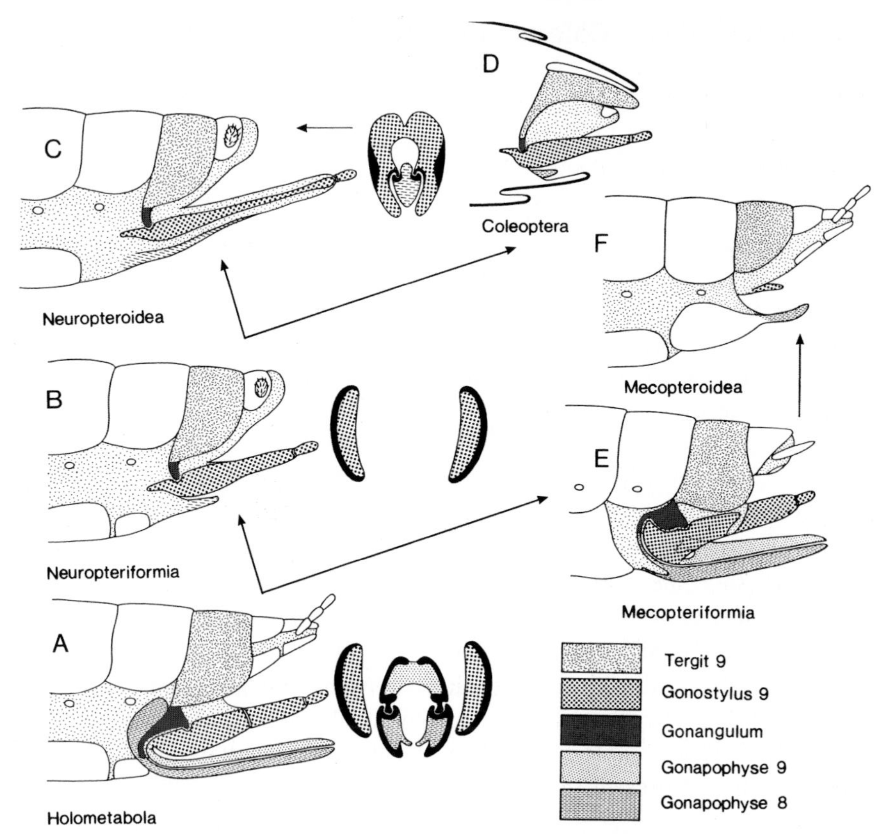

Fig. 140. Evolution of the ovipositor within the Holometabola. **A** Ground pattern Holometabola (*left* side view; *right* cross section). The ovipositor is formed from the paired gonapophyses of abdominal segments 8 and 9, which are connected to each other. The gonapophyses originate with basal gonocoxites from the sternal region of the two segments. Gonocoxites 8 have a jointed connection with the tergum of segment 9 over the gonangulum – the latter being a triangular sclerite plate. Gonocoxites 9 bear long gonostyli which flank the ovipositor laterally. **B–D** Neuropteriformia. **B** Ground pattern Neuropteriformia. Gonapophyses 8 are greatly shortened and fused. Gonapophyses 9 are completely reduced. Gonostyli of the ninth segment are transformed into vaginal palps (sensory palps). The cerci are greatly reduced; they have no joints. These ground pattern elements are realized in the Coleoptera. **C** Ground pattern Neuropteroidea. Evolution of a new ovipositor in the stem lineage of the unity. Gonostyli 9 (vaginal palps) unite dorsally and form together a half-pipe. The floor of the pipe originates from the fused gonapophyses 8. The long ovipositor of the Raphidioptera represents this ground pattern. In the Megaloptera, and also within the Planipennia, the morphologically identical egg-laying apparatus is shortened. **D** Ground pattern Coleopteroidea. With the two free gonostyli 9 (vaginal palps) the Coleoptera remain more primitive than the Neuropteroidea. In contrast the retraction of the end of the body into abdominal segment 8 is a distinct apomorphy of the beetles. **E–F** Mecopteriformia. **E** Ground pattern Mecopteriformia. In the stem lineage of this unity the ovipositor underwent no essential modifications. With paired gonapophyses 8 and 9, gonostyli 9 and the jointed gonangulum the plesiomorphic egg-laying apparatus of the Hymenoptera corresponds to the ground pattern of the Holometabola. The transformation into a poison sting first took place within the Hymenoptera in the stem lineage of the Aculeata. **F** Ground pattern of the Mecopteroidea. The whole ovipositor is greatly reduced and only vestigial. The ovipositor disappears completely within the unity. (Mickoleit 1973)

Neuropteroidea – Coleopteroidea

Neuropteroidea

Autapomorphies (Fig. 139 → 3)

- New ovipositor.
  The vaginal palps (gonostyli 9) unite dorsally to form a new ovipositor (Fig. 140 C). The fused gonapophyses 8 form the floor of the tube. Also unique is an inner ovipositor musculature; the new ovipositor is interspersed transversally with short muscle cords.
- Reduction of the paraglossae in the labium of the imago.
- Raptorial feeding in the larva.

In traditional classifications the three taxa Planipennia, Megaloptera and Raphidioptera are all given the same rank as "orders". Within the Holometabola they are distinguished by a number of symplesiomorphies. To these belong the roof-like resting position of the large wings and their rich network of veins, as well as the long multiarticulated antennae and the construction of the tarsi from five articles (Figs. 141, 142).

Let us examine the kinship relations between the three unities of the Neuropteroidea. The Planipennia form the adelphotaxon of an, as yet, unnamed unity N. N., consisting of the Megaloptera and Raphidioptera; this hypothesis has at present the best arguments.

Planipennia – N. N. (Megaloptera + Raphidioptera)

Planipennia

Autapomorphies (larva) (Fig. 139 → 4)

- Sucking out of prey.
  Special mode of predacious feeding, the structural elements of which form the two following apomorphic peculiarities.
- Sucking tube formed from the mandibles and maxillae.
  Long mandibles and laciniae (maxillae) are folded into one another to form a functional unity. A tube for the introduction of enzymes into the prey and the uptake of dissolved nutrients is formed between them (Fig. 141 C).
  A poison canal in the maxillae also belongs in the ground pattern.
- Closure of the mouth.
  The paired suction tubes open laterally in the preoral cavity. The primary oral opening is closed, only recognizable in the form of a slit.

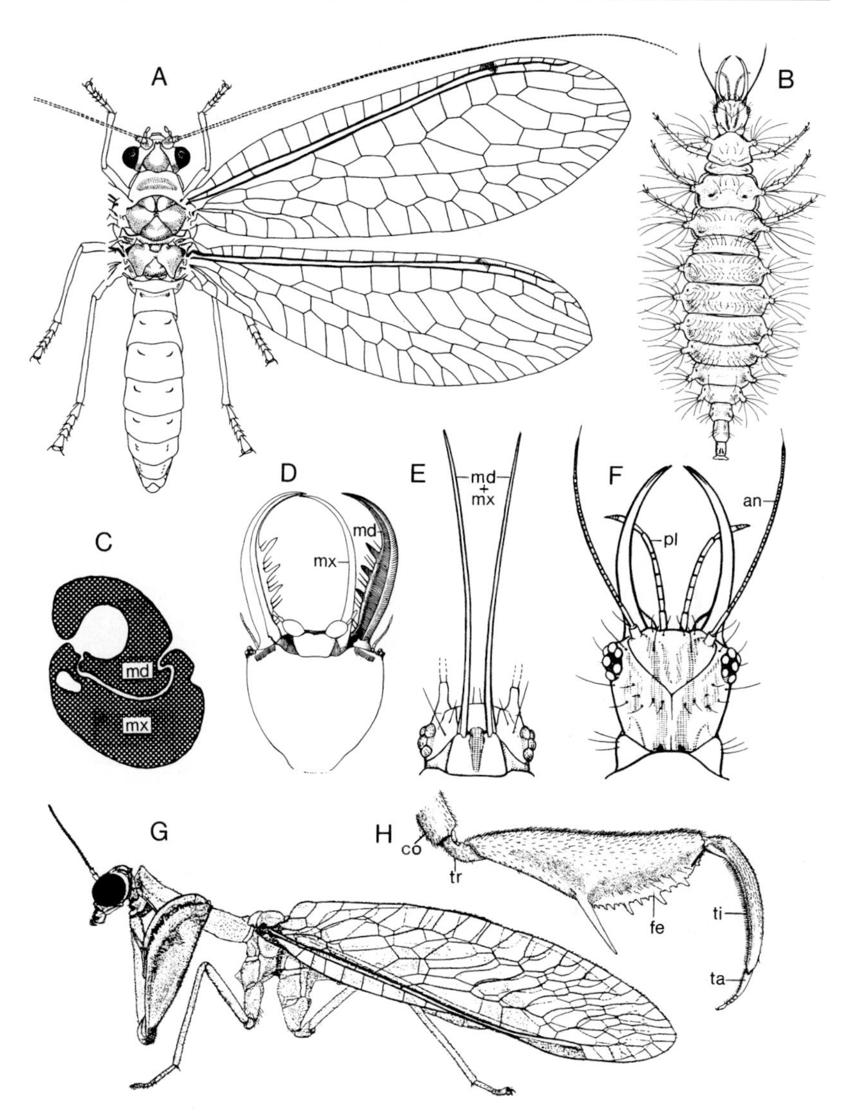

Fig. 141. Planipennia (= Neuroptera). **A** *Chrysopa.* Male. Original features: network of wing veins; long, multiarticulated antennae. Tarsi of five articles. **B** *Chrysopa.* Larva. **C** *Osmylus chrysops.* Larva. Cross section through the connected mandibles and maxilla with feeding canal between the two and poison canal in the maxilla. **D** *Myrmeleon formicarius.* Ventral view of the larval head. The maxilla lies to the right beneath the mandible; it is bent inwards to the left. **E** *Sisyra.* Larval head (ventral). Mandibles+maxillae needle-shaped. **F** *Chrysopa.* Larval head, dorsal. Mandibles lie above the maxillae. **G** *Mantispa styriaca.* Resting position with the wings held roof-like over the body. **H** *Mantispa styriaca.* Left capturing leg from the inner side. *an* Antenna; *co* coxa; *fe* femur; *md* mandible; *mx* maxilla; *pl* palpus labialis; *ta* tarsus; *ti* tibia. (**A, B, E, F** New 1991; **C** Kaestner 1973; **D** Weber 1933; **G, H** Ulrich 1965)

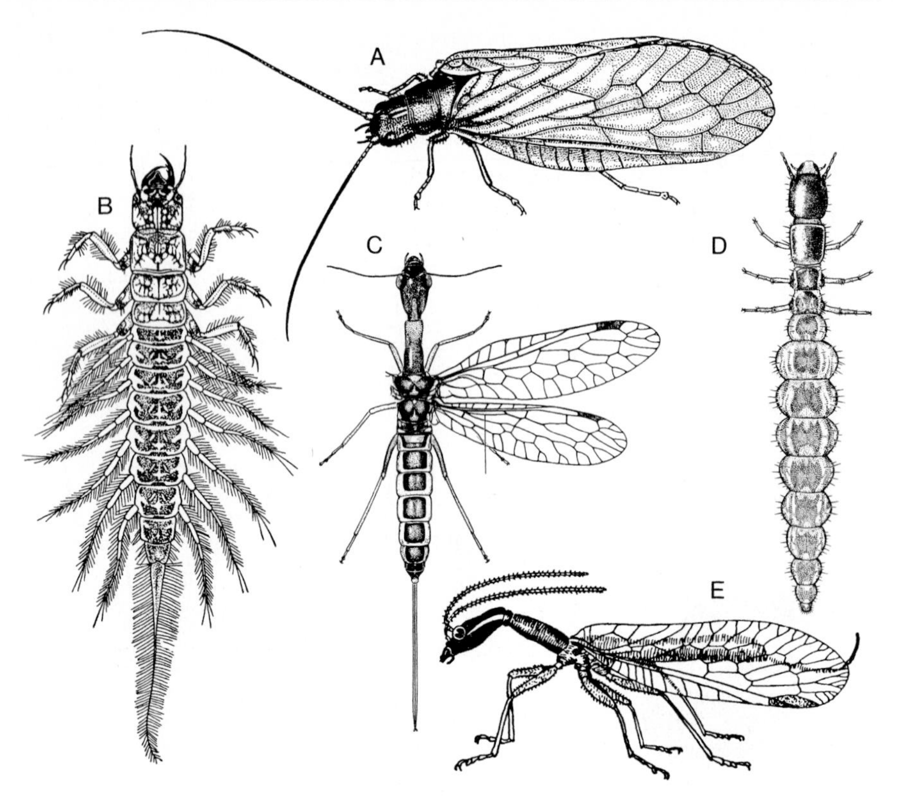

Fig. 142. Megaloptera and Raphidioptera. **A–B** Megaloptera. **A** *Sialis*. Imago. **B** *Sialis*. Aquatic larva with tracheal gills on the abdomen. **C–E** Raphidioptera. **C** *Raphidia palaeformis*. Female. Aerial view. Cylinder-shaped prothorax between head and mesothorax. **D** *Phaeostigma robusta*. Larva. **E** *Raphidia oblita*. Imago in side view with camel-neck profile. (**A, B** Engelhardt 1959; **C, D** Aspöck & Aspöck 1991; **E** Eidmann & Kühlhorn 1970)

- Reduction of the maxillary palps.
- Closure of the midgut.
  Mid- and endgut only connected by a lumenless tissue cord. Excrement cannot be deposited during the larval phase.
- Opening of the majority of the Malpighian tubules into the proctodaeum (rectum).
- Production of cocoon silk from the Malpighian tubules.
  In the last larval stage the Malpighian tubules secrete a silk, which is delivered from the rectum. The end of the abdomen functions as a spool in shaping the pupal cocoon.
- Reduction of the tarsi to one article.

A considerable variability in the structure of the suction tubes is connected with the suction mechanism of the Planipennia larva. As well as medially bent jaws (*Myrmeleon*, *Chrysopa*), there are needle- or skewer-like extended implements (*Osmylus*, *Sisyra*) that can even be bent outwards (Fig. 141 D–F).

This leads us to some remarks about the diverse biology of the Planipennia (Neuroptera). The larvae of the ant lion *Myrmeleon* bury themselves in sandy substrates and lie in wait at the bottom of funnels for their prey. Insects that happen on the steep slope are pelted with sand. While ant lions are not particularly selective, the larvae of the lacewing *Chrysopa* prefer colonies of aphids (Aphidina). *Sisyra* which attaches to sponges documents a totally different specialty. The secondary aquatic larvae suck in the tissues of freshwater sponges; the maxillary poison canal is missing. In the imago of *Mantispa* the anterior appendages have evolved into capturing legs; it thereby forms a striking likeness to the mantes (Mantodea) (Fig. 141 G, H).

N. N.: Megaloptera + Raphidioptera

Autapomorphies (Fig. 139 → 5)

- Modified tergite of the second abdominal segment. Attachment point for a strong intersegmental musculature between tergites 2–5 (ACHTELIG 1981).
- Racemose branching of the ovaries and a high number of ovarioles (ACHTELIG 1981).
- Cleaning behavior.
  Both antennae are cleaned at the same time with the two front legs. In contrast the corresponding mechanism in the adelphotaxon Planipennia is plesiomorphous; here only one antenna at a time is stroked by the two front legs. Within the Holometabola the behavior of the Megaloptera + Raphidioptera is seen again only in the Trichoptera from the Mecopteriformia (convergence) (JANDER 1966).

Megaloptera

Autapomorphy (Fig. 139 → 6)

- Aquatic larvae.
  Long lateral tracheal gills on abdominal segments 1–8 (or 1–7) (Fig. 142 B). Stipes of the maxilla divided.

Species-poor unity of the Neuropteroidea. The larvae of the dobsonflies or alder flies colonize still and flowing fresh waters. As predators they feed on a broad spectrum of invertebrate animals. Pupation takes place on land.

The imago is characterized by two pairs of large, brown to dark gray wings (Fig. 142 A). Imagines are often found near water. Only a limited nutrient uptake. The female lays eggs on plants on the banks or above the water level (*Phragmites*).

Sialis, *Corydalus*.

Raphidioptera

Autapomorphies (Fig. 139 → 7)

- Rostral tapering of the prognathous head capsule.
- Elongation of the prothorax to a cylinder (Fig. 142 C, E).
  The pronotum (dorsal plate) is pulled downwards over the sides; thereby the ventral prosternum is only visible as a narrow slit.
  The anterior appendages are attached far back on the prothorax.
- Third article of the (originally five-articled) tarsus of the imago broadened into a heart-shape.
- Pretarsus without arolium.

Snakeflies or camel-neck flies are only found in the northern hemisphere.

The characteristic apomorphous "camel-neck" profile results from the oblique position of the prothorax in comparison with the horizontally held body and the light downwards bending of the head. Raptorial feeding.

Slim terrestrial larvae with biting mouthparts represent a plesiomorphic state (Fig. 142 D).

Raphidia, *Inocellia*.

Coleopteroidea

We come to the Coleopteroidea as the sister group of the Neuropteroidea and examine first the characteristic complex female genital apparatus.

The egg-laying apparatus of the Coleoptera comprises essential ground pattern features of the Neuropteroidea with the shortening of gonapophyses 8, the reduction of gonapophyses 9 and the evolution of gonostyli 9 to vaginal palps. Compared with the Neuropteroidea these are without doubt plesiomorphic states. In contrast, an autapomorphy of the **Coleoptera** is the construction of the posterior body; the **end of the abdomen is withdrawn into the eighth segment** and in the female is only extended to lay eggs (Fig. 140 D).

This comparison Neuropteroidea – Coleopteroidea is for the latter limited to the Coleoptera. In the Strepsiptera, the possible sister group of the Coleoptera, the egg-laying apparatus is fully reduced.

Autapomorphy (Fig. 139 → 8)

- ? Derived flight mechanism.
  Metathoracic wings alone used for propulsion in flight (p. 331).

Coleoptera – ? Strepsiptera

Coleoptera

In the ground pattern of the beetles the whole surface of the body forms an enclosed **shell**, between the sections of which very few spaces or membranous zones can be found (HENNIG 1969). We present the essential characteristic features with the three autapomorphies which are required for this unique construction of the Coleoptera first.

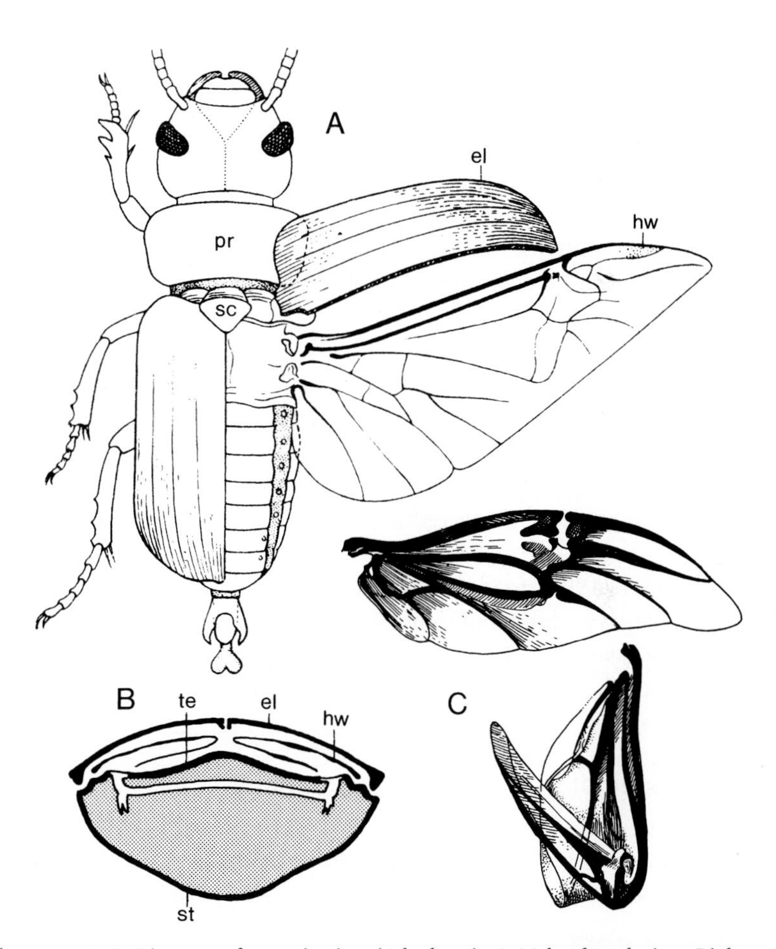

Fig. 143. Coleoptera. **A–B** Diagram of organization (Adephaga). **A** Male, dorsal view. Right wings out-spread; left wings in the resting position. Male copulatory apparatus extended to its maximum. **B** Cross section through the middle of the abdomen. Elytra rest laterally on the margins of the sternum. Folded hindwings between elytra and tergum. **C** *Melolontha melolontha* (cockchafer). Hindwing in outspread state (*top*) and after folding (*bottom*). *el* elytra; *hw* hindwing; *pr* prothorax; *sc* scutellum (derivative of the mesothorax); *st* sternum; *te* tergum. (**A, B** Weber & Weidner 1974; Hennig 1994; **C** Jeannel 1949)

Autapomorphies (Fig. 139 → 9)

- Elytra.
  The forewings are transformed into strongly sclerotized protective covers without veins under which the primary longer hindwings are hidden when at rest (Fig. 143 A, C).
- Protective casing over the abdomen.
  The elytra which meet medially form a firm covering for the first six abdominal segments. Thickened edges of the elytra (epipleura) lie on the margins of the greatly broadened and upward-projecting sternites of the abdominal segments (Fig. 143 B).
- Cryptopleury on the prothorax.
  Paranota as lateral surfaces of the tergum bend ventralwards, cover the pleura of the prothorax and make a firm connection with the sternum of the segment. The prothorax thus becomes a stiff capsule.
- Soft abdominal tergites.
  In connection with the evolution of elytra the tergites of the first six abdominal segments are transformed into soft structures. Only abdominal tergites 7 and 8 remain in the plesiomorphic state of strongly sclerotized covering plates.
- Withdrawal of the posterior end.
  In both sexes the last body section with the ninth abdominal segment and the remains of the tenth segment is withdrawn into the eighth segment. Thereby the exterior genital appendages – as already discussed for the female (p. 326) – are retracted into the body interior.
- Hindwings with strong reduction of vein branching in the remigium.

BEUTEL (1997) made a comparative analysis of concurrent hypotheses concerning the high-ranking adelphotaxa relationships within the Coleoptera; he favors the following phylogenetic systematization.

Refer to the original paper for further information.

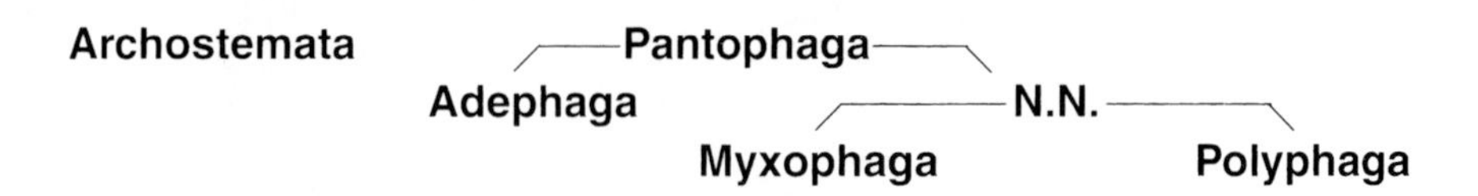

Strepsiptera

As larval parasites whose hosts are various insects, generally only the mm-long Strepsiptera (twisted-wing parasites) are characterized by a mass of derived features. We list here the most conspicuous autapomorphies.

Autapomorphies (Fig. 139 → 10)

- Larval parasitism.
  Endoparasites from the second to the last larval stage.
- Adult in both sexes without nutrient uptake. Corresponding reduction of mouthparts.

Males.

- Antenna of seven articles; third article always with a ventrolateral process; fourth article with an olfactory pit.
- Compound eyes of the imago taken over from the larva as stemmata (? paedomorphosis).
- Short mesothorax, long metathorax.
- "Halteres" (Fig. 144 A). Wings of the mesothorax greatly shortened; only the remnants of venation found. Proximally furnished with sensory structures, distally thickened to club-like structures. The forewings represent in structure and function a convergence to the halteres on the metathorax of the Diptera.
- Fan-like hindwings (Fig. 144 A). The wings of the metathorax are greatly broadened and foldable lengthways. They have a simple longitudinal venation without cross-connections.
- Trochanter and femur of the pro- and mesothorax appendages fused.
- No cerci, no gonopods.

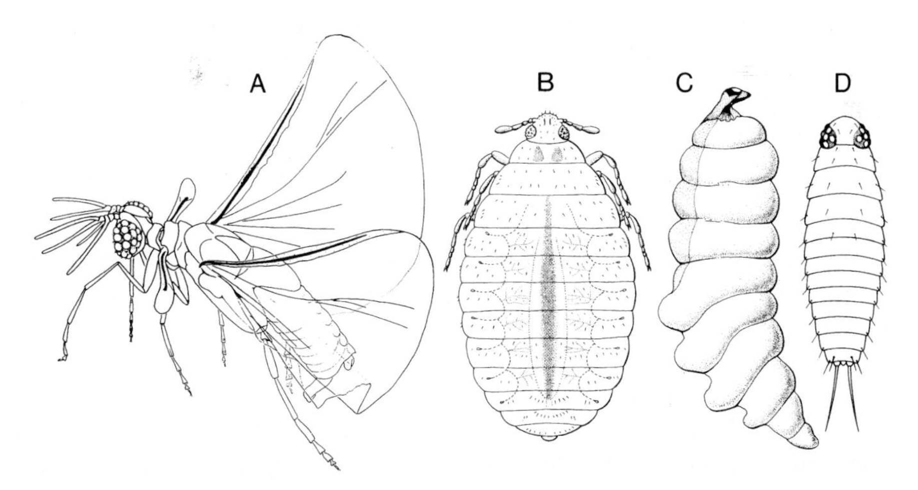

Fig. 144. Strepsiptera. **A** *Dundoxenos vilhenai* (Stylopidia). Male with club-like forewings and fan-like hindwings. **B** *Eoxenos laboulbenei* (Mengenillidia). Dorsoventrally flattened female with ten abdominal segments. **C** *Coriophagus rieki* (Stylopidia). Endoparasitic maggot-like female. Only the dark pigmented anterior part of head, thorax and first abdominal segment projects out of the hosts body. **D** *Hylecthrus* (Stylopidia). First larva (primary larva). (**A** Kinzelbach 1966; **B–D** Kathirithamby 1991)

- Nervous system. Apart from the supraesophageal ganglion two concentrations of ganglia on the ventral side – a thoracic mass consisting of the subesophageal ganglion, the thoracic ganglia and some abdominal ganglia, and a smaller abdominal mass from the rest of the abdominal ganglia.
- Tracheal system with two thoracic and seven abdominal spiracle pairs.

Females.

- Without wings.
- Without exterior genitalia.
- Viviparous.
- Primitively free-living ♀♀ (Mengenillidia) – these, however, have the following derived features: Body compressed dorsoventrally; antennae have a maximum of five articles; number of ommatidia in the compound eyes less than in the male; thorax without spiracles (Fig. 144 B).

The Strepsiptera are divided into the subtaxa Mengenillidia and Stylopidia (KINZELBACH 1971). The **Mengenillidia** (? monophylum) comprises the two supraspecific unities *Mengenilla* and *Eoxenos* consisting of just a few species. Only these species have free-living females with thoracic appendages (Fig. 144 B) – a clear plesiomorphic state.

The species-rich **Stylopidia** are characterized by the extremely derived behavior of the female. The posterior body remains as a whitish, maggot-like formation without appendages in the interior of the host organism during its entire life. Only a short anterior end, comprising the head, thorax and first abdominal segment with spiracles, is found to the exterior; it projects out of the host as a curved, darkly pigmented and strongly sclerotized formation (Fig. 144 C).

In what way are the different states in the two taxa incorporated into the life cycle of the Strepsiptera?

We have already mentioned vivipary in the autapomorphies. The female of the Mengenillidia and Stylopidia discharge tiny larvae only a fraction of a mm long, with three leg pairs. This is the primary larva (first larva) which represents the infection stage of the cycle (Fig. 144 D). This invades species-specific insect hosts and transforms into the secondary larva. This undergoes several molts, whereby the Mengenillidia larva retains its legs, while the Stylopidia larva becomes a legless maggot. The secondary larva forms the feeding stage of the Strepsiptera; it takes up soft nutrients from the host.

The cuticle of the last endoparasitic secondary larva hardens to a puparium in which pupation takes place. In the Mengenillidia this process takes place externally. The last larva migrates out of the host, forms the puparium and molts inside this to the pupa; primitively the male and female adults leave the puparium. In contrast, in the Stylopidia, the puparium is formed inside the host. Only the males hatch, the females remain fixed in the puparium.

The **adelphotaxa of the Strepsiptera** has not yet been incontestably identified (KINZELBACH 1981; N.P. KRISTENSEN 1995). Even the placement within the Holometabola is questioned; the male larva has exterior wingbuds. The larval stemmata do not degenerate, but are taken over into the imago; this could, however, be the result of paedomorphosis. Within the Holometabola it is only in the Coleoptera and Strepsiptera that the hindwings alone function as flight organs. This is without doubt an apomorphy in both taxa. There are, however, no convincing arguments to trace this back to the feature pattern of a common stem species. Special congruencies in wing venation have been referred to (KUKALOVÁ-PECK & LAWRENCE 1993; KUKALOVÁ-PECK 1997). Opposing this is the hypothesis of an adelphotaxa kinship between the Strepsiptera and Diptera (WHITING et al. 1997; WHITING 1998).

Mecopteriformia [28]

Autapomorphies (Fig. 139 → 11)

- Formation of the pupal cocoon from silk secreted by labial glands.
- Appendages of the larvae with an unpaired pretarsal claw.

Cocoon formation of silk from the labial glands is lacking in the sister group Neuropteriformia and can therefore, with some confidence, be interpreted as a derived feature of the Mecopteriformia. It is possible that this mode could already belong in the ground pattern of the Holometabola; it must then, however, have been lost early in the stem lineage of the Neuropteriformia – and for this there is no evidence.

Hymenoptera – Mecopteroidea

Hymenoptera

Autapomorphies (Fig. 139 → 12)

- Forewing main flight organ.
  The wings of the mesothorax form the principal flight apparatus. The wings of the metathorax are comparatively small; venation is strongly reduced (Fig. 145 A).

[28] tax. nov.

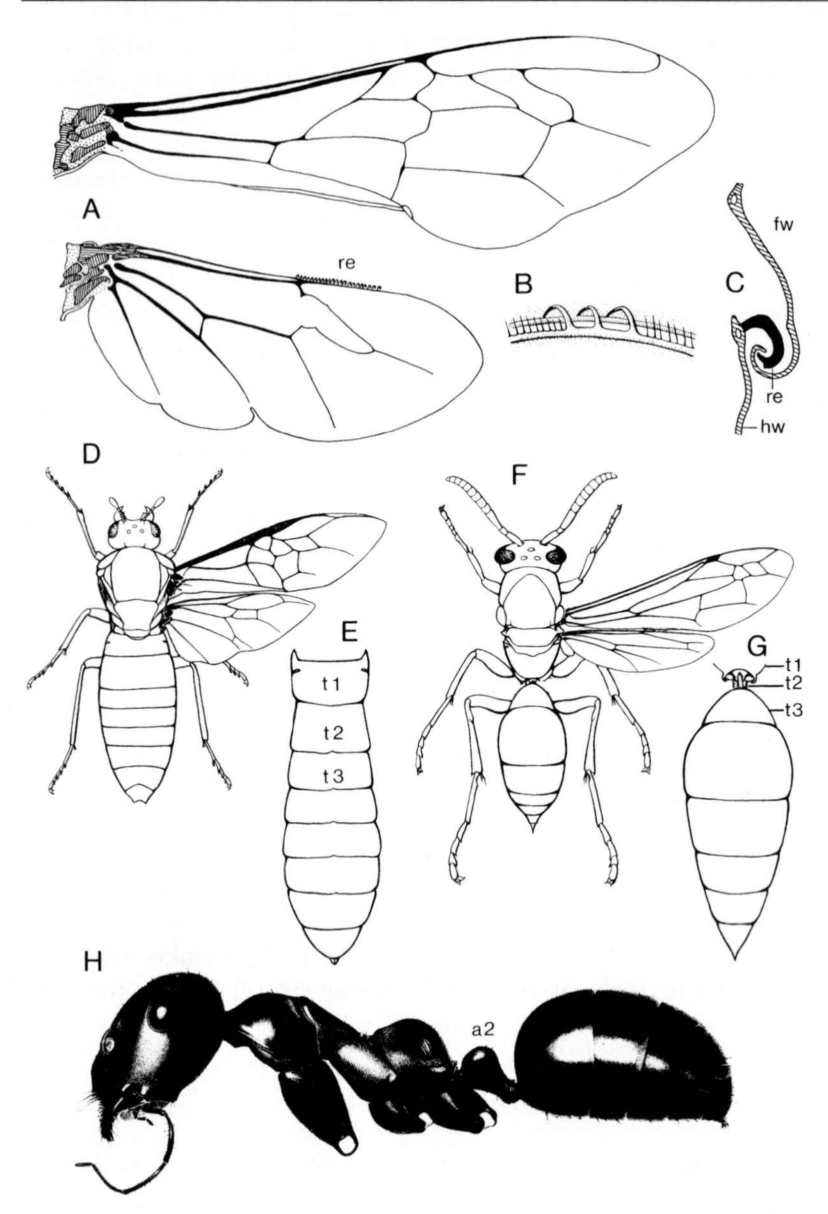

Fig. 145. Hymenoptera. **A** *Apis mellifera* (Apocrita). Wings of the right-hand side of the body; hindwing with retinaculum. **B** Braconidae (Apocrita). Hooks from the retinaculum which grasp over the forewing. **C** Bees. Coupling mechanism between fore- and hindwing in longitudinal section. **D** *Perga*. **E** *Guiglia* ("Symphyta"). Broad connection between abdominal segments 1 and 2. **F–G** *Polistes* (Apocrita). The tergum of abdominal segment 1 (propodeum) forms the back wall of the metathorax. Abdominal segment 2 (petiolus) as a stalk forms the connection between the anterior and posterior body. **H** *Cataglyphis savignyi*. Worker (Apocrita, Formicidae). Side view with prominent bulging abdominal segment 2. *a2* Abdominal segment 2 (petiolus); *fw* forewing; *hw* hindwing; *re* retinaculum; *t* terga of abdominal segments. (**A,C** Seifert 1995; **B,D–G** Naumann 1991; **H** Wehner, Wehner & Agosti 1994)

- Mesothorax bears flight musculature.
  The main part of the flight musculature lies in the mesothorax. This segment is correspondingly larger than the following metathorax.
- Retinaculum (hamuli) for connecting the wings.
  The front margin of the hindwing carries a row of inwardly bent setae. These hook from underneath onto the upturned posterior margin of the forewing. In this fashion fore- and hindwing form together one functional unit (Fig. 145 B, C).
- Anchoring of the forewing in the resting position.
  On the dorsal side of the metathorax there are bubble-like swellings (cenchri), in which spines from the anal field of the forewing mesh. Together they form an anchor apparatus which fixes the wings on the back while at rest. Only realized in the "Symphyta" with the exception of the Cephidae.
  According to VILHELMSEN (1997) independently reduced in the stem lineages of the Cephidae and Apocrita.
- Pronotum firmly connected to the mesothorax.
- Abdominal segment 1 fused with the mesothorax; its sternite greatly reduced.
- Tergites and sternites of the abdomen overlap like scales.
- Larval phytophagy. Larvae rely on living plants for food.
- Eyes of the larvae with a single corneal lens. This covers several separate ommatidia.
- Haplo-diploid sex determination.
  Only the diploid females are the product of the union of egg and sperm. The haploid males originate from unfertilized egg cells.

Further apomorphic peculiarities can be found in an overview by VILHELMSEN (1997). The monophyly of the Hymenoptera is incontestably established.

The **ground pattern** of the Hymenoptera also consists, however, of a whole series of original features that first underwent wide-ranging transformations within the unity. To these belong the broad connection between all abdominal segments and the biting-chewing orthopteroid mouthparts. Licking-sucking mouthparts with the elongation of the glossae of the labium to a suction pipe characterize only the Apoidea (bees) within the Apocrita. The plesiomorphic ovipositor with gonapophyses 8 and 9, gonostyli 9 and the jointed gonangulum represents the ground pattern of the Mecopteriformia (Fig. 140 E) – as taken over from the stem species of the Holometabola.

The additive evolution of new features has led to certain feature pairs within the Hymenoptera which are used in traditional classifications to construct artificial groups, in which yet again paraphyla and monophyla are opposed. These are:

- **Hymenoptera: "Symphyta" – Apocrita.**
- **Apocrita: "Terebrantes" – Aculeata**

KÖNIGSMANN (1976–1978, 1981) laid the foundations for a phylogenetic system of the Hymenoptera. Although the non-monophyletic nature of the "Symphyta" and "Terebrantes" was clearly emphasized, they are still found in many textbooks as valid system unities. Thus we take a methodologically oriented position.

"Symphyta" – Apocrita

In conventional classifications about 10% of the 115,000 recent Hymenoptera are grouped together under the name "**Symphyta**". Well-known taxa are the Siricidae, the Tenthredinidae and the Cephidae. An essential diagnostic feature is the broad connection between the thorax and abdomen and likewise between the first and second abdominal segments (Fig. 145 D, E). This, however, is a common plesiomorphy for members of the "Symphyta" – and, as such, has no value for phylogenetic kinship research. VILHELMSEN (1997) definitively eliminated the paraphylum, divided it into monophyletic subgroups, and discussed their kinship with one another and to the Apocrita.

The **Apocrita** have a "**wasp-waist**". Here, the first abdominal segment is incorporated into the thorax; the tergum forms the sloping back wall of the metathorax, the sternum disappears. The second abdominal segment (petiole) experiences a radical restriction and a joint originates over which the rest of the posterior body is separated from the anterior body (Fig. 145 F–H).

We hypothesize a single evolution of this apomorphous construction within the Hymenoptera (principle of parsimony). Without conflict with other feature assessments the "wasp-waist" can be used to establish a large monophylum Apocrita within the Hymenoptera.

"Terebrantes" – Aculeata

The "**Terebrantes**", with taxa such as the Ichneumonidae, Chalcididae and Cynipidae, are Hymenoptera with wasp-waists. They – like the "Symphyta" – carry over the plesiomorphic egg-laying apparatus from the ground pattern of the Mecopteriformia. Once again, we can see that a symplesiomorphy is of no use when investigating phylogeny.

In contrast, the ovipositor of the **Aculeata** (wasps, ants, bees) is transformed into a **sting with poison gland** – as seen in the Chrysididae, Formicidae, Vespidae or the Apoidea. We now have a derived feature at our disposal. Assuming a single evolution in the stem lineage of a unity including all wasps, ants and bees, the monophyly of the Aculeata within the Apocrita is established with the autapomorphy "sting with poison gland".

Mecopteroidea

The caddis flies and the butterflies on the one hand (Amphiesmenoptera), and scorpion flies, fleas and dipterans (Antliophora) on the other, form together the last large unity of the Holometabola.

Autapomorphies (Fig. 139 → 13)

- Stipes of the larval maxillae transversely divided into basistipes and distipes. The distal section has a special extensor muscle (cranial extensor) (Fig. 146 A).
- Labial palps of the larvae without muscles.
- Ovipositor greatly reduced; in the ground pattern only vestigial (Fig. 140 F).

Amphiesmenoptera – Antliophora

Amphiesmenoptera

Between the Trichoptera and the Lepidoptera around 20 synapomorphies have been documented (N.P. KRISTENSEN 1984). As autapomorphies they establish without doubt a monophylum Amphiesmenoptera consisting of these two taxa. The following list includes a selection.

Autapomorphies (Fig. 139 → 14)

- Prelabium and hypopharynx fused to a lobe from which the salivary glands open apically in the larva.
  (In the haustellum of the Trichoptera the lobe evolved to an organ for the uptake of fluids. The fusion of prelabium and hypopharynx belongs, however, in the ground pattern of the Amphiesmenoptera.)
- Wings densely covered with hairs (p. 337).
- Anal vein of the forewing with loop formation from which results a double Y figure.
- Pretarsus over the claws with a pseudempodium – a strong hair with a socket (Fig. 146 B).
- Paired glands in the abdomen opening in the fifth sternite, probably defense glands.
- Insertion of the ventral diaphragm muscles.
  The abdominal ventral diaphragm which consists of ventral muscle wings inserts into the ventral nerve cord. This is a unique phenomenon within the insects.

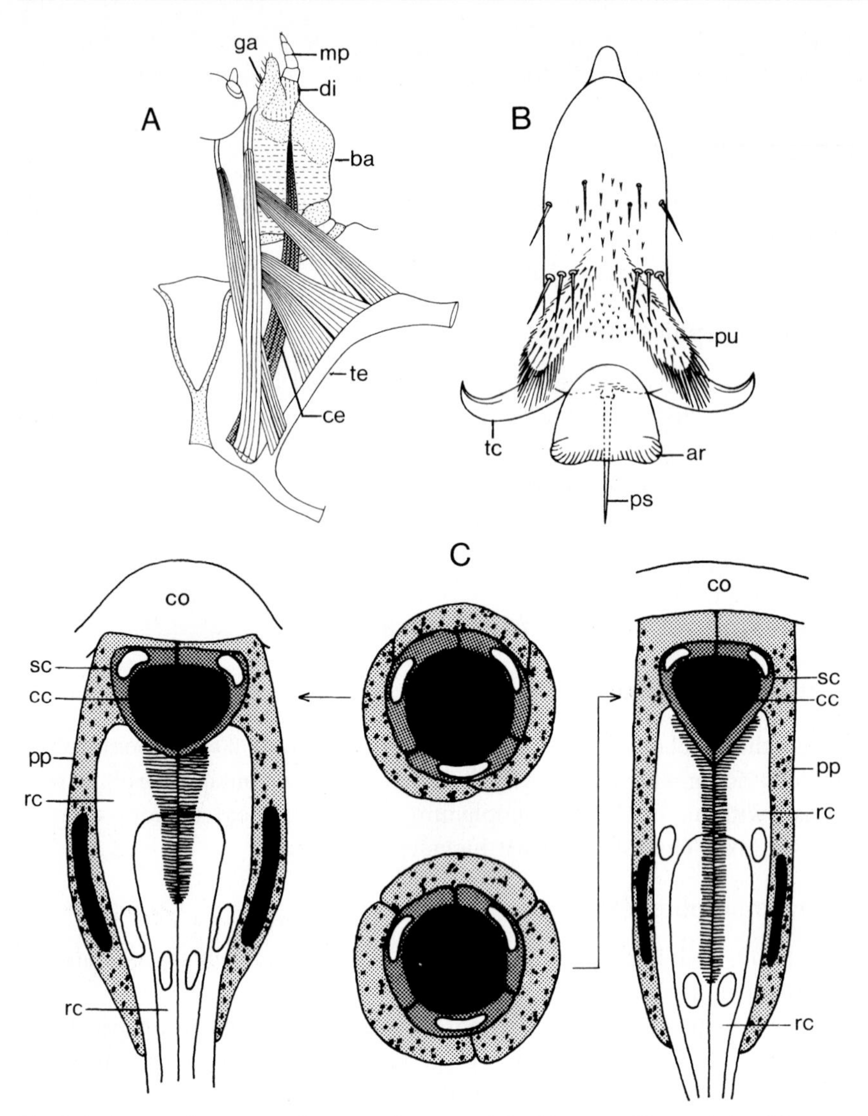

Fig. 146. Apomorphous ground pattern features of the Mecopteroidea and Amphiesmenoptera. **A** Mecopteroidea. Larval maxilla with transverse division of the stipes into basistipes and dististipes shown on an example of a caddis fly larva (Limnephilidae, Trichoptera). **B–C** Amphiesmenoptera. **B** Pretarsus with pseudempodium – a socketed seta over the claws (*Epiphyas*, Lepidoptera). **C** Ommatidia of the larval stemmata in longitudinal sections and cross sections (middle). Left *Operophthera brumata* (Lepidoptera); right *Stenophylax* (Trichoptera). The identical equipment with three semper cells and three primary pigment cells is judged to be a synapomorphy of the caddis flies and the butterflies and moths. *ar* Arolium; *ba* basistipes; *cc* crystalline cone; *ce* cranial extensor of the dististipes; *co* cornea; *di* dististipes; *ga* galea; *mp* maxillary palp; *pp* primary pigment cell (mantle cell); *ps* pseudempodium; *pu* pulvillus; *rc* retinula cell; *sc* semper cell; *tc* tarsal claw; *te* tentorium. (**A** N.P. Kristensen 1991; **B** Nielsen & Common 1991; **C** Paulus & Schmidt 1978)

The plesiomorphous alternative of the Antliophora is known from *Panorpa* (Mecoptera). The ventral diaphragm runs continuously above the nerve cord and remains separate from it. This state is the normal case in the Pterygota.
- Heterogametic female – a classic cytological synapomorphy between the Trichoptera and Lepidoptera.
- Stemmata of the larvae with a three-part crystalline cone.
The crystalline cone is the secretion product of three semper cells (crystalline cone cells). The crystalline cone and usually also the retinula cells are surrounded by three primary pigment cells (mantle cells, main pigment cells) (Fig. 146C) (PAULUS & SCHMIDT 1978).
In the ground pattern of the Insecta the ommatidia have four crystalline cone cells and two mantle cells. In the stem lineage of the Amphiesmenoptera one crystalline cone cell is transformed into a primary pigment cell (mantle cell).

Trichoptera – Lepidoptera

Trichoptera

Autapomorphies (Fig. 139 → 15)

- Haustellum.
A lobe of the prelabium (prementum) and hypopharynx forms an organ with a cuticular canal system which can be extruded for the uptake of fluids (Fig. 147G–J).
- Clypeolabrum.
Fusion of clypeus and labrum (convergence to the Mecoptera).
- Mandibles strongly reduced; without articulation.
- Aquatic larvae (Fig. 147B, C).
Evolution of aquatic larvae in the stem lineage of the Trichoptera with closure of the spiracles (apneustic). The larvae have a closed tracheal system.
- Larval antennae reduced to papillae; without musculature.
- Larval tentorium only weakly developed.

The dense **covering of the wings with hairs or setae** is the most characteristic feature of the Trichoptera, and has given the unity their name. Nevertheless, for the following reasons, this feature is not an autapomorphy of the caddis flies. The hairs of the Trichoptera and the scales on the wings of the Lepidoptera are interpreted as homologous formations. In comparing hairs and scales the latter are without doubt the apomorphous manifestation of the homologous feature. Consequently, possession of

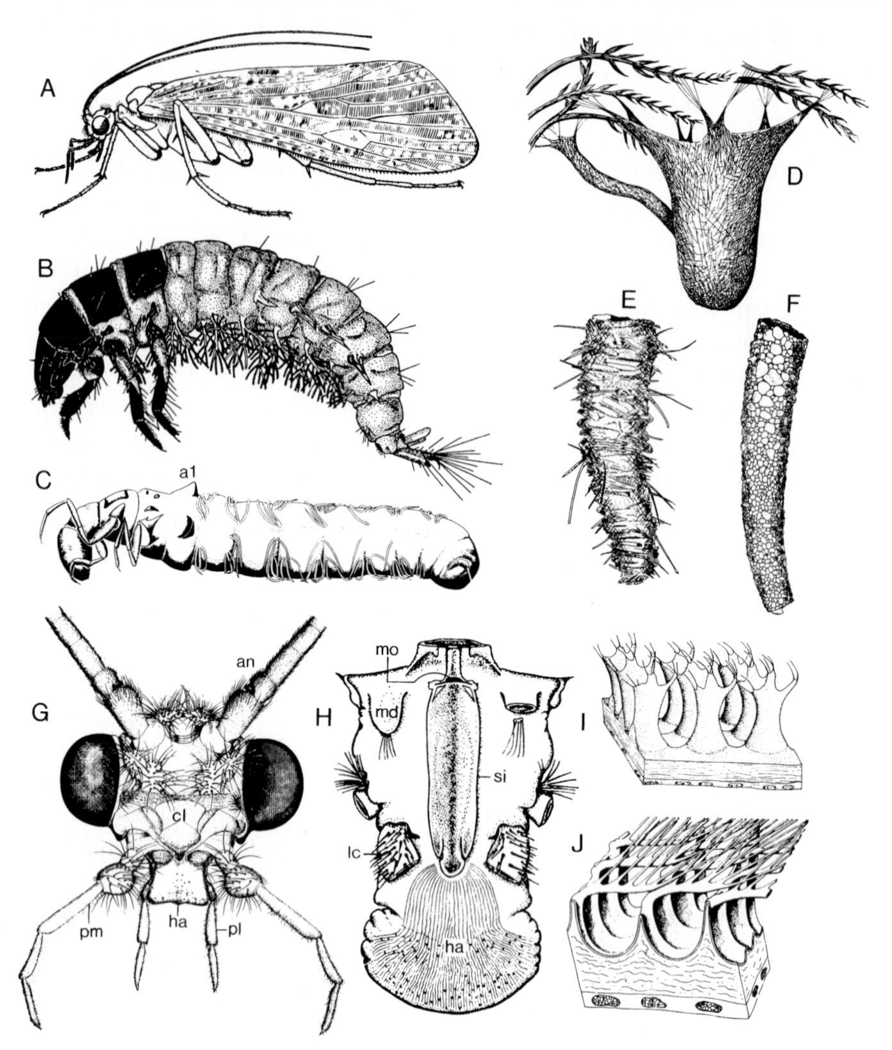

Fig. 147. Trichoptera. **A** *Hydropsyche*. Imago. **B** *Hydropsyche*. Larva with prognathous mandibles. Without a case. Forms a capturing net. With segmental tracheal gills. **C** *Limnephilus flavicornis*. Larva with orthognathous mouthparts. With case. The cones on the first abdominal segment serve for anchoring within the case. Likewise with tracheal gills. **D** *Plectrocnemia conspersa*. Capturing net and tube of the larva; formed from secretion of the salivary glands. **E** *Plectrotarsus gravenhorstii*. Case of the larva using bits of plants. **F** *Antipodoecia turneri*. Case of the larva using small stones. **G** *Rhyacophila septentrionalis*. Male. Frontal view of the head with haustellum. Species with "free-living" predacious larvae without capturing net or case. **H** *Phryganea bipunctata*. View of the head after removal of labrum, maxillary palps, lacinia and left mandible. The microtrichia canals converge on the haustellum in the direction of the sitophore (back wall of the cibarium). **I** *Rhyacophila dorsalis*. Microtrichia of the cuticle of the haustellum with filaments. **J** *Phryganea bipunctata*. The microtrichia on the upper surface of the haustellum form closed canals. *a1* Abdominal segment 1; *an* antenna; *cl* clypeolabrum; *ha* haustellum; *lc* lacinia; *md* mandible; *mo* mouth; *pl* palpus labialis; *pm* palpus maxillaris; *si* sitophore. (**A,C** Kaestner 1973; **B,D** Despax 1951; **E,F** Neboiss 1991; **G** Klemm 1966; **H–J** Malicky 1973)

wings covered with hairs can be placed in the feature mosaic of the last common stem species of the Trichoptera and Lepidoptera and postulated as an autapomorphy in the ground pattern of the Amphiesmenoptera comprising these two unities. In other words, wing hairs are not a new formation in the stem lineage of the Trichoptera; rather they are carried over unchanged from the ground pattern of the Amphiesmenoptera into the stem species of the Trichoptera. In the stem lineage of the sister group Lepidoptera the wing hairs are transformed into scales.

The **haustellum** is the outstanding imaginal autapomorphy in the ground pattern of the Trichoptera. The tongue-shaped mainly membranous organ is retracted when at rest. Increase in hemolymph pressure extrudes the organ to the front. The cuticle of the haustellum forms a unique conducting structure for fluid nutrients; it projects out in finger-like branched microtrichia, which themselves form a system of connected net-like canals. This micro-canal system serves in the capillary uptake and conduction of fluids, which through suction-pumping are led over the preoral cavity, the cibarium and the mouth into the pharynx.

The Holometabola have primary terrestrial larvae. From this hypothesis it follows convincingly that the **aquatic larvae** of the Trichoptera must have developed secondarily – convergently to the aquatic larvae of the Megaloptera. It is, however, not clear whether a larva with or without a portable bag (Fig. 147D–F) belongs in the ground pattern of the Trichoptera and whether the larva of the stem species was equipped with tracheal gills or not.

Lepidoptera

The Lepidoptera (moths and butterflies), like the Amphiesmenoptera, are established as a monophylum through a large number of autapomorphies (N.P. KRISTENSEN 1984; N.P. KRISTENSEN & SKALSKI 1998). From the two dozen or so derived features I list a number of the most impressive which can be easily represented.

Autapomorphies (Fig. 139 → 16)

- Wing scales.
  The covering of the wings with broad scales is the classic and perhaps most striking apomorphic peculiarity of moths and butterflies. The single scale is anchored by its pedicel in a basal ring (scale socket) of the wings. Wing scales evolved in the stem lineage of the Lepidoptera from wing hairs.
  In the Micropterigoidea and a few other taxa scales in the form of solid plates form the only wing covering. Upper and lower lamellae are here

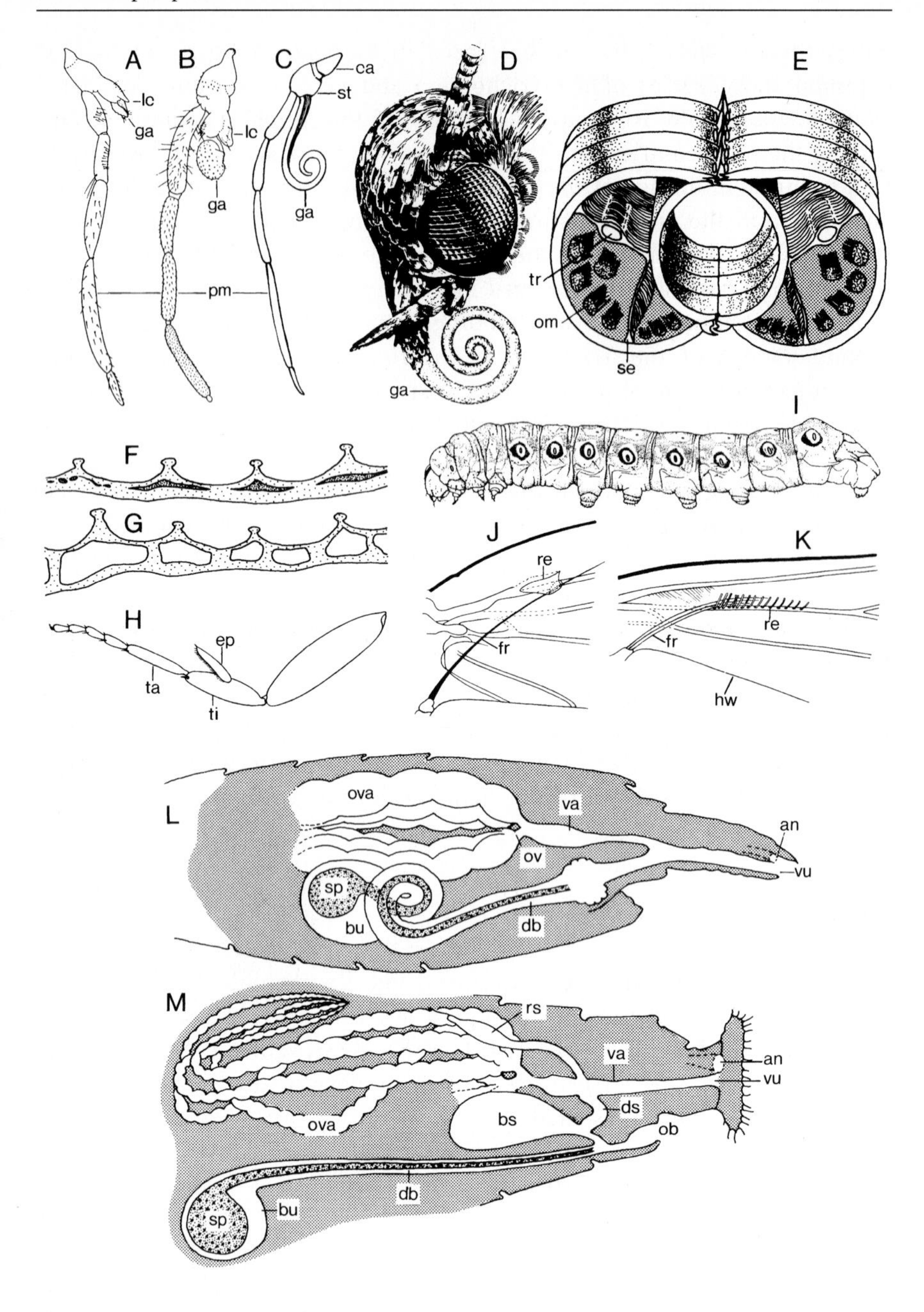
A
B
C
D
E
lc
ga
lc
ga
ca
st
ga
tr
om
se
pm
ga
F
G
H
ep
ta
ti
I
J
re
fr
K
re
fr
hw
L
ova
va
an
ov
vu
sp
bu
db
M
rs
va
an
vu
ds
bs
ob
ova
db
sp
bu

completely united (Fig. 148 F). These plesiomorphic scales belong in the ground pattern of the Lepidoptera.
In contrast apomorphic scales have an internal lacunal system between the two lamellae (Fig. 148 G); this first evolved within the Lepidoptera in the stem lineage of the Myoglossata (p. 344).

- Loss of the median ocellus in the adult.
- Corpotentorium with posteromedian process.
  Outgrowth on the ventral termination of the occipital foramen. The ventrolongitudinal throat muscles insert here. In the sister group Trichoptera the process is not found.
- Maxillary palps of the imago.
  With flexion points between articles 1/2 and 3/4. Article 4 is the longest section.
- Labial palps of the imago.
  End article with a group of sensilla in a depression.

◀ **Fig. 148.** Lepidoptera. Features from the ground pattern of the Lepidoptera and autapomorphies from various stem lineages within the Lepidoptera. **A–E** Mouthparts. **A** *Sabatinca* (Micropterigoidea). **B** *Agathiphaga* (Agathiphagoidea) with plesiomorphies from the ground pattern of the Lepidoptera. Maxilla with galea and lacinia as chewing structures and long multi-articulated palp. **C** *Mnemonica aurycyanea* (Eriocranioidea) with short muscle-free proboscis. First evolutionary step from the stem lineage of the Glossata. **D** *Heliozela* (Incurvarioidea) with simple longitudinal muscles in the proboscis. Second evolutionary step in the stem lineage of the Myoglossata. **E** *Pieris brassicae* (Ditrysia). Cross section of the proboscis. Short, densely packed oblique muscles in the half-pipes. Third evolutionary step in the stem lineage of the Ditrysia. **F–G** Wing scales using *Tinea monophthalma* (Incurvarioidea) as an example. Ultrastructure in cross section. **F** Plesiomorphous, solid scales from the ground pattern of the Lepidoptera. **G** Apomorphous scales with internal lacunae from the ground pattern of the Myoglossata. **H** *Epiphyas* (Ditrysia). Anterior appendage with epiphyse on the tibia. Autapomorphy of the Lepidoptera. **I** *Carthaea* (Ditrysia). Larva. Short appendages with hooks on abdominal segments 3–6 and 10. Autapomorphy of the Neolepidoptera. **J–K** Retinaculofrenulate coupling of the wings. Evolved in the stem lineage of the Heteroneura. Ventral view. **J** *Barea* (Ditrysia). Male. Membranous retinaculum on the forewing. Frenulum of the hindwing a single seta (ground pattern of the Heteroneura). **K** *Phthorimaea* (Ditrysia). Female. Retinaculum a row of stiff scales. Frenulum a few setae grouped together. **L–M** Female genital tract. **L** Monotrysic state as a plesiomorphic ground pattern feature of the Lepidoptera. A single terminal genital opening beneath the anus functions for both copulation and egg laying. In copulation a spermatophore enters into the stalked sac-shaped bursa copulatrix. The sperm move directly from the bursa into the vagina. **M** Ditrysic state as an autapomorphy of the Ditrysia. Ventral of the egg-laying opening (vulva) there is a separate copulation opening (ostium bursae) in the region of abdominal segments 7 and 8. The sperm migrate out of the bursa copulatrix through a new connection (ductus seminalis) into the vagina and further in a receptaculum seminis. From this storage organ they come back to fertilize the egg in the vagina. *an* Anus; *bs* bulla seminalis; *bu* bursa copulatrix; *ca* cardo; *db* ductus bursae; *ds* ductus seminalis; *ep* epiphyse; *fr* frenulum; *ga* galea; *hw* hindwing; *la* lacinia; *ob* ostium bursae; *om* oblique muscle; *ov* oviduct; *ova* ovariole; *pm* palpus maxillaris; *re* retinaculum; *rs* receptaculum seminis; *se* septum; *sp* spermatophore; *st* stipes; *ta* tarsus; *ti* tibia; *tr* trachea; *va* vagina; *vu* vulva (**A, B, D, F–M** Nielsen & Common 1991; **C** Weber 1933; **E** Kaestner 1973)

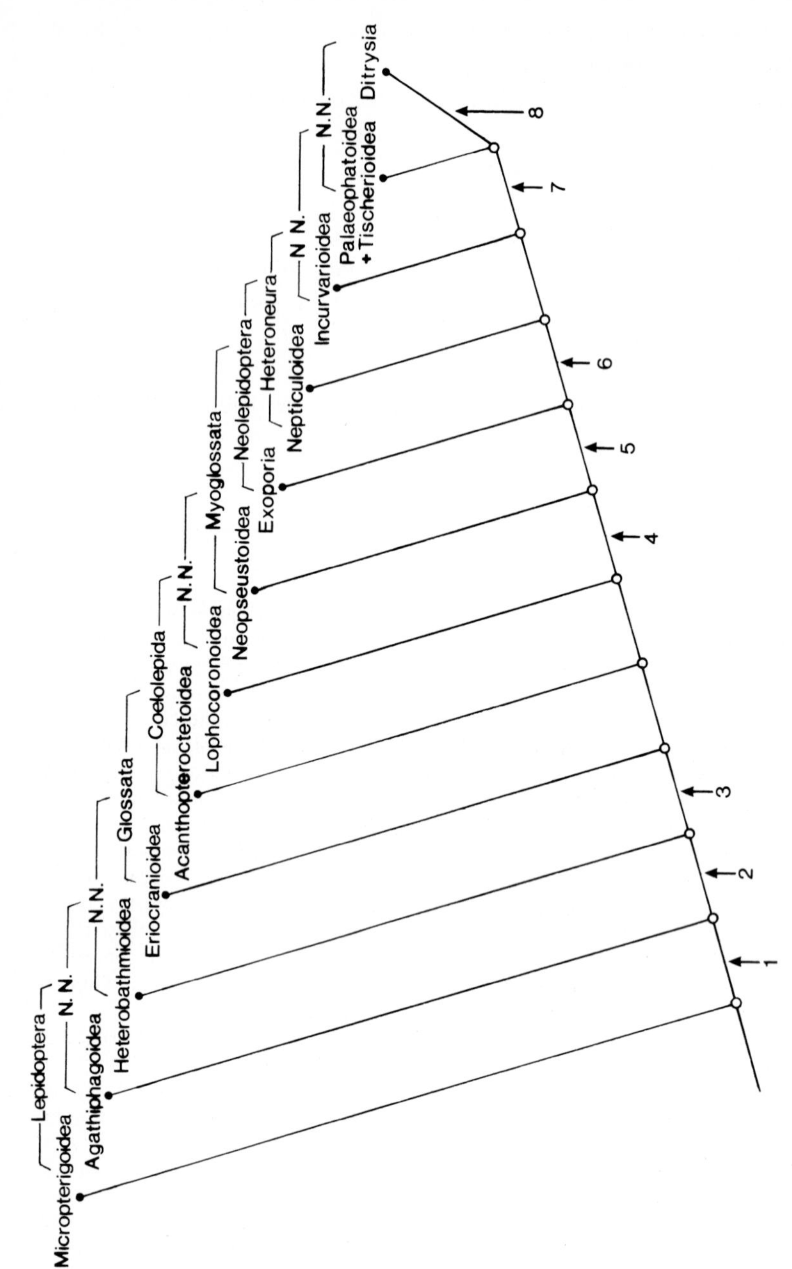

Fig. 149. Basal section of the phylogenetic system of the Lepidoptera from the Micropterigoidea to the Ditrysia. Demonstration of the additive evolution of apomorphous features from the ground pattern of the Lepidoptera up to the ground pattern of the Ditrysia. The *arrows* mark the stem lineages in which the autapomorphies explained in the text under numbers 1–8 evolved. (Redrawn from N.P. Kristensen & Skalski 1998)

- Tibia of the prothorax appendages.
  With a process (epiphyse) for cleaning the antennae or the proboscis (Fig. 148 H).
- Cerci absent in both sexes.
- Maxillary palps of the larvae with less than five articles.

In an impressive analysis of the phylogeny of the Lepidoptera a research team based around N.P. KRISTENSEN has established roughly 50 high-ranking subtaxa as monophyla and explained their kinship to one another (N.P. KRISTENSEN 1984, 1997b; N.P. KRISTENSEN & SKALSKI 1998; NIELSEN & COMMON 1991; NIELSEN & N.P. KRISTENSEN 1996 amongst others). A detailed representation of the arguments is not possible within the confines of this book. However, I particularly wish to look at a few aspects which seem to me to be of general importance.

The Ditrysia form a very large monophylum, which possess separate genital openings for copulation and egg laying (Fig. 148 M). The Ditrysia comprise almost 99% of all the 150,000 recent butterfly and moth species – from the small cosmopolitan clothes moth *Tineola bisselliella* to the large, blue shimmering *Morpho* butterflies in South America.

Of special interest is the remaining one percent. I reproduce the corresponding basal section of the phylogenetic system of the Lepidoptera from the Micropterigoidea to the Ditrysia (Fig. 149) with the aim of demonstrating the worth of phylogenetic systematics with good clarification of kinship relations in a monophyletic species group. The justification for all the branching steps lie in the cited literature. We will demonstrate here the sequence of the additive evolution of some chosen features – as they developed one after the other from the stem species of recent Lepidoptera to the formation of the ground pattern of the Ditrysia. The arrows in the kinship diagram refer to the stem lineage sections in which apomorphies 1–8 described below evolved.

1. After division into the lineage of the **Micropterigoidea** and the **stem lineage of all other Lepidoptera**, only minor changes took place at first in the latter. To these belong the reduction of the paraglossae on the labium, the division of the spermatheca duct into two sections and the loss of spiracles as respiratory openings on the metathorax of the larvae.
2. For the stem lineage after the branching off of the Agathiphagoidea N.P. KRISTENSEN recorded the evolution of new sensilla on the antennae (sensilla auricilla). The Micropterigoidea, Agathiphagoidea and also the Heterobathmioidea have plesiomorphic orthopteroid mouthparts. In these three taxa the galeae and laciniae of the maxillae are primary chewing structures (Figs. 109 D, 148 A, B).
3. A central evolutionary step in the **stem lineage of the Glossata** is the transformation of the maxillary galeae into a coiled proboscis; the

galeae develop in the form of half-pipes which are anchored dorsally and ventrally to one another. In the stem lineage of the Glossata a new organ originated for the uptake of water and fluid nutrients from narrow slits. The evolution of the proboscis took place over several stages in three stem lineages. After its origin the proboscis is at first free of musculature (Eriocranioidea); coiling is effected simply through the elasticity of the pipe wall (Fig. 148C). Correlated to the evolution of the proboscis in the stem lineage of the Glossata further mouthpart elements were transformed. The maxillary laciniae are reduced and are no longer sclerotized. The mandibles are eliminated as chewing structures in the imago; here, they lack articulation.

4. In the **stem lineage of the Myoglossata** in a second evolutionary step an inner musculature developed in the proboscis – in the manifestation of a longitudinal muscle or a few longitudinal bands (Fig. 148D; N.P. KRISTENSEN & NIELSEN 1981). Furthermore, in this stem lineage the apomorphous wing scales evolved with a lacunal system between the upper and lower lamellae (Fig. 148G). As well as these, plesiomorphic scales from the ground pattern of the Lepidoptera were retained.
5. In the **stem lineage of the Neolepidoptera** brain and subesophageal ganglion are united to form a compact mass; only a narrow opening remains for the stomodaeum and aorta to pass through.
   In the larvae, muscular appendages equipped with crochets develop on the abdomen. These are the prolegs on segments 3–6 and the anal legs on segment 10 (Fig. 148I). From the plesiomorphous state of the pupa dectica with freely movable mandibles for biting through the cocoon, the pupa adectica with immobile mandibles unsuitable for biting evolved.
6. In the **stem lineage of the Heteroneura** a new connection between the fore- and hindwings developed – the retinaculofrenulate binding. In the female the frenulum of the hindwing consists primarily of numerous setae, in the male of a single compound seta (Fig. 148J). The setae develop on the wings from a thickened frenulum base. The retinaculum of the forewing in the male is a membranous, primitively short and broad hook; in the female it usually consists of one or two rows of stiff, erect scales (NIELSEN & COMMON 1991).
7. In the **Ditrysia and their sister group** (Palaephatoidea + Tischerioidea) the female possesses an apomorphic frenulum of a few closely set setae (Fig. 148K). This change took place in the last but one stage of the Lepidoptera phylogeny depicted here, before the origin of the Ditrysia.
8. Up to the Ditrysia, a common opening existed in the female genital tract for copulation and egg laying. This is the plesiomorphic monotrysian state (Fig. 148L). In the **stem lineage of the Ditrysia** the two passageways were separated from one another; an independent copulatory

opening evolved ventral to the egg-laying opening – the apomorphic ditrysian state was realized. Furthermore a seminal duct developed as a new interior connection for transporting sperm between the bursa copulatrix and vagina (Fig. 148 M). Finally, in the stem lineage of the Ditrysia the third evolutionary step in the formation of the proboscis took place. In place of the simple longitudinal muscles from the ground pattern of the Myoglossata, a complicated system of numerous short oblique muscles appeared, which are arranged in sets with different orientations in the half-pipes of the proboscis (Fig. 148 E).

Antliophora

HENNIG (1969) grouped the Mecoptera, Siphonaptera and Diptera together as the Antliophora. However, the sperm pump in the male genital tract from which the group derives its name is obviously not an homologous organ of the Mecoptera and Diptera (MICKOLEIT 1973; WILLMANN 1981 b). "More work is needed to elucidate the distribution and structures of sperm pumps" within the Antliophora (N.P. KRISTENSEN 1991, p.139).

Autapomorphies (Fig. 139 → 17)

- Dagger-like mandibles in the imago; foremost condyle weakly developed (in the Diptera reduced). In the stem lineage of the Siphonaptera the mandibles were completely reduced.
- Winged segments with a "posterior notal wing process" as an attachment point for pleural muscles.
- Larval mouthparts without lateral labral retractor, without hypopharynx retractor and without ventral salivarium opener.

In the debate about the kinship relations between the three Antliophora taxa there are two competing hypotheses. Sister groups are either the Mecoptera and Siphonaptera or the Diptera and Siphonaptera. N.P. KRISTENSEN (1991) presents convincing arguments for the first alternative. The Mecoptera and Siphonaptera form together as adelphotaxa an as yet unnamed monophylum, to which the Diptera is the sister group.

N. N. (Mecoptera + Siphonaptera) – Diptera

N. N.: Mecoptera + Siphonaptera

Autapomorphies (Fig. 139 → 18)

- Acanthae in the proventriculus (Fig. 150 E).
  The proventriculus is a spherical foregut section in front of the midgut. The lumen is provided with an enclosing mantle of tightly packed setae; each seta is formed from one cell (RICHARDS & RICHARDS 1969). The remarkably similar arrangement of these tightly packed long "acanthae" in the Mecoptera and Siphonaptera is the foremost of their presumed synapomorphies (N.P. KRISTENSEN 1991).
- Sperm ultrastructure.
  The axoneme of the sperm is coiled into a spiral around the mitochondrial body.
- Absence of external labral muscles in the imago of scorpion flies and fleas.
- Sexual dimorphism in the concentration of ventral ganglia. The male possess one pair of ganglia more than the female.

Mecoptera

Autapomorphies (Fig. 139 → 19)

- Clypeus and labrum fused to a uniform appendage (clypeolabrum), i.e., the primitive jointed connection between these two parts of the head is given up. Thereby, except for the epipharyngeal compressor, the musculature of the labrum is reduced.
- Reduction of all muscles of the hypopharynx.
- Genital apparatus on the ninth abdominal segment of the male (Fig. 150 A, D, F, G). External elements of the genitals form the paired basistyli (gonobases) and dististyli (gonostyli); they articulate with one another. The basistyli grow together to form a capsule. The abdomen of the female is held fast during copulation with the forceps-like dististyli (WILLMANN 1981 a, b).
- Perineustic larva.
  The spiracles on the larval metathorax are closed.

With the above-mentioned apomorphous peculiarities, the monophyly of a taxon Mecoptera (scorpion flies) with 500 recent species is assured (WILLMANN 1989; N.P. KRISTENSEN 1991).

Within the framework of the Antliophora – i.e. compared to the Diptera and Siphonaptera – the Mecoptera are characterized by various original

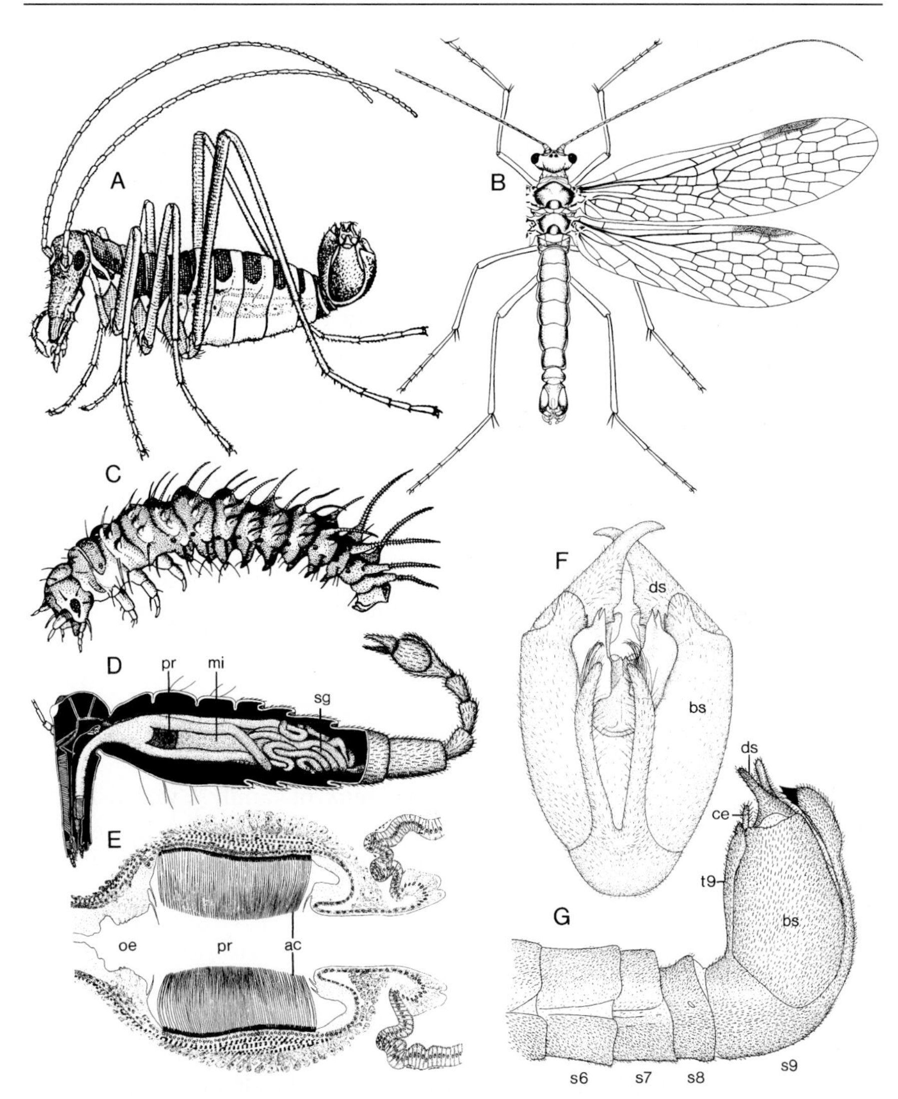

Fig. 150. Mecoptera. **A** *Apteropanorpa tasmanica*. Male. Head with long rostrum. Posterior end with upright genital apparatus. **B** *Chorista*. Male. Abdominal segment 9 with the genital apparatus is thickened. **C** *Panorpa communis*. Larva. The existence of appendages on the thorax and on abdominal segments 1–8 is plesiomorphous. **D** *Panorpa communis*. Male. Sagittal section to demonstrate the proventriculus and the salivary glands. The bow-shaped end of the abdomen directed forwards is reminiscent of the opisthosoma with poison spine in the Scorpiones; hence the name scorpion flies. **E** *Panorpa communis*. Proventriculus as an oval foregut section with tightly packed setae (acanthae). Synapomorphy of the Mecoptera and Siphonaptera. **F** *Panorpa communis*. Male. Caudoventral view of the genitalia. **G** *Neopanorpa furcata*. Male. Lateral view of the end of the abdomen with genitalia. *ac* Acanthae (setae in the proventriculus); *bs* basistylus; *ce* cercus; *ds* dististylus; *mi* midgut; *oe* esophagus; *pr* proventriculus; *s* sternite; *sg* salivary gland; *t* tergite. (**A, B** Byers 1991; **C** Grassé 1951; **D, E** Grell 1938; **F, G** Willmann 1981 a)

features. Only in the Mecoptera do two pairs of wings belong as a plesiomorphy in the ground pattern. Only the larvae of the Mecoptera possess appendages on the thorax and abdomen (Fig. 150 C). According to the hypothesis of an adelphotaxa relationship between the Mecoptera and Siphonaptera, legless larvae must have evolved convergently in the stem lineages of the Siphonaptera and Diptera.

A phylogenetic system of the Mecoptera has been produced (WILLMANN 1987, 1989). A taxon Nannomecoptera with only a few species in southern continents is placed opposite a unity Pistillifera consisting of all other Mecoptera.

Nannomecoptera

The Nannomecoptera are interpreted as a monophylum through some fusions and reductions. ♂♂ with greatly shortened cerci; tergum and sternum of abdominal segment 8 united to a single sclerital ring. In the ♀♀ base articles of the cerci fused with the pleural region of segment 11. In both sexes the mandibles are reduced to functionless appendages.

Pistillifera

For the adelphotaxon Pistillifera a special piston (pistillum) in the sperm pump is a conspicuous evolutionary novelty.

Of the eight unities with the traditional rank of family I would like to mention, as well as the Panorpidae (*Panorpa*) with strongly derived features, the Boreidae with the winter-active species *Boreus hiemalis*.

Siphonaptera

Ectoparasitism with feeding on the blood of homoiothermic vertebrates is responsible for the numerous characteristic features of the fleas, by means of which the Siphonaptera (Aphaniptera) can without doubt be established as a monophylum.

Compared with the sister group Mecoptera the ground pattern of the Siphonaptera also has some original features. These include a dorsal opener of the salivarium and metathoracic spiracles in the last larval stage (N.P. KRISTENSEN 1991). These plesiomorphies prevent an interpretation of the fleas as a subordinate subtaxon of the Mecoptera.

Autapomorphies (Fig. 139 → 20)

- Temporary ectoparasite on warm-blooded vertebrates.
- Piercing-sucking mouthparts for the uptake of blood with three long stylets: Paired maxillary laciniae and an unpaired epipharyngeal stylet from

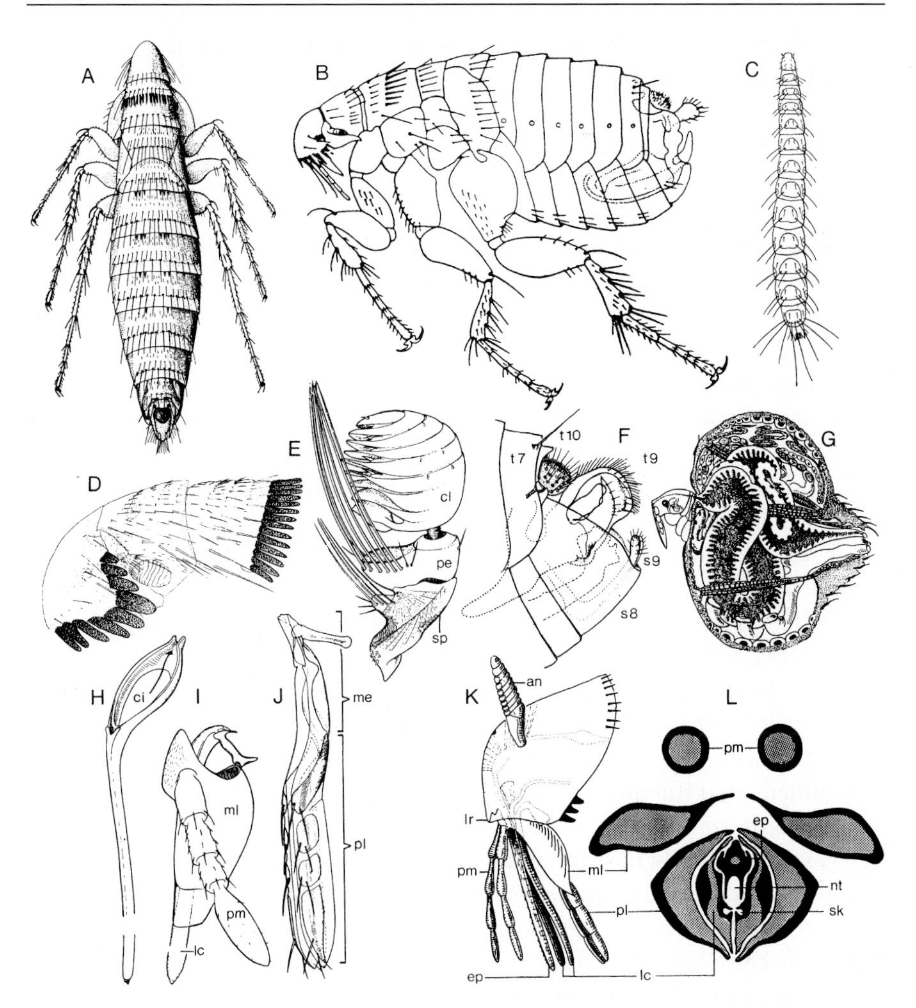

Fig. 151. Siphonaptera (Aphaniptera). **A** *Ceratophyllus gallinae*. Dorsal view. Female. **B** *Ctenocephalides canis*. Side view. Male. **C** *Xenopsylla cheopis*. Larva. **D** *Macropsylla hercules*. Head and pronotum. Male. Head with uniform ctenidium that overlaps to the cheeks in front of the antennae (ground pattern feature). Further ctenidium at the posterior of the pronotum. **E** *Ctenocephalides canis*. Antenna. Female. Scapus, pedicellus and club-shaped clava of ten articles. The clava inserts over a short petiole (first article) on the pedicellus. **F** *Pulex irritans*. End of the abdomen. Tergite 8 is largely hidden. Various sclerites of the genital apparatus originate at the posterior of tergite 9 as grasping structures. Tergite 10 forms the sensory pygidial plate (autapomorphy of the fleas). A further grasping organ is formed from sternite 9. **G** *Tunga penetrans*. Round swollen female. **H–J** *Ctenocephalides canis*. Sclerotized portions of the following mouthparts: **H** Epipharyngeal piercing seta and cibarium. **I** Maxilla with wing-like maxillary lobe, the lacinia differentiated as piercing seta and the maxillary palp of four articles. **J** Labium with labial palp of four articles. **K** Diagram of a flea head. **L** Cross section through the mouthparts. *an* Antenna; *ci* cibarium; *cl* clava; *ep* epipharyngeal piercing seta; *lc* lacinia; *lr* labrum; *me* mentum; *ml* maxillary lobe; *nt* nutrient tube; *pe* pedicellus; *pl* labial palp; *pm* maxillary palp; *s* sternite; *sk* salivary canal; *sp* scapus; *t* tergite. (**A** Kaestner 1973; **B, C** Séguy 1951c; **D** Dunnet & Mardon 1991; **E, H–J** Wenk 1953; **F, G** Piekarski 1954; **K, L** Seifert 1995)

the inner wall of the clypeolabrum. Piercing apparatus surrounded by labium and labial palps (Fig. 151 K, L).
- Mandibles completely reduced.
- Wings completely reduced.
- Body laterally compressed. Segmental tergites and sternites overlap caudalwards. Setae of the thorax and abdomen directed backwards (Fig. 151 A, B). (Decrease in resistance when moving through feathers or hair).
- Ocelli absent.
- Compound eyes small, strongly modified. With one lens.
- Short antennae. Flagella following the pedicellus swollen to a club (clava) (Fig. 151 E).
- Lateral pits on the head capsule into which the antennae can be retracted while in hair or plumage.
- Ctenidium – a crown of strong setae on the head (Fig. 151 D) (within the Siphonaptera reduced: *Pulex irritans*).
- Metathorax appendages differentiated into jumping legs; trochanter greatly enlarged (Fig. 151 B).
- Pretarsi of the appendages with very long claws (for anchoring in hair, plumage), but without pulvilli and empodium.
- Ten abdominal segments, the last of which forms a small sensory pygidial plate with trichobothria (Fig. 151 F).
- Larva without appendages or eyes (Fig. 151 C). A pair of anal legs on the rudimentary 11th abdominal segment (? appendage derivatives).

Fleas are primarily **nest inhabitants** with remarkable adaptations to different nest structures. The roomy structures of large mammals accommodate species with large jumping powers; inhabitants of nests of small mammals and birds can only jump short distances. Mammal and bird nests form the specific biotope of the apodous flea larvae. Their basic nutrient supply is the numerous secretions of the host organism as well as the blood of the animal, which the flea imago, after uptake, discards liberally with the feces into the nest. Single synanthropic animals such as *Pulex irritans* and *Ctenocephalides canis* find a nest substitute in the gaps in wooden buildings that collect organic waste. Fleas are not found on certain mammals and birds without nests, such as hoofed animals.

The **life cycle** is only seldom completed in the nest of a single host species. More usually the nests of different hosts with similar conditions are used by single flea species. Correspondingly there are multiple possibilities for the adult to acquire blood – also in species with common but inappropriate names such as the human flea (*Pulex irritans*), dog flea (*Ctenocephalides canis*) or poultry flea (*Ceratophyllus gallinae*). The species *Pygiopsylla hoplia* has been found on 35 different hosts among a broad spectrum of marsupials and rodents (DUNNET & MARDON 1991). Even a change between mammals and birds as blood donor is possible. However, there are also monophagous fleas; *Bradyopsylla echidnae* sucks blood exclusively from the egg-laying echidna *Tachyglossus aculeatus*.

Fleas are originally **temporary parasites**; they make contact with the host primarily for nutrient uptake. Differing stages of increasing temporal connection find their extreme in the tropical sandflea *Tunga penetrans*. Mated females drill into the skin of diverse mammals and here swell to form spherical stationary parasites (Fig. 151 G). Only the tip of the abdomen projects out of the skin for respiration, to lay eggs and for excrementation.

From the large number of Siphonaptera autapomorphies the **construction of the piercing apparatus** is a characteristic that is not found in other insects. Fleas have three stylets (Fig. 151 H–L). On the wing-like basal article of the maxillae (maxillary lobes) a lacinia is inserted in the form of a long stylet. Longitudinal canals are recessed in the laciniae through which the secretion of the salivary glands is introduced into the host. An epipharyngeal piercing seta, an outgrowth from the inner wall of the clypeolabrum, lies between the laciniae as the third element. The three setae surround a feeding tube that is connected in the head with a cibarial sucking pump. The piercing apparatus is conducted from the labium which bears the setae in a furrow. The distal multiarticulated labial palps surround the epipharynx and laciniae in the form of a sheath.

Diptera

With the last high-ranking unity of the Insecta we come again to an extremely well-established monophylum. HENNIG (1973) described more than 30 derived peculiarities of which the following overview gives a flavor.

Autapomorphies (Fig. 139 → 21)

Imago

- Halteres, wings, thorax.
  The transformation of the hindwings into minute halteres with a sensory function is the outstanding apomorphic feature of the Diptera. With their evolution the forewings became the only flight device of the flies and the mesothorax segment which carries them became the strongest thoracic segment. Prothorax and metathorax are decreased in size (Fig. 152 A).
- Piercing and sucking proboscis.
  Labrum, mandibles, hypopharynx and the maxillary laciniae are elongated into setae-like devices (Fig. 152 B).

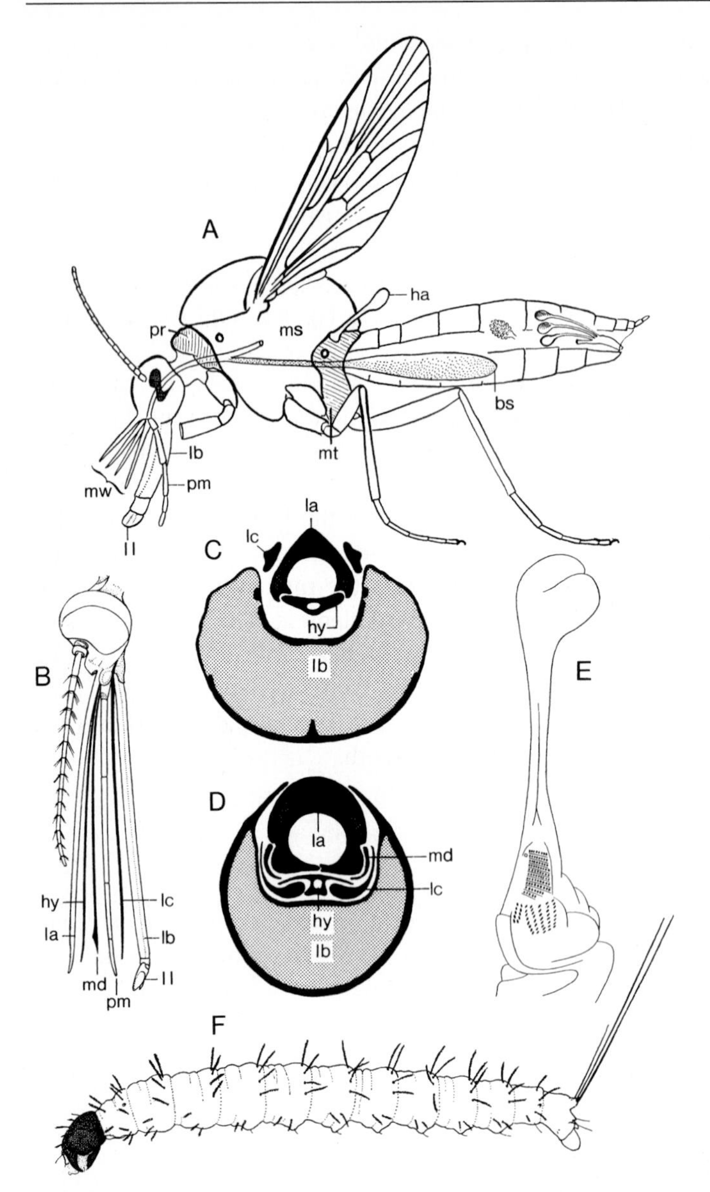

Fig. 152. Diptera. **A** Organization of the imago emphasizing derived ground pattern features. **B** *Anopheles* (Culicidae). Mouthparts in side view. **C** *Eristalis arbustorum* (Syrphidae). Cross section of the proboscis. Plesiomorphous manifestation of the sucking pipe in its formation from labrum and hypopharynx. **D** *Culex pipiens* (Culicidae). Apomorphous manifestation of the sucking pipe. Closed nutrient tube formed from the ventral curling of the lateral edges of the labrum. **E** *Lucilia* (Calliphoridae). Haltere in dorsal view. **F** *Phlebotomus papatasii* (Psychodidae). Habitus with ground pattern features of the Diptera larva. Separated prognathous head. Lack of thoracic appendages. *bs* Closed sac of the esophagus; *ha* haltere; *hy* hypopharynx; *la* labrum; *lb* labium; *lc* lacinia; *ll* labella; *md* mandible; *ms* mesothorax; *mt* metathorax; *mw* mouthparts; *pm* palpus maxillaris; *pr* prothorax. (**A, C–F** Hennig 1973; **B** Hennig 1994)

The labrum is primarily half-pipe shaped. Together with the hypopharynx as a ventral covering it forms the suction pipe of the proboscis apparatus (Fig. 152C). The curling of the labrum to a closed feeding tube is strongly derived (Fig. 152D).
On the mandibles the fore-joint is reduced; on the maxillae the galea has disappeared. With a furrow-like depression on its dorsal side the labium (prementum) forms the sheath for the conduction of the piercing stylets. Glossae and paraglossae are reduced. The two-part palps are differentiated into cushion-like labellae.
- Absence of spiracles on the eighth abdominal segment in the male.
- Cerci in the female of only two articles.
- Esophagus with long ventral diverticulum (crop).

Larva
- Prognathous head.
- Loss of thoracic legs (Fig. 152F).
- Closing apparatus absent on the spiracles.

For the interpretation of the **halteres** as an autapomorphy of the flies and therefore for the establishment of the Diptera as a monophylum we need not necessarily know their subordinate sister groups – whether these are mosquitoes and flies or not (Vol. I, p. 38). If by this we mean the traditional division of the flies into the Nematocera and Brachycera, then mosquitoes and flies are indeed not adelphotaxa. Phylogenetic systematical research has shown the "Nematocera" to be a paraphylum, while the monophyletic Brachycera form only a subtaxon of the recently newly constructed unity Neodiptera.

MICHELSEN (1996) differentiated eight high-ranking monophyletic taxa, which are distributed as follows among the two sister groups of the Diptera.

Taxon 1. **Polyneura:** Tipuloidea + Trichoceridae + Tanyderidae + Ptychopteridae.

Taxon 2. **Oligoneura:** Psychodida + Blephariceroidea + Culicumorpha + Neodiptera.
Thereby the Neodiptera comprise 75% of the 100,000 recent Diptera – these are the "Nematocera"-taxa Scatopsoidea, Anisopodidae, Perissommatidae, Axymyiidae, Pachyneuridae, Sciaroidea, Bibionoidea and the Brachycera.

This systematization is based essentially on the assessment of a number of special features of the skeleton and musculature. Their listing here would only be meaningful in a detailed examination of the anatomy of the Diptera – and that is not the aim of this book. Moreover, the justifications for the new consistent phylogenetic system of the Diptera must be seen as temporary (MICHELSEN 1996).

Independent of this FRIEDRICH & TAUTZ (1997) have presented a molecular analysis of the phylogeny of the Diptera.

Tardigrada – a Taxon with Unclarified Kinship

Tardigrada, which inhabit the interstitial system of marine sand, belong to the smallest metazoa. With body lengths of only 100–150 µm, bizarre organisms have developed in this biotope – such as *Tanarctus tauricus* with enormous appendages or *Actinarctus doryphorus* with a wide protective shield (Fig. 153). The 300 µm long giant *Batillipes mirus* bears in the place of claws tiny adhesive plates on the appendages; these are excellent adhesion structures in this unsettled environment. If one transfers a population of *Batillipes mirus* with sediment to a Petri dish the tiny bears free themselves and descend leisurely from the sand grains when all is still.

In conventional classifications marine species are all placed in one taxon "**Heterotardigrada**" for which no derived peculiarities exist. In contrast the **Eutardigrada** obviously form a monophylum (Fig. 154C–E). Only these have Malpighian tubules (apomorphy); furthermore, in the Eutardigrada the gonoducts open in the endgut (apomorphy) and in the "Heterotardigrada" separately before the anus (plesiomorphy). To the Eutardigrada belong species from the well-known taxa *Macrobiotus*, *Hypsibius* and *Milnesium*, which are widespread in limnic-terrestrial biotopes. Here, body lengths of 1 mm are often reached; *Macrobiotus hufelandi* can even attain 1.4 mm.

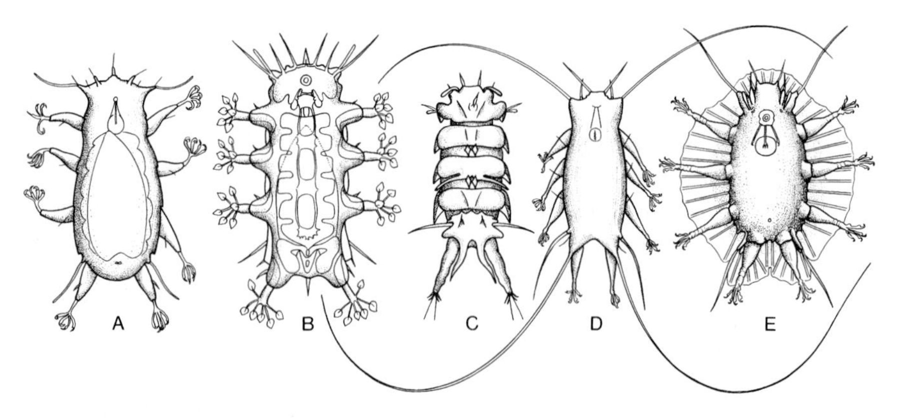

Fig. 153. Tardigrada. Marine species from the interstitial system of sand. **A** *Halechiniscus subterraneus*. **B** *Batillipes mirus*. **C** *Stygarctus bradypus*. **D** *Tanarctus tauricus*. **E** *Actinarctus doryphorus*. (Ax 1966)

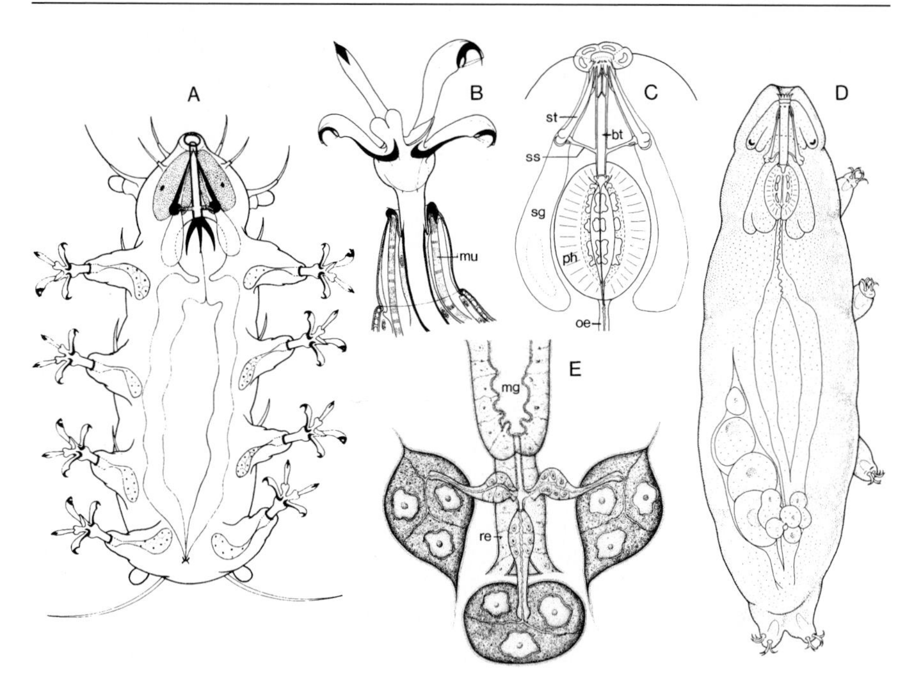

Fig. 154. Tardigrada. **A–B** *Styraconyx hallasi* ("Heterotardigrada"). **A** Ventral view. **B** Ventral view of third leg. **C–E** Eutardigrada. **C** *Halobiotus crispae.* Oral pipe – pharyngeal apparatus with two stylets, stylet supporters and stylet glands. **D** *Macrobiotus dianeae.* Dorsal view. Female. Unpaired dorsal ovary displaced to the side. **E** *Halobiotus crispae.* Malpighian tubules. Two lateral and one dorsal organ with three voluminous sections each of three cells; three tube-shaped sections lead into the intestine. *bt* Buccal tube (oral pipe); *mg* midgut; *mu* oblique muscle; *oe* esophagus; *ph* pharynx; *re* rectum; *sg* stylet gland; *ss* stylet supporter; *st* stylet. (**A, B** R.M. Kristensen 1977; **C** R.M. Kristensen 1982a; **D** R.M. Kristensen 1982b; **E** Møbjerg & Dahl 1996)

Autapomorphies

- Appendages.
  Four pairs of short uniramous appendages; each with four claws.
- Combination of the pharynx with a stylet apparatus (Fig. 154C).
  From a long buccal tube follows an oval pharynx with a triradial lumen and ectodermal myoepithelium (EIBYE-JACOBSEN 1996); this may be a plesiomorphy. Unique, however, is the functional coupling with two stylets, which lie lateral to the pharynx and can be protruded through the buccal tube to pierce objects of prey. The stylets rest on stylet supporters which are themselves anchored onto the buccal tube.
- ? Unpaired gonads dorsal of the gut in both sexes.

The **problem of the phylogenetic kinship** of the Tardigrada remains **unsolved.** The obvious difficulties are often explained by the hypothesis of an evolutionary miniaturization of the body and a secondary impoverishment of the feature pattern. The Tardigrada have in the course of progenesis undergone a process of **miniaturization** (DEWEL & DEWEL 1997). If one wants to justify this assumption, the "miniaturized" Tardigrada must be shown to be a subtaxon of a more comprehensive monophylum with larger organisms and a corresponding macroscopic stem species.

Let us examine this concept in the framework of the following argument and ask **three precise questions**. Do the Tardigrada have certain features that can be assessed as synapomorphies with corresponding features in the (1) Articulata, (2) Arthropoda or (3) subgroups of these – and which therefore could be seen as the basis for the establishment of discrete adelphotaxa relations?

1. Metamerism with segmental coelomic sacs, nephridia and a ladder-like nervous system; teloblastic formation of new segments; prostomium and pygidium and differentiation of longitudinal muscle bands from the body wall musculature (p. 42) belong in the ground pattern of the **Articulata.** Apart from debatable similarities in the nervous system (see below) no ground pattern feature of the Articulata is realized in the Tardigrada. The famous five coelomic sac pairs of entodermal origin (MARCUS 1929a, b) have not been confirmed (EIBYE-JACOBSEN 1997a); possibly, in early ontogeny, paired lateral mesodermal strips develop that expand into cavities; however, in the current state of research this is unsure. In contrast, it is certain that during development of the Tardigrada no serially arranged coelomic sacs appear. The adult has an undivided, uniform body cavity that is surrounded by the basal lamina of adjoining organs. According to our definition this is a primary body cavity (Vol. I, p. 113). For the hypothesis of a division of the Tardigrada body into segments – three head segments and four to five trunk segments (DEWEL & DEWEL 1997) – ontogeny does not provide any basis. There is, of course, no post-embryonic teloblastic growth.
2. Without coelom and without nephridia the Tardigrada understandably lack the characteristics of the **Arthropoda** coelom, such as the pericardial septum or the segmental sacculi of the nephridia (p. 75); the Tardigrada have neither a blood circulatory system nor a heart with ostia.
   How about the cuticle with chitin and lack of epidermal cilia and the correlated molting connected with this? These phenomena are not limited to the Tardigrada and Arthropoda. Chitin is widespread in the Metazoa; the question of whether we are dealing with α-chitin in the Tardigrada remains unanswered (DEWEL & DEWEL 1997).
   A pair of antennae innervated from the brain belongs in the ground pattern of the Arthropoda; comparable appendage derivatives are absent in the Tardigrada. Diverse sensory structures at the front of mar-

ine species cannot be traced back to antennae. From the ultrastructure of the sensilla several support cells have been described (KRISTENSEN 1981) and discussed as a possible synapomorphy with the Arthropoda; this finding is, however, disputed (DEWEL & DEWEL 1997).

3. Within the Arthropoda we begin with the **Onychophora**. Here a comparison of the four leg pairs of the Tardigrada with the uniramous segmental trunk appendages of the Onychophora is suggested. However, grave differences quickly become clear. The four claws per Tardigrada appendage (Fig. 154B) are opposed by two claws in the Onychophora. The ventral transverse row of spiny bulges, on which the appendages of the Onychophora rest during locomotion, are without counterparts. The Tardigrada have no body wall musculature; isolated muscles intersperse the body cavity as stem and appendage muscles (Fig. 154B). In the Onychophora ring and diagonal muscles from the body wall musculature extend into the appendages (BIRKET-SMITH 1974). To innervate four pairs of highly mobile appendages four pairs of ganglia with a connection to the brain are to be expected for the Tardigrada. In their uniting to four ganglia with close-lying connectives the organization of the nervous system is fundamentally different to the widely spaced ventral cords of the Onychophora with 9–10 commissures per neuromere.

No congruencies exist which can support the hypothesis of a homology between the appendages of the Tardigrada and the Arthropoda. Comparing the two unities I interpret the legs of the Tardigrada as the product of an independent evolution and hypothesize that they are an autapomorphy of the taxon.

With this we must return to the nervous system. If one still wishes to consider a homology of the "ventral ladder" nervous system of the Tardigrada and Arthropoda, despite a non-provable homology of the appendages, then one would expect congruencies in the brain with the, at least in the Euarthropoda, clear division into proto-, deuto- and tritocerebrum. This, however, is not the case. According to DEWEL & DEWEL (1996), the brain of the Tardigrada corresponds "approximately" to the protocerebrum of the Arthropoda – provided that the model concepts of these authors withstand an empirical examination.

In a comparison with the **Euarthropoda** the Malpighian tubules at the mid-endgut boundary come into question (Fig. 154E). Such excretory organs evolved convergently in the stem lineages of the Arachnida and Tracheata with the transfer to land. If one wishes to hypothesize a synapomorphic congruence between the Malpighian tubules of the Eutardigrada and of certain Euarthropoda, then the Tardigrada must be proven to be representatives of the Arachnida or Tracheata; for this moreover a reduction in the Malpighian tubules in the "Heterotardigrada" would be necessary. I find no grounds for such considerations. We must assume an independent evolution of excretory organs given the label Malpighian tubules within the Tardigrada (GREVEN 1982) even when the similarities with the Malpighian tubules of the Protura (Insecta) are impressive (MØBJERG & DAHL 1996).

What is the result of this discussion for the assertion of a "miniaturization" of the Tardigrada?

One cannot prove a sister group kinship either to the monophylum Arthropoda as a whole or to a certain subtaxon of the Arthropoda that, a posteriori, could make plausible the hypothesis of a regressive evolution with miniaturization. The alternative is to be emphatically emphasized. **The Tardigrada can be primary small organisms.**

Every single species and every monophyletic unity of species has an adelphotaxon among recent organisms – independent of whether we can hypothesize this with good justification or not. The adelphotaxon of the Tardigrada has up to now not been provable.

Molecular results appear to be of little help at present. Analyses of 18S ribosomal RNA gene sequences have both supported (GAREY et al. 1996) and, at the same time, undermined (MOON & KIM 1996) a kinship between the Tardigrada and Arthropoda.

The hypothesis of a kinship between the Tardigrada and Arthropoda, in whatever form, may at present be without serious competition – but using the given feature pattern the instrument of phylogenetic systematics is unable to demonstrate this.

References

ABELE, L. G. & B. E. FELGENHAUER (1986). Phylogenetic and phenetic relationships among lower Decapoda. Journal of Crustacean Biology **6**, 385–400.

ABELE, L. G., KIM. W. & B. E. FELGENHAUER (1989). Molecular evidence for inclusion of the phylum Pentastomida in the Crustacea. Mol. Biol. Evol. **6**, 685–691.

ACHTELIG, M. (1981). Neuropteroidea. Revisionary notes. 286–287, 290, 293–295, 299–300, 307–308. In W. HENNIG. Insect Phylogeny. Wiley & Sons. Chichester.

ALBERTI, G. (1979). Zur Feinstruktur der Spermien und Spermiocytogenese von Prokoenenia wheeleri (Rucker, 1901) (Palpigradi, Arachnida). Zoomorphologie **94**, 111–120.

ALBERTI, G. (1995). Comparative spermatology of Chelicerata: review and perspective. In B. G. JAMIESON, J. AUSIO & J.-L. JUSTINE (Eds.). Advances in spermatozoal phylogeny and taxonomy. Mem. Mus. natn. Hist. nat. **166**, 203–230. Paris.

ALBERTI, G. & C. WEINMANN (1985). Fine structure of spermatozoa of some labidognath spiders (Filistatidae, Segestriidae, Dysderidae, Oonopidae, Scytodidae, Pholcidae; Araneae; Arachnida) with remarks on spermiogenesis. J. Morph. **185**, 1–35.

ANKEL, W. E. (1958). Begegnung mit Limulus. Natur und Volk **88**, 101–110, 153–162. Frankfurt a.M.

ASPÖCK, H. & U. ASPÖCK (1991). Raphidioptera. In: The Insects of Australia. Vol. I, 521–524. Melbourne University Press. Melbourne.

AVEL, M. (1959). Classe des Annélides Oligochètes (Oligochaeta Huxley, 1875). Traité de Zoologie V, 224–470. Paris.

AX, P. (1968). Die Bedeutung der interstitiellen Sandfauna für allgemeine Probleme der Systematik, Ökologie und Biologie. Veröffentl. Inst. Meeresforsch. Bremerhaven. Sonderband II, 15–65.

AX, P. (1969). Populationsdynamik, Lebenszyklen und Fortpflanzungsbiologie der Mikrofauna des Meeressandes. Verh. Dtsch. Zool. Ges. Innsbruck 1968, 65–113.

AX, P. (1984). Das Phylogenetische System. Systematisierung der lebenden Natur aufgrund ihrer Phylogenese. G. Fischer. Stuttgart, New York.

AX, P. (1987). The Phylogenetic System. The systematization of organisms on the basis of their phylogenesis. J. Wiley & Sons. Chichester, New York, Brisbane, Toronto, Singapore.

AX, P. (1994). Japanoplana insolita n. sp. – eine neue Organisation der Lithophora (Seriata, Plathelminthes) aus Japan. Microfauna Marina **9**, 7–23.

BADONNEL, A. (1951). Ordre des Psocopteres. Traité de Zoologie **X**, 1301–1340. Paris.

BÄHR, R. (1971). Die Ultrastruktur der Photorezeptoren von Lithobius forficatus L. (Chilopoda: Lithobiidae). Z. Zellforsch. **116**, 70–93.

BÄHR, R. (1974). Contribution to the morphology of chilopod eyes. Symp. Zool. Soc. London **32**, 383–404.

BALDERSON, J. (1991). Mantodes. In: The Insects of Australia. Vol. I, 348–356. Melbourne University Press. Melbourne.

BANG, F. B. & B. G. BANG (1962). Studies on sipunculid blood: immunologic properties of coelomic fluid and morphology of "urn cells". Cah. Biol. Mar **3**, 363–374.

BARTH, F. G. (1971). Der sensorische Apparat der Spaltsinnesorgane (Cupiennius salei Keys. Araneae). Z. Zellforsch. **112**, 212–246.

BARTH, F. G. (1972). Die Physiologie der Spaltsinnesorgane I. Modellversuche zur Rolle des cuticularen Spaltes beim Reiztransport. J. Comp. Physiol. **78**, 315–336.

BARTH, F. G. & W. LIBERA (1970). Ein Atlas der Spaltsinnesorgane von Cupiennius salei Keys. Chelicerata (Araneae). Z. Morph. Tiere **68**, 343–369.

BARTOLOMAEUS, T. (1992). Ultrastructure of the photoreceptor in the larvae of Lepidochiton cinereus (Mollusca, Polyplacophora) and Lacuna divaricata (Mollusca, Gastropoda). Microfauna Marina **7**, 215–236.

BARTOLOMAEUS, T. (1993 a). Die Leibeshöhlenverhältnisse und Nephridialorgane der Bilateria – Ultrastruktur, Entwicklung und Evolution. Habilitationsschrift Univ. Göttingen.

BARTOLOMAEUS, T. (1993 b). Die Leibeshöhlenverhältnisse und Verwandtschaftsbeziehungen der Spiralia. Verh. Deutsch. Zool. Ges. **88**, 1, 42.

BARTOLOMAEUS, T. (1994). On the ultrastructure of the coelomic lining in the Annelida, Sipuncula and Echiura. Microfauna Marina **9**, 171–220.

BARTOLOMAEUS, T. (1995). Structure and formation of the uncini in Pectinaria koreni, Pectinaria auricoma (Terebellida) and Spirorbis spirorbis (Sabellida): implications for annelid phylogeny and the position of the Pogonophora. Zoomorphology **115**, 161–177.

BARTOLOMAEUS, T. (1997). Ultrastructure of the renopericardial complex of the interstitial gastropod Philinoglossa helgolandica Hertfing, 1932 (Mollusca: Opisthobranchia). Zool. Anz. **235**, 165–176.

BARTOLOMAEUS, T. (1998). Chaetogenesis in polychaetous Annelida – significance for annelid systematics and the position of the Pogonophora. Zoology **100**, 348–364.

BARTOLOMAEUS, T. & W. AHLRICHS (1998). Morphologie und Phylogenie der Mollusca. Shaker Verlag. Aachen.

BARTOLOMAEUS, T., MEYER, K., HAUSEN, H., SCHWEIGKOFLER, M. & S. SCHULZ (1997). Die Anneliden-Borste Morphogenese und Bedeutung für die Systematik (Teil I und II). Mikrokosmos **88**, 91–98 und 159–167.

BARTOLOMAEUS, T. & H. RUHBERG (1999). Ultrastructure of the body cavity lining in embryos of Epiperipatus biolleyi (Onychophora, Peripatidae) – a comparison with annelid larvae. Invertebrate Biology **118**, 165–174.

BAUM, S. (1973). Zum "Cribellaten-Problem": Die Genitalstrukturen der Oecobiinae und Urocteinae (Arach.: Aran.: Oecobiidae). Abh. Verh. naturwiss. Verein Hamburg (NF) **16**, 101–153.

BEKLEMISCHEW, W. N. (1958). Grundlagen der vergleichenden Anatomie der Wirbellosen. Bd. 1. VEB Deutscher Verlag der Wissenschaften. Berlin.

BERTHOLD, T. & T. ENGESER (1987). Phylogenetic analysis and systematization of the Cephalopoda (Mollusca). Verh. naturwiss. Ver. Hamburg (NF) **29**, 187–220.

BEUTEL, R. (1997). Über Phylogenese und Evolution der Coleoptera (Insecta). insbesondere der Adephaga. Abh. naturwiss. Ver. Hamburg (NF) **31**, 1–164.

BIRKET-SMITH, S. J. R. (1974). The anatomy of the body wall of Onychophora. Zool. Jb. Anat. **93**, 123–154.

BITSCH, J. (1973). Voies digestives céphaliques des Insectes. Traité de Zoologie **VIII**, 1, 42–60. Paris.

BITSCH, J. (1979). Morphologie abdominale des Insects. Traité de Zoologie **VIII**, 2, 291–578. Paris.

BITSCH, J. (1994). The morphological groundplan of Hexapoda: critical review of recent concepts. Ann. Soc. Entomol. Fr. (N. S.) **30**, 103–129.

BLOWER, J. G. (1985). Millipedes. Synopses of the British Fauna (New Series) **35**, 1–242.

BOROFFKA, I. & R. HAMP (1969). Topographie des Kreislaufsystems und Zirkulation bei Hirudo medicinalis (Annelida, Hirudinea). Z. Morph. Tiere **64**, 59–76.

BORUCKI, H. (1996). Evolution und Phylogenetisches System der Chilopoda. Verh. naturw. Ver. Hamburg (NF) **35**, 95–226.

BOTNARIUC, N. & N. VIÑA BAYES (1977). Contribution a la connaissance de la biologie de Cyclestheria hislopi (BAIRD), (Conchostraca: Crustacea) de Cuba. In: T. ORGHIDAN & A. NUNEZ (Eds.). Résultats des Expeditions Biospéologique Cubano-Roumaines a Cuba. Bucaresti. Vol. **2**, 257–262.

BOUDREAUX, H. B. (1979). Arthropod phylogeny with special reference to Insects. J. Wiley & Sons. New York, Chichester, Brisbane, Toronto.

BOXSHALL, G. A. (1983). A comparative functional analysis of the major maxillopodan groups. In: F. R. SCHRAM (Ed.). Crustacean Issues 1. Crustacean Phylogeny. 121–143. A. A. Balkema. Rotterdam.

BOXSHALL, G. A. (1991). A review on the biology and phylogenetic relationships of the Tantulocarida, a subclass of Crustacea recognized in 1983. Verh. Dtsch. Zool. Ges. **84**, 271–279.

BOXSHALL, G. A., F. D. FERRARI & H. TIEMANN (1984). The ancestral copepod: towards a consensus of opinion at the First International Conference on Copepoda 1981. Crustaceana. Suppl. 7, 68–84.

BOXSHALL, G. A. & R. HUYS (1989). New tantulocarid, Stygotantulus stocki, parasitic on harpacticoid copepods, with an analysis of the phylogenetic relationships within the Maxillopoda. J. Crust. Biol. **9**, 126–140.

BOXSHALL, G. A. & R. J. LINCOLN (1983). Tantulocarida. A new class of Crustacea ectoparasitic on other crustaceans. J. Crust. Biol. **3**, 1–16.

BOXSHALL, G. A. & R. J. LINCOLN (1987). The life cycle of the Tantulocarida (Crustacea). Phil. Trans. Roy. Soc. London **315**, 267—303.

BOYLE, P. R. (1969). Fine Structure of the eyes of Onithochiton neglectus (Mollusca: Polyplacophora). Z. Zellforsch. **102**, 313–322.

BOYLE, P. R. (1974). The Aesthetes of Chitons. II. Fine structure in Lepidochiton cinereus (L.). Cell. Tiss. Res. **153**, 383–398.

BRIEN, P. (1959). Classe des Endoproctes ou Kamptozoaires (Endoprocta Nitsche 1870, Kamptozoa Cori 1929). Traité de Zoologie **V**, 1, 927–1007. Paris.

BRINKHURST, R. O. (1994). Evolutionary relationships within the Clitellata: an update. Megadrilogica **5**, 109–112.

BRINKHURST, R. O. & S. R. GELDER (1989). Did the lumbriculids provide the ancestor of the branchiobdellidans, acanthobdellidans and leeches? Hydrobiologia **180**, 7–15.

BRINKHURST, R. O. & A. F. L. NEMEC (1987). A comparison of phenetic and phylogenetic methods applied to the systematics of Oligochaeta. Hydrobiologia **155**, 65–74.

BROCCO, S.L., O'CLAIR, R. M. & R. A. CLONEY (1974). Cephalopod integument: the ultrastructure of Kölliker's organs and their relationship to setae. Cell. Tiss. Res. **151**, 293–308.

BUCHHOLZ, H. (1951). Die Larvenformen von Balanus improvisus. Kieler Meeresforsch. **8**, 49–57.
BUDELMANN, B. U., SCHIPP, R. & S. VON BOLETZKY (1997). Cephalopoda. In F. W. HARRISON & A. J. KOHN (Eds.). Microscopic Anatomy of Invertebrates 6A, 119–414. Wiley-Liss, Inc. New York.
BUDELMANN, B. U. & J. Z. YOUNG (1985). Central pathways of the nerves of the arms and mantle of Octopus. Phil. Trans. Royal Soc. London B **310**, 109–122.
BÜNING, J. (1994). The Insect Ovary. Ultrastructure, previtellogenic growth and evolution. Chapman & Hall. London.
BÜNING, J. (1998). Ovariole: Structure, Types and Phylogeny. In M. LOCKE & F. W. HARRISON (Eds). Microscopic Anatomy of Invertebrates **11**, 897–932. Wiley-Liss, Inc. New York.
BUNKE, D. (1986). Ultrastructural investigations in the spermatozoon and its genesis in Aeolosoma litorale with considerations on the phylogenetic implications for the Aeolosomatidae (Annelida). J. Ultrastr. Mol. Str. Res. **95**, 113–130.
BYERS, G. W. (1991). Mecoptera. In: The Insects of Australia. Vol. II, 696–704. Melbourne University Press. Melbourne,
CALABY, J. H. & M. D. MURRAY (1991). Phthiraptera. In: The Insects of Australia. Vol. I, 421–428. Melbourne University Press. Melbourne.
CALS, P., DELAMARE DEBOUTTEVILLE, C. & J. RENAUD-MORNANT (1968). Nature et adaptations à un mode de vie en milieu interstitiel des structures cuticulaires céphaliques chez Derocheilocaris remanei Delamare Deboutteville et Chappuis (Crustacea-Mystacocarida). C. R. Acad. Sc. Paris **266**, 126–129.
CARVER, M., GROSS, G. F. & T. E. WOODWARD (1991). Hemiptera. In: The Insects of Australia. Vol. I, 429–509. Melbourne University Press. Melbourne.
CASANOVA, B. (1991). Origine protocéphalique antennaire de la carapace chez les Leptostracés, Mysidacés et Eucaridés (Crustacés). Compt. rend. hebd. Seánc. l'Acad. Scienc. **312** (III), 461–468.
CASANOVA, B. (1993). Ĺorigine protocéphalique de la carapace chez les Thermosbaenacés, Tanaidacés, Cumacés et Stomatopodes. Crustaceana **65**, 144–150.
CAULLERY, M. (1914). Sur les Siboglinidae, type nouveau d'invertébrés receuillis par l'expedition du Siboga. Bull. Soc Zool. France **39**, 350–353.
CHAPMAN, R. F. (1991). General Anatomy and Function. In: The Insects of Australia. Vol. I, 33–67. Melbourne University Press. Melbourne.
CHOPARD, L. (1949). Ordre des Dictyoptères Leach, 1818. Traité de Zoologie **IX**, 355–407.
CHRISTIAN, U. H. (1973). Trichobothrien, ein Mechanorezeptor bei Spinnen. Elektronenmikroskopische Befunde bei der Winkelspinne Tegenaria derhami (Scopoli), (Agelenidae, Araneae). Verh. Dtsch. Zool. Ges. Mainz, 1972, 31–36.
CHRISTOFFERSEN, M. L. (1988). Phylogenetic systematics of the Eucarida (Crustacea, Malacostraca). Revista brasileira de Zoologia **5**, 325–351.
CLARK, R. B. (1964). Dynamics in metazoan evolution. The origin of the coelom and segments. Oxford University Press. London.
CODDINGTON, J. A. (1990). Cladistics and spider classification: araneomorph phylogeny and the monophyly of orbweavers (Araneae: Araneomorphae; Orbiculariae). Acta Zool. Fennica **190**, 75–87.
CODDINGTON, J. A. & H. W. LEVI (1991). Systematics and Evolution of spiders (Araneae). Annu. Rev. Ecol. Syst. **22**, 565–592.
CONDÉ, B. & J. PAGÉS (1991). Diplura. In: The Insects of Australia. Vol. I, 269–271. Melbourne University Press. Melbourne.
CUÉNOT, L. (1949). Les Onychophores. Traité de Zoologie **VI**, 3–37. Paris.
DAHL, E. (1952). Mystacocarida. Reports of the Lund University Chile Expedition 1948–1949. 7. Lunds. Univ. Arsskr. N. F. 2., **48**, 6, 1–41.
DAHL, E. (1956). Some crustacean relationships. In K. G. WINGSTRAND (Ed.). Bertil Hanström. Zoological papers in honour of his sixty fifth birthday. Zool. Inst. Lund, 138–147.
DALLAI, R. & B. A. AFZELIUS (1984). Paired spermatozoa in Thermobia (Insecta, Thysanura). J. Ultrastruct. Res. **86**, 67–74.

DELAMARE DEBOUTTEVILLE, C. (1954). Recherches sur l'écologie et la repartition du Mystacocaride Derocheilocaris remanei Delamare et Chappuis en Méditerranee. Vie et Milieu **4**, 321–380.

DELAMARE DEBOUTTEVILLE, C. & P.-A. CHAPPUIS (1954). Morphologie des Mystacocarides. Arch. de Zool. exp. et gen. **91**, 7–24.

DESPAX, R. (1949). Ordre des Plécoptères. Traité de Zoologie **IX**, 557–586. Paris.

DESPAX, R. (1951). Ordre des Trichoptères. Traité de Zoologie **X**, 125–173. Paris.

DEWEL, R. A. & W. C. DEWEL (1996). The brain of Echiniscus viridissimus Peterfi, 1956 (Heterotardigrada): a key to understanding the phylogenetic position of tardigrades and the evolution of the arthropod head. Zool. J. Linn. Soc. **116**, 35–49.

DEWEL, R. A. & W. C. DEWEL (1997). The place of tardigrades in arthropod evolution. In: R. A. FORTEY & R. H. THOMAS (Eds). Arthropod relationships. Syst. Ass. Spec. Vol. **55**, 109–123. London.

DOHLE, W. (1964). Die Embryonalentwicklung von Glomeris marginata (VILLIERS) im Vergleich zur Entwicklung anderer Diplopoden. Zool. Jb. Anat. **81**, 241–310.

DOHLE, W. (1965). Über die Stellung der Diplopoden im System. Verh. Dtsch. Zool. Ges. Kiel 1964, 597–606.

DOHLE, W. (1970). Über Eiablage und Entwicklung von Scutigera coleoptrata (Chilopoda). Bull. Mus. natn. Hist. nat. **41**(2), 53–57.

DOHLE, W. (1972). Über die Bildung und Differenzierung des postnauplialen Keimstreifs von Leptochalia spec. (Crustacea, Tanaidacea). Zool. Jb. Anat. **89**, 503–566.

DOHLE, W. (1974). The segmentation of the germ band of Diplopoda compared with other classes of Arthropoda. Symp. Zool. Soc. London **32**, 143–161.

DOHLE, W. (1976). Zur Frage des Nachweises von Homologien durch die komplexen Zell- und Teilungsmuster in der embryonalen Entwicklung höherer Krebse (Crustacea, Malacostraca, Peracarida). Sitzungsber. Ges. Naturf. Freunde Berlin. N. F. **16**, 125–144.

DOHLE, W. (1980). Sind die Myriapoden eine monophyletische Gruppe? Eine Diskussion der Verwandtschaftsbeziehungen der Antennata. Abh. naturw. Ver. Hamburg (NF) **23**, 45–104.

DOHLE, W. (1985). Phylogenetic pathways in the Chilopoda. Bijdr. Dierk. **55**, 55–66.

DOHLE, W. (1988). Myriapoda and the ancestry of insects. The Manchester Polytechnic, 1–28. Manchester.

DOHLE, W. (1990). Some observations of morphology and affinities of Craterostigmus tasmanianus (Chilopoda). In A. MINELLI (Ed.). Proc. 7th internat. Congr. Myriapodology. 69–79. Brill. Leiden.

DOHLE, W. (1996). Antennata (Tracheata, Monantennata, Atelocerata). In W. WESTHEIDE & R. RIEGER (Hrsg.). Spezielle Zoologie. Teil 1. 582–600. G. Fischer. Stuttgart, Jena, New York.

DOHLE, W. (1997). Myriapod-insect relationships as opposed to an insect-crustacean sister group relationship. In: R. A. FORTEY & R. H. THOMAS (Eds.). Arthropod relationships. Syst. Ass. Spec. Vol. **55**, 305–315. London.

DUNGER, W. (1993). Antennata. In H.-E. GRUNER (Ed.). Lehrbuch der Speziellen Zoologie. 4. Aufl. Band I, 4. Teil: Arthropoda (ohne Insecta), 1031–1160 G. Fischer. Jena, Stuttgart, New York.

DUNLOP, J. A. & P. A. SELDEN (1997). The early history and phylogeny of the chelicerates. In R. A. FORTEY & R. H.THOMAS (Eds.). Arthropod relationships. Syst. Ass. Spec. Vol. **55**, 221–235. London.

DUNNET, G. M. & D. K. MARDON (1991). Siphonaptera. In: The Insects of Australia. Vol. II, 705–716. Melbourne University Press. Melbourne.

EAKIN, R. M. & J. A. WESTFALL (1965). Fine structure of the eye of Peripatus (Onychophora). Z. Zellforsch. **68**, 278–300.

EERNISSE, D. J. (1997). Arthropod and annelid relationships reexamined. In: R. A. FORTEY & R. H. THOMAS (Eds.). Arthropod relationships. Syst. Ass. Spec. Vol. **55**, 43–56. London.

EERNISSE, D. J., ALBERT, J. S. & F. E. ANDERSEN (1992). Annelida and Arthropoda are not sister taxa: a phylogenetic analysis of spiralian metazoan morphology. Syst. Biol. **41**, 305–330.

EERNISSE, D. J. & P. D. REYNOLDS (1994). Polyplacophora. In: F. W. HARRISON & A. J. KUHN (Eds.). Microscopic Anatomy of Invertebrates **5**, Mollusca I, 55–110. Wiley-Liss, Inc. New York.

EHLERS, U. & B. SOPOTT-EHLERS (1997). Ultrastructure of the subepidermal musculature of Xenoturbella bocki, the adelphotaxon of the Bilateria. Zoomorphology **117**, 71–79.

EIBYE-JAGOBSEN, D. & C. NIELSEN (1996). Rearticulation of annelids. Zool. Scr. **25**, 275–282.

EIBYE-JACOBSEN, J. (1996). On the nature of pharyngeal muscle cells in the Tardigrada. Zool. J. Linn. Soc. **116**, 123–138.

EIBYE-JACOBSEN, J. (1997a). New observations on the embryology of the Tardigrada. Zool. Anz. **235**, 201–216.

EIBYE-JACOBSEN, J. (1997b). Development, ultrastructure and function of the pharynx of Halobiotus crispae Kristensen, 1982 (Eutardigrada). Acta Zool. **78**, 329–347.

EIDMANN, H. & F. KÜHLHORN (1970). Lehrbuch der Entomologie. Parey. Hamburg, Berlin.

EISENBEIS, G. & W. WICHARD (1985). Atlas zur Biologie der Bodenarthropoden. G. Fischer. Stuttgart, New York.

EMSCHERMANN, P (1965). Das Protonephridiensystem von Urnatella gracilis Leidy (Kamptozoa). Z. Morph. Ökol. Tiere **55**, 859–914.

EMSCHERMANN, P. (1972). Loxokalypus socialis gen. et sp. nov. (Kamptozoa, Loxokalypodidae fam. nov.), ein neuer Kamptozoentyp aus dem nördlichen Pazifischen Ozean. Ein Vorschlag zur Neufassung der Kamptozoensystematik. Marine Biology **12**, 237–254.

EMSCHERMANN, P. (1996). Kamptozoa (Entoprocta), Kelchwürmer. In: W. WESTHEIDE & R. RIEGER (Hrsg.). Spezielle Zoologie. Teil 1. 337–344. G. Fischer. Stuttgart, Jena, New York.

ENGELHARDT, W. (1959). Was lebt in Tümpel, Bach und Weiher. Franckh'sche Verlagshandlung. Stuttgart.

ENGHOFF, H. (1984). Phylogeny of millipedes – a cladistic analysis. Z. zool. Syst. Evolut.-forsch. **22**, 8–26.

ENGHOFF, H. (1990). The groundplan of the chilognathan millipedes (external morphology). In A. MINELLI (Ed.). Proc. 7th internat. Congr. Myriapodology. 1–21. Brill. Leiden.

ENGHOFF, H., DOHLE, W. & J. G. BLOWER (1993). Anamorphosis in millipeds (Diplopoda) – the present state of knowledge with some developmental and phylogenetic considerations. Zool. J. Linn. Soc. **109**, 103–234.

ERIKSSON, S. (1934). Studien über die Fangapparate der Branchiopoden nebst einigen phylogenetischen Bemerkungen. Zool. Bidr. Uppsala **15**, 23–287.

ERSEUS, C. (1987). Phylogenetic analysis of the aquatic Oligochaeta under the principle of parsimony. Hydrobiologia **155**, 75–89.

FAGE, L. (1949a). Classe des Pygnogonides Latreille 1810. Traité de Zoologie **VI**, 906–941. Paris.

FAGE, L. (1949b). Classe des Mérostomacés. Traité de Zoologie **VI**, 217–262. Paris.

FAUCHALD, K. & G. ROUSE (1997). Polychaete systematics: past and present. Zool. Scripta **26**, 71–138.

FECHTER, H. (1961). Anatomie und Funktion der Kopfmuskulatur von Cylindroiulus teutonicus (Pocock). Zool. Jb. Anat. **79**, 479–528.

FELGENHAUER, B. E. & L. G. ABELE (1983). Phylogenetic relationships among shrimp-like decapods. In F. R. SCHRAM (Ed.). Crustacean Phylogeny, 291–311. A. A. Balkema. Rotterdam.

FERRAGUTI, M. (1983). Annelida-Clitellata. In: K. G. ADIYODI & R. G. ADIYODI (Eds.). Reproductive Biology of Invertebrates, Vol. II: Spermatogenesis and sperm function. 343–385. John Wiley & Sons Ltd. Chichester.

FERRAGUTI, M. & G. LANZAVECCHIA (1977). Comparative electron microscopic studies of muscle and sperm cells in Branchiobdella pentodonta Whitman and Bythonomus lemani Grube (Annelida Clitellata). Zoomorphologie **88**, 19–36.

FISCHER-PIETTE, E. & A. FRANC (1960). Classe des Polyplacophores. Traité de Zoologie **V**, 1701–1785. Paris.

FLÖSSNER, D. (1972). Krebstiere, Crustacea. Kiemen- und Blattfüßer, Branchiopoda. Fischläuse, Branchiura. Die Tierwelt Deutschlands **60**, 1–501. G. Fischer. Jena.
FOELIX, R. F. (1992). Biologie der Spinnen. Thieme. Stuttgart, New York.
FORSTER, R. R., PLATNICK, N. I, & M. R. GRAY (1987). A review of the spider superfamilies Hypochiloidea and Austrochiloidea (Araneae, Araneomorpha). Bull. Amer. Mus. Nat. Hist. **185**, 1–116.
FRANC, A. (1960). Classe des Bivalves. Traité de Zoologie V, 1845–2164. Paris.
FRANKE, M. (1993). Ultrastructure of the protonephridia in Loxosomella fauveli, Barentsia matsushimana and Pedicellina cernua. Implications for the protonephridia in the ground pattern of the Entoprocta (Kamptozoa). Microfauna Marina **8**, 7–38.
FRANZÉN, A. (1962). Notes on the morphology and histology of Xironogiton instabilia (Moore 1893) (Fam. Branchiobdellidae) with special reference to the muscle cells. Zool. Bidr. Uppsala **35**, 369–384.
FRIEDRICH, M. & D. TAUTZ (1995). Ribosomal DNA phylogeny of the major extant arthropod classes and the evolution of myriapods. Nature **376**, 165—167.
FRIEDRICH, M. & D. TAUTZ (1997). Evolution and phylogeny of the Diptera: molecular phylogenetic analysis using 28S rDNA sequences. Syst. Biol. **46**, 674–698.
FUNCH, P. & R. M. KRISTENSEN (1995). Cycliophora is a new phylum with affinities to Entoprocta and Ectoprocta. Nature **378**, 711–714.
FRYER, G. (1987 a). Morphology and classification of the so-called Cladocera. Hydrobiologia **145**, 19–28.
FRYER, G. (1987 b). A new classification of the branchiopod Crustacea. Zool. J. Linn. Soc. London **91**, 357–383.
GAREY, J. R., KROTEC, M., NELSON, D. R. & J. BROOKS (1996). Molecular analysis supports a tardigrade-arthropod association. Invertebrate Biology **115**, 79–88.
GERHARDT, U. (1941). Merostomata. Handb. Zool. **3** (2), 1; 11–96.
GERTSCH, W. J. (1958). The spider family Hypochilidae. Amer. Mus. Novitates 1912, 1–28.
GEWECKE, M. (1995). Motorik. In: M. GEWECKE (Hrsg.). Physiologie der Insekten. G. Fischer. Stuttgart, Jena, New York.
GHISELIN, M. T. (1981). Categories, life, and thinking. The Behavioral and Brain Sciences **4**, 269–313.
GHISELIN, M. T. (1988). The origin of molluscs in the light of molecular evidence. Oxford Surveys in Evolutionary Biology **5**, 66–95.
GLATZ, L. (1972). Der Spinnapparat haplogyner Spinnen (Arachnida, Araneae). Z. Morph. Tiere **72**, 1–25.
GLATZ. L. (1973). Der Spinnapparat der Orthognatha (Arachnida, Araneae). Z. Morph. Tiere **75**, 1–50.
GLENNER, H., GRYGIER, M. J., HOEG, J. T., JENSEN, P. G. & F. R. SCHRAM (1995). Cladistic analysis of the Cirripedia Thoracica. Zool. J. Linn. Soc. London **114**, 365–404.
GRASSÉ, P.-P. (1951). Ordre des Mécoptères. Traité de Zoologie **X**, 71–124. Paris.
GREENSLADE, P. J. (1991). Collembola. In: The Insects of Australia. Vol. I, 265–268. Melbourne University Press. Melbourne.
GRELL, K. G. (1938). Der Darmtraktus von Panorpa communis L. und seine Anhänge bei Larve und Imago. (Ein Beitrag zur Anatomie und Histologie der Mecopteren.). Zool. Jb. Anat. **64**, 1–86.
GREVEN, H. (1982). Homologues or analogues? A survey of some structural patterns in Tardigrada. Proc. 3. Int. Symp. Tardigrada. 55–76. East Tennessee University Press. Johnson City, Tennessee.
GRUNER, H.-E. (1969). Klasse Crustacea – Krebstiere. In: Urania Tierreich. Wirbellose Tiere 2, 262–420. H. Deutsch. Frankfurt/M., Zürich.
GRUNER, H.-E. (1993). Klasse Crustacea, Krebse. In: H.-E. GRUNER (Ed.). Lehrbuch der Speziellen Zoologie. 4. Aufl. Band I. 4.Teil: Arthropoda (ohne lnsecta), 448–1030. G. Fischer. Jena, Stuttgart, New York.

GRYGIER, M. J. (1985). Comparative morphology and ontogeny of the Ascothoracida, a step towards a phylogeny of the Maxillopoda. Dissertation Abstracts International, Section B, **45**, 2466B–2467B.

GRYGIER, M. J. (1987). New records, external and internal anatomy, and systematic position of Hansen's Y-larvae (Crustacea: Maxillopoda Facetotecta). Sarsia **72**, 261–278.

GÜNTHER, K. (1969 a). Ordnung Ephemeroptera – Eintagsfliegen. In: Urania Tierreich. Insekten, 36–43. H. Deutsch. Frankfurt/M., Zürich.

GÜNTHER, K. (1969 b). Ordnung Heteroptera – Wanzen. In: Urania Tierreich. Insekten, 140–165. H. Deutsch. Frankfurt/M., Zürich.

GÜNTHER, K. & K. HERTER (1974). Dermaptera (Ohrwürmer). Handb. Zool. **4** (2). 2/11. W. de Gruyter. Berlin, New York.

GUSTUS, R. M. & R. A. CLONEY (1972). Ultrastructural similarities between setae of brachiopods and polychaetes. Acta Zool. **53**, 229–233.

HAAS, F. (1935). Bivalvia. Teil 1. Bronns Klassen u. Ordnungen d. Tierreichs. 3. Bd. Mollusca. III. Abt., 1–984.

HAAS. W. (1972). Untersuchungen über die Mikro- und Ultrastruktur der Polyplacophorenschale. Biomineralisation **5**, 1–52.

HAAS, W. (1981). Evolution of calcareous hardparts in primitive molluscs. Malacologia **21**, 403–418.

HAAS, W. (1989). Suckers and arm hooks in Coleoidea (Cephalopoda, Mollusca) and their bearing for Phylogenetic Systematics. Abh. naturw. Verein Hamburg **28**, 165–185.

HALL, J. R. (1972). Aspects of the biology of Derocheilocaris typica (Crustacea: Mystacocarida). II. Distribution. Mar. Biol. **12**, 42–52.

HALLBERG, E. (1982). The fine structure of the compound eye of Argulus foliaceus (Crustacea: Branchiura). Zool. Anz. **208**, 227–236.

HALLBERG, E. & R. ELOFSSON (1983). The larval compound eye of barnacles. J. Crust. Biol. **3**, 17–24.

HALLBERG, E., ELOFSSON, R. & M. J. GRYGIER (1985). An ascothoracid compound eye (Crustacea). Sarsia **70**, 167–171.

HARANT, H. & P.-P. GRASSÉ (1959). Classe des annélides achètes ou hirudinées ou sangsues. Traité de Zoologie **V**, 1, 471–593. Paris.

HARTMANN, G. (1966–1989). Ostracoda. In: Bronns Klassen u. Ordn. Tierreich 5 (1), 2(4), 1–1067.

HARTWICH, G. (1993). Stamm Entoprocta (syn. Kamptozoa), Kelchwürmer. In H.-E. GRUNER (Hrsg.). Lehrbuch der Speziellen Zoologie. 5. Aufl. Band I, 5. Teil. 455–462. G. Fischer. Jena, Stuttgart, New York.

HASZPRUNAR, G. (1988). On the origin and evolution of major gastropod groups, with special reference to the Streptoneura. J. Mollus. Stud. **54**, 367–441.

HASZPRUNAR, G. (1992). The first molluscs – small animals. Boll. Zool. **59**, 1–16.

HASZPRUNAR, G. (1996). The Mollusca: Coelomate turbellarians or mesenchymate annelids? In: J. TAYLOR (Ed.). Origin and evolutionary radiation of the Mollusca. 1–28. Oxford University Press. Oxford, New York, Tokyo.

HASZPRUNAR, G. & K. SCHAEFER (1997). Anatomy and phylogenetic significance of Micropilina arntzi (Mollusca, Monoplacophora, Micropilinidae fam. nov.) Acta Zool. **77**, 315–334.

HAUB, F. (1972). Das Cibarialsklerit der Mallophaga-Amblycera und der Mallophaga-lschnocera (Kellogg) (Insecta). Z. Morph. Tiere **73**, 249–261.

HAUPT, J. (1973). Die Ultrastruktur des Pseudoculus von Allopauropus (Pauropoda) und die Homologie der Schläfenorgane. Z. Morph. Tiere **76**, 173–191.

HAUPT, J. (1979). Phylogenetic aspects of recent studies on Myriapod sense organs. In: M. CAMATINI (Ed.). Myriapod biology. 391–406. Academic Press Inc. London.

HAUPT, J. (1983). Vergleichende Morphologie der Genitalorgane und Phylogenie der liphistiomorphen Webspinnen (Araneae: Mesothelae). Z. zool. Syst. Evolut.-forsch. **21**, 275–293.

HENNIG, W. (1950). Grundzüge einer Theorie der phylogenetischen Systematik. Deutscher Zentralverlag. Berlin.

HENNIG, W. (1953). Kritische Bemerkungen zum phylogenetischsn System der Insekten. Beitr. Ent. **3**, 1–85.

HENNIG, W. (1966). Phylogenetic systematics. University of Illinois Press. Urbana, Chicago, London.

HENNIG, W. (1969). Die Stammesgeschichte der Insekten. W. Kramer. Frankfurt am Main.

HENNIG, W. (1973). Diptera (Zweiflügler). Handb. Zool. **4**(2), 2/31, 1–337. Berlin.

HENNIG, W. (1981). Insect phylogeny. J. Wiley and Sons. Chichester, New York, Brisbane, Toronto.

HENNIG, W. (1994). Wirbellose II. Gliedertiere. 5. Aufl. (Nachdruck). G. Fischer, Jena.

HENTSCHEL, E. J. & H. WAGNER (1996). Zoologisches Wörterbuch. G. Fischer. Jena.

HESSLER, R. R. (1964). The Cephalocarida. Comparative skeleton musculature. Mem. Connecticut Acad. Arts Sci. **16**, 1–97.

HESSLER, R. R. (1969 a). Mystacocarida. In: R. C. MOORE (Ed.). Treatise on Invertebrate Palaeontology. Part R. Arthropoda 4, Vol. l, 192–195.

HESSLER. R. R. (1969 b). Branchiura. In: R. C. MOORE (Ed.). Treatise on Invertebrate Palaeontology. Part R. Arthropoda 4. Vol 1, 203–206.

HESSLER, R. R. (1969 c). Euphausiacea. In: R. C. MOORE (Ed.). Treatise on Invertebrate Palaeontology, Part R. Arthropoda 4. Vol. 1, 394–398.

HESSLER, R. R. (1983). A defense of the caridoid facies: Wherein the early evolution of the Eumalacostraca is discussed. In: F. R. SCHRAM (Ed.). Crustacean Phylogeny, 145–164. A. A. Balkema. Rotterdam.

HESSLER, R. R. (1992). Reflections on the phylogenetic position of the Cephalocarida. Acta Zool. **73**, 315–316.

HESSLER, R. R. & W. A. NEWMAN (1975). A trilobitomorph origin for the Crustacea. Fossils and Strata **4**, 437–459.

HILKEN, G. (1994). Struktur und phylogenetischer Ursprung der Tracheensysteme von Scutigera coleoptrata (Chilopoda, Notostigmophora). Verh. Dtsch. Zool. Ges. Jena. **87**(1), 104.

HILKEN, G. (1997). Tracheal systems in Chilopoda: a comparison under phylogenetic aspects. Ent. Scand. Suppl. **51**, 49–60.

HILKEN, G. (1998). Vergleich von Tracheensystemen unter phylogenetischem Aspekt. Verh. naturwiss. Verein Hamburg (NF) **37**, 5–94.

HILKEN, G. & O. KRAUS (1994). Struktur und Homologie der Komponenten des Gnathochilariums der Chilognatha. Verh. naturw. Ver. Hamburg (NF) **34**, 33–50.

HOEG, J. T. (1987). The relation between cypris ultrastructure and metamorphosis in male and female Sacculina carcini (Crustacea, Cirripedia). Zoomorphology **107**, 299–311.

HOEG, J. T. (1991). Functional and evolutionary aspects of the sexual systems in the Rhizocephala (Thecostraca: Cirripedia). In: R. T. BAUER & J. M. MARTIN (Eds). Crustacean sexual biology, 208–227. Columbia University Press. New York.

HOLT, P. C. (1965). The systematic position of the Branchiobdellidae (Annelida: Clitellata). Syst. Zool. **14**, 25–32.

HOLT, P. C. (1989). Comments on the classification of the Clitellata. Hydrobiologia **180**, l-5.

HOPKIN, S. P, & H. S. ANGER (1992). On the structure and function of the glue-secreting glands of Henia vesuviana (Newport, 1845) (Chilopoda: Geophilomorpha). Ber. nat.-med. Verein Innsbruck. Suppl. **10**, 71–79.

HOPKIN, S. P & H. J. READ (1992). The Biology of Millipedes. Oxford University Press. Oxford, New York, Tokyo.

HÜTHER, W. (1968). Erstnachweis der Pauropoda Hexamerocerata für Südamerika, mit Beschreibung einer neuen Art. Rev. Ecol. Biol. Sol. **5**, 561–567.

HUYS, R. (1991). Tantulocarida (Crustacea: Maxillopoda): A new taxon from the temporary Meiobenthos. Marine Ecology **12**, 1–34.

HUYS, R. & G. A. BOXSHALL (1991). Copepod Evolution. The Ray Society, London.

HUYS, R. BOXSHALL, G. A. & R. J. LINCOLN (1993). The tantulocaridan life cycle: the circle closed? J. Crust. Biol. **13**, 432–442.

ILLIES, J. (1968). Ephemeroptera (Eintagsfliegen). Handb. Zool. **4** (2), 2/5, 1–63. Berlin.

IMADATÉ, G. (1991). Protura. In: The Insects of Australia. Vol. I, 265–268. Melbourne University Press. Melbourne
IMMS, A. D. (1957). A General Textbook of Entomology. Methuen & Co. Ltd. London.
IVANOV, A. V. (1957). Neue Pogonophora aus dem nordwestlichen Teil des Stillen Ozeans. Zool. Jb. Syst. **85**, 431–500,
IVANOV, A. V. (1963). Pogonophora. Academic Press Inc. London.
IVANOV, D. L. (1996). Origin of Aculifera and problems of monophyly of higher taxa in molluscs. In: J. TAYLOR (Ed.). Origin and evolutionary radiation of the Mollusca. 59–65. Oxford University Press. Oxford, New York, Tokyo.
JACOBS, W. & M. RENNER (1974). Taschenlexikon zur Biologie der Insekten. G. Fischer. Stuttgart.
JACOBS, W. & F. SEIDEL (1975). Systematische Zoologie: Insekten. Uni-Taschenbücher 368. G. Fischer. Stuttgart.
JÄGERSTEN, G. (1972). Evolution of the metazoan life cycle. Academic Press. London, New York.
JAHN, I., LÖTHER, R. & K. SENGLAUB (1982). Geschichte der Biologie. Theorien, Methoden, Institutionen, Kurzbiographien. VEB Gustav Fischer Verlag. Jena.
JAMIESON, B. G. M. (1983 a). The ultrastructure of the spermatozoon of the oligochaetoid polychaete Questa spec. (Questidae, Annelida) and its phylogenetic significance. J. Ultrastruct. Res. **84**, 238—251.
JAMIESON, B. G. M. (1983 b). Spermatozoal ultrastructure: evolution and congruence with a holomorphological phylogeny of the Oligochaeta (Annelida). Zool. Scr. **12**, 107–114.
JAMIESON, B. G. M. (1987). The ultrastruclure and phylogeny of insect spermatozoa. Cambridge University Press. Cambridge.
JAMIESON, B. G. M. (1988). On the phylogeny and higher classification of fhe Oligochaeta. Cladistics **4**, 364–410.
JAMIESON, B. G. M. (1991). Ultrastructure and phylogeny of crustacean spermatozoa. Memoirs of the Queensland Museum **31**, 109–142.
JAMIESON, B. G. M. (1992). Oligochaeta. In: F. W. HARRISON & S. L. GARDINER (Eds). Microscopic Anatomy of Invertebrates 7, 217–322. Wiley-Liss, Inc. New York.
JANDER, U. (1966). Untersuchungen zur Stammesgeschichte von Putzbewegungen von Tracheaten. Z. Tierpsychol. **23**, 799–844.
JANSSON, B.-O. (1966). On the ecology of Derocheilocaris remanei Delamare and Chappuis (Crustacea, Mystacocarida). Vie et Milieu **17**, 143–186.
JEANNEL, R. (1949). Ordre des Coléoptères. Partie générale. Traité de Zoologie **IX**, 771–891. Paris.
KABATA, Z. (1979). Parasitic Copepoda of British Fishes. The Ray Society. London.
KAESTNER, A. (1941). 1. Ordnung der Arachnida: Scorpiones. Handb. Zool. **3** (2), 1; 117–240. Berlin.
KAESTNER, A. (1965). Lehrbuch der Speziellen Zoologie. Band I. Wirbellose, 1. Teil. G. Fischer. Jena.
KAESTNER, A. (1973). Lehrbuch der Speziellen Zoologie. Band I. Wirbellose. 3. Teil. Insecta: B. Spezieller Teil. G. Fischer. Jena.
KALTENBACH, A. (1968). Embioidea (Spinnfüßer). Handb. Zool. **4** (2), 2/8, 1–29. Berlin.
KARG, W. (1971). Die freilebenden Gamasina (Gamasides), Raubmilben. Die Tierwelt Deutschlands **59**, 1–475. G. Fischer. Jena.
KATHIRITHAMBY, J. (1991). Strepsiptera. In: The Insects of Australia. Vol. II, 684–695. Melbourne University Press. Melbourne.
KEY, K. H. L. (1991). Phasmatodea. In: The Insects of Australia. Vol. I, 394–404. Melbourne University Press. Melbourne.
KILIAS, R. (1993 a). Stamm Mollusca, Weichtiere. In: H.-E. GRUNER (Hrsg.). Lehrbuch der Speziellen Zoologie. 5. Aufl. Band I, 3. Teil: Mollusca, Sipunculida, Echiurida, Annelida, Onychophora, Tardigrada, Pentastomida. 9–245. G. Fischer. Jena, Stuttgart, New York.
KILIAS, R. (1993 b). Stamm Mollusca-Weichtiere. Urania Tierreich. Wirbellose 1 (Protozoa-Echiurida). 390–631. Urania-Verlag. Leipzig, Jena, Berlin.

KINCHIN, I. M. (1994).The biology of tardigrades. Portland Press. London.
KINZELBACH, R. (1966). Zur Kopfmorphologie der Fächerflügler (Strepsiptera, Insecta). Zool. Jb. Anat. **84**, 559–684.
KINZELBACH, R. (1971). Strepsiptera (Fächedlügler). Handb. Zool. **4** (2), 2/24, 1–68. Berlin.
KINZELBACH, R. (1981). Strepsiptera. Revisionary notes. 311–313. In: W. HENNIG. Insect phylogeny. Wiley & Sons. Chichester.
KJELLESVIG-WAERING, E. N. (1986). A restudy of the fossil Scorpionida of the world. Palaeontographica Americana **55**, 1–287.
KLASS, K.-D. (1995). Die Phylogenie der Dictyoptera. Cuvillier Verlag. Göttingen.
KLASS, K.-D. (1997). The external male genitalia and the pylogeny of Blattaria and Mantodea. Bonn. Zool. Monogr. **42**,1–341.
KLAUSNITZER, B. (1996). lnsecta (Hexapoda), Insekten. In: W. WESTHEIDE & R. RIEGER (Hrsg.). Spezielle Zoologie. Teil 1. 601–681. G. Fischer. Stuttgart, Jena, New York.
KLAUSNITZER, B. & K. RICHTER (1981). Stammesgeschichte der Gliedertiere (Articulata). Die Neue Brehm-Bücherei. A. Ziemsen. Wittenberg Lutherstadt.
KLEINOW, W. (1966). Untersuchungen zum Flügelmechanismus der Dermapteren. Z. Morph. Ökol. Tiere **56**, 363–416.
KLEMM, N. (1966). Die Morphologie des Kopfes von Rhyacophila Pict. (Trichoptera). Zool. Jb. Anat. **83**, 1–51.
KLUGE, A. G. (1997). Testability and the refutation and corroboration of cladistic hypotheses. Cladistics **13**, 81–96.
KOCH, M. (1997). Monophyly and phylogenetic position of the Diplura (Hexapoda) Pedobiologia **41**, 9–12.
KÖNIGSMANN, E. (1976). Das phylogenetische System der Hymenoptera. Teil 1: Einführung, Grundplanmerkmale, Schwestergruppe und Fossilfunde. Dt. ent. Z. **23**, 253–279.
KÖNIGSMANN, E. (1977). Das phylogenetische System der Hymenoptera. Teil 2: ."Symphyta". Dt. ent. Z. **24**, 1–40.
KÖNIGSMANN, E. (1978a). Das phylogenetische System der Hymenoptera. Teil 3: "Terebrantes" (Unterordnung Apocrita). Dt. ent. Z. **25**, 1–55.
KÖNIGSMANN, E. (1978b). Das phylogenetieche System der Hymenoptera. Teil 4: Aculeata (Unterordnung Apocrita). Dt. ent. Z., **25**, 365–435.
KÖNIGSMANN, E. (1981). Hymenoptera. Revisonary notes. 318, 403–406. In: W. HENNIG. Insect phylogeny. Wiley & Sons. Chichester.
KORN, H. (1960). Ergänzende Beobachtungen zur Struktur der Larve von Echiurus abyssalis Skor. Z. wiss. Zool. **164**, 199–237.
KRAUS, O. (1978). Liphistius and the evolution of spider genitalia. Symp. Zool. Soc. London **42**, 235–254.
KRAUS, O. (1984). Male spider genitalia: Evolutionary changes in structure and function. Verh. naturwiss. Ver. Hamburg (NF) **27**, 373–382.
KRAUS, O. (1990). On the so-called thoracic segments in Diplopoda. In: A. MINELLI (Ed.). Proc. 7th internat. Congr. Myriapodology. 63–68. Brill. Leiden.
KRAUS, O. (1997). Phylogenetic relationships between higher taxa of tracheate arthropods. In: R. A. FORTEY & R. H. THOMAS (Eds). Arthropod relationships. Syst. Ass. Spec. Vol. **55**, 295–303. London.
KRAUS, O. & M. KRAUS (1993a). Divergent transformation of chelicerae and original arrangement of eyes in spiders (Arachnida, Araneae). Mem. Queensland Mus. **33**, 579–584.
KRAUS, O. & M. KRAUS (1993b). Phylogenetisches System der Tracheata: Die Frage nach der Schwestergruppe der Insecta. Mitt. Dtsch. Ges. Allg. Angew. Ent. **8**, 441–446.
KRAUS, O. & M. KRAUS (1994). Phylogenetic System of the Tracheata (Mandibulata): on "Myriapoda"-Insecta relationships, phylogenetic age and primary ecological niches. Verh. naturw. Ver. Hamburg (NF) **34**, 5–31.
KRAUS, O. & M. KRAUS (1996). On myriapod-insect interrelationhips. In: J. J. GEOFFROY, J.-P. MAURIÈS & M. NGUYEN DUY-JACQUEMIN (Eds). Acta Myriapodologica. Mem. Mus. natn. Hist. nat. **169**, 283–290. Paris.

KRISTENSEN, N. P. (1975). The phylogeny of hexapod "orders". A critical review of recent accounts. Z. zool. Syst. Evolut.-forsch. **13**, 1–44.

KRISTENSEN, N. P (1981). Phylogeny of insect orders. Ann. Rev. Entomol. **26**, 135–157.

KRISTENSEN, N. P. (1984). Studies on the morphology and systematics of primitive Lepidoptera. Steenstrupia **10**, 141–191.

KRISTENSEN, N. P. (1989). Insect phylogeny based on morphological evidence. In: B. FERNHOLM, K. BREMER & H. JÖRNVALL (Eds.). The Hierarchy of Life. Nobel Symp. **70**, 295–306. Elsevier Science Publishers. Amsterdam.

KRISTENSEN, N. P. (1991). Phylogeny of extant hexapods. In: The lnsects of Australia. Vol. I, 125–140. Melbourne University Press. Melbourne.

KRISTENSEN, N. P. (1994). Phylogeny of extant hexapods. In: I. D. NAUMANN (Ed.). Systematic and applied Entomology, 117–132. Melbourne University Press. Melbourne (Reproduction of the paper of 1991).

KRISTENSEN, N. P. (1995). Forty years' Insect Phylogenetic Systematics. Zool. Beitr,. N. F. **36**, 83–124.

KRISTENSEN, N. P. (1997a). The ground plan and basal diversification of the hexapods. In: R. A. FORTEY & R. H. THOMAS (Eds.). Arthropod relationships. Syst. Ass. Spec.Vol. **55**, 281–293. London.

KRISTENSEN. N. P (1997b). Early evolution of the Lepidoptera + Trichoptera lineage: phylogeny and ecological scenario. In: P. GRANDCOLAS (Ed.). The origin of biodiversity in insects: phylogenetic tests of evolutionary scenarios. Mem. Mus. natn. Hist. nat. **173**, 253–271. Paris.

KRISTENSEN, N. P & E. S. NIELSEN (1981). Intrinsic proboscis musculature in non-ditrysian Lepidoptera-Glossata: Structure and phylogenetic significance. Ent. scand. Suppl. **15**, 299–304.

KRISTENSEN, N. P & A. W. SKALSKI (1998). Phylogeny and palaeontology. In: N. P. KRISTENSEN (Ed.). Lepidoptera: Moths and butterflies. 1. Handbuch der Zoologie. Handbook of Zoology **IV**, 35, 7–25. De Gruyter. Berlin & New York.

KRISTENSEN, R. M. (1976). On the fine structure of Batillipes noerrevangi Kristensen 1976. 1. Tegument and moulting cycle. Zool. Anz. **197**, 129–150.

KRISTENSEN, R. M. (1977). On the marine genus Styraconyx (Tardigrada, Heterotardigrada, Halechiniscidae) with description of a new species from a warm spring on Disco Island, West Greenland. Astarte **10**, 87–91.

KRISTENSEN, R. M. (1978). On the structure of Batillipes noerrevangi Kristensen 1978. 2. The muscle-attachements and the true cross-striated muscles. Zool. Anz. **200**, 173–184.

KRISTENSEN, R. M. (1981). Sense organs of two marine arthrotardigrades (Heterotardigrada. Tardigrada). Act. Zool. **62**, 27–41.

KRISTENSEN, R. M. (1982a). The first record of cyclomorphosis in Tardigrada based on a new genus and species from Arctic meiobenthos. Z. zool. Syst. Evolut.-forsch. **20**, 249–270.

KRISTENSEN, R. M. (1982b). New aberrant eutardigrades from homothermic springs of Disco Island, West Greenland. Proc. 3. Int. Symp. Tardigrada. 203–220. East Tennessee State University Press. Johnson City, Tennessee.

KRÜGER, P. (1940a). Ascothoracida. Bronns Klassen und Ordnungen des Tierreichs **5**, 1. Abt., 4. Buch. 1–46.

KRÜGER, P. (1940b). Cirripedia. Bronns Klassen und Ordnungen des Tierreichs **5**, 1. Abt., 3. Buch, 1–560.

KUKALOVÁ-PECK, J. (1983). Origin of insect wing and wing venation from the arthropodan leg. Can. J. Zool. **61**, 1618–1669.

KUKALOVÁ-PECK, J. (1985). Ephemeroid wing venation based upon gigantic Carboniferous mayflies and basic morphology, phylogeny and metamorphosis of pterygote insects (Insecta, Ephemerida). Can. J. Zool. **63**, 933–955.

KUKALOVÁ-PECK, J. (1987). New Carboniferous Diplura, Monura, and Thysanura, the hexapod ground plan and the role of thoracic side lobes in the origin of wings (Insecta). Can. J. Zool. **65**, 2327–2345.

KUKALOVÁ-PECK, J. (1991). Fossil history and the evolution of hexapod structures. In: The Insects of Australia. Vol. I, 141–179. Melbourne University Press. Melbourne.

KUKALOVÁ-PECK, J. (1992). The "Uniramia" do not exist: the ground plan of the Pterygota as revealed by Permian Diaphanapterodea from Russia (Insecta: Palaeodictyopteroidea). Can. J. Zool. **70**, 236–255.

KUKALOVÁ-PECK, J. (1997). Arthropod phylogeny and "basal" morphological structures. In: R. A. FORTEY & R. H. THOMAS (Eds.). Arthropod relationships. Syst. Ass. Spec. Vol. **55**, 249–268. London.

KUKALOVÁ-PECK, J. & J. F. LAWRENCE (1993). Evolution of the hindwing in Coleoptera. Can. Ent. **125**, 181–258.

LANZAVECCHIA, G. & M. DE EGUILEOR (1976). Studies on the helical and paramyosinic muscles.V. Ultrastructural morphology and contraction speed of muscular fibres of Erpobdella octoculata and Erpobdella testacea (Annelida, Hirudinea). J. Submicr. Cytol. **8**, 69–88.

LARSEN, O. (1945). Das thorakale Skelettmuskelsystem der Heteropteren. Ein Beitrag zur vergleichenden Morphologie des Insektenthorax. Acta Univ. Lund NF **41** (11), 1–83.

LAUTERBACH, K.-E. (1973). Schlüsselereignisse in der Evolution der Stammgruppe der Euarthropoda. Zool. Beitr. **19**, 251–299.

LAUTERBACH, K.-E. (1974). Über die Herkunft des Carapax der Crustaceen. Zool. Beitr. **20**, 273–327.

LAUTERBACH, K.-E. (1975). Über die Herkunft der Malacostraca (Crustacea). Zool. Anz. **194**, 165–179.

LAUTERBACH, K.-E. (1980 a). Schlüsselereignisse in der Evolution des Grundplans der Mandibulata (Arthropoda). Abh. naturwiss. Ver. Hamburg (NF) **23**, 105–161.

LAUTERBACH, K.-E. (1980 b). Schlüsselereignisse in der Evolution des Grundplans der Arachnata (Arthropoda). Abh. naturwiss. Ver. Hamburg (NF) **23**, 103–327

LAUTERBACH, K.-E. (1983 a). Zum Problem der Monophylie der Crustacea. Verh. naturwiss. Ver. Hamburg (NF) **26**, 293–320.

LAUTERBACH, K.-E. (1983 b). Gedanken zur Entstehung der mehrfach paarigen Exkretionsorgane von Neopilina (Mollusca, Conchifera). Z. zool. Syst. Evolut.-forsch. **21**, 38–52.

LAUTERBACH, K.-E. (1983 c). Erörterungen zur Stammesgeschichte der Mollusca, insbesondere der Conchifera. Z. zool. Syst. Evolut.-forsch. **21**, 201–216.

LAUTERBACH, K.-E. (1984 a). Das phylogenetische System der Mollusca. Mitt. dtsch. malak. Ges. **37**, 66–81. Frankfurt a. M.

LAUTERBACH, K.-E. (1984 b). Homologieverhältnisse der Molluskenherzen. Zool. Anz. **212**, 163–179.

LAUTERBACH, K.-E. (1986). Zum Grundplan der Crustacea. Verh. naturwiss. Ver. Hamburg (NF) **28**, 27–63.

LAUTERBACH, K.-E. (1996). Grabwespen (Hymenoptera-Sphecidae) in Bielefeld und Umgegend I: Sandwespen (Ammophilomorpha). Mit Anmerkungen zum Phylogenetischen System und Problemen der Phylogenetischen Systematik. Ber. Naturw. Verein Bielefeld u. Umgegend **37**, 127–152.

LEE, M. S. Y. (1996). The phylogenetic approach to biological taxonomy: practical aspects. Zool. Scripta **25**, 187–190.

LEMCHE, H. (1957). A new living deep-sea mollusc of the Cambro-Devonian class Monoplacophora. Nature **179**, 413–416.

LEMCHE, H. & K. G. WINGSTRAND (1959). The anatomy of Neopilina galatheae Lemche, 1957. Galathea Rept. **3**, 9–71.

LEWIS, J. G. E. (1981). The biology of centipedes. Cambridge University Press. Cambridge.

LIDEN, M. & B. OXELMAN (1996). Do we need phylogenetic taxonomy. Zool. Scripta **25**, 183–185.

LINDER, F. (1946). Affinities within the Branchiopoda, with notes on some dubious fossils. Arkiv f. Zoologi **37A**, 4, 1–28.

LINNÉ, C. (1735). Systema naturae, sive regna tria naturae systematice proposita per classes, ordines, genera, et species. Lugduni Batavorum.

LIVANOW, N. (1931). Die Organisation der Hirudineen und die Beziehungen dieser Gruppe zu den Oligochaeten. Ergebn. Fortschr. Zoologie **7**, 378–484.

LOHMANN, H. (1996). Das phylogenetische System der Anisoptera (Odonata). Entomol. Z. **106**, 209–252.

LOMBARDI, J. & E. E. RUPPERT (1982). Functional morphology and locomotion in Derocheilocaris typica (Crustacea, Mystacocarida). Zoomorphology **100**, 1–10.

LÜTER, C. (1998). Zur Ultrastruktur, Ontogenese und Phylogenie der Brachiopoda. Cuvillier Verlag. Göttingen.

LYAL, C. H. C. (1985). Phylogeny and classification of the Psocodea, with special reference to the lice (Psocodea: Phthiraptera). Syst. Ent. **10**, 145–165.

LYAL, C. H. C. (1986). External genitalia of Pscocodea, with particular reference to Lice (Phthiraptera). Zool. Jb. Anat. **114**, 277–292.

MACKENSTEDT, U. & K. MÄRKEL (1987). Experimental and comparative morphology of radula renewal in pulmonates (Mollusca, Gastropoda). Zoomorphology **107**, 209–239.

MAHNER, M. (1993). Systema Cryptoceratorum Phylogeneticum (Insecta, Heteroptera). Zoologica **48**, 143, 1–302. Schweizerbart'sche Verlagsbuchhandlung. Stuttgart.

MALICKY, H. (1973). Trichoptera (Köcherfliegen). Handb. Zool. **4** (2), 2/29, 1–114. Berlin.

MANGOLD-WIRZ, K. & P. FIORONI (1970). Die Sonderstellung der Cephalopoda. Zool. Jb. Syst. **97**, 522–631.

MANTON, S. M. (1977). The Arthropoda. Habits, Functional Morphology and Evolution. Clarendon Press. Oxford.

MARCUS, E. (1929 a). Tardigrada. Bronn's Klassen und Ordnungen des Tierreichs **5**, 4. Abt., 3. Buch. 1–608. Akademische Verlagsgesellschaft Leipzig.

MARCUS, E. (1929 b). Zur Embryologie der Tardigraden. Zool. Jb. Anat. **50**, 333–384.

MARPLES, B. J. (1968). The hypochilomorph spiders. Proc. Linn. Soc. London **179**, 11–31.

MARTENS, J. (1978). Weberknechte, Opiliones. Die Tierwelt Deutschlands **64**, 1–464.

MARTENS, J., HOHEISEL, U. & M. GÖTZE (1981). Vergleichende Anatomie der Legeröhren der Opiliones als Beitrag zur Phylogenie der Ordnung. Zool. Jb. Anat. **105**, 13–76.

MARTIN, J. W. & C. E. CASH-CLARK (1995). The extemal morphology of the onychopod "cladoceran" genus Bythotrephes (Crustacea, Branchiopoda, Onychopoda, Cercopagididae), with notes on the morphology and phylogeny of the order Onychopoda. Zool. Scripta **24**, 61–90.

MARTINI, E. (1952). Lehrbuch der medizinischen Entomologie. G. Fischer. Jena.

MEGLITSCH, P.A. & F. R. SCHRAM (1991). Invertebrate Zoology. Oxford University Press. Oxford, New York.

MICHELSEN, V. (1996). Neodiptera: New insights into the adult morphology and higher level phylogeny of Diptera (Insecta). Zool. J. Linn. Soc. **117**, 71–102.

MICKOLEIT, G. (1973). Über den Ovipositor der Neuropteroidea und Coleoptera und seine phylogenetische Bedeutung (Insecta, Holometabola). Z. Morph. Tiere **74**, 37–64.

MILLOT, J. (1949 a). Ordre des Uropyges. Traité de Zoologie **VI**, 533–562. Paris.

MILLOT, J. (1949 b). Ordre des Amblypyges. Traité de Zoologie **VI**, 563–588. Paris.

MILLOT, J. (I949 c). Ordre des Aranéides. Traité de Zoologie **VI**, 589–743. Paris.

MILLOT, J. (1949 d). Ordre des Palpigrades. Traité de Zoologie **VI**, 520–532. Paris.

MILLOT, J. & M. VACHON (1949 a). Ordre des Scorpions. Traité de Zoologie **VI**, 386–436 Paris.

MILLOT, J. & M. VACHON (1949 b). Ordre des Solifuges. Traité de Zoologie **VI**, 482–519. Paris.

MINELLI, A. & S. BORTOLETTO (1988). Myriapod metamerism and arthropod segmentation. Biol. Journ. Linn. Soc. **33**, 323–343.

MINELLI, A. & S. BORTOLETTO (1990). Segmentation in centipedes and its taxonomic implications. In: A. MINELLI (Ed.). Proc. 7th Intern. Congr. Myriapodology. 81–88. Brill. Leiden.

MØBJERG, N. & C. DAHL (1996). Studies on the morphology and ultrastructure of the malpighian tubules of Halobiotus crispae Kristensen, 1982 (Eutardigrada). Zool. J. Linn. Soc. **116**, 85–95.

MONOD, T (1940). Thermosbaenacea. Bronns Klassen und Ordnungen des Tierreiches **5**, 1. Abt., 4. Buch, 1–24.

MOON, S.Y. & W. KIM (1996). Phylogenetic position of the Tardigrada based on the 18S ribosomal RNA gene sequences. Zool. J. Linn. Soc. **116**, 61–69.

MORITZ, M. (1993a). Stamm Onychophora, Stummelfüßer. In: H.-E. GRUNER (Hrsg.). Lehrbuch der Speziellen Zoologie. 5. Aufl. Band I, 3. Teil. 470–495. G. Fischer. Jena, Stuttgart, New York.

MORITZ, M. (1993b). Arachnata. In H.-E. GRUNER (Hrsg.). Lehrbuch der Speziellen Zoologie. 4. Aufl. Band I, 4. Teil. 64–442. G. Fischer. Jena, Stuttgart, New York.

MOUND, L. A. & B. S. HEMING (1991). Thysanoptera. In: The Insects of Australia. Vol. I, 510–515. Melbourne University Press. Melbourne.

MOURA, G. & M. L. CHRISTOFFERSEN (1996). The system of the mandibulate arthropods: Tracheata and Remipedia as sister groups, "Crustacea" non-monophyletic. J. Comp. Biol. **1**, 95–113.

MÜLLER, A. H. (1994). Lehrbuch der Paläozoologie Bd. II, Teil 2. 4. Aufl. G. Fischer. Jena, Stuttgart.

MÜLLER, K. J. & D. WALOSSEK (1987). Morphology, ontogeny, and life habit of Agnostus pisiformis from the Upper Cambrium of Sweden. Fossils and Strata **19**, 1–124.

NAUMANN, I. D. (1991). Hymenoptera. In: The Insects of Australia. Vol. II, 916–1000. Melbourne University Press. Melbourne.

NAUMANN, I. D. (Ed.). (1994). Systematic and applied Entomology. Melbourne University Press. Melbourne.

NEBOISS, A. (1991). Trichoptera. In: The Insects of Australia. Vol. II, 787–816. Melbourne University Press. Melbourne.

NEW, T. R. (1991). Neuroptera. In: The Insects of Australia. Vol. I, 525–542. Melbourne University Press. Melbourne.

NEWMAN, W. A. (1974). Two new deep-sea Cirripedia (Ascothoracida and Acrothoracica) from the Atlantic. J. Mar. Biol. Ass. U.K. **54**, 437–456.

NEWMAN, W. A. (1992). Origin of Maxillopoda. Acta Zool. **73**, 319–322.

NEWMAN, W. A., ZULLO, V.A. & T. H. WITHERS (1969). Cirripedia. In: R. C. MOORE (Ed.). Treatise on Invertebrate Palaeontology. Part R. Arthropoda 4. Vol. **1**, 206–295.

NIELSEN, C. (1971). Entoproct life-cycles and the entoproct/ectoproct relationship. Ophelia **9**, 209–341.

NIELSEN, C. (1987). Structure and function of metazoan ciliary bands and their phylogenetic significance. Acta Zool. **68**, 205–262.

NIELSEN, C. (1995). Animal evolution. Interrelationships of the living Phyla. Oxford University Press. Oxford, New York, Tokyo.

NIELSEN, C. (1997). The phylogenetic position of the Arthropoda. In: R. A. FORTEY & R. H.THOMAS (Eds.). Arthropod relationships. Syst. Ass. Spec. Vol. **55**, 11–22. London.

NIELSEN, C. & A. JESPERSEN (1997). Entoprocta. In: F. W. HARRISON & R. M. WOOLLACOTT (Eds.). Microscopic Anatomy of Invertebrates **13**, 13–43. Wiley-Liss, Inc. New York.

NIELSEN, C. & J. ROSTGAARD (1976). Structure and function of an entoproct tentacle with a discussion of ciliary feeding types. Ophelia **15**, 115–140.

NIELSEN, E. S. & I. F. B. COMMON (1991). Lepidoptera. In: The Insects of Australia.Vol. II, 817–915. Melbourne University Press. Melbourne.

NIELSEN, E. S. & N. P. KRISTENSEN (1996). The Australian moth family Lophocoronidae and the basal phylogeny of the Lepidoptera-Glossata. Invertebrate Taxonomy **10**, 1199–1302.

NILSSON, D. E. & D. OSARIO (1997). Homology and parallelism in arthropod sensory processing. In: R. A. FORTEY & R. M. THOMAS (Eds). Arthropod relationships. Syst. Ass. Spec. Vol. **55**, 333–347.

NOODT, W. (1954). Crustacea Mystacocarida von Süd-Afrika. Kieler Meeresforsch. **10**, 243–246.

NOODT, W. (1964). Natürliches System und Biogeographie der Syncarida (Crustacea, Malacostraca). Gewässer und Abwässer **37/38**, 77–186.

NOTT, J. A. & B. A. FOSTER (1969). On the structure of the antennular attachment organ of the Cypris larva of Balanus balanoides (L.). Phil. Trans. Royal Soc. London B, **256**, 115–134.

OLESEN, J. (1998). A phylogenetic analysis of the Conchostraca and Cladocera (Crustacea. Branchiopoda, Diplostraca). Zool. J. Linn. Soc. London **122**, 491–536.

OLESEN, J., MARTIN, J. W. & E. W. ROESSLER (1997). External morphology of the male of Cyclestheria hislopi (Baird, 1859) (Crustacea, Branchiopoda, Spinicaudata), with a comparison of male claspers among the Conchostraca and Cladocera and its bearing on phylogeny of the "bivalved" Branchiopoda. Zool. Scripta 25, 291–316.

ORRHAGE, L. (1973). Light and electron microscope studies of some brachiopod and pogonophoran setae? With a discussion of the "annelid seta" as a phylogenetic-systematic character. Z. Morph. Tiere **74**, 253–270.

PACLT, J. (1956). Biologie der primär flügellosen Insekten. G. Fischer. Jena.

PALMER, J. M. (1985). The silk and silk production system of the funnel-web mygalomorph spider Euagrus (Araneae, Dipluridae). J. Morph. **186**, 195–207.

PAULUS, H. F. (1972 a). Zum Feinbau der Komplexaugen einiger Collembolen. Eine vergleichend-anatomische Untersuchung (Insecta, Apterygota). Zool. Jb. Anat. **89**, 1–116

PAULUS, H. F. (1972 b). Die Feinstruktur der Stirnaugen einiger Collembolen (Insecta, Entognatha) und ihre Bedeutung für die Stammesgeschichte der Insekten. Z. zool. Syst. Evolut.-forsch. **10**, 81–122.

PAULUS, H. F. (1979). Eye structure and the monophyly of the Arthropoda. In: A. P. GUPTA (Ed.). Arthropod phylogeny. 299–383. Van Nostrand Reinhold Comp. New York.

PAULUS, H. F. (1985). Euarthropoda, Allgemeines. In R. SIEWING (Hrsg.). Lehrbuch der Zoologie. Systematik. 768–787. G. Fischer. Stuttgart. New York.

PAULUS, H. F. (1986). Evolutionswege zum Larvalauge der Insekten – ein Modell für die Entstehung und die Ableitung der ozellären Lateralaugen der Myriapoda von Fazettenaugen. Zool. Jb. Syst. **113**, 353–371.

PAULUS, H. (1996). Euarthropoda, Gliederfüßer i. e. S. In: W. WESTHEIDE & R. RIEGER (Hrsg.). Spezielle Zoologie. Teil 1. 435–444. G. Fischer. Stuttgart, Jena, New York.

PAULUS, H. F. & M. SCHMIDT (1978). Evolutionswege zum Larvalauge der Insekten: Die Stemmata der Trichoptera und Lepidoptera. Z. zool. Syst. Evolut.-forsch. **16**, 188–216.

PAULUS, H. & P. WEYGOLDT (1996). Arthropoda, Gliederfüßer. In W. WESTHEIDE & R. RIEGER (Hrsg.). Spezielle Zoologie. Teil 1. 411–419. G. Fischer. Stuttgart, Jena, New York.

PENNAK, R. W. (1978). Fresh-water invertebrates of the United States. J. Wiley & Sons, New York

PESSON, P. (1951). Ordre des Homoptères. Traité de Zoologie **X**, II, 1390–1647. Paris.

PETERS, H. M. (1982). Wie Spinnen der Familie Uloboridae ihre Beute einspinnen und verzehren. Verh. naturwiss. Ver. Hamburg (NF) **25**, 147–167.

PETERS, W. L. & I. C. CAMPBELL (1991). Ephemeroptera. In: The Insects of Australia. Vol. I, 279–293. Melbourne University Press. Melbourne.

PFAU, H. K. (1986). Untersuchungen zur Konstruktion, Funktion und Evolution des Flugapparates der Libellen (Insecta, Odonata). Tijdschr. Ent. **129**, 35–123.

PFAU, H. K. (1991). Contributions of functional morphology to the phylogenetic systematics of Odonata. Adv. Odonatol. **5**, 109–141.

PFLUGFELDER, O. (1948). Entwicklung von Paraperipatus amboinensis n. sp. Zool. Jb. Anat. **69**, 443–492.

PFLUGFELDER, O. (1968). Onychophora. G. Fischer. Stuttgart.

PIEKARSKI, G. (1954). Lehrbuch der Parasitologie. Springer. Berlin, Göttingen. Heidelberg.

PILGER, J. F. (1993). Echiura. In: F.W. HARRISON & M. E. RICE (Eds.). Microscopic Anatomy of Invertebrates **12**, 185–236. Wiley-Liss, Inc. New York.

PLATNICK, N. I. (1977). The hypochiloid spiders: a cladistic analysis, with notes on the Atypoidea (Arachnida, Araneae). Amer. Mus. Novitates 2627, 1–23.

PLATNICK. N. I., CODDINGTON, J. A., FORSTER, R. R. & C. E. GRISWOLD (1991). Spinneret morphology and the phylogeny of haplogyne spiders (Araneae, Araneomorphae). Amer. Mus. Novitates 3016, 1–73.

PLATNICK, N. I. & W. J. GERTSCH (1976). The suborders of spiders: a cladistic analysis. Amer. Mus. Novitates 2607, 1–15.

PLATNICK, N. I. & P. A. GOLOBOFF (1985). On the monophyly of the spider suborder Mesothelae (Arachnida: Araneae). J. N.Y. Entomol. Soc. **93**, 1265–1270.

POCOCK, R. 1. (1893). On the classification of the tracheate Arthropoda. Zool. Anz. **16**, 271–275.

POISSON, R. (1951). Ordre des Heteroptères. Traité de Zoologie **X**, II, 1657–1803. Paris.

PORTMANN, A. (1961). Die Tiergestalt. F. Reinhardt AG. Basel.

PRIESNER, H. (1968). Thysanoptera (Physopoda, Blasenfüßer). Handb. Zool. **4** (2), 2/19, 1–32. Berlin.

PURSCHKE, G. (1996). Pogonophora, Bartwürmer. In: W. WESTHEIDE & R. RIEGER (Hrsg.). Spezielle Zoologie. Teil 1. 383–395. G. Fischer. Stuttgart. Jena, New York.

PURSCHKE, G. (1997). Ultrastructure of nuchal organs in polychaetes (Annelida) – new results and review. Acta Zool. **78**, 123–143.

PURSCHKE, G. & A. B. TZETLIN (1996). Dorsolateral ciliary folds in the polychaete foregut: structure, prevalence and phylogenetic significance. Acta Zool. **77**, 33–49.

PURSCHKE, G., WESTHEIDE, W., ROHDE, D. & R. O. BRINKHURST (1993). Morphological reinvestigation and phytogenetic relationship of Acanthobdella peledina (Annelida, Clitellata). Zoomorphology **113**, 91–101.

PURSCHKE, G., WOLFRATH, F. & W. WESTHEIDE (1997). Ultrastructure of the nuchal organ and cerebral organ in Onchnesoma squamatum (Sipuncula, Phascolionidae). Zoomorphology **117**, 23–31.

DE QUEIROZ, K. (1992). Phylogenetic definitions and taxonomic philosophy. Biol. Philos. **7**, 295–313.

DE QUEIROZ, K. (1994). Replacement of an essentialistic perspective on taxonomic definitions as exemplified by the definition of Mammalia. Syst. Biol. **43**, 497–510.

DE QUEIROZ, K. (1997a). The Linnean Hierarchy and the evolutionization of taxonomy, with emphasis on the problem of nomenclature. Aliso **15**, 125–144.

DE QUEIROZ, K. (1997b). Misunderstandings about the phylogenetic approach to biological nomenclature: a reply to Lidén and Oxelman. Zool. Scripta **26**, 67–70.

DE QUEIROZ, K. & J. GAUTHIER (1992). Phylogenetic taxonomy. Annu. Rev. Ecol. Syst. **23**, 449–480.

RAMCKE, J. (1965). Kopf der Schweinelaus (Haematopinus suis L., Anoplura). Zool. Jb. Anat. **82**, 547–663.

RAVEN, R. J. (1985). The spider infraorder Mygalomorphae (Araneae): cladistics and systematics. Bull. Amer. Mus. Nat. Hist. **182**, 1–180.

RAVOUX, P. (1975). Endosquelette et musculature céphaliques de Scutigerella immaculata Newport (Symphyla: Scutigerellidae). Bull. Mus. natn. Hist. nat., Paris, 3[e]sér, no332, Zoologie **234**, 1189–1238.

REMANE, A. (1957). Arthropoda – Gliedertiere. Handbuch der Biologie **4**, 209–310.

REMY, P (1950). Les Millotauropus, types d'un nouveau groupe des Pauropodes. C. R. Acad. Sci. Paris, **230**, 472–474.

REMY, P. (1953). Description de nouveaux types de Pauropodes: "Millotauropus" et "Rabaudauropus". Mem. Inst. Sci. Madagascar, A, **8**, 25–41.

RENAUD-DEBYSER, J. (1963). Recherches écologiques sur la faune interstitielle des sables (Bassin d'Arcachon, Ile de Bimini, Bahamas). Vie et Milieu (Suppl.) **15**, 1–157.

RENAUD-MORNANT, J. (1976). Un nouveau genre de Crustacé Mystacocaride de la zone neotropicale: Ctenocheilocaris claudiae n. g. n. sp. C. R. Sc. Paris. **282**, 863–866.

RENTZ, D. C. F. (1991a). Grylloblattodea. In: The Insects of Australia. Vol. I, 357–359. Melbourne University Press. Melbourne.

RENTZ, D.C. F. (1991b). Orthoptera. In: The Insects of Australia. Vol. I, 369–391. Melbourne University Press. Melbourne.

RENTZ, D. C. F. & D. K. McE. KEVAN (1991). Dermaptera. In: The Insects of Australia. Vol. I, 360–368. Melbourne University Press. Melbourne.

RICE, M. E. (1993). Sipuncula. In: F. W. HARRISON & M. E. RICE (Eds.). Microscopic Anatomy of Invertebrates **12**, 237–325. Wiley-Liss, Inc. New York.
RICHARDS, P. A. & A. G. RICHARDS (1969). Acanthae: a new type of cuticular process in the proventriculus of Mecoptera and Siphonaptera. Zool. Jb. Anat. **86**, 158–176.
RICHTER, S. (1993). Die Eucarida sind nicht monophyletisch – Ommatidienstruktur deutet auf nähere Verwandtschaft von Euphausiacea, Syncarida und Peracarida (Crustacea). Verh. Dtsch. Zool. Ges. **86**, 1, 144.
RICHTER, S. (1994a). Monophylie der Lophogastrida und phylogenetisches System der Peracarida (Crustacea). Verh. Dtsch. Zool. Ges. **87**, 1, 231.
RICHTER, S. (1994b). Aspekte der phylogenetischen Systematik der Malacostraca (Crustacea) unter besonderer Berücksichtigung der Caridoida auf der Grundlage vergleichend-morphologischer Untersuchungen der Komplexaugen und anderer Merkmale. Dissertation. Fachbereich Biologie. Freie Universität Berlin. 171.
RICHTER, S. (1999). The structure of the ommatidia of the Malacostraca (Crustacea) – a plylogenetic approach. Verh. naturwiss. Ver. Hamburg (NF) **38**, 161–204.
RIEDL, R. (1963). Fauna und Flora der Adria. P. Parey. Hamburg und Berlin.
RILEY, J., BANAJA, A. A. & J. L. JAMES (1978). The phylogenetic relationships of the Pentastomida: The case for their inclusion within the Crustacea. Int. J. Parasit. **8**, 245–254.
RILLING, G. (1960). Zur Anatomie des braunen Steinläufers Lithobius forficatus L. (Chilopoda). Skelettmuskelsystem, peripheres Nervensystem und Sinnesorgane des Rumpfes. Zool. Jb. Anat. **138**, 303–369.
RILLING, G. (1968). Lithobius forficatus. G. Fischer. Stuttgart.
RISLER. H. (1957). Der Kopf von Thrips physapus L. (Thysanoptera, Terebrantia). Zool. Jb. Anat. **76**, 251–302.
ROESSLER, E. W, (1995). Review of Colombian Conchostraca (Crustacea) – ecological aspects and life cycles – family Cyclestheriidae. Hydrobiologia **298**, 113–124.
ROSS, E. S. (1991). Embioptera. In: The Insects of Australia. Vol. I, 405–409. Melbourne University Press. Melbourne.
ROTH, L. M. (1991). Blattodea. In: The Insects of Australia. Vol. I, 320–329. Melbourne University Press. Melbourne.
ROUSE, G. W & K. FAUCHALD (1995). The articulation of annelids. Zool. Scripta **24**, 269–301.
ROUSE, G. W. & K. FAUCHALD (1997). Cladistics and polychaetes. Zool. Scripta **26**, 139–204.
RUDOLPH, D. (1982). Site, process and mechanisms of active uptake of water vapour from the atmosphere in the Psocoptera. J. Insect Physiol. **28**, 205–212.
RUDOLPH, D. (1983). The water-vapour uptake system of the Phthiraptera. J. Insect Physiol. **29**, 15–25.
RUDOLPH, D. & W. KNÜLLE (1982). Novel uptake systems for atmospheric water vapour among insects. J. exp. Zool. **222**, 321–333.
RUHBERG, H. (1985). Die Peripatopsidae (Onychophora). Zoologica **137**, 1–184.
RUHBERG, H. (1996). Onychophora, Stummelfüßer. In: W. WESTHEIDE & R. RIEGER (Hrsg.). Spezielle Zoologie. Teil 1. 420–428. G. Fischer. Stuttgart, Jena, New York.
SALVINI-PLAWEN, L. von (1971). Schild- und Furchenfüßer (Caudofoveata und Solenogastres), verkannte Weichtiere am Meeresgrund. Die Neue Brehm-Bücherei. Ziemsen. Wittenberg Lutherstadt.
SALVINI-PLAWEN, L. von (1981). On the origin and evolution of the Mollusca. In: Origine dei grandi Phyla dei Metazoi. Acc. Naz. Lincei Roma. Atti dei Convegni Lincei **49**, 235–293.
SALVINI-PLAWEN, L. von (1985). Early evolution and the primitive groups. The Mollusca. Vol. **10**, 59–150.
SALVINI-PLAWEN, L. von (1990). Origin, phylogeny and classification of the Phylum Mollusca. Iberus **9**, 1–33.
SALVINI-PLAWEN, L. von & T. BARTOLOMAEUS (1995). Mollusca: Mesenchymata with a "coelom". In: G. LANZAVECCHIA, R. VALVASSORI & M. D. CANDIA CARNEVALI (Eds). Body cavities: function and phylogeny. Selected Symposia and Monographs U. Z. I, **8**, 75–92. Mucchi, Modena.

SALVINI-PLAWEN, L. von & G. HASZPRUNAR (1987). The Vetigastropoda and the systematics of streptoneurous Gastropada (Mollusca). J. Zool. Lond. **211**, 747–770.

SALVINI-PLAWEN, L. von & G. STEINER (1996). Synapomorphies and plesiomorphies in the higher classification of Mollusca. In: J. TAYLOR (Ed.). Origin and evolutionary radiation of the Mollusca. 29–51. Oxford University Press. Oxford, New York, Tokyo.

SANDERS, H. L. (1955). The Cephalocarida, a new subclass of Crustacea from Long Island Sound. Proc. Nat. Acad. Sci. **41**, 61–66.

SANDERS, H. L. (1963). The Cephalocarida. Functional morphology, larval development, comparative external anatomy. Mem. Connecticut Acad. Arts Sci. **15**, 1–80.

SARS, G. O. (1887). On Cyclestheria hislopi (BAIRD), a new generic type of bivalve Phyllopoda. Forh.Vidensk. Selsk. Krist. 1887. 1–65.

SAWYER, R. T. (1986). Leech biology and behavior. Vol. I–III. Clarendon Press, Oxford.

SCHAEFER, K. & G. HASZPRUNAR (1997). Anatomy of Laevipilina antarctica, a monoplacophoran limpet (Mollusca) from antarctic waters. Acta Zool. **77**, 295–314

SCHALLER, F. (1962). Die Unterwelt des Tierreiches. Kleine Biologie der Bodentiere. Springer. Berlin, Göttingen, Heidelberg.

SCHEDL, W. (1991). Hymenoptera: Unterordnung Symphyta (Pflanzenwespen). Handb. Zool. **4**, 31. Berlin, New York.

SCHELTEMA, A. H. (1996). Phylogenetic position of Sipuncula, Mollusca and the progenetic Aplacophora. In: J. TAYLOR (Ed.). Origin and evolutionary radiation of the Mollusca. 53–58. Oxford University Press. Oxford, New York, Tokyo.

SCHLEE, D. (1969a). Sperma-Übertragung (und andere Merkmale) in ihrer Bedeutung für das phylogenetische System der Sternorrhyncha (Insecta, Hemiptera). Phylogenetische Studien an Hemiptera I: Psylliformes (Psyllina und Aleyrodina) als monophyletische Gruppe. Z. Morph. Tiere **64**, 95–138.

SCHLEE, D. (1969b). Die Verwandtschaftsbeziehungen innerhalb der Sternorrhyncha aufgrund synapomorpher Merkmale. Phylogenetische Studien an Hemiptera II: Aphidiformes (Aphidina+Coccina) als monophyletische Gruppe. Stuttgart. Beitr. Naturkde. **199**, 1–19.

SCHLEE, D. (1969c). Morphologie und Symbiose; ihre Beweiskraft für die Verwandtschaftsbeziehungen der Coleorrhyncha (Insecta, Hemiptera). Phylogenetische Studien an Hemiptera IV: Heteropteroidea (Heteroptera+Coleorrhyncha) als monophyletische Gruppe. Stuttgart. Beitr. Naturkde. **210**, 1–27.

SCHMINKE, H. K. (1973). Evolution, System und Verbreitungsgeschichte der Familie Parabathynellidae (Bathynellacea. Malacostraca). Mikrofauna des Meeresbodens **24**, 1–192.

SCHMINKE, H. K. (1978a). Die phylogenetische Stellung der Stygocarididae (Crustacea, Syncarida) - unter besonderer Berücksichtigung morphologischer Ähnlichkeiten mit Larvenformen der Eucarida. Z. zool. Syst. Evolut.-forsch. **16**, 225–239.

SCHMINKE, H. K. (1978b). Notobathynella longipes sp. n. and new records of other Bathynellacea (Crustacea, Syncarida) from New Zealand. N. Z. Journal of Marine and Freshwater Research **12**, 457–462.

SCHMINKE, H. K. (1981). Adaptation of Bathynellacea (Crustacea, Syncarida) to life in the interstitial ("Zoea-Theory"). Int. Revue ges. Hydrobiol. **66**, 575–637.

SCHMINKE, H. K. (1996). Crustacea. Krebse. In: W. WESTHEIDE & R. RIEGER (Hrsg.). Spezielle Zoologie. Teil 1. 501–581. G. Fischer. Stuttgart, Jena, New York.

SCHOLTZ, G. (1995). Head segmentation in Crustacea – an immunocytochemical study. Zoology **98**, 104–114.

SCHOLTZ, G. (1997). Cleavage, germ band formation and head segmentation: the ground pattern of the Euarthropada. In: R. A. FORTEY & R. H. THOMAS (Eds). Arthropod relationships. Syst. Ass. Spec. Vol. **55**, 317–332. London.

SCHOLTZ, G. & W. DOHLE (1996). Cell lineage and cell fate in crustacean embryos – a comparative approach. Int. J. Dev. Biol. **40**, 211–220.

SCHOLTZ, G. & S. RICHTER (1995). Phylogenetic systematics of the reptantian Decapoda. Zool. J. Linn. Soc. **113**, 289–328.

SCHRAM, F. R. (1986). Crustacea. Oxford University Press. New York, Oxford.

SCHRAM, F. R., YAGER, J. & M. J. EMERSON (1986). Remipedia. Part I. Systematics. San Diego Society Natural History. Memoir **15**, 1–60.

SCHUBART, O. (1934). Tausendfüßler oder Myriapoda I. Diplopoda. In: F. DAHL. Die Tierwelt Deutschlands **28**, 1–318. G. Fischer, Jena.

SCHÜRMANN, F.-W. (1987). Histology and ultrastructure of the onychophoran brain. In: A. P. GUPTA (Ed.). Arthropod brain, development, structure, and functions. 159–180. John Wiley & Sons, Inc. New York.

SCHÜRMANN. F.-W. (1995). Common and special features of the nervous system of Onychophora: a comparison with Arthropoda, Annelida and some other invertebrates. In: O. BREIDBACH & W. KUTSCH (Eds.). The nervous system of invertebrates: an evolutionary and comparative approach. 139–159. Birkhäuser Verlag. Basel.

SCHULT J. (1983). Taster haplogyner Spinnen unter phylogenetischem Aspekt (Arachnida: Araneae). Verh. naturwiss. Ver. Hamburg (NF) **26**, 69–84.

SCHULZ, K. (1975). The chitinous skeleton and its bearing on taxonomy and biology of ostracodes. Bull. Amer. Palaeontol. **65**, 587–599.

SCHULZ, K. (1976). Das Chitinskelett der Podocopida (Ostracoda, Crustacea) und die Frage der Metamerie dieser Gruppe. Dissertation, Universität Hamburg.

SCRIBAN, I. A. & H. AUTRUM (1934). Hirudinea-Egel. Handb. Zool. **2** (2), 119–352. Berlin.

SEELINGER, G. & U. SEELINGER (1983). On the social organisation, alarm and fighting in the primitive cockroach Cryptocercus punctulatus Scudder. Z. Tierpsychol. **61**, 315–333.

SEGUY, E. (1951 a). Ordre des Mallophages. Traité de Zoologie **X**, 1341–1364. Paris.

SEGUY, E. (1951 b). Ordre des Anoploures ou Poux. Traité de Zoologie **X**, 1365–1384. Paris.

SEGUY, E. (1951 c). Ordre des Siphonaptères. Traité de Zoologie **X**, 745–769. Paris.

SEIFERT, G. (1995). Entomologisches Praktikum. G. Thieme. Stuttgart, New York.

SHULTZ, J. W. (1989). Morphology of locomotory appendages in Arachnida: evolutionary trends and phylogenetic implications. Zool. J. Linn. Soc. **97**, 1–56.

SHULTZ, J. W. (1990). Evolutionary morphology and phylogeny of Arachnida. Cladistics **6**, 1–38

SIDDALL, M. E. & E. M. BURRESON (1995). Phylogeny of the Euhirudinea: Independent evolution of blood feeding by leeches? Can. J. Zool. **73**, 1048–1064.

SIEG, J. (1984). Neuere Erkenntnisse zum natürlichen System der Tanaidacea. Zoologica **136**, 1–132.

SIEWING, R. (Hrsg.). (1985). Lehrbuch der Zoologie. Systematik. G. Fischer. Stuttgart, New York.

SMITHERS, C. N. (1991a). Zoraptera. In: The Insects of Australia. Vol. I, 410–411. Melbourne University Press. Melbourne.

SMITHERS, C. N. (1991 b). Psocoptera. In: The Insects of Australia.Vol. I, 412–420. Melbourne University Press. Melbourne.

SNODGRASS, R. E. (1965). A textbook of Arthropod Anatomy. Hafner Publ. Comp. New-York, London.

SOPOTT-EHLERS, B. (1985). The phylogenetic relationships within the Seriata (Platyhelminthes). In: S. CONWAY MORRIS, J. D. GEORGE, R. GIBSON & H. M. PLATT (Eds.). The origins and relationships of lower invertebrates. The Systematics Association. Spec. Vol. **28**, 159–167. Clarendon Press. Oxford.

SPEARS, T. & L. G. ABELE (1997). Crustacean phylogeny inferred from 18S rDNA. In: R. A. FORTEY & R. H. THOMAS (Eds). Arthropod relationships. Syst. Ass. Spec.Vol. **55**, 169–187. London.

SPIES, T. (1981). Structure and phylogenetic interpretation of diplopod eyes (Diplopoda). Zoomorphology **98**, 241–260.

SPENGEL, J. W. (1881). Die Geruchsorgane und das Nervensystem der Mollusken. Z. wiss. Zool. **35**, 333–383.

STEPHEN, A. C. & S. J. EDMONDS (1972). The phyla Sipuncula and Echiura. The British Museum (Natural History). London.

STORCH, V. (1984). Echiura and Sipuncula. In: J. BEREITER-HAHN, A. G. MATOLTSY & K. SYLVIA RICHARDS (Eds). Biology of the Integument. 1 Invertebrates, 368–375. Springer. Berlin, Heidelberg, New York, Tokyo.

STORCH, V, & B. G. M. JAMIESON (1992). Further spermatological evidence for including the Pentastomida (Tongue worms) in the Crustacea. Int. J. Parasit. **22**, 95—108.

STORCH, V. & H. RUHBERG (1993). Onychophora. In: F. W. HARRISON & M. E. RICE (Eds.). Microscopic Anatomy of Invertebrates **12**, 11–56. Wiley-Liss, Inc. New York.

STORCH, V. & U. WELSCH (1972). Über Bau und Entstehung der Mantelstacheln von Lingula unguis L. (Brachiopoda). Zool. Jb. Anat. **94**, 441–452.

STORCH, V. & U. WELSCH (1993). Kükenthals Leitfaden für das Zoologische Praktikum. G. Fischer. Stuttgart, Jena, New York.

STØRMER, L. (1970). Arthropods from the lower Devonian (lower Emsion) of Alken an der Mosel, Germany. Part. 1: Arachnida. Senckenbergiana Lethaea **51**, 335–369.

SUNDBERG, P. & F. PLEIJEL (1994). Phylogenetic classification and the definition of taxon names. Zool. Scripta **23**, 19–25.

TÉTRY, A. (1959). Classe des Sipunculiens (Sipunculidea de Quatrefages, 1866). Traité de Zoologie **V**, 1, 785–854. Paris.

THEISCHINGER, G. (1991). Plecoptera. In: The Insects of Australia. Vol. I, 311–319. Melbourne University Press. Melbourne.

TIEGS, O. W. (1947). The development and affinities of the Pauropoda, based on a study of Pauropus silvaticus Quart. 3. microsc. Sci. **88**, 165–267.

TIEMANN, H. (1984). Is the taxon Harpacticoida a monophyletic one? First International Conference on Copepoda. Crustaceana suppl. **7**, 47–59.

TURCATO, A., FUSCO, G. & A. MINELLI (1995). The sternal pore areas of geophilomorph centipedes (Chilopoda: Geophilomorpha). Zool. J. Linn. Soc. **115**, 185–209.

ULRICH, H. (1965). Der Fang- und Greifapparat von Mantispa - ein Vergleich mit Mantis. Natur u. Museum **95**, 499–508.

VACHON, M. (1949). Ordre des Pseudoscorpions. Traité de Zoologie **VI**, 437–481. Paris.

VERHOEFF, K. W. (1925). Klasse Chilopoda. Bronns Klassen Ordnungen Tier. **5** (II). VII + 725pp.

VILHELMSEN, L. (1997). The phylogeny of lower Hymenoptera (Insecta). with a summary of the early evolutionary history of the order. J. Zool. Syst. Evol. Research **35**, 49–70.

VOLLMER, C. (1952). Kiemenfuß, Hüpferling und Muschelkrebs. Die Neue Brehm-Bücherei. Akad. Verlagsgesellschaft, Geest & Portig K. G. Leipzig.

VOLLMER, C. (1960). Wasserflöhe. Die Neue Brehm-Bücherei. A. Ziemsen Verlag. Wittenberg Lutherstadt.

WÄGELE, J. W. (1989). Evolution und phylogenetisches System der Isopoda. Stand der Forschung und neue Erkenntnisse. Zoologica **140**, 1–262.

WÄGELE, J. W. (1993). Rejection of the "Uniramia" hypothesis and implications of the Mandibulata concept. Zool. Jb. Syst. **120**, 253–288.

WÄGELE, J. W. (1994). Review of methodological problems of "Computer cladistics" exemplified with a case study on isopod phylogeny (Crustacea: Isopoda). Z. zool. Syst. Evolut.-forsch. **32**, 81–107.

WALOSSEK, D. (1993). The Upper Cambrian Rehbachiella and the phylogeny of Branchiopoda and Crustacea. Fossils and strata **32**, 1–202.

WALOSSEK, D. (1995). The Upper Cambrian Rehbachiella, its larval development, morphology and significance for the phylogeny of Branchiopoda and Crustacea. Hydrobiologia **298**, 1–13.

WALOSSEK, D. & K. J. MÜLLER (1990). Upper Cambrian stem-lineage crustaceans and their bearing upon the monophyletic origin of Crustacea and the position of Agnostus. Lethaia **23**, 409–427.

WALOSSEK, D. & K. J. MÜLLER (1994). Pentastomid parasites from the Lower Palaeozoic of Sweden. Trans. Royal. Soc. Edinburgh: Earth Sciences **85**, 1–37.

WALOSSEK, D. & K. J. MÜLLER (1997). Cambrian "Orsten"-type arthropods and the phylogeny of Crustacea. In: R. A. FORTEY & R. H. THOMAS (Eds.). Arthropod relationships. Syst. Ass. Spec. Vol. **55**, 139–153. London.

WARÉN, A. & S. GOFAS (1996). A new species of Monoplacophora, redescription of the genera Veteropilina and Rokopella, and new information of three species of the class. Zool. Scripta **25**, 215–232.

WASSON, K. (1997). Systematic revision oi colonial kamptozoans (entoprocts) of the Pacific coast of North America. Zool. J. Linn. Soc. **121**, 1–63.

WATSON, J. A. L. & F. J. GAY (1991). Isoptera. In: The Insects of Australia. Vol. I, 330–347. Melbourne University Press. Melbourne.

WATSON, J. A. L. & A. F. O'FARREL (1991). Odonata. In: The Insects of Australia. Vol. I. 294–310. Melbourne University Press. Melbourne.

WATSON, J. A. L. & G. B. SMITH (1991 a). Archaeognatha. In: The Insects of Australia. Vol. I, 272–274. Melbourne University Press. Melbourne.

WATSON, J. A. L. & G. B. SMITH (1991 b). Thysanura. Zygentoma. In: The Insects of Australia. Vol. I, 275–278. Melbourne University Press. Melbourne.

WEBER, H. (1933). Lehrbuch der Entomologie. G. Fischer. Jena.

WEBER, H. & H. WEIDNER (1974). Grundriß der Insektenkunde. G. Fischer. Stuttgart.

WEHNER, R., WEHNER, S. & D. AGOSTI (1994). Patterns of biogeographic distribution within the bicolor species group of the North African desert ant, Cataglyphis Foerster 1850. Senckenbergiana biologica **74**, 163–191.

WEIDNER, H. (1972). Copeognatha (Staubläuse). Handb. Zool. **4** (2), 2/9, 1–94. Berlin.

WEISBLAT, D. A., WEDEEN, C. J. & R. KOSTRIKEN (1993). Evolutionary conservation of developmental mechanisms: comparison of annelids and arthropods. In: A. C. SPRADLING (Ed.). Evolutionary conservation and developmental mechanisms. 125–140. Wiley-Liss, Inc. New York.

WENK, P. (1953). Der Kopf von Ctenocephalus canis (Curt.) (Aphaniptera). Zool. Jb. Anat. **73**, 103–164.

WESENBERG-LUND, C. (1943). Biologie der Süßwasserinsekten. Springer. Berlin. Wien.

WESTHEIDE, W. (1980). Erpobdella octoculata (Hirudinea). Spermatophorenübertragung, Kokonablage, Schlüpfen der Jungtiere. Publ. Wiss. Film., Sekt. Biol., Ser. 13, Nr. 27, 1–12.

WESTHEIDE, W. (1981). Nahrungsaufnahme bei Egeln (Hirudinea). Publ. Wiss. Film., Sekt. Biol., Ser. 14, Nr. 20. 1–17.

WESTHEIDE, W. (1996 a). Articulata. In: W. WESTHEIDE & R. RIEGER (Hrsg.). Spezielle Zoologie. Teil 1. 350–352. G. Fischer. Stuttgart, Jena, New York.

WESTHEIDE, W. (1996 b). Annelida. Ringelwürmer. In: W. WESTHEIDE & R. RIEGER (Hrsg,). Spezielle Zoologie. Teil 1. 353–410. G. Fischer. Stuttgart, Jena, New York.

WESTHEIDE, W. (1997). The direction of evolution within the Polychaeta. J. Nat. Hist. **31**, 1–15.

WEYGOLDT, P (1960). Embryologische Untersuchungen an Ostrakoden: Die Entwicklung von Cyprideis litoralis (G. S. BRADY) (Ostracoda, Podocopa, Cytheridae). Zool. Jb. Anat. **78**, 369–426.

WEYGOLDT, P. (1964). Vergleichend-embryologische Untersuchungen an Pseudoscorpionen (Chelonethi). Z. Morph. Ökol. Tiere **54**, 1–106.

WEYGOLDT, P. (1965). Vergleichend-embryologische Untersuchungen an Pseudoscorpionen III. Die Entwicklung von Neobisium muscorum Leach (Neobisiinae, Neobisiidae). Mit dem Versuch einer Deutung der Evolution des embryonalen Pumporgans. Z. Morph. Ökol. Tiere **55**, 321–382.

WEYGOLDT, P. (1966). Moos- und Bücherskorpione. Die Neue Brehm-Bücherei. Ziemsen. Wittenberg Lutherstadt.

WEYGOLDT, P. (1971). Vergleichend-embryologische Untersuchungen an Pseudoskorpionen V. Das Embryonalstadium mit seinem Pumporgan bei verschiedenen Arten und sein Wert als taxonomisches Merkmal. Z. zool. Syst. Evolut.-forsch. **9**, 3–29.

WEYGOLDT, P. (1972). Geißelskorpione und Geißelspinnen (Uropygi und Amblypygi). Z. Kölner Zoo **15**, 95–107.

WEYGOLDT, P. (1996). Chelicerata. In: W. WESTHEIDE & R. RIEGER (Hrsg.). Spezielle Zoologie. Teil 1. 449–497. G. Fischer. Stuttgart, Jena, New York.

WEYGOLDT, P. & H. F. PAULUS (1979). Untersuchungen zur Morphologie, Taxonomie und Phylogenie der Chelicerata, I. Morphologische Untersuchungen. II. Cladogramme und Entfaltung der Chelicerata. Z. zool. Syst. Evolut.-forsch. **17**, 85–116 u. 177–200.

WHEELER, W. C. (1997). Sampling, groundplans, total evidence and the systematics of arthropods. In: R. A. FORTEY & R. H. THOMAS (Eds.). Arthropod relationships. Syst. Ass. Spec. Vol. **55**, 88–96. London.

WHEELER, W. C., CARTWRIGHT, P. & C. Y. AAYASHI (1993). Arthropod phylogeny: a combined approach. Cladistics **9**, 1–39.

WHEELER, W. C., SCHUA, C. & R. BANG (1993). Cladistic relationships among higher groups of Heteroptera: congruence between morphological and molecular data sets. Ent. Scand. **24**, 121–137.

WHITING, M. F. (1998). Long branch distraction and the Strepsiptera. Syst. Biol. **47**, 134–138.

WHITING, M. F., CARPENTER, J. C., WHEELER, Q. D. & W. C. WHEELER (1997). The Strepsiptera problem: Phylogeny of the holometabolous insect orders inferred from 18S and 28S ribosomal DNA sequences and morphology. Syst. Biol. **46**, 1–68.

WHITINGTON, P. M. & J. P. BACON (1997). The organization and development of the arthropod ventral nerve cord: insights into arthropod relationships. In: R. A. FORTEY & R. H. THOMAS (Eds.). Arthropod relationships. Syst. Ass. Spec. Vol. **55**, 349–367. London

WILLMANN, R. (1981 a). Das Exoskelett der männlichen Genitalien der Mecoptera (lnsecta). 1. Morphologie. Z. zool. Syst. Evolut.-forsch, **19**, 96–150.

WILLMANN, R. (1981 b). Das Exoskelett der männlichen Genitalien der Mecoptera (lnsecta). II. Die phylogenetischen Beziehungen der Schnabelfliegen-Familien. Z. zool. Syst. Evolut.-forsch. **19**, 153–174.

WILLMANN, R. (1987). The phylogenetic system of the Mecoptera. Syst. Ent. **12**, 519–524.

WILLMANN, R. (1989). Evolution und Phylogenetisches System der Mecoptera (Insecta: Holometabola). Abh. Senckenberg. naturforsch. Ges. **544**, 1–153.

WILLMANN, R. (1997 a). Phylogeny and the consequences of phylogenetic systematics. In: P. LANDOLDT & M. SARTORI (Eds.). Ephemeroptera & Plecoptera: Biology-Ecology-Systematics. 499–510. MTL. Fribourg.

WILLMANN, R. (1997 b). Advances and Problems in insect phylogeny. In: R. A. FORTEY & R. H. THOMAS (Eds.). Arthropod relationships. Syst. Ass. Spec. Vol. **55**, 269–279. London.

WILLS, M. A. (1997). A phylogeny of recent and fossil Crustacea derived from morphological characters. In: R. A. FORTEY & R. H. THOMAS (Eds.). Arthropod relationships. Syst. Ass. Spec. Vol. **55**, 189–209. London.

WINGSTRAND, K. G. (1972). Comparative spermatology of a pentastomid, Raillietiella hemidactyli, and a branchiuran crustacean, Argulus foliaceus, with a discussion of pentastomid relationships. Kongl. Danske Vid. Selskab. Biolog. Skr. **19**, (4), 1–72.

WINGSTRAND, K. G. (1985). On the anatomy and relationships of recent Monoplacophora. Galathea Rept. **16**, 7–94.

WINTER, G. (1980). Beiträge zur Morphologie und Embryologie des vorderen Körperabschnitts (Cephalosoma) der Pantopoda Gerstaecker, 1863. Z. zool. Syst. Evolut.-forsch. **18**, 27–61.

WOLTER, H. (1963). Vergleichende Untersuchungen zur Anatomie und Funktionsmorphologie der stechend-saugenden Mundwerkzeuge der Collembolen. Zool. Jb. Anat. **81**, 27–100.

WOOTTON, R. J. & C. R. BETTS (1986). Homology and function in the wings of Heteroptera. Syst. Ent. **11**, 389–400.

YAGER, J. (1987). Speleonectes tulumensis n. sp. (Crustacea: Remipedia) from two anchialine cenotes of the Yucatan peninsula, Mexico. Stygologia **3**, 160–166.

YAGER, J. (1991). The Remipedia (Crustacea): Recent investigations of their biology and phylogeny. Verh. Dtsch. Zool. Ges. **84**, 261–269.

YAGER, J. & F. SCHRAM (1986). Lasionectes entrichoma, new genus, new species (Crustacea: Remipedia) from anchialine caves in the Turks and Caicos, British West Indies. Proc. Biol. Soc. Wash. **99**, 65–70.

ZIEGLER, B. (1983). Spezielle Paläontologie. E. Schweizerbart. Stuttgart.

ZILCH, R. (1972). Beitrag zur Verbreitung und Entwicklungsbiologie der Thermosbaenacea. Int. Revue ges. Hydrobiol. **57**, 75–107.

ZRZAVÝ, J. HYPŠA, V. & M. VLÁŠKOVÁ (1997). Arthropod phylogeny: taxonomic congruence, total evidence and conditional combination approaches to morphological and molecular data sets. In: R. A. FORTEY & R. H. THOMAS (Eds.). Arthropod relationships. Syst. Ass. Spec. Vol. **55**, 97–107. London.

ZWICK, P. (1981). Plecoptera. Revisionary notes. 172–178. In W. HENNIG. Insect Phylogeny. Wiley & Sons. Chichester.

Index

The names of various taxa can be found on many pages. Quoting every page number for each taxon seems to be a needless padding for the reader. Relevant information is given in the following structure.

- Standard print of page numbers is leading to statements in the continual text.
- Page numbers printed in bold refer to autapomorphies of the mentioned taxa.
- Page numbers printed in italics are pointing to names in a diagram of phylogenetic kinship as well as in the lines under the illustrations.
- Names of non-monophyletic taxa are in quotation marks.
- An obelisk marks fossil taxa.

Printing: Mercedes-Druck, Berlin
Binding: Stürtz AG, Würzburg